Essentials of Nucleosynthesis and Theoretical Nuclear Astrophysics

AAS AMERICAN ASTRONOMICAL SOCIETY IOP | ebooks™

About the program:

AAS-IOP Astronomy ebooks is the official book program of the American Astronomical Society (AAS), and aims to share in depth the most fascinating areas of astronomy, astrophysics, solar physics and planetary science. The program includes publications in the following topics:

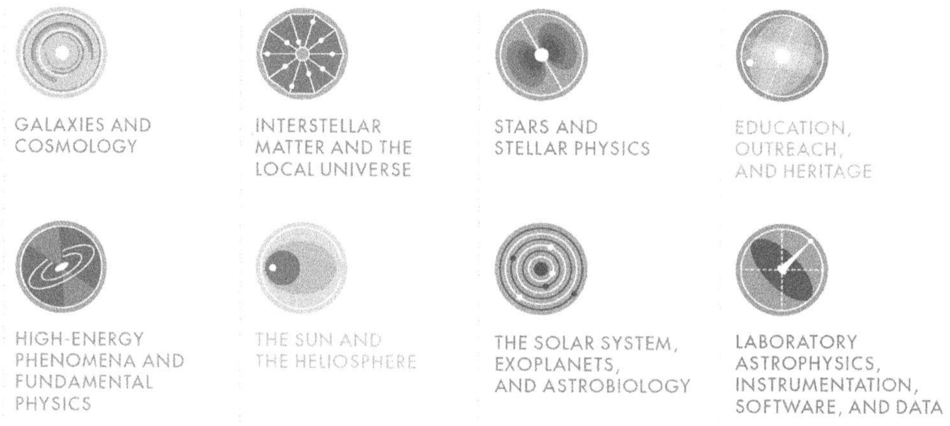

Books in the program range in level from short introductory texts on fast-moving areas, graduate and upper-level undergraduate textbooks, research monographs and practical handbooks.

For a complete list of published and forthcoming titles, please visit iopscience.org/books/aas.

About the American Astronomical Society

The American Astronomical Society (aas.org), established 1899, is the major organization of professional astronomers in North America. The membership (~7,000) also includes physicists, mathematicians, geologists, engineers and others whose research interests lie within the broad spectrum of subjects now comprising the contemporary astronomical sciences. The mission of the Society is to enhance and share humanity's scientific understanding of the universe.

Editorial Advisory Board

Steve Kawaler
Iowa State University, USA

Ethan Vishniac
John Hopkins University, USA

Dieter Hartmann
Clemson University, USA

Piet Martens
Georgia State University, USA

Dawn Gelino
NASA Exoplanet Science Institute, Caltech, USA

Joan Najita
National Optical Astronomy Observatory, USA

Bradley M. Peterson
The Ohio State University / Space Telescope Science Institute, USA

Scott Kenyon
Smithsonian Astrophysical Observatory, USA

Daniel Savin
Columbia University, USA

Stacy Palen
Weber State University, USA

Jason Barnes
University of Idaho, USA

James Cordes
Cornell University, USA

Essentials of Nucleosynthesis and Theoretical Nuclear Astrophysics

Thomas Rauscher
Department of Physics, University of Basel, Switzerland

IOP Publishing, Bristol, UK

Multimedia content is available for this book from http://iopscience.iop.org/book/978-0-7503-1149-6.

ISBN 978-0-7503-1149-6 (ebook)
ISBN 978-0-7503-1150-2 (print)
ISBN 978-0-7503-1828-0 (myPrint)
ISBN 978-0-7503-1151-9 (mobi)

DOI 10.1088/2514-3433/ab8737

Version: 20200701

AAS–IOP Astronomy
ISSN 2514-3433 (online)
ISSN 2515-141X (print)

British Library Cataloguing-in-Publication Data: A catalogue record for this book is available from the British Library.

Published by IOP Publishing, wholly owned by The Institute of Physics, London

IOP Publishing, Temple Circus, Temple Way, Bristol, BS1 6HG, UK

US Office: IOP Publishing, Inc., 190 North Independence Mall West, Suite 601, Philadelphia, PA 19106, USA

Contents

Preface

It is always a special challenge to write about interdisciplinary topics. Experts in any of the involved fields may find their topic not given adequate coverage. At the same time, parts of the potential audience may find some other fields outside their expertise not accessible enough. Often, there may not even be agreement on which research areas are sufficiently closely related to form an interdisciplinary field of research warranting treatment of its own.

Introducing nuclear astrophysics in that sense involves all these pitfalls. This becomes already apparent in the fact that there does not seem to be a unique definition of the field and various groups or research communities appear to have slightly different interpretations of the term. Some people emphasize the nuclear physics aspect and thus view the meaning as being "nuclear physics in astrophysical context." Others favor a different perspective, realizing that "nuclear" would just be an adjective, a modifier of the main noun astrophysics and thus emphasize the astrophysical modeling in need of some nuclear physics input and addressing the behavior and synthesis of nuclides. In the choice of the title of this book, I tried to unambiguously clarify that both the nuclear physics aspect and the astrophysical modeling of nucleosynthesis should be covered when presenting this research area. It is the nature of such an interdisciplinary field that there is an intimate connection between the respective research areas, informing each other, and that the special requirements of the connected fields cannot be fully understood by only perusing an isolated research area.

The choice of title also reflects that it was not my goal to provide full coverage of all details but rather to present the essential definitions and approaches which enable the reader to enter the research field and to get a basic understanding of the specialties and specific problems. The book also should equip the reader with the knowledge necessary to follow discussions of current open questions in the understanding of the origin of the elements. Beyond the decision to focus on theory, I obviously had to make choices on which topics to include in what detail. These were partly based on the usual selection in established introductory courses of the field but also partly on personal preference and expertise. My research experience also informed the choices, including frequently encountered or discussed topics. The basic goal was to provide a toolbox for working on nucleosynthesis and nuclear astrophysics and a formulary for quick reference, also for the more experienced researcher. For this reason, derivations or illustrative examples are not always given, only when they serve to illustrate a particular point or to convey a better basic understanding. In that sense, this work was not conceived as a classical textbook. It can also be used in the classroom, of course, but may require additional guidance and a selection or re-ordering of sections to fit the particular requirements. The contained material can be adapted for an introductory as well as a more specialized course but having completed introductory courses on quantum mechanics, thermal physics, and nuclear physics certainly would be advantageous to make full use of the content of this book.

It is inherent in an active field of research that some information may become quickly outdated. Certain fundamental equations and aspects, on the other hand, always remain untainted by time, as they provide the foundation for progress. In order to avoid the fate of many textbooks, which after some time contain an entwined mix of fundamental and outdated information, I have divided the book into two main parts. Part I is supposed to present fundamental equations and definitions, which may stand the test of time without the need for frequent revision. Part II addresses the current understanding of nucleosynthesis in various astrophysical sites in a more qualitative, descriptive manner as this information is expected to be more volatile and prone to change when our understanding of astrophysical objects and the evolution of the Universe advances.

Obviously, there is a history of textbooks in the involved fields having contributed to my own education, both while being a student and also later as an experienced scientist. I have drawn inspiration from many of them, in addition to using original research papers and lecture notes from my own teaching at the University of Basel, Switzerland, University of Hertfordshire, UK, and in various summer or winter schools. The most educating, publicly available sources are specifically mentioned in the Further Reading sections at the end of each chapter. They should be consulted for additional details and further references. I consciously avoided to provide extensive references in the main text, as this impacts legibility and distracts from the presentation of content.

Essentially, this book is designed as a concise summary of the most important concepts and equations related to nucleosynthesis and theoretical nuclear astrophysics. It came to life out of the desire to create a one-stop resource providing quick access to a combination of information otherwise provided at various level of detail in different books and original articles. I know that it will be useful for my students and myself, and I sincerely hope that it will also appeal to a wider audience.

Thomas Rauscher, Basel, November 2019

About the Author

Thomas Rauscher

Thomas Rauscher studied physics at the University of Technology in Vienna, Austria, and astronomy at the University of Vienna. After obtaining his PhD at the University of Technology with work on nuclear reaction rates for primordial nucleosynthesis, he embarked on an international academic career, with research stays at the Harvard-Smithsonian Center for Astrophysics in Cambridge, MA, USA, at the Institute for Nuclear Chemistry of the University of Mainz, Germany, at the University of Basel, Switzerland, at UC Santa Cruz, CA, USA, and at the Hungarian Institute for Nuclear Research (ATOMKI) in Debrecen, Hungary. He held academic positions as Assistant Professor in the Department of Physics of the University of Basel, Switzerland, and as tenured Reader in Astrophysics at the University of Hertfordshire, UK. Currently he is Adjunct Lecturer at the University of Basel, Visiting Research Fellow at the University of Hertfordshire, and also employed by the Swiss Federal Institute for Intellectual Property in Berne, Switzerland.

He is the recipient of several prestigious funded fellowships, having been Erwin Schrödinger fellow (Austrian National Science Foundation), Alexander von Humboldt fellow (Germany), APART fellow (Austrian Programme for Advanced Research and Technology, Austrian Academy of Sciences), PROFIL professor (Swiss National Science Foundation), and MTA fellow (Hungarian Academy of Sciences). In 1999, he received the Ludwig Boltzmann Prize from the ÖPG (Austrian Physical Society) for his work in theoretical physics. Further, he received the Outstanding Reviewer Award from the IOP (UK) in 2016 and a similar award, the Outstanding Referee Award, from the APS (American Physical Society) in 2017. He is Elected Fellow of the Institute of Physics (FinstP), Elected Fellow of the UK Higher Education Academy (FHEA), and Fellow of the Royal Astronomical Society (FRAS), as well as member of several learned societies, including the ÖPG, APS, AAAS, and IAU.

Thomas Rauscher is the author of more than 650 scientific publications with over 15,000 citations.

Symbols

a_{LE}	Lane–Emden scaling parameter
a_{NLD}	Nuclear level density parameter
a_{nuc}	Diffuseness parameter of optical potential
A	Nuclear mass number
$\hat{A}$	Antisymmetrization operator
$\mathcal{A}$	Relative isotopic mass
b_{G}	Gamow factor
B_{ν}	Planck's function by frequency
$\mathcal{B}$	Nuclear binding energy
C_{P}	Heat capacity at constant pressure
C_{V}	Heat capacity at constant volume
$\mathcal{C}_{\mathrm{scr}}$	Electron screening factor
D_{ℓ}^{u}	Spatial derivative of the nuclear wave function
D^{lev}	Nuclear level spacing
δ_{def}	Nuclear deformation parameter
δ_{ℓ}	Nuclear phase shift
$\delta_{\ell}^{\mathrm{C}}$	Coulomb phase shift
$\tilde{\delta}$	Kronecker symbol
$\mathcal{D}_{\mathrm{e}}$	Electron degeneracy
Δ_{ato}	Atomic mass excess
Δ_{break}	Nuclear pairing gap
Δ_{G}	Gamow width
Δ_{n}	Neutron number to seed ratio (νp-process)
Δ_{np}	Neutron–proton mass difference
Δ_{pair}	Nuclear pairing term
Δ_{shell}	Shell correction in the liquid drop mass formula
E_{C}	Compound formation energy
E_{F}	Fermi energy
E_{G}	Gravitational binding energy
E_{kin}	Kinetic energy
E_{nuc}	Nuclear energy
E_{pot}	Potential energy
E_{res}	Resonance energy
$E^{\mathrm{s.p.}}$	Single-particle energy
E_{rot}	Rotational energy
E_{T}	Thermal energy
E^{x}	Nuclear excitation energy
E_{γ}	Photon energy
ε	Net energy generation per volume
ε	Mass energy density
η_{b}	Baryon-to-photon ratio (SBBN)
η_{n}	Neutron excess
η_{S}	Sommerfeld parameter
f_{prod}	Stellar production factor
$f_{\ell}^{\mathrm{match}}$	Matching factor
$\tilde{f}^{\gamma}$	Photon strength function

$\bar{ft}$	ft-value in β-decay
$F_{\rm G}$	Gravitational force
$F_{\rm inc}$	Particle number current density
$F_{\rm SEF}$	Stellar enhancement factor
F_γ	Radiation flux
$\tilde{F}_{\mu\nu}$	Energy–momentum tensor
$\mathcal{F}_{\rm form}$	Form factor
$\mathcal{F}_{\rm IMF}$	Initial mass function
$\mathcal{F}^{\rm y}$	Nuclide yield per blocked mass in simple GCE
$\mathcal{F}_{\mathcal{K}}$	Scattering amplitude
$\bar{\mathcal{F}}$	Flow
g	Statistical number of degrees of freedom
$g_{\rm V}$	Coupling constant for Fermi decays
$g_{\rm A}$	Coupling constant for Gamow–Teller decays
$g_{\mu\nu}$	Metric tensor
G	Nuclear partition function
$G_{\rm pair}$	Nuclear pairing strength
$G_{\rm vir}$	Virial function
$\mathcal{G}$	Gamow energy
$\gamma_{\rm BW}$	Breit–Wigner width ratio
$\hat{\gamma}$	Reduced width
$\hat{\gamma}_\ell^{\rm s.p.}$	Dimensionless single-particle reduced width
Γ	Partial resonance width
$\tilde{\Gamma}$	Polytropic exponent (adiabatic exponent)
$\hat{\Gamma}_{\rm tot}$	Total resonance width
H	Hubble parameter
$\underline{h}$	Partial Hamiltonian operator
$H_{\rm P}$	Pressure scale height
$\mathcal{H}$	Total Hamiltonian operator
I_ν	Intensity of blackbody radiation by frequency
$I_{\rm iso}$	Total isospin
I_3	z-component of isospin
$\mathcal{I}$	Shift factor
$j_{\rm nuc}$	Spin quantum number of nucleon
J	Total spin quantum number
$\mathcal{J}_\ell$	Regular Coulomb wave function
$\hat{\mathcal{J}}_\ell$	Irregular Coulomb wave function
$\tilde{k}$	GR curvature constant
$\mathcal{K}$	Wave number
κ	Opacity
$\kappa_{\rm Thom}$	Opacity from Thomson scattering
κ_ν	Opacity by frequency
$l_{\rm mix}$	Mixing length
$l_{\rm part}$	Particle mean free path
l_γ	Photon mean free path
L	Luminosity
$\mathcal{L}^*$	Stellar surface luminosity
$\mathcal{L}^*_{\rm Ed}$	Eddington luminosity
ℓ	Orbital angular momentum quantum number, partial wave

ℓ_z	Magnetic quantum number
$\mathcal{L}$	Legendre polynomial
λ_{dec}	Decay rate
λ_E	Emission rate of nuclear electric γ-radiation
λ_{fold}	Renormalization of folding potential
λ_M	Emission rate of nuclear magnetic γ-radiation
λ_{trans}	Transformation rate
λ_{vib}	Vibrational mode quantum number
$\lambdabar$	deBroglie wave length divided by 2π
Λ	Cosnological constant
m	Particle mass
m_{nuc}	Nuclear mass
m_{ato}	Atomic mass
$m_{nucleon}$	mass of a nucleon
M_{Ch}	Chandrasekhar mass
M_{ej}	Mass ejection rate in GCE models
M_F	Nuclear matrix element for Fermi decays
M_{GT}	Nuclear matrix element for Gamow–Teller decays
M_{Jeans}	Jeans mass (mass limit for stability of a self-gravitating cloud)
$\mathcal{M}$	Multipole order, multipolarity
$\bar{\mu}$	Mean molecular weight
μ_A	Reduced mass number
μ_c	Chemical potential
μ_c^{tot}	Chemical potential including rest energy ($\mu_c + m_0 c^2$)
μ^{mag}	Magnetic moment
μ_{red}	Reduced mass
n_{net}	Size of reaction network
n_q	Nodal quantum number
$\tilde{n}$	Polytropic index
$\bar{n}$	Number density
$\bar{n}_n$	Neutron number density
$\bar{n}_p$	Proton number density
$\bar{n}_\gamma$	Photon number density
N	Neutron number
$\mathcal{N}$	Particle number
ν_e	Electron neutrino (similar for muon- and tau-neutrino with index μ, τ, respectively)
$\bar{\nu}_e$	Electron antineutrino (similar for muon- and tau-antineutrinos)
$o(p)$	Occupation number by momentum
$o(E)$	Occupation number by energy
$\mathcal{O}_\ell$	Penetration factor
$\tilde{\mathcal{O}}^{E1}$	Electric dipole operator
$\tilde{\mathcal{O}}^{E2}$	Electric quadrupole operator
$\tilde{\mathcal{O}}^{M1}$	Magnetic dipole operator
ω_{BW}	Breit–Wigner spin factor
Ω	Solid angle
Ω_b	Ratio of baryon density to critical density (SBBN)
Ω_{tot}	GR density parameter (total density to critical density)
$\tilde{\Omega}$	Grand canonical potential (Landau potential)

ω	Angular frequency
$\tilde{\omega}$	GR ideal fluid coefficient
p	Momentum
p_{F}	Fermi momentum
P	Pressure
P_{γ}	Radiation pressure
$\hat{P}$	Permutation operator
$\mathcal{P}$	Excited state population factor
$\tilde{\mathcal{P}}$	Isotope population factor
$\mathcal{P}_{\pi_{\mathrm{q}}}$	Parity projection factor in the nuclear level density
$\mathcal{P}_{\mathrm{J}}$	Spin projection factor in the nuclear level density
π_{q}	Parity quantum number
Ψ	Total quantum wave function
Ψ_{SFR}	Star formation rate
q	Net energy generation per mass
q_{dec}	Deceleration parameter
Q	Heat content
Q_{nuc}	Reaction Q–value
$\mathcal{Q}$	Quantum number (generic)
$\mathcal{Q}_4$	Electric quadrupole moment
$\hat{\mathcal{Q}}$	Quantum operator (generic)
r_{seed}	Neutron-to-seed abundance ratio ($Y_{\mathrm{n}}/Y_{\mathrm{seed}}$)
$\tilde{r}$	Scale factor in GR metric
R_{nuc}	Nuclear radius
R_{GR}	Scalar curvature
R_{S}	Schwarzschild radius
$R_{\mu\nu}$	Ricci curvature tensor
$\mathcal{R}$	R-function, R-matrix
$\tilde{\mathcal{R}}$	Return function in simple GCE model
ρ	Mass density
ρ_{nex}	Classical s-process neutron exposure
ρ_{nuc}	Nuclear matter density
$\tilde{\rho}$	Density of states
$\tilde{\rho}_{\mathrm{nuc}}$	Nuclear level density
s_{q}	Spin quantum number
s_{z}	Quantum number of spin projection onto z-axis
$\hat{s}_{\ell}$	Nuclear strength function
S	Entropy
S^{b}	Entropy per baryon
S_{n}	Neutron separation energy
S_{p}	Proton separation energy
S_{pro}	Projectile separation energy
S_{α}	α-particle separation energy
S_{γ}	Radiation entropy
$S_{\ell}^{\alpha\alpha}$	Diagonal elements of the scattering matrix
$S_{\ell}^{\alpha\beta}$	Scattering matrix
$\tilde{S}$	Astrophysical S-factor
$\mathcal{S}$	Spectroscopic factor
σ	Interaction cross section

σ^*	Stellar cross section
$\sigma_{\rm abs}$	Nuclear absorption cross section
$\sigma_{\rm BW}$	Breit–Wigner cross section
$\sigma_{\rm cut}$	Spin cut-off parameter in the nuclear level density
$\sigma_{\rm el}$	Elastic scattering cross section
$\sigma^{\rm eff}$	Effective cross section
$\sigma_{\rm lab}$	Laboratory cross section
$\sigma_{\rm reac}$	Reaction cross section
$\sigma_{\rm Thom}$	Thomson cross section
$\sigma_{\rm tot}$	Total cross section
$\langle \sigma v \rangle$	Reactivity
$\langle \sigma^* v \rangle$	Stellar reactivity
t	Time
$t_{1/2}$	Half-life
T_ℓ	Transmission coefficient for partial wave ℓ
$\mathcal{T}$	Total transmission coefficient
$\hat{T}$	Partial transmission coefficient
$\tau_{\rm ex}$	Expansion timescale
$\tau_{\rm ff}$	Free-fall timescale
$\tau_{\rm KH}$	Kelvin–Helmholtz timescale
$\tau_{\rm life}$	Lifetime of a nuclide with respect to reactions and/or decays
$\bar{\tau}_{\rm life}$	Average lifetime for a given reaction or decay
$\tau_{\rm nex}$	Time-integrated neutron flux
$\tau_{\rm nuc}$	Nuclear timescale
$\tau_{\rm S}$	Reaction rate parameter
τ_Δ	Process timescale
Θ	Moment of inertia
$\tilde{\Theta}$	Heaviside function
u_ℓ	Radial quantum wave function
$u^2_{j_{\rm nuc}}$	Occupation of nuclear hole states
$\tilde{u}$	Internal energy per volume
$\tilde{u}_\gamma$	Internal energy per volume for the radiation field
U	Internal energy
$U_{\rm NLD}$	Shifted excitation energy in the nuclear level density
$U_{\rm scr}$	Electron screening potential
$\mathcal{U}$	Thermodynamic velocity distribution
v	Velocity
$v_{\rm ex}$	Expansion velocity
$v^2_{j_{\rm nuc}}$	Occupation of nuclear particle states
V	Volume
$V_{\rm C}$	Coulomb potential
$V_{\rm G}$	Gravitational potential
$V_{\rm nuc}$	Nuclear interaction potential (negative for attractive potential)
$V_{\rm Roche}$	Roche potential
$v_{\rm T}$	Thermal velocity
$\tilde{V}$	Specific volume
$\mathcal{V}_{\rm fold}$	Effective nucleon–nucleon interaction for folding potential
$\tilde{W}$	Width fluctuation correction

$\mathcal{W}$	Weight of transition in stellar cross section
$\mathcal{W}_{\mathrm{R}}$	Racah coefficient for angular momentum coupling
x_{fiss}	Fissionability parameter
X_i	Mass fraction of nuclide i
${}^{*}X$	Contribution to the stellar rate
$\mathcal{X}$	Clebsch–Gordan coefficient
ξ	Lane–Emden scaled radius
Y	Total composition
Y_i	Abundance of nuclide i
Y_{e}	Electron abundance
$\mathcal{Y}$	Spherical harmonic function
Z	Charge number, proton number
$\mathcal{Z}$	Metallicity

Physical Constants

$\tilde{a}$	7.565×10^{-16} J K^{-4} m^{-3}	Radiation density constant
c	$2.997\,924\,58 \times 10^{8}$ m s^{-1}	Speed of light in vacuum
e_C	$1.602\,176\,634 \times 10^{-19}$ C	Elementary charge
G	$6.674\,3 \times 10^{-11}$ m^{3} kg^{-1} s^{-2}	Gravitational constant
h	$6.626\,070\,15 \times 10^{-34}$ J s	Planck's constant
$\hbar$	$1.054\,571\,817 \times 10^{-34}$ J s	$h/(2\pi)$
H_0	67.74 ± 0.46 km s^{-1} Mpc^{-1}	Hubble constant
k_B	$1.380\,649 \times 10^{-23}$ J K^{-1}	Boltzmann constant
$L_\odot$	3.846×10^{26} W	Solar luminosity
m_e	$9.109\,383\,701\,5 \times 10^{-31}$ kg	Electron mass
m_n	$1.674\,927\,498\,04 \times 10^{-27}$ kg	Neutron mass
m_p	$1.672\,621\,923\,69 \times 10^{-27}$ kg	Proton mass
m_u	$1.660\,5 \times 10^{-27}$ kg	Atomic mass unit
m_α	$6.644\,657\,335\,7 \times 10^{-27}$ kg	Mass of α-particle (^{4}He)
M_u	$9.999\,999\,996\,5 \times 10^{-4}$ kg mole^{-1}	Molar mass constant
$M_\odot$	$1.988\,4 \times 10^{30}$ kg	Solar mass
N_A	$6.022\,140\,76 \times 10^{23}$ mole^{-1}	Avogadro's number
ρ_0	2.3×10^{17} kg m^{-3}	Nuclear density
$R_\odot$	$6.963\,42 \times 10^{5}$ km	Solar radius (equatorial)
$\tilde{\sigma}_S$	5.67×10^{-8} W m^{-2} K^{-4}	Stefan's constant

Unit Conversions

Length, area	
	1 lightyear = 1 ly = $9.460\,730\,472\,580\,8 \times 10^{15}$ m
	1 parsec = 1 pc = 3.261 56 ly = $3.085\,677\,581\,491\,37 \times 10^{16}$ m
	1 fermi = 1 fm = 10^{-15} m
	1 Ångström = 1 Å = 10^{-10} m
	1 barn = 1 b = 0.01 fm^2
Mass	
	$m_\mathrm{n} = 1.008\,664\,915\,6 m_\mathrm{u} = 939.567\,36$ MeV/c^2
	$m_\mathrm{p} = 1.007\,276\,466\,88 m_\mathrm{u} = 938.272\,029$ MeV/c^2
Energy	
	1 erg = 1 g $\mathrm{cm}^2\,\mathrm{s}^{-2} = 10^{-7}$ J
	1 electron volt = 1 eV = $1.602\,19 \times 10^{-12}$ erg
	$m_\mathrm{u}c^2 = 1.492\,43 \times 10^{-3}$ erg = 931.494 043 MeV
	$m_\mathrm{e}c^2 = 8.187\,27 \times 10^{-7}$ erg = 0.510 998 918 MeV
	1 Bethe = 1 B = 1 foe = 10^{51} erg = 10^{44} J
Power	
	1 erg $\mathrm{s}^{-1} = 10^{-7}$ W
Time	
	1 min = 60 s
	1 h = 60 min = 3600 s
	1 d = 24 h = 86,400 s
	1 yr = 365.256 d = $3.155\,81 \times 10^7$ s

Part I

Essentials

Essentials of Nucleosynthesis and Theoretical Nuclear Astrophysics

Thomas Rauscher

Chapter 1

Basic Properties of Stars and the Stellar Plasma

1.1 Introduction

While stellar models (see Chapter 2) allow to study the interior of a star and its changes during the lifetime of a star, they also have to be able to describe observable, universal quantities of stars. Among these are

- total mass M,
- radius R,
- effective surface temperature T_{eff},
- total energy output per time, surface luminosity $\mathcal{L}^*$,
- composition Y (of outmost layers).

Note that all of them may change with time, partly quite drastically, as illustrated in Section 2.1 for M and R. Usually only a tiny fraction of a star's life can be observed during a human life span, though.

1.2 Blackbody Radiation and Luminosity

The electromagnetic radiation emitted by a star in the visible and near-visible wavelength range can be approximated by *blackbody radiation*. The ideal blackbody is in thermal equilibrium with its surroundings, is a perfect absorber and emitter of radiation at any wavelength, and emits a continuous spectrum whose shape (as a function of wavelength) is defined by its temperature alone. This ideal can be realized in nature by a short mean free path of photons within an opaque medium, leading to a larger number of absorptions and re-emissions of each photon before it can escape. This is fulfilled well in the stellar interior, not so well in stellar atmospheres. Nevertheless, the global properties of stellar electromagnetic radiation retain the ones of a blackbody.

doi:10.1088/2514-3433/ab8737ch1

The electromagnetic spectrum (as a function of wavelength λ) of a blackbody at temperature T is related to the Planck function (see Section 1.5.3) and described by

$$I_\lambda(T) = \frac{4\pi\hbar c^2}{\lambda^5}\frac{1}{\exp\left(\frac{2\pi\hbar c}{\lambda kT}\right) - 1}. \tag{1.1}$$

This is the power output per unit area per wavelength interval per solid angle and thus would, e.g., be given in units of W m^{-3} $sterad^{-1}$. For a derivation, see Section 1.5.3. Two well-known radiation laws can be derived from this function. It shows a pronounced emission peak at a wavelength λ_{peak} depending on T. This can be found by differentiation of Equation (1.1) and leads to *Wien's law*

$$\lambda_{peak}(T) \times T = C_{Wien}, \tag{1.2}$$

with the constant $C_{Wien} = 2.898 \times 10^{-3}$ m · K, when the wavelength is given in meters and the temperature in K.

The total power flux F_A (total power output) per unit area can be found by integrating Equation (1.1) over all wavelengths and solid angles. This yields

$$F_A = \tilde{\sigma}_S T^4, \tag{1.3}$$

which is the *Stefan–Boltzmann law*. Stefan's constant is given by $\tilde{\sigma}_S = 5.67 \times 10^{-8}$ W m^{-2} K^{-4} for temperature measured in K. The total power radiated from the surface of a sphere with radius R is surface area times flux, $4\pi R^2 F_A$. This power output of a star can be measured. It is called *luminosity* $\mathcal{L}^*$. For a blackbody radiator we can use the Stefan–Boltzmann law to derive a relation between luminosity, radius and surface temperature

$$\mathcal{L}^* = 4\pi R^2 \tilde{\sigma}_S T^4. \tag{1.4}$$

Given a measured stellar luminosity $\mathcal{L}^*_{measured}$, the *effective (surface) temperature* T_{eff} of the star is defined as the temperature of a blackbody radiating the same energy per time,

$$T_{eff} = \left(\frac{\mathcal{L}^*_{measured}}{4\pi R^2 \tilde{\sigma}_S}\right)^{1/4}. \tag{1.5}$$

Astronomical literature often quotes the luminosity in *absolute bolometric magnitudes* M_{bol}. The term *bolometric* refers to an integration over all wavelengths instead of measuring just in a wavelength band. Absolute magnitudes provide a logarithmic scale for luminosities.

$$\mathcal{L}^*_{measured} = \mathcal{L}^*_0 10^{-0.4 M_{bol}}, \tag{1.6}$$

with the normalization $\mathcal{L}^*_0 = 3.0128 \times 10^{28}$ W. This is the definition adopted by IAU Resolution B2 in 2015, replacing older definitions which were based on solar luminosity and magnitude. The absolute bolometric magnitude of the Sun in this scale is $M_{bol,\odot} = 4.74$. Note that in this magnitude scale, a brighter star has a smaller magnitude. For example, the red giant star Antares radiates about 65,000 times the energy of the Sun and thus has an absolute bolometric magnitude of −7.29. (Antares emits most light in the infrared and so its absolute magnitude in the visual range is −5.28.)

An important way to demonstrate the connection between $\mathcal{L}^*$, T_{eff}, and R is the *Hertzsprung–Russell diagram (HRD)*. In the original diagram, absolute magnitudes were plotted versus spectral types. In the modern versions of the HRD, luminosity is shown as a function of effective temperature in a doubly logarithmic plot, with T_{eff} increasing from right to left on the horizontal axis (see Figure 6.1 in Chapter 6). Another frequently used option is the photometric HRD, which links absolute magnitudes with photometric "colors," i.e., wavelength bands (see Figure 1.1). The most common system for defining such bands is the U(ltraviolet)–B(lue)–V(isual) system, with the central wavelengths defined as: 364 nm (U), 442 nm (B), 540 nm (V). According to Equation (1.4), a higher luminosity at given T_{eff} implies larger radius. With the commonly adopted form of the HRD (with flipped temperature axis), hotter stars are toward the left end and cooler stars toward the right. For a specific T_{eff}, smaller stars are at lower luminosities and larger stars at higher luminosities. For example, white dwarfs (see Section 2.5) are found in the bottom left of this diagram,

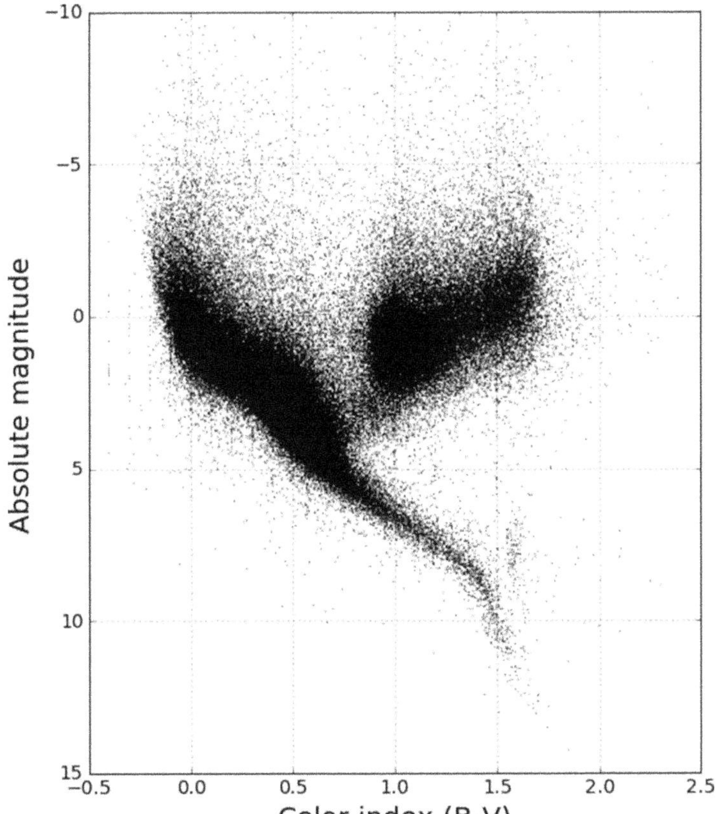

Figure 1.1. A real HRD created using data for about 114,000 stars from the HIPPARCOS main catalog I/239.

whereas red supergiants are in the top right. A star's position in the diagram will change over its lifetime, as its surface temperature and radius will change. Entering a large number of stars in this diagram, it was found early on that they populate several separated regions in the plot. These regions indicate long phases of stability in a star's life and can be connected to hydrostatic burning phases or stable endpoints of stellar evolution (such as white dwarfs). For example, most stars fall on a relatively narrow band extending from hot giant stars to cool dwarf stars, the *main sequence*. Stars on the main sequence are in the hydrogen-burning phase. Toward the upper right with respect to the main sequence another region identifies red giants burning helium to uphold hydrostatic equilibrium. Since H- and He-burning comprise by far the largest fraction of a star's life, most stars will be in one or the other phase when a census of the current stellar distribution is taken. The life of a star and its track in the HRD are briefly described in Chapter 6.

1.3 Abundances and Mass Fractions

While the laws derived for blackbody radiation describe the global emission spectra, deviations from this blackbody spectrum allow to extract information on the composition of the stellar plasma. The prerequisites for the blackbody approximation, short mean free paths for protons and thermal equilibrium, are not met in the outer layer of a star and in the stellar atmosphere. The spectra of escaping photons are therefore modified from a pure blackbody spectrum. The most important modification is the appearance of discrete emission or absorption lines at specific wavelengths. These are due to partially ionized atoms, which emit or absorb photons at discrete energies corresponding to the energy differences of electron levels in their atomic shells. Analyzing the stellar spectra allows to derive the composition of the layers through which the photons have passed. Explaining the details of abundance analysis with stellar spectra is beyond the scope of this book. It is a difficult task involving identification of absorption lines and determination of line widths in measured stellar spectra, using stellar model atmospheres and atomic physics. Here, only a few basic definitions are given, which are also needed when studying the change of abundances through nuclear reactions in Section 4.2 and, of course, when comparing the results of model calculations to observed abundances. It should also be mentioned that, with few exceptions, only elemental abundances can be observed in most stars, whereas isotopic abundances are central to nucleosynthesis investigations.

In thermodynamics, the number $\mathcal{N}$ of particles in a volume V is given by the *number density*

$$\bar{n} = \frac{\mathcal{N}}{V}. \tag{1.7}$$

For gases or plasmas, being pure or mixtures of i components, often the *specific volume* is used, which is the volume per mass unit and directly connected to the matter density ρ:

$$\overline{V} = \frac{1}{\rho} = \frac{1}{\sum_i \overline{n}_i m_i}, \tag{1.8}$$

where m_i is the mass of the ith component. By definition, 1 mole of a substance contains N_A particles and N_A is *Avogadro's constant*. According to this definition, the mass of N_A atoms with *relative isotopic mass* $\mathcal{A}_i$ is

$$m_i N_A = \mathcal{A}_i M_u, \tag{1.9}$$

where the *molar mass constant* is denoted by M_u. Note that the relative isotopic mass $\mathcal{A}_i = m_i/m_u$ is a dimensionless quantity and should not be confused with the atomic or nuclear mass to be used for m_i (see also Section 3.2). The molar mass constant has the dimension of mass per mole and is very close to 1 g mole^{-1}. (Before the 2019 redefinition of the SI base units it was exactly 10^{-3} kg mole^{-1} by definition.) Therefore the mass of 1 mole of a substance is (almost) $\mathcal{A}$ grams. Using Equation (1.9), density ρ can be expressed as

$$\rho = \sum_i \rho_i = \sum_i \rho X_i = \sum_i \overline{n}_i m_i = \sum_i \frac{\overline{n}_i M_u}{N_A} \mathcal{A}_i. \tag{1.10}$$

In the context of nuclear reaction rates (see Section 4.2), the approximation $\mathcal{A} \approx A$ is used for nuclides, with the *nuclear mass number* $A = Z + N$ being the sum of atomic number (proton number) Z and neutron number N. The mass number therefore counts the constituent particles in an atomic nucleus. Assuming that all particles have mass m_u and using A implies that $m_i \approx A_i m_u$ and that the density is defined as a nucleon density $\rho_{nuc} = \sum_i \overline{n}_i M_u A_i / N_A$. Then the mass fraction (see below) actually is a nucleon fraction. While speaking of nucleon densities may be natural in the nuclear physics context, astrophysical models describe stellar plasmas using the matter density as defined in Equation (1.8). The numerical differences arising from exchanging matter density with nucleon density, however, are tiny and usually negligible, especially in plasmas composed of only mono-atomic species (no molecules). The ratio $\mathcal{A}/A$ for stable nuclides ranges from 0.99884 (for ^{56}Fe) to 1.00782505 (for ^{1}H). Therefore $\mathcal{A}$ and A are used interchangeably in later sections while the $\approx$ symbol is used in this section to emphasize the distinction between them.

The contribution of the ith component to the total density ρ is called the *mass fraction*

$$X_i = \frac{\rho_i}{\rho}. \tag{1.11}$$

Obviously, mass fractions are normalized as

$$\sum_i X_i = \sum_i \frac{\overline{n}_i M_u}{\rho N_A} \mathcal{A}_i = 1. \tag{1.12}$$

For historical reasons, elements heavier than helium are called "metals" in astronomy. That is why the mass fraction of plasma content beyond hydrogen and helium is called *metallicity*,

$$\mathcal{Z} = \sum_{i \in \text{"metals"}} Y_i \mathcal{A}_i = 1 - X_{\text{H}} - X_{\text{He}} \approx 1 - Y_{^1\text{H}} - 4Y_{^4\text{He}}. \tag{1.13}$$

For comparison, the solar metallicity is $\mathcal{Z}_\odot \approx 0.0196$.

Density fluctuations (e.g., shock waves in a stellar plasma) influence $\bar{n}$. To obtain a measure of particles per volume that is independent of such density changes, the *relative abundance* Y_i (sometimes called *mole fraction*) is introduced,

$$Y_i = \frac{\bar{n}_i \text{M}_\text{u}}{\rho \text{N}_\text{A}} = \frac{X_i}{\mathcal{A}_i} \approx \frac{X_i}{A_i}. \tag{1.14}$$

It is important to note that, contrary to mass fractions, abundances Y_i are not normalized and thus only allow to specify relative amounts of material by comparing their individual abundances. The sum of relative abundances

$$Y = \sum_i Y_i \tag{1.15}$$

is called *total composition* of the material. For comparison, the solar total composition is $Y_\odot \approx 1.639$. This notation for mass fractions X_i, abundances Y_i, and total composition Y should not be confused with the symbols X and Y sometimes used in literature for the mass fractions of hydrogen $X_{^1\text{H}}$ and of helium $X_{^4\text{He}}$ (as used, for example, in Big Bang nucleosynthesis, see Chapter 10).

The abundances of species with opposite electric charges are not independent of each other if charge neutrality is required. The standard plasma in the deep interior of a star (where nuclear reactions are occurring) is fully ionized and consists of a mix of free, negatively charged electrons and positively charged protons and heavier nuclei. Since each positive charge (free or inside a nucleus) is balanced by an electron, the *electron abundance* Y_e is given by

$$Y_\text{e} = \sum_j Z_j Y_j = \sum_j \frac{Z_j}{\mathcal{A}_j} X_j \approx \sum_j \frac{Z_j}{A_i} X_j, \tag{1.16}$$

where j is enumerating the nuclear species (not including electrons) and Z is the nuclear charge number (sometimes also called atomic number). With this relation, the composition of a neutral plasma can also be given as

$$Y = Y_\text{e} + \sum_j Y_j = \sum_j (Z_j + 1) Y_j \approx \sum_j \frac{Z_j + 1}{A_j} X_j. \tag{1.17}$$

In a neutral plasma, the electron abundance is also a measure of the neutron-to-proton ratio and thus specifies how neutron-rich a plasma is. Equation (1.16) shows that $Y_e \propto Z/(Z + N)$ and thus $Y_e = 1/2$ implies an equal number of neutrons and protons in the plasma. A similar measure is the *neutron excess* defined as

$$\eta_n = \sum_i (N_i - Z_i) Y_i \approx 1 - 2Y_e. \tag{1.18}$$

Table 1.1 illustrates the meaning of η_n and Y_e for a few exemplary values.

As an alternative to the abundance, one can use the *mean molecular weight* $\bar{\mu}$, which is just the reciprocal of the abundance,

$$\bar{\mu}_i = \frac{1}{Y_i}. \tag{1.19}$$

It is obvious that abundances can be simply added to arrive at the ion composition whereas mean molecular weights cannot be added directly, which makes working with them cumbersome.

Astronomical abundance data are usually given as ratios of two elements $El1$ and $El2$ relative to their solar ratio, using the *bracket notation*

$$\left[\frac{El1}{El2}\right] = \log_{10}\left(\frac{(\bar{n}_{El1}/\bar{n}_{El2})_*}{(\bar{n}_{El1}/\bar{n}_{El2})_\odot}\right) = \log_{10}\left(\frac{\bar{n}_{El1}}{\bar{n}_{El2}}\right)_* - \log_{10}\left(\frac{\bar{n}_{El1}}{\bar{n}_{El2}}\right)_\odot. \tag{1.20}$$

For example, for the Fe abundance of a star:

- Solar: $\left[\frac{\text{Fe}}{\text{H}}\right] = 0$,
- Twice solar: $\left[\frac{\text{Fe}}{\text{H}}\right] = \log_{10}(2) \approx 0.3$,
- Half solar: $\left[\frac{\text{Fe}}{\text{H}}\right] = \log_{10}(0.5) \approx -0.3$.

The Fe abundance is often used as a marker for the metallicity of a star. More precisely, the relation between metallicity and Fe abundance is usually assumed to be

$$\begin{aligned}\left[\frac{\text{“metals”}}{\text{H}}\right] &= \log_{10}\left(\frac{\sum_{i\in\text{“metals”}}\bar{n}_i}{\bar{n}_\text{H}}\right)_* - \log_{10}\left(\frac{\bar{n}_{i\in\text{“metals”}}}{\bar{n}_\text{H}}\right)_\odot \\ &= C_\text{metallicity} \times \left[\frac{\text{Fe}}{\text{H}}\right],\end{aligned} \tag{1.21}$$

where the proportionality constant is assigned values $0.9 \leqslant C_\text{metallicity} \leqslant 1$.

Table 1.1. Some Examples Illustrating the Relation between Neutron Excess η_n and Electron Abundance Y_e

η_n	Y_e	Comment
1	0	Pure neutrons
0	0.5	Equal number of protons and neutrons
−1	1	Pure protons

Often the abundance of an element is compared to the hydrogen abundance in the same object and therefore the *ε-notation* was introduced as a shorthand:

$$\varepsilon(El1) = \left(\frac{\overline{n}_{El1}}{\overline{n}_{\mathrm{H}}}\right)_{*}. \tag{1.22}$$

To avoid small numbers, sometimes $\log_{10}(\varepsilon(El1)) + 12$ is tabulated for objects with low metallicity (implying $\log_{10}\overline{n}_{\mathrm{H}} \approx 12$).

1.4 Further Thermodynamic Relations

The equilibrium distribution of particles in a gas or plasma is determined by thermodynamic quantities such as temperature T, pressure P, and *chemical potential* $\mu_{\mathrm{c}}^{\mathrm{tot}}$. In thermodynamics, these physical quantities are viewed as parameters determining how the internal energy U of a gas is changed by a change in entropy S, heat content Q, volume V, or particle number $\mathcal{N}$,

$$\mathrm{d}U = T\,\mathrm{d}S - P\,\mathrm{d}V + \sum_i \mu_{\mathrm{c},i}^{\mathrm{tot}}\,\mathrm{d}\mathcal{N}_i = dQ - P\,\mathrm{d}V + \sum_i \mu_{\mathrm{c},i}^{\mathrm{tot}}\,\mathrm{d}\mathcal{N}_i, \tag{1.23}$$

where i enumerates the different particle species. The partial derivatives of the heat Q with respect to temperature are the *heat content at constant pressure*

$$C_{\mathrm{P}} = \left.\frac{\mathrm{d}Q}{\mathrm{d}T}\right|_{\mathrm{P}} \tag{1.24}$$

and the *heat capacity at constant volume*

$$C_{\mathrm{V}} = \left.\frac{\mathrm{d}Q}{\mathrm{d}T}\right|_{\mathrm{V}}, \tag{1.25}$$

describing the relation between heat and temperature.

Another useful thermodynamic relation is the *grand canonical potential (Landau potential)*

$$\widetilde{\Omega} = U - TS - \sum_i \mu_{\mathrm{c},i}^{\mathrm{tot}}\mathcal{N}_i \tag{1.26}$$

which simplifies to

$$\widetilde{\Omega} = -P\overline{V} = -\frac{P}{\rho} \tag{1.27}$$

for a homogeneous gas with negligible temperature and pressure fluctuations across the specific volume $\overline{V}$. This condition is fulfilled for astrophysical plasmas in nucleosynthesis studies. Equation (1.27) provides a simple means to calculate all other thermodynamic properties once the pressure has been determined. For example, entropy is given by

$$S = -\left(\frac{\partial \widetilde{\Omega}}{\partial T}\right)_{\mathcal{V},\mu_{c,i}^{\text{tot}}}. \tag{1.28}$$

The chemical potential μ_c^{tot} accounts for the establishment of a chemical equilibrium in processes affecting the particle number, e.g., by diffusion or by production or destruction of a particle species due to chemical or nuclear reactions. For example, in a system with only one species, particles move from regions with high chemical potential to regions with lower chemical potential until equilibration is reached and the chemical potential has the same value everywhere. Thus, in the case of chemical equilibrium

$$\sum_i \mu_{c,i}^{\text{tot}} \, d\mathcal{N}_i = 0. \tag{1.29}$$

The total chemical potential

$$\mu_c^{\text{tot}} = \mu_c + \mu_c^0 \tag{1.30}$$

has the dimension of an energy and includes a kinetic part μ_c (often also called chemical potential) and a rest energy μ_c^0 at absolute zero temperature. The latter is, for example, the energy equivalent of the particle rest mass.

This also allows to derive equilibrium concentrations. If there are, e.g., four types of particles connected by forward and reverse reactions $a_1 + a_2 \leftrightarrow b_1 + b_2$, in chemical (thermodynamic) equilibrium

$$\mathcal{N}_{a_1}\mu_{c,a_1}^{\text{tot}} + \mathcal{N}_{a_2}\mu_{c,a_2}^{\text{tot}} = \mathcal{N}_{b_1}\mu_{c,b_1}^{\text{tot}} + \mathcal{N}_{b_2}\mu_{c,b_2}^{\text{tot}}. \tag{1.31}$$

In equilibrium, $\Delta\mu_c^{\text{tot}} = \mu_{c,a_1}^{\text{tot}} + \mu_{c,a_2}^{\text{tot}} - \mu_{c,b_1}^{\text{tot}} - \mu_{c,b_2}^{\text{tot}} = 0$ and the particle numbers $\mathcal{N}$ are constant. Without equilibrium, the sign of

$$\Delta\mu_c^{\text{tot}} = \left(\mu_{c,a_1}^0 + \mu_{c,a_2}^0 - \mu_{c,b_1}^0 - \mu_{c,b_2}^0\right) + \left(\mu_{c,a_1} + \mu_{c,a_2} - \mu_{c,b_1} - \mu_{c,b_2}\right) \tag{1.32}$$

determines in which direction the reaction will proceed. Note that the first bracketed sum is the difference in rest mass before and after the reaction. This is the binding energy difference, also called reaction Q–value. For nuclear reactions, masses $m_{\text{nuc}}c^2$ are of the order of several MeV and thus dominate $\Delta\mu_c^{\text{tot}}$ at low temperature. The kinetic contribution μ_c is determined by the properties of the particle gas as discussed in Section 1.5. For nuclear reactions, the kinetic contribution becomes comparable to the nuclear reaction Q–value Q_{nuc} only for temperatures exceeding approximately 40 MK. Higher temperatures allow reactions with $Q_{\text{nuc}} < 0$ to occur. Thus, the chemical potential and also the equilibrium abundances depend on the temperature. See Sections 3.6 and 4.1 for further details.

1.5 Equations-of-state for Stellar Plasmas

1.5.1 Thermodynamic Derivation

1.5.1.1 Pressure Integral

An equation-of-state (EOS) is describing the behavior of matter when changing certain thermodynamic variables. In the usual formulation, an EOS relates pressure to other quantities, such as density and temperature. In general, for a gas of non-interacting particles occupying a spherical symmetric region of space the pressure can be derived from the *pressure integral*

$$P = \frac{1}{3}\int_0^\infty vpw(p)\,\mathrm{d}p, \tag{1.33}$$

using the velocity v and the momentum $p = m_{\text{part}}v$ of particles with mass m_{part}. The density function $w(p)$ describes how many particles carry a momentum in the interval $[p, p + \mathrm{d}p]$. The factor 1/3 comes from the average over the three spatial directions.

The pressure integral can be derived from the description of elastic particle collisions with a wall, thus exerting pressure. Consider a beam of particles impinging at a velocity v and angle θ with the normal of the surface (Figure 1.2). The momentum transferred by the collision of one particle with the wall is twice the momentum component normal to the surface,

$$\Delta p = 2p\cos\theta. \tag{1.34}$$

Taking $w(\theta, p)$ to be the number density of particles with momenta in the interval $[p, p + \mathrm{d}p]$ and directions in the cone $\theta + \mathrm{d}\theta$. Since the particle distribution is isotropic, the number of particles is proportional to the solid angle $\mathrm{d}\Omega$ and thus

$$\frac{w(\theta, p)\,\mathrm{d}\theta\mathrm{d}p}{w(p)\,\mathrm{d}p} = \frac{\mathrm{d}\Omega}{4\pi} = \frac{2\pi\sin\theta\,\mathrm{d}\theta}{4\pi} = \frac{1}{2}\sin\theta\,\mathrm{d}\theta. \tag{1.35}$$

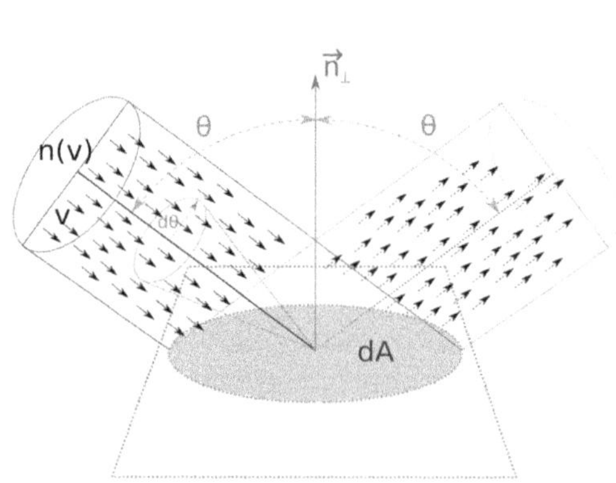

Figure 1.2. Particle beam colliding elastically with a wall. (See text for further description.)

The number of particles from this beam striking the wall in a time interval Δt is the number density multiplied by the volume $v\Delta t\ \mathrm{d}A\cos\theta$ with $\mathrm{d}A$ being the area of incidence of the beam on the wall,

$$w(\theta, p)v\Delta t\ \mathrm{d}A\cos\theta. \tag{1.36}$$

The momentum transferred to the wall by these particles is

$$\Delta p_\theta = w(\theta, p)\ \mathrm{d}\theta\ \mathrm{d}pv\Delta t\ \mathrm{d}A\cos\theta\Delta p. \tag{1.37}$$

Since pressure is transferred momentum per unit and unit area, Equation (1.37) has to be divided by $\Delta A\Delta t$ to obtain the pressure contribution of these particles,

$$\mathrm{d}P = w(\theta, p)\ \mathrm{d}\theta\ \mathrm{d}pv\cos\theta\Delta p. \tag{1.38}$$

Using Equations (1.34) and (1.35) we can rewrite the above equation as

$$\begin{aligned} \mathrm{d}P &= w(\theta, p)\ \mathrm{d}\theta\ \mathrm{d}pv\cos\theta 2p\cos\theta \\ &= \frac{1}{2}\sin\theta\ \mathrm{d}\theta\ w(p)\ \mathrm{d}pv\cos\theta 2p\cos\theta \\ &= vpn(p)\ \mathrm{d}p\cos^2\theta\sin\theta\ \mathrm{d}\theta. \end{aligned} \tag{1.39}$$

Integration over all particles, i.e., over all momenta p and all angles θ finally yields the pressure

$$\begin{aligned} P &= \int_0^\infty vpw(p)\ \mathrm{d}p \int_0^{\pi/2} \cos^2\theta\sin\theta\ \mathrm{d}\theta \\ &= \int_0^\infty vpw(p)\ \mathrm{d}p \int_0^1 \cos^2\theta\ \mathrm{d}\cos\theta \\ &= \frac{1}{3}\int_0^\infty vpw(p)\ \mathrm{d}p. \end{aligned} \tag{1.40}$$

The pressure integral can also be viewed as an average of $\vec{p}\cdot\vec{v}$ over the distribution function,

$$P = \frac{w}{3}\langle\vec{p}\cdot\vec{v}\rangle. \tag{1.41}$$

Although the derivation of the pressure integral seems to have been assuming classical particles impinging on a wall, no assumptions concerning specific properties of the gas were made. It is a general theorem, applying to all sorts of "gas," including, for example, a "gas" of photons. Application of the pressure integral only requires an adequately defined momentum distribution function $w(p)$, which can also be derived from quantum mechanical properties, see Section 1.5.1.2. It also equally applies to relativistic motion of particles.

1.5.1.2 Quantum Mechanical Expression

Classical particles can have a continuum of positions and momenta and thus occupy a phase space continuum (x, y, z, p_x, p_y, p_z) with total momentum

$p = \sqrt{p_x^2 + p_y^2 + p_z^2}$. Thus, also the momentum distribution $w(p)$ is continuous. For a quantum mechanical treatment, a quantized phase space and quantized distribution function have to be introduced. In order to do this, the function $w(p) = \tilde{\rho}(p)o(p)$ is constructed as a product of a density of discrete quantum states $\tilde{\rho}(p)$ and their occupation $o(p)$. The density of states follows from solving the quantum mechanical Schrödinger equation (see also Sections 1.7 and 3.3) for particles confined to a cubic box with side length L and volume $V = L^3$. The obtained solutions are standing waves $\sin k_x x \times \sin k_y y \times \sin k_z z$ with the wave vector $\vec{k}$ given by

$$\vec{k} = \begin{pmatrix} k_x \\ k_y \\ k_z \end{pmatrix} = \frac{\pi}{L}\begin{pmatrix} n_x \\ n_y \\ n_z \end{pmatrix} = \frac{1}{\hbar}\vec{p}\,. \tag{1.42}$$

The quantum numbers n_x, n_y, n_z are positive integers enumerating the solutions. If counting starts at 1, they give the number of half wavelengths being accommodated between two opposite sides of the box in each spatial direction. In other words, the number of nodes of the standing wave inside the box would be one less than the corresponding quantum number. The correspondence of the wave vector with the momentum vector $\vec{p} = \hbar\vec{k}$ can be derived from *Heisenberg's uncertainty principle* $\Delta x \Delta p_x = 2\pi\hbar$ and the *deBroglie wavelength* $\lambda_{\mathrm{deB}} = 2\pi\hbar/p$. They also imply that the quantum phase space consists of "cells" with volume $(2\pi\hbar)^3$,

$$\mathrm{d}p_x\,\mathrm{d}p_y\,\mathrm{d}p_z\,\mathrm{d}x\,\mathrm{d}y\,\mathrm{d}z = \mathrm{d}p_x\,\mathrm{d}p_y\,\mathrm{d}p_z\,\mathrm{d}V = (2\pi\hbar)^3. \tag{1.43}$$

For particles with intrinsic angular momentum, i.e., spin, each cell in the six-dimensional quantum phase space is degenerate in the additional quantum number of the spin projection (see Section 1.7) or polarization and the number of states has to be multiplied by a statistical factor g (depending on spin) to account for this. Thus the number of states per unit phase space is $g/(2\pi\hbar)^3$. Then the number of states enclosed by a "sphere" in momentum space for momentum p is

$$\frac{4\pi}{3}\frac{g}{(2\pi\hbar)^3}p^3. \tag{1.44}$$

Since the phase space volume is found by integrating over the position and momentum space

$$V_{\mathrm{phase}} = \int \mathrm{d}^3p\,\mathrm{d}^3r = \int 4\pi p^2\,\mathrm{d}p\,4\pi r^2\,\mathrm{d}r, \tag{1.45}$$

it is obvious that an increase by $\mathrm{d}p$ increases the phase space volume by $4\pi p^2\,\mathrm{d}p$. Therefore the state density at momentum p is

$$\tilde{\rho}(p) = 4\pi\frac{g}{(2\pi\hbar)^3}p^2. \tag{1.46}$$

This expression is also valid for a relativistic gas.

To allow the calculation of the pressure integral as defined in Equation (1.33) and using $w(p) = \tilde{\rho}(p)\, o(p)$, the occupation probability $o(p)$ of the available momentum (or energy) states for momentum $p(E)$ has to be known. The equilibrium distribution of the particles in the quantum states can be described by general thermodynamic equations, as given earlier in Equations (1.23)–(1.28) and thus determination of $o(p)$ requires the knowledge of the chemical potential. The occupation probability $o(p)$ also depends on the type of the particles, whether they are bosons or fermions, which determines the statistics to be applied. Without derivation, the relevant particle statistics are given below. For identical bosons (particles with zero or integer spin) any number of particles can be in a given quantum state. This leads to the *Bose–Einstein statistics*

$$o(E)_{\mathrm{BE}} = \frac{1}{\mathrm{e}^{(E_{\mathrm{part}}(p)-\mu_{\mathrm{c}})/(k_{\mathrm{B}}T)} - 1}, \tag{1.47}$$

where we have rewritten $o(p)$ as $o(E)$, which is the average occupation number of states with energy $E_{\mathrm{part}}(p)$. The energy of a particle with mass m in a quantum state with momentum p obeys the relativistic relation

$$E^2_{\mathrm{part}}(p) = p^2\mathrm{c}^2 + m^2\mathrm{c}^4. \tag{1.48}$$

Although the denominator of $o(E)_{\mathrm{BE}}$ is zero at $E_{\mathrm{part}}(p) = \mu_{\mathrm{c}}$, integrating over the occupation number still yields a finite number, see Equation (1.53). The accumulation of particles at energies close to μ_{c} is a consequence of the possibility of having several particles sharing the same quantum state.

On the other hand, the *Pauli exclusion principle* limits the number of fermions (particles with half-integer spin) in each state by stating that no two such particles can have an identical set of quantum numbers. Due to Pauli's exclusion principle, at $T = 0$ the lowest energy states would be filled with g particles each. Since only a finite number N of particles is available, quantum states up to an energy $E_{\mathrm{F}} = \mu_{\mathrm{c}}$ are filled. This energy of the last occupied state is called *Fermi energy* E_{F}. The corresponding momentum is the *Fermi momentum* p_{F}. Above E_{F} no states are occupied. Using Equation (1.44) and the total number of particles N, the Fermi momentum can be determined as

$$p_{\mathrm{F}} = \left(\frac{3(2\pi\hbar)^3 N}{4\pi g V}\right)^{1/3} = 2\pi\left(\frac{3\bar{n}}{4\pi g}\right)^{1/3}\hbar. \tag{1.49}$$

Since the state density up to p_{F} is given by Equation (1.46) and is zero above p_{F}, the resulting occupation distribution must be a step function valued zero above p_{F}. This situation changes when the temperature is increased. For a gas of fermions with finite temperature $T > 0$, the resulting *Fermi–Dirac statistics* yields

$$o(E)_{\mathrm{FD}} = \frac{1}{\mathrm{e}^{(E_{\mathrm{part}}(p)-\mu_{\mathrm{c}})/(k_{\mathrm{B}}T)} + 1}. \tag{1.50}$$

Physically, this describes the excitation above E_F (or p_F) of an increasing fraction of particles with increasing temperature. An increase in T removes particles from states with lower momentum and adds them into states at higher momentum, thus softening the hard edge of the step function, as illustrated in Figure 1.3. A gas at $T = 0$, with all particles occupying only states at or below p_F, is called a fully degenerate gas.

For both distributions, Equations (1.47) and (1.50), the average occupation of each quantum state becomes comparable when the exponential function in the denominator becomes sufficiently large so that the ±1 in the denominator can be neglected. This is the case when

$$\exp\left[(mc^2 - \mu_c)/(k_B T)\right] \gg 1, \tag{1.51}$$

implying that even the states at lowest momentum are hardly occupied. This will occur at high T or at the very low μ_c of a dilute gas. When the average occupation is very low, $o(E)_{BE} \approx o(E)_{FD} \ll 1$, a distinction between bosons and fermions is not necessary anymore and both behave like classical particles. The average occupation of a state becomes

$$o(E)_{MB} \approx o(E)_{FD} \approx \frac{1}{e^{(E_{part}(p) - \mu_c)/(k_B T)}}, \tag{1.52}$$

which is the *Maxwell–Boltzmann statistics*, also found in the classical derivation of the properties of the ideal gas (see also Section 1.5.2).

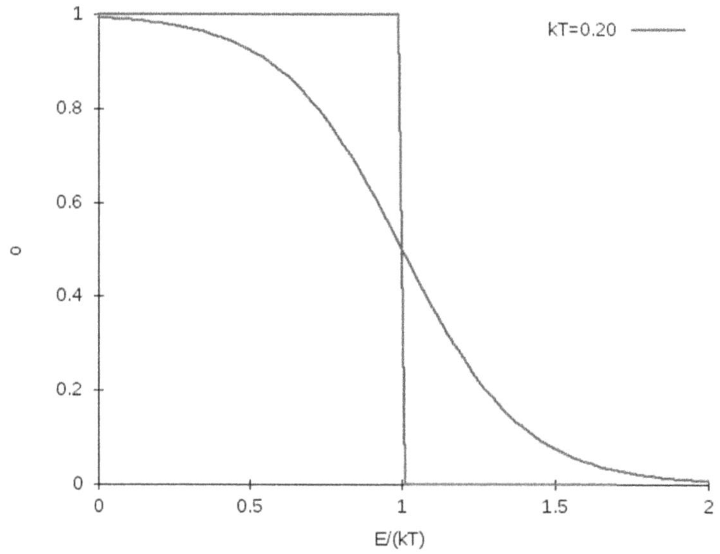

Figure 1.3. (Animation) Fermi–Dirac distribution $o(E)_{FD}$ with $\mu_c/(k_B T) = 1$ as function of energy per $k_B T$ for varying $k_B T$. The step function at E_F resulting at $T = 0$ is also shown. Animation available online at https://doi.org/10.1088/978-0-7503-1149-6.

Not only can the pressure integral of Equation (1.33) be expressed with the distributions $o(E)_{\mathrm{BE}}$, $o(E)_{\mathrm{FD}}$, or $o(E)_{\mathrm{MB}}$, but also the total number of particles must be recovered when integrating over the state density and its occupation,

$$N = \int_0^\infty o(p)\tilde{\rho}(p)\,\mathrm{d}p. \tag{1.53}$$

This equation can be used to determine μ_c when N is known, see Section 1.5.2 for an example. Likewise, the internal energy of a gas can be written as

$$U = \int_0^\infty E_{\mathrm{part}}(p)o(p)\tilde{\rho}(p)\,\mathrm{d}p. \tag{1.54}$$

1.5.2 Ideal (Maxwell–Boltzmann) Gas

With the help of Equations (1.46) and (1.52), Equation (1.53) can be specified as

$$\bar{n} = \frac{N}{V} = 4\pi\frac{g}{(2\pi\hbar)^3}\int_0^\infty \frac{p^2}{\mathrm{e}^{(E_{\mathrm{part}}(p)-\mu_\mathrm{c})/(\mathrm{k_B}T)}}\,\mathrm{d}p, \tag{1.55}$$

with the particle number density $\bar{n}$. In the non-relativistic case we have $E(p) = p^2/(2m)$ and thus

$$\bar{n} = 4\pi\frac{g}{(2\pi\hbar)^3}\mathrm{e}^{\mu_\mathrm{c}/(\mathrm{k_B}T)}\int_0^\infty p^2\mathrm{e}^{-p^2/(2m\mathrm{k_B}T)}\,\mathrm{d}p = \frac{g}{(2\pi\hbar)^3}\mathrm{e}^{\mu_\mathrm{c}/(\mathrm{k_B}T)}(2\pi m\mathrm{k_B}T)^{3/2}, \tag{1.56}$$

since the remaining integral can be solved analytically. This determines the chemical potential to be

$$\mu_\mathrm{c} = \mathrm{k_B}T\ln\left[\frac{\bar{n}(2\pi\hbar)^3}{g}(2\pi m\mathrm{k_B}T)^{-3/2}\right] \tag{1.57}$$

and thus yields the Maxwell–Boltzmann distribution

$$w(p) = \mathrm{d}\bar{n} = \frac{4\pi\bar{n}p^2}{(2\pi m\mathrm{k_B}T)^{3/2}}\mathrm{e}^{-p^2/(2m\mathrm{k_B}T)} \tag{1.58}$$

to be used in Equation (1.33). It is worth noting that although the peak of distribution, specifying the most probable momentum, is at $\sqrt{2m\mathrm{k_B}T}$, the average momentum is higher, $\sqrt{3m\mathrm{k_B}T}$. This is due to the asymptotically extending tail to higher momentum.

Alternatively, to avoid having to solve the integral of Equation (1.33), the expression for the average kinetic energy of classical, non-relativistic particles in an ideal gas can be used:

$$U = \langle E_{\mathrm{kin}}\rangle = \frac{3\mathrm{k_B}T}{2} = \frac{m}{2}\langle v^2\rangle = \frac{m}{2}\left\langle\frac{p^2}{m^2}\right\rangle = \frac{m}{2}\frac{\langle p^2\rangle}{m^2} = \frac{1}{2m}\langle p^2\rangle. \tag{1.59}$$

Furthermore, because of $p = mv$ and using the above relation, we have

$$\langle pv \rangle = \frac{1}{m}\langle p^2 \rangle = 3\mathrm{k_B}T. \tag{1.60}$$

This can be readily inserted into Equation (1.41) to obtain the pressure

$$P = \frac{w}{3}\langle p \cdot v \rangle = \frac{\bar{n}}{3}\langle p \cdot v \rangle = \frac{\bar{n}}{3}3\mathrm{k_B}T = \bar{n}\mathrm{k_B}T. \tag{1.61}$$

In the above derivation it was assumed that $U = \langle E_{\mathrm{kin}} \rangle = 3\mathrm{k_B}T/2$ is already known. If this is not the case, U and P can be derived by inserting Equation (1.58) into Equations (1.54) and (1.33), respectively.

The entropy of the ideal gas can be simply derived by applying Equation (1.28). According to Equation (1.27), the required grand canonical potential for a Maxwell–Boltzmann gas is given by

$$\widetilde{\Omega}_{\mathrm{MB}} = -\frac{\bar{n}\mathrm{k_B}T}{\rho}. \tag{1.62}$$

Finally, the entropy is obtained as

$$S_{\mathrm{MB}} = \frac{\bar{n}\mathrm{k_B}}{\rho}\left(\frac{5}{2} - \frac{\mu_{\mathrm{c}}}{\mathrm{k_B}T}\right) = \frac{\bar{n}\mathrm{k_B}}{\rho}\left\{\frac{5}{2} + \ln\left[\frac{g}{\bar{n}(2\pi\hbar)^3}(2\pi m\mathrm{k_B}T)^{3/2}\right]\right\}. \tag{1.63}$$

This is the entropy per unit volume, which also is the entropy per mole. Often, the *entropy per baryon* is used, which is obtained by dividing the above entropy by the ideal gas constant. The ideal gas constant is nothing else than $\mathrm{N_A k_B}$ and thus

$$S_{\mathrm{MB}}^{\mathrm{b}} = \frac{\bar{n}}{\rho \mathrm{N_A}}\left(\frac{5}{2} - \frac{\mu_{\mathrm{c}}}{\mathrm{k_B}T}\right) = Y_x\left(\frac{5}{2} - \frac{\mu_{\mathrm{c}}}{\mathrm{k_B}T}\right), \tag{1.64}$$

where Y_x is the abundance of the particle x with mass m and chemical potential μ_{c}.

Equations (1.56)–(1.61) only apply to classical particles moving at non-relativistic speeds. Inserting the relativistic momentum of Equation (1.48) into Equation (1.55) the derivation can be generalized to account for particles reaching relativistic speeds. In this case, however, the integral in Equation (1.56) cannot be solved analytically. Instead, its solution includes a modified Bessel function of the second kind K_2. The full expression for the momentum distribution is called the *Maxwell–Jüttner distribution,*

$$w_{\mathrm{MJ}}(p) = \left(4\pi m^3 \mathrm{c}^3 \frac{\mathrm{k_B}T}{mc^2} K_2(mc^2/(\mathrm{k_B}T))\right)^{-1} \mathrm{e}^{-\gamma_{\mathrm{p}}(p)mc^2/(\mathrm{k_B}T)}, \tag{1.65}$$

with the relativistic γ factor for the momentum

$$\gamma_{\mathrm{p}}(p) = \sqrt{1 + \left(\frac{p}{mc}\right)^2}. \tag{1.66}$$

It should be noted that the Maxwell–Jüttner distribution does not account for pair creation of particles and antiparticles, which will occur when $k_B T$ approaches mc^2. This affects total particle number and chemical potential and requires the introduction of quantum statistics. Despite of elevated temperatures in later burning stages, however, atomic nuclei never attain relativistic speeds in stellar burning and thus it is sufficient to use the standard Maxwell–Boltzmann distribution to account for their pressure contribution (see Section 2.6). Electrons, on the other hand, may become relativistic. Their pressure is given by the relations given in Section 1.5.4.

1.5.3 Photon Gas

The pressure integral derived in Section 1.5.1.1 can also be applied to electromagnetic blackbody radiation, treating it as a "gas" with photons as "particles" with mass zero and all moving at the same speed c. Although photons are bosons with intrinsic spin $s_q = 1$, their degrees of freedom g are not $2s_q + 1$ but rather $g = 2$ for the two helicity states of relativistic particles (spin orientation remains fixedly parallel or anti-parallel to the momentum direction). In contrast to other particle gases, the number of photons N_γ can change because they are massless particles and thus photons of zero energy can be created and destroyed without violating energy conservation. Rearranging Equation (1.23) to

$$\mathrm{d}Q = \mathrm{d}U + P\,\mathrm{d}V - \mu_c\,\mathrm{d}N_\gamma \tag{1.67}$$

shows that with $\mathrm{d}U = 0$ and $\mathrm{d}V = 0$, but $\mathrm{d}N_\gamma \neq 0$, it is only possible to uphold $\mathrm{d}Q = 0$ with a vanishing chemical potential. Using the density of states given by Equation (1.46) and the distribution function of Equation (1.47) with $\mu_c = 0$, the photon number density is obtained as

$$\bar{n}_\gamma = \frac{N_\gamma}{V} = \frac{8\pi}{(2\pi\hbar)^3}\int_0^\infty p^2(\mathrm{e}^{pc/(k_B T)} - 1)^{-1}\,\mathrm{d}p = \frac{8\pi}{c^3}\int_0^\infty \nu^2(\mathrm{e}^{h\nu/(k_B T)} - 1)^{-1}\,\mathrm{d}\nu. \tag{1.68}$$

Here, $E = 2\pi\hbar\nu = pc$ is used, with ν being the photon frequency. Similarly, the internal energy per volume is given by

$$\begin{aligned}\tilde{u}_\gamma = \frac{U}{V} &= \frac{8\pi c}{(2\pi\hbar)^3}\int_0^\infty p^3(\mathrm{e}^{pc/(k_B T)} - 1)^{-1}\,\mathrm{d}p \\ &= \frac{16\pi^2\hbar}{c^3}\int_0^\infty \nu^3(\mathrm{e}^{2\pi\hbar\nu/(k_B T)} - 1)^{-1}\,\mathrm{d}\nu = \int_0^\infty B_\nu\,\mathrm{d}\nu.\end{aligned} \tag{1.69}$$

The expression for the energy density at a given frequency B_ν is also called the *Planck function*. The integrals in the above two equations are not easily solvable. Their solutions include the *Riemann zeta function* ζ. The solution of Equation (1.68) is proportional to $2\zeta(3)$ with

$$\zeta(3) = \frac{7\pi^3}{180} - 2\sum_{n=1}^{\infty}\frac{1}{n^3[\exp(2\pi n) - 1]} \approx 1.202. \tag{1.70}$$

The solution of Equation (1.69) is proportional to $6\zeta(4)$ with

$$\zeta(4) = \frac{\pi^4}{90}. \tag{1.71}$$

The full solutions are

$$\bar{n}_\gamma = \zeta(3)\frac{16\pi k_B^3}{(2\pi\hbar c)^3}T^3 \tag{1.72}$$

and

$$\tilde{u}_\gamma = \frac{8\pi^5 k_B^4}{15(2\pi\hbar c)^3}T^4 = \tilde{a}T^4, \tag{1.73}$$

defining the *radiation density constant* $\tilde{a}$. A numerical value for this constant would be $\tilde{a} = 7.565 \times 10^{-16}$ J K^{-4} m^{-3}. It is related to Stefan's constant $\tilde{\sigma}_S = \tilde{a}c/4$ appearing in the Stefan–Boltzmann law, see Equation (1.3). Comparing Equations (1.72) and (1.73) it follows that $\tilde{u}_\gamma = 2.7\bar{n}_\gamma k_B T$. Hence the average energy of a photon in the gas is $2.7k_B T$, which is to be compared to the average energy $1.5k_B T$ of a particle in a classical, non-relativistic ideal gas (see Equation (1.59)). The maximum of the function in photon frequency shown in Equation (1.69), however, is at $2.82k_B T/h$, showing that the most probable energy $2.82k_B T$ of a photon is slightly higher than the average energy of all photons. Using Equation (1.73) with the pressure integral, the pressure of the photon gas is derived as

$$P_\gamma = \frac{\tilde{a}}{3}T^4. \tag{1.74}$$

A comparison of basic properties of the ideal gas and the photon gas is given in Table 1.2.

The relation to Stefan's constant implies that the intensity I_ν radiated by a blackbody at a given frequency is $c/4$ times the photon energy density at this frequency, given by the Planck function. From Equation (1.69) it can easily be seen that

$$I_\nu\, d\nu = \frac{c}{4}B_\nu = \frac{c}{4}\frac{16\pi^2\hbar}{c^3}\nu^3(e^{2\pi\hbar\nu/(k_B T)} - 1)^{-1}\, d\nu = \frac{2\pi h}{c^2}\nu^3(e^{h\nu/(k_B T)} - 1)^{-1}\, d\nu \tag{1.75}$$

Table 1.2. Comparison of a Maxwell–Boltzmann Gas (Ideal Gas) and a Photon Gas (Radiation)

Quantity	Ideal Gas	Photon Gas
Distribution	Maxwell–Boltzmann, Equation (1.52)	Bose–Einstein, Equation (1.47)
Average energy	$1.5k_B T$	$2.7k_B T$
Most probable energy	$0.5k_B T$	$2.82k_B T$
Pressure	$\bar{n}k_B T = YN_A\rho k_B T$	$\tilde{a}/3T^4$

which is similar to Equation (1.1) but using frequency ν instead of wavelength λ. (Note that $|d\nu| = c/\lambda^2 |d\lambda|$.)

As already described for the ideal gas in Section 1.5.2, also for a photon gas the entropy can be determined by application of Equations (1.27) and (1.28). This leads to

$$S_\gamma = \frac{4\tilde{a}}{3} \frac{T^3}{\rho} \tag{1.76}$$

for the entropy per mole. The entropy per baryon is obtained by dividing Equation (1.76) by $N_A k_B$,

$$S_\gamma^b = \frac{4\tilde{a}}{3} \frac{T^3}{\rho N_A k_B}. \tag{1.77}$$

1.5.4 Degenerate Fermi Gas

Fermions in an astrophysical plasma are electrons and nucleons (neutrons, protons). They have the spin quantum number $s_q = 1/2$ (see also Section 1.7) and $g = 2s_q + 1 = 2$ degrees of freedom. As pointed out in Section 1.5.1.2, the Maxwell–Boltzmann distribution, leading to the relations for the ideal gas, is the limit of the Fermi–Dirac distribution for high temperature and/or a very dilute gas. At lower T, states at low energy become more and more occupied. When only states at or below the Fermi momentum p_F (Equation (1.49)) are occupied, we speak of a degenerate gas. At what temperature this occurs depends on the gas density and the mass of the gas particles because the Fermi energy is given by (see Section 1.5.1.2)

$$E_F = p_F c \frac{(2\pi\hbar)^2}{2m} \left(\frac{3\bar{n}}{4\pi g} \right)^{2/3}. \tag{1.78}$$

Thus, at similar number density $\bar{n}$, the Fermi energy of nucleons is a factor $m_e/m_{\text{nucleon}} \approx 1/1836$ lower than the one of electrons. For heavier atomic nuclei, the Fermi energies are even lower by one to two orders of magnitude. This ensures that the ion part of the plasma (composed of nucleons and nuclei) in stars can be treated as an ideal gas even when the electron component is already degenerate. Therefore, the main application of the relations for a non-relativistic and relativistic Fermi gas, which are derived in the following, is to the electrons in a dense stellar plasma. In terms of electron abundance Y_e, the Fermi energy for degenerate, relativistic electrons in MeV can be estimated by

$$E_F^{e^-} \simeq 4.642 \times 10^{-3} (\rho Y_e)^{1/3}, \tag{1.79}$$

where ρ is the matter density in g cm^{-3}. In the equations following below in this section, $g = 2$ is assumed and g is not specified explicitly anymore.

Since there are no momenta above p_F in a fully degenerate gas, the upper limit for the integration in Equation (1.33) is reduced to

$$
\begin{aligned}
P &= \frac{1}{3}\int_0^\infty vpw(p)\,\mathrm{d}p = \frac{1}{3}\int_0^\infty vp\frac{8\pi p^2}{(2\pi\hbar)^3}\widetilde{\Theta}(p_F - p)\,\mathrm{d}p \\
&= \frac{8\pi}{3(2\pi\hbar)^3}\int_0^{p_F} p^3 v\,\mathrm{d}p,
\end{aligned}
\tag{1.80}
$$

where $\widetilde{\Theta}$ is the *Heaviside step function*, defined as

$$
\widetilde{\Theta}(x) = \begin{cases} 0 & \text{for } x < 0, \\ 1 & \text{for } x \geqslant 0. \end{cases}
\tag{1.81}
$$

1.5.4.1 Non-relativistic Case

For non-relativistic fermions $v = p/m$ and thus

$$
P = \frac{8\pi}{3(2\pi\hbar)^3}\int_0^{p_F}\frac{p^4}{m}\,\mathrm{d}p = \frac{8\pi}{15m(2\pi\hbar)^3}p_F^5.
\tag{1.82}
$$

Since

$$
p_F = \frac{2\pi\hbar}{2}\left(\frac{3\bar{n}}{\pi}\right)^{1/3},
\tag{1.83}
$$

the final expression for the pressure is

$$
P = \frac{(2\pi\hbar)^2}{20m}\left(\frac{3}{\pi}\right)^{2/3}\bar{n}^{5/3} = \frac{(2\pi\hbar)^2}{20m}\left(\frac{3}{\pi}\right)^{2/3}(Y_x \mathrm{N_A}\rho)^{5/3},
\tag{1.84}
$$

where Equation (1.14) was used and Y_x is the abundance of the fermions in the gas, e.g., Y_e for electrons. Note that the pressure is independent of temperature, in contrast to what was found for the ideal gas and the photon gas.

1.5.4.2 Relativistic Case

At particle speeds approaching sizeable fractions of c, the relativistic expression $p = m\gamma_{RT}v$ for the momentum has to be used, with the *relativistic* γ *factor*

$$
\gamma_{RT} = \left(\sqrt{1 - \frac{v^2}{c^2}}\right)^{-1}.
\tag{1.85}
$$

Thus, the expression

$$
v = \frac{p/m}{\sqrt{1 + \frac{p^2}{m^2c^2}}} = \frac{x_{\mathrm{rel}}c}{\sqrt{1 + x_{\mathrm{rel}}^2}}
\tag{1.86}
$$

can be used in Equation (1.80), leading to the general expression

$$P = \frac{8\pi}{3(2\pi\hbar)^3}\int_0^{p_F} p^3 \frac{x_{rel}c}{\sqrt{1+x_{rel}^2}}\,dp \tag{1.87}$$

for the pressure. The factor

$$x_{rel} = \frac{p}{mc} \tag{1.88}$$

indicates the importance of relativistic effects. For $x_{rel} \to 0$, relativistic effects can be neglected, whereas they dominate for $x_{rel} \gg 1$.

For a more general treatment, the integral in Equation (1.87) can be solved analytically by transformation of the integration variable, yielding

$$P = \frac{m^4c^5\pi}{3(2\pi\hbar)^3}\left[{}_Fx_{rel}\sqrt{{}_Fx_{rel}^2+1}\,(2{}_Fx_{rel}^2-3) + \frac{3}{\sinh({}_Fx_{rel})}\right], \tag{1.89}$$

with ${}_Fx_{rel} = p_F/(mc)$. This expression is valid for $0 < x_{rel} < \infty$.

To obtain the pressure in the *ultra-relativistic limit* $v \approx c$, v is replaced by c in Equation (1.80). This results in

$$\begin{aligned} P &= \frac{8\pi c}{3(2\pi\hbar)^3}\int_0^{p_F} p^3\,dp = \frac{8\pi c}{3(2\pi\hbar)^3}\frac{p_F^4}{4} \\ &= \frac{2\pi\hbar c}{8}\left(\frac{3}{\pi}\right)^{1/3}\bar{n}^{4/3} = \frac{2\pi\hbar c}{8}\left(\frac{3}{\pi}\right)^{1/3}(Y_x N_A \rho)^{4/3}. \end{aligned} \tag{1.90}$$

A degenerate, initially non-relativistic Fermi-gas becomes relativistic when the density is increased, since p_F depends on the density. A numerical value for the density beyond which relativistic effects are not negligible anymore is obtained from assuming ${}_Fx_{rel} \approx 1$ and solving for the number density. This yields

$$\begin{aligned} \bar{n} &\approx \frac{8\pi}{3}\left(\frac{mc}{2\pi\hbar}\right)^3, \\ \rho &\approx \frac{8\pi}{3N_A Y_x}\left(\frac{mc}{2\pi\hbar}\right)^3. \end{aligned} \tag{1.91}$$

For a degenerate electron gas, the boundary density is $\rho \approx 9.74 \times 10^8/Y_e$ kg m^{-3}. It should be noted that ${}_Fx_{rel} \approx 1$ implies $v \approx c/\sqrt{2}$.

1.5.5 Ultra-relativistic Gas

For ultra-relativistic particles in a gas, approaching the speed of light, the contribution of the rest mass to their energy is negligible and the relativistic energy equation (see also Equation (1.48))

$$E = \sqrt{p^2c^2 + m^2c^4} \tag{1.92}$$

simplifies to

$$E = pc. \tag{1.93}$$

Two examples for gases involving ultra-relativistic particles were already discussed above, the photon gas in Section 1.5.3 and the pressure of a degenerate, ultra-relativistic Fermi-gas given in Equation (1.90). The photon gas is a boson gas and the properties of all ultra-relativistic boson gases, also with finite particle masses, can be related to it, as can be shown by solving Equations (1.33), (1.53), and (1.54) in the ultra-relativistic limit and comparing to the solutions for the photon gas:

$$\begin{aligned} \text{Number density} \quad & \bar{n}_{\mathrm{UB}} = \frac{g}{2}\bar{n}_\gamma, \\ \text{Internal energy/mol} \quad & u_{\mathrm{UB}} = \frac{g}{2}u_\gamma, \\ \text{Pressure} \quad & P_{\mathrm{UB}} = \frac{g}{2}P_\gamma, \\ \text{Entropy/mol} \quad & S_{\mathrm{UB}} = \frac{g}{2}S_\gamma, \end{aligned} \tag{1.94}$$

where the subscript γ identifies the quantities obtained for the photon gas and g is the number of degrees of freedom, as defined previously.

Also Fermi-gas quantities can be related to the properties of the photon gas in the ultra-relativistic limit. An additional, constant factor appears due to the difference between Bose–Einstein and Fermi–Dirac statistics, see Equations (1.47), (1.50):

$$\begin{aligned} \text{Number density} \quad & \bar{n}_{\mathrm{UF}} = \frac{3g}{8}\bar{n}_\gamma = \frac{3}{4}\bar{n}_{\mathrm{UB}}, \\ \text{Internal energy/mol} \quad & u_{\mathrm{UF}} = \frac{7g}{16}u_\gamma = \frac{7}{8}u_{\mathrm{UB}}, \\ \text{Pressure} \quad & P_{\mathrm{UF}} = \frac{7g}{16}P_\gamma = \frac{7}{8}P_{\mathrm{UB}}, \\ \text{Entropy/mol} \quad & S_{\mathrm{UF}} = \frac{7g}{16}S_\gamma = \frac{7}{8}S_{\mathrm{UB}}. \end{aligned} \tag{1.95}$$

Plasmas at high temperature and comparatively low density are encountered in the advanced burning phases of massive stars (Chapter 6) and in stellar explosions (see, for example, Section 8.3 and Chapter 9). In this case, also leptons, such as electrons and positrons, can contribute. Then effective quantities have to be constructed, accounting for the pressure and entropy contributions of all species in the plasma. For a mixture of a non-relativistic particle gas and a photon gas, this is discussed in Section 2.6. At sufficiently high temperature, electron–positron pairs are created (see Chapter 6 and Figure 6.2). This is also relevant for explosive environments. Here, an example for an effective entropy is given, bridging the two extreme cases of having only radiation entropy S_γ or having an additional contribution by ultra-relativistic e^-, e^+, each contributing to the entropy as given by Equation (1.95):

$$S = S_\gamma + 2f_{\mathrm{ent}}(T)S_{\mathrm{UF}} = S_\gamma\left[1 + \frac{7}{4}f_{\mathrm{ent}}(T)\right]. \tag{1.96}$$

The function $0 \leqslant f_{\mathrm{ent}}(T) \leqslant 1$ is smoothly varying with T and gives the contribution of the ultra-relativistic particles. It has to be fit to the actual environmental conditions. For the innermost layers of a star ejected in a core-collapse supernova explosion (see Section 9.3) it can be approximated by

$$f_{\mathrm{ent}}(T_9) = \frac{T_9^2}{T_9^2 + 5.3}, \tag{1.97}$$

where T_9 is the temperature in GK.

1.6 Basics of Scattering

1.6.1 Interaction Cross Sections

Scattering theory provides a description of particles or waves moving through a medium containing *scattering centers*, e.g., other particles, with which they interact. Kinetic energy and direction of motion may be affected in interactions with the scattering centers. The aim is to obtain a statistical distribution of energy and direction after the scattering process by solving differential equations of motion. This is called the *direct scattering problem*. The *inverse scattering problem* concerns the reconstruction of properties of the scattering centers when knowing the final distribution of scattered particles or waves. Both problems can be approached in a classical framework (where the statistical component arises from unknown or uncertain scattering parameters) or in a quantum mechanical treatment, depending on the underlying physics.

Within the context of this book, (atomic or ion) scattering processes appear in photon and particle motion carrying energy in a stellar plasma, and in the description of nuclear reactions occurring when two atomic nuclei interact. Details are provided in Section 2.7 and Chapter 3, respectively.

Different types of scattering processes can be distinguished. The two main categories are elastic and inelastic scattering. In *inelastic scattering* a fraction of or all kinetic energy in the system of particle and scattering center is used to excite internal degrees of freedom of particle, scatterer, or both. Thus, the kinetic energy of the particle is altered after the interaction. In *elastic scattering*, the kinetic energy before and after the scattering process remains the same. Note that the direction of motion still may have changed. Inelastic scattering can be further subdivided into categories assigned to various microscopic processes using the kinetic energy previously present in the system. Examples are also given in Sections 2.7 and 3.5.

A central quantity in the treatment of scattering processes is the interaction *cross section*. Each type of scattering process can be assigned its specific cross section, their relative sizes measuring the relative probability that the process occurs. Cross sections of individual processes sum up to the *total cross section*, providing a measure of the probability that some interaction happens. (Note that sometimes the term "total cross section" is used in a different manner, to address the integrated differential cross section, see Equations (1.99) and (1.100) below. In this use, it would refer to a single process. The term total cross section is used in this book to denote the cross section given by the sum of all interaction processes at a scattering center.)

The term "cross section" is inspired by the classical notion of the area of interacting particles perpendicular to the direction of their relative motion. In classical physics and without forces acting at a distance, the particles must meet within the areas defined by their geometric cross sections to scatter from each other. For example, the total geometric cross section of two hard spheres with radii r_1, r_2 and only interacting at contact would be the sum of their individual cross sections

$$\sigma_{\text{tot,sph}} = \pi(r_1 + r_2)^2. \tag{1.98}$$

This also illustrates why an interaction cross section carries the dimension of an area. Even in this simple, classical picture the interaction cross section is larger than the actual geometric cross section of either interaction partner (unless one of them is reduced to point size). Interaction cross sections are even less related to the geometric size of a body when taking into account the range of various forces, such as the electromagnetic, gravitational, and nuclear forces. Furthermore, they may depend on the interaction energy.

A more general definition of the cross section is based on the comparison of incident particle flux (current density) F_{inc} and outgoing scattered flux $F_{\text{out}}(\theta, \phi)$ as a function of scattering angle θ and azimuthal angle ϕ (the solid angle Ω, see Figure 1.4). The scattered flux into the solid angle element $\text{d}\Omega$ around (θ, ϕ), for example, can be measured experimentally. The cross section describing the interaction with a single interacting center and resulting particles moving in this solid angle $\text{d}\Omega$ is called *differential cross section* and given by

$$\frac{\text{d}\sigma}{\text{d}\Omega}(\theta, \phi) = \frac{F_{\text{out}}(\theta, \phi)}{F_{\text{inc}}}, \tag{1.99}$$

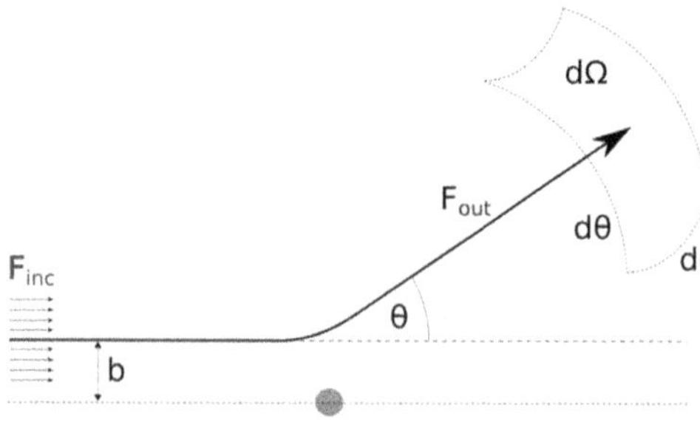

Figure 1.4. Sketch of a scattering process at a scattering center (red dot) with the definition of important quantities.

with the dimension of area per solid angle.

The usual, angle-independent cross section is then given by integrating over the full solid angle,

$$\sigma = \oint_{4\pi} \frac{\mathrm{d}\sigma}{\mathrm{d}\Omega}\,\mathrm{d}\Omega = \int_0^{2\pi} \int_0^{\pi} \frac{\mathrm{d}\sigma}{\mathrm{d}\Omega} \sin\theta\,\mathrm{d}\theta\,\mathrm{d}\phi. \tag{1.100}$$

The integrated cross section σ as well as the differential cross section $\mathrm{d}\sigma/\mathrm{d}\Omega$ may depend on the kinetic energy of the incoming particles.

The integrated cross section of Equation (1.100) can be described as "number of scatterings per scattering center per time divided by incoming particle flux." In classical physics, a particle flux is given in particles per area per time, for instance $\bar{n}v$ with v being the particle velocity and $\bar{n}$ the number density of particles in the beam. The area would be the interaction cross section perpendicular to the beam direction. Further geometric factors would appear if the cross section varies depending on the geometrical orientation of the beam.

In the definition Equation (1.99) it can also be specified that all interactions are considered that scatter particles at an angle ϕ into the region between θ and $\theta + \mathrm{d}\theta$. Then the outgoing particles are moving into an annular solid angle element. The cross section for this is

$$\frac{\mathrm{d}\sigma}{\mathrm{d}\theta} = 2\pi \sin\theta \left(\frac{\mathrm{d}\sigma}{\mathrm{d}\Omega}\right)_\theta \tag{1.101}$$

for constant azimuthal angle θ. Assuming there is axial symmetry in the scattering process it is sufficient to consider the dependence on θ. Figure 1.4 shows the definition of the *impact parameter* b, which is the distance of the asymptote of the particle trajectory from the symmetry axis going through the center of the scatterer. In a classical scattering process the scattering angle depends on b and the energy of the incoming particles, $\theta = \theta(b, E_{\mathrm{kin}})$. Particle number conservation demands that all particles coming through the annular zone around the symmetry axis limited by b and $b + \mathrm{d}b$ are scattered into the annular solid angle element $2\pi \sin\theta\,\mathrm{d}\theta$ and therefore

$$\left(\frac{\mathrm{d}\sigma}{\mathrm{d}\Omega}\right)_\theta = \frac{b}{\sin\theta}\left|\frac{\mathrm{d}b}{\mathrm{d}\theta}\right|. \tag{1.102}$$

An important example for an axially symmetric scattering process is Coulomb scattering, in which a charged particle is scattered on a charged scattering center. This means that their interaction is given by the Coulomb potential, which falls off with $1/r$, where r is the distance from the center. The differential scattering cross section for Coulomb scattering into angle θ as defined in Equation (1.102) is given by

$$\left(\frac{\mathrm{d}\sigma}{\mathrm{d}\Omega}\right)_\theta^{\mathrm{Rutherford}} = \left(\frac{Z_1 Z_2 e_{\mathrm{C}}^2}{4E_{\mathrm{kin}}}\right)^2 \frac{1}{\sin^4(\theta/2)}, \tag{1.103}$$

with the charge numbers Z_1, Z_2 and the interaction energy E_{kin}. This is *Rutherford's scattering formula*. In this form the equation is valid when the scattering center is much more massive than the projectile and E_{kin} is the projectile energy. It is also valid for not largely differing masses m_1, m_2 of projectile and target when using the center-of-mass system. Then E_{kin} is taken to be the *center-of-mass energy*

$$E_{\mathrm{kin}} = \frac{\mu_{\mathrm{red}} v_{\mathrm{rel}}^2}{2}, \tag{1.104}$$

with the relative velocity v_{rel} and the *reduced mass*

$$\mu_{\mathrm{red}} = \frac{m_1 m_2}{m_1 + m_2}. \tag{1.105}$$

A useful relation to convert between laboratory energy E_{lab} and center-of-mass energy E_{cm} is

$$E_{\mathrm{cm}} = \frac{m_2}{m_1 + m_2} E_{\mathrm{lab}}, \tag{1.106}$$

where again m_1 is the mass of the moving projectile and m_2 the mass of the resting target, and $m_2 \geqslant m_1$.

It is interesting to note that the Rutherford cross section can be rewritten to show that the scattering cross section is inversely proportional to the fourth power of the transferred momentum $q = |\vec{p}_{\mathrm{before}} - \vec{p}_{\mathrm{after}}|$,

$$\left(\frac{\mathrm{d}\sigma}{\mathrm{d}\Omega}\right)_\theta^{\mathrm{Rutherford}} = (2Z_1 Z_2 m_1 e_C^2)^2 \frac{1}{q^4}. \tag{1.107}$$

The Rutherford cross section for Coulomb scattering is only correct for point charges. It has to be modified for a spatially extended target with a charge distribution $\rho_C(r)$. It can be shown that in this case the cross section from Equations (1.103), (1.107) is modified only by a factor,

$$\left(\frac{\mathrm{d}\sigma}{\mathrm{d}\Omega}\right)_\theta^{\mathrm{extended}} = \left(\frac{\mathrm{d}\sigma}{\mathrm{d}\Omega}\right)_\theta^{\mathrm{Rutherford}} \mathcal{F}_{\mathrm{form}}^2(q). \tag{1.108}$$

The *form factor* $\mathcal{F}_{\mathrm{form}}$ depends on the shape of the charge distribution, which therefore can be inferred from observed scattering angles. (Note that this does not always uniquely determine the charge distribution.) For example, electrons scattered on atomic nuclei are used to probe the charge density distributions inside atomic nuclei, see Section 3.2. Although this only determines proton distributions inside the nucleus and not the ones of neutrons, it is an important step to learn about the nuclear matter density distribution, see also Equation (3.4). Equation (1.103) assumed spinless projectiles and targets. A further correction has to be applied to account for electrons being fermions with spin 1/2. Furthermore, depending on their energy, they can travel at relativistic speeds. The extension of the Rutherford cross section to electron scattering is called *Mott cross section*. It is given by

$$\left(\frac{\mathrm{d}\sigma}{\mathrm{d}\Omega}\right)_\theta^{\mathrm{Mott}} = \left(\frac{\mathrm{d}\sigma}{\mathrm{d}\Omega}\right)_\theta^{\mathrm{Rutherford}}\left[1 - \frac{v_\mathrm{e}}{\mathrm{c}}\sin\left(\frac{\theta}{2}\right)\right], \tag{1.109}$$

using the electron speed v_e. An extended charge distribution can be probed by electrons using

$$\left(\frac{\mathrm{d}\sigma}{\mathrm{d}\Omega}\right)_\theta = \left(\frac{\mathrm{d}\sigma}{\mathrm{d}\Omega}\right)_\theta^{\mathrm{Mott}} \mathcal{F}^2_{\mathrm{form}}(q). \tag{1.110}$$

Although the Rutherford and Mott cross sections were originally derived from classical mechanics, it was shown later that a quantum mechanical treatment leads to the same result.

Cross sections can also be treated in wave mechanics and therefore also in quantum mechanics. The particle motion is described by the wave functions appearing as the solution of a Schrödinger equation in quantum mechanics (see also Section 3.3.2 for details on the Schrödinger equation). The quantum mechanical flux is defined as

$$\vec{F}_{\mathrm{qu}} = \frac{\hbar}{2im}(\psi^*\vec{\nabla}\psi - \psi\vec{\nabla}\psi^*) \tag{1.111}$$

where m is the particle mass and ψ, ψ^* the particle wave function and its complex conjugate, respectively. In the one-dimensional case of plane waves, this reduces to

$$F_{\mathrm{qu}} = \pm|\psi|^2\frac{\hbar\mathcal{K}}{m}. \tag{1.112}$$

For a non-relativistic particle the momentum p is related to wave number $\mathcal{K}$ and velocity as $p = \hbar\mathcal{K} = mv$. Equation (1.112) then reduces to $F_{\mathrm{qu}} = \pm v|\psi|^2$. Thus, the number density $\bar{n}$ is replaced by $|\psi|^2$ in quantum mechanics. It is interesting to note that in the stationary case of a continuously incoming and fully elastically scattered particle beam, the right-hand side of Equation (1.112) is zero. This is because in this case the superposition of incoming and outgoing waves yields a *standing wave*. Figure 1.5 shows a sketch of such wave scattering.

The stationary scattering problem in quantum mechanics is to determine the wave function constructed from the incoming wave (for the continuous flow of incoming particles) $\psi_\mathcal{K}^-$ and the outgoing wave (having been scattered at the interaction center) $\psi_\mathcal{K}^+$. The asymptotic solution (far away from the interaction center) can be written as

$$\lim_{r\to\infty}\psi_\mathcal{K}(r) = \psi_\mathcal{K}^-(r) + \psi_\mathcal{K}^+(r) = \mathrm{e}^{ikr\cos\theta} + \mathcal{F}_\mathcal{K}(\theta, \phi)\frac{\mathrm{e}^{ikr}}{r}, \tag{1.113}$$

where r is the distance to the interaction center and $\psi_\mathcal{K}^- = \mathrm{e}^{ikz} = \mathrm{e}^{ikr\cos\theta}$ is a plane wave. The function $\mathcal{F}_\mathcal{K}(\theta, \phi)$ is called *scattering amplitude* and describes the

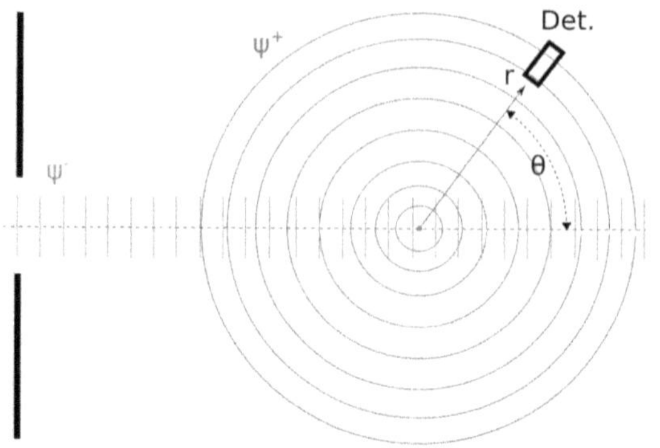

Figure 1.5. Wave scattering: A plane wave ψ^- is emitted from a collimator and scatters at a scattering center (red dot), creating an outgoing spherical wave ψ^+, which is registered in a detector placed at angle θ and distance r. The detector is sufficiently far away from the incident beam so that interference between incoming and outgoing waves can be neglected.

alteration of the wave amplitude by the interaction process. It is directly related to the differential cross section:

$$\frac{\mathrm{d}\sigma}{\mathrm{d}\Omega}(\theta, \phi) = |\mathcal{F}_K(\theta, \phi)|^2. \tag{1.114}$$

This equation is valid for elastic scattering of particles. A scattering amplitude will also appear for inelastic processes, including reactions (see Sections 3.5 and 3.6).

This wave mechanical treatment of cross sections can be applied to all quantum interaction processes. Cross sections for photons interacting with matter (atoms, ions) are used in Section 2.7. Cross sections for interactions of atomic nuclei (nuclear scattering, nuclear reactions) will be discussed in detail in Sections 3.5 and 3.6.

1.6.2 Mean Free Path and Random Walk

The general expression for the *mean free path* of a particle traveling in a medium is

$$l = \frac{1}{\bar{n}\sigma_{\mathrm{int}}}, \tag{1.115}$$

where $\bar{n}$ is the number density of the scattering centers with which the particle can interact and σ_{int} is an interaction cross section as defined in Section 1.6.1, quantifying the probability that an interaction will occur. Assuming that all interaction cross sections are equal, the mean free path defines an average length over which a particle can travel before it will be scattered. The mean free path is characteristic for a medium but can also dependent on the energy (or velocity) of the traveling particle because σ_{int} may depend on it. An important application of this concept is shown in Section 2.7.2, where the paths of photons determine the energy flow in the interior of a star. Another field of interest is in nuclear radiation and other nuclear processes, for which, for example, the penetration depth of radiation or a particle beam into a material is an important quantity to predict material damage and to construct shielding.

At each scattering center the particle will be deflected from a straight path. Therefore it becomes unlikely that travel over a distance longer than the mean free path will take the shortest, straight path. Assuming that the direction change at each scattering is random and that the scattering is elastic, i.e., without loss of kinetic energy of the particle, the resulting path is described by a *random walk*. The assumption of randomness is not only applicable to quantum mechanical interactions but also to classical scattering when the actual collision parameter is unknown.

The distance $D_{\text{mfp}}(N_{\text{int}})$ of the particle from a starting point after N_{int} interactions between which the particle traveled a distance l is simply given by the vector addition

$$D_{\text{mfp}}(N_{\text{int}}) = \left|\vec{D}_{\text{mfp}}(N_{\text{int}})\right| = \left|\sum_{i=1}^{N_{\text{int}}} \vec{l}_i\right|. \tag{1.116}$$

Then the average squared distance $\langle \vec{D}^2_{\text{mfp}} \rangle$ is

$$\langle \vec{D}^2_{\text{mfp}} \rangle = \left(\sum_{i}^{N_{\text{int}}} \langle \vec{l}_i^{\,2} \rangle\right) + 2\left(\sum_{i=1}^{N_{\text{int}}} \sum_{j=i+1}^{N_{\text{int}}} \langle \vec{l}_i \cdot \vec{l}_j \rangle\right) + \cdots \tag{1.117}$$

Further assuming isotropic scattering, i.e., equal probability to be scattered in any direction, and equal mean free paths between interactions, Equation (1.117) simplifies to

$$\langle \vec{D}^2_{\text{mfp}} \rangle = \sum_{i}^{N_{\text{int}}} \langle \vec{l}_i^{\,2} \rangle = N_{\text{int}} \langle \vec{l}^{\,2} \rangle = N_{\text{int}} l^2 \tag{1.118}$$

because all scalar products of the path vectors will average to zero.

The time t_D it takes to cover a distance $D_{\text{mfp}} \approx \sqrt{N_{\text{int}}}\, l$ in such a random walk is proportional to the time taken to cover the length of the mean free path, which depends on the velocity v_{part} of the particle,

$$t_D \approx \sqrt{N_{\text{int}}} \frac{l}{v_{\text{part}}}. \tag{1.119}$$

1.7 Spin, Parity, and Selection Rules in Quantum Mechanics

Atoms, atomic nuclei, and fundamental particles in the stellar plasma have to be treated in quantum physical approaches. These usually use the Schrödinger picture of quantum mechanics, which describes particle behavior by wave mechanics of an abstract probability wave function obtained by solving the Schrödinger wave equation. Such approaches have already been referred to in Sections 1.5 and

1.6.1. Schrödinger equations and their solutions will be further discussed in Section 3.3.2 and later. Here, the focus is on features of angular momentum addition that do not appear in classical mechanics and restrict the emission of particles or radiation in interaction processes. Such selection rules play an important role also in nuclear reactions as discussed in Chapter 3.

The wave solutions of a Schrödinger equation can be labeled by *quantum numbers*. A complete set of quantum numbers identifies a particular solution of the equation with all observables as accurately determined as allowed by quantum physics. The observables connected to the quantum numbers are eigenvalues (stable solutions) of the equation. (For derivations and further details, see any textbook on quantum mechanics, for example Messiah 2014.)

Some of these quantum numbers, like orbital angular momentum, correspond to the quantum mechanical equivalent to a classical angular momentum. Others, such as the spin quantum number, also behave like angular momenta although they do not have a classical counterpart. In quantum physics, all angular momenta are quantized. A direct visualization as a classical vector that is quantized in length as well as in orientation, however, is only possible in exceptional cases.

Assume that $\hat{\vec{\mathcal{Q}}}$ is a quantum mechanical angular momentum operator. Then the eigenvalue (solution to the equation) of the square of the angular momentum operator, which is an observable, is given by

$$\left| \hat{\vec{\mathcal{Q}}}^2 \right| = \mathcal{Q}(\mathcal{Q} + 1)\hbar^2, \tag{1.120}$$

where $\mathcal{Q}$ is the angular momentum quantum number. Quantum numbers can assume integer or half-integer values. Allowed values have to be spaced by one and are $\mathcal{Q} = 0, 1, 2, \ldots$ or $\mathcal{Q} = 1/2, 3/2, 5/2, \ldots$, respectively. Another observable would be the projection of the angular momentum onto one of the coordinate axes (by definition usually the z-axis is used). Its quantum number $\mathcal{Q}_z$ can assume the values $\mathcal{Q}_z = -\mathcal{Q}, -\mathcal{Q} + 1, \ldots, \mathcal{Q} - 1, \mathcal{Q}$ due to the quantization of the orientation. These are $2\mathcal{Q} + 1$ values. (Simultaneous projection onto another axis does not yield an eigenvalue and thus does not result in a good quantum number.) This projection quantum number is often called *magnetic quantum number* as it can be determined by applying a magnetic field. Without magnetic field, the energy eigenvalues of each solution differing only in $\mathcal{Q}_z$ are identical. Then the state (i.e., the solution) is called *degenerate* (see also Section 1.5.4). Application of an external magnetic field forces the energy eigenvalues to be different because of the interaction of the spins with the magnetic field and thus is a way to remove the degeneracy.

The result of an addition of two angular momenta again is an angular momentum and again has to obey the quantization rules. Quantum mechanically this is understood by composing a quantum system ψ_{add} from two wavefunctions ψ_1, ψ_2, which are eigenfunctions, i.e., solutions of a Schrödinger equation, and have angular momentum quantum numbers $\mathcal{Q}_1$, $\mathcal{Q}_{z,1}$ and $\mathcal{Q}_2$, $\mathcal{Q}_{z,2}$, respectively. The eigenfunction of the composite system with quantum numbers $\mathcal{Q}_{\rm add}$, $\mathcal{Q}_{\rm z,add}$ can be expanded in angular momentum space as

$$\psi_{\rm add} = \sum_{1,2} \mathcal{X}^{\mathcal{Q}_1\mathcal{Q}_{\rm z,1}\mathcal{Q}_2\mathcal{Q}_{\rm z,2}}_{\mathcal{Q}_{\rm add},\,\mathcal{Q}_{\rm z,add}} \psi_1\psi_2. \tag{1.121}$$

The coefficients $\mathcal{X}^{\mathcal{Q}_1\mathcal{Q}_{\rm z,1}\mathcal{Q}_2\mathcal{Q}_{\rm z,2}}_{\mathcal{Q}_{\rm add},\,\mathcal{Q}_{\rm z,add}}$ are called *Clebsch–Gordan coefficients* and vanish when $\mathcal{Q}_{\rm add}$, $\mathcal{Q}_{\rm z,add}$ is not a quantum mechanical addition of $\mathcal{Q}_1$, $\mathcal{Q}_{\rm z,1}$ and $\mathcal{Q}_2$, $\mathcal{Q}_{\rm z,2}$. The square of the Clebsch–Gordan coefficient represents the probability to find the state $\psi_{\rm add}$ as a product state of ψ_1 and ψ_2. The coefficients are tabulated or can be calculated with available computer codes. Sometimes a bracket notation is used for the Clebsch–Gordan coefficient, indicating which spin quantum numbers are added (to the left side of the vertical bar inside the curly brackets) to obtain which total spin quantum numbers,

$$\mathcal{X}^{\mathcal{Q}_1\mathcal{Q}_{\rm z,1}\mathcal{Q}_2\mathcal{Q}_{\rm z,2}}_{\mathcal{Q}_{\rm add},\,\mathcal{Q}_{\rm z,add}} = \{\mathcal{Q}_1\mathcal{Q}_{\rm z,1}\mathcal{Q}_2\mathcal{Q}_{\rm z,2}|\mathcal{Q}_{\rm add},\,\mathcal{Q}_{\rm z,add}\}. \tag{1.122}$$

Other widely used notations for spin addition include the 3j- and 6j-symbols and the Racah coefficients. The *3j-symbols* are related to the Clebsch–Gordan coefficients by

$$\begin{Bmatrix} \mathcal{Q}_1 & \mathcal{Q}_2 & \mathcal{Q}_{\rm add} \\ \mathcal{Q}_{\rm z,1} & \mathcal{Q}_{\rm z,2} & \mathcal{Q}_{\rm z,add} \end{Bmatrix}_{\rm 3j} = \frac{(-1)^{\mathcal{Q}_1-\mathcal{Q}_2-\mathcal{Q}_{\rm z,add}}}{\sqrt{2\mathcal{Q}_{\rm add}+1}}\{\mathcal{Q}_1\mathcal{Q}_{\rm z,1}\mathcal{Q}_2\mathcal{Q}_{\rm z,2}|\mathcal{Q}_{\rm add},\,(-\mathcal{Q}_{\rm z,add})\}. \tag{1.123}$$

The *6j-symbols* are defined as a sum over a product of four 3j-symbols,

$$\begin{aligned}\begin{Bmatrix} \mathcal{Q}_1 & \mathcal{Q}_2 & \mathcal{Q}_3 \\ \mathcal{Q}_4 & \mathcal{Q}_5 & \mathcal{Q}_6 \end{Bmatrix}_{\rm 6j} = &\sum_{\mathcal{Q}_{\rm z,1}\mathcal{Q}_{\rm z,2}\dots\mathcal{Q}_{\rm z,6}} (-1)^{\sum_{k=1}^{6}(\mathcal{Q}_k-\mathcal{Q}_{{\rm z},k})} \\ &\times \begin{Bmatrix} \mathcal{Q}_1 & \mathcal{Q}_2 & \mathcal{Q}_3 \\ -\mathcal{Q}_{\rm z,1} & -\mathcal{Q}_{\rm z,2} & -\mathcal{Q}_{\rm z,3} \end{Bmatrix}_{\rm 3j} \\ &\times \begin{Bmatrix} \mathcal{Q}_1 & \mathcal{Q}_5 & \mathcal{Q}_6 \\ \mathcal{Q}_{\rm z,1} & -\mathcal{Q}_{\rm z,5} & \mathcal{Q}_{\rm z,6} \end{Bmatrix}_{\rm 3j} \begin{Bmatrix} \mathcal{Q}_4 & \mathcal{Q}_2 & \mathcal{Q}_6 \\ \mathcal{Q}_{\rm z,4} & \mathcal{Q}_{\rm z,2} & -\mathcal{Q}_{\rm z,6} \end{Bmatrix}_{\rm 3j} \\ &\times \begin{Bmatrix} \mathcal{Q}_4 & \mathcal{Q}_5 & \mathcal{Q}_3 \\ -\mathcal{Q}_{\rm z,4} & \mathcal{Q}_{\rm z,5} & \mathcal{Q}_{\rm z,3} \end{Bmatrix}_{\rm 3j}. \end{aligned} \tag{1.124}$$

The summation is over all six magnetic quantum numbers, as allowed by the 3j-symbols. Finally, the 6j-symbols are related to the *Racah coefficients* $\mathcal{W}_{\rm R}$ used for recoupling spins by

$$\begin{Bmatrix} \mathcal{Q}_1 & \mathcal{Q}_2 & \mathcal{Q}_3 \\ \mathcal{Q}_4 & \mathcal{Q}_5 & \mathcal{Q}_6 \end{Bmatrix}_{\rm 6j} = (-1)^{\mathcal{Q}_1+\mathcal{Q}_2+\mathcal{Q}_4+\mathcal{Q}_5}\mathcal{W}_{\rm R}(\mathcal{Q}_1\mathcal{Q}_2\mathcal{Q}_5\mathcal{Q}_4;\,\mathcal{Q}_3\mathcal{Q}_6). \tag{1.125}$$

All the above coefficients and symbols are extensively used in nuclear physics to express spin addition rules in a concise manner.

The addition rules can be summarized as

$$|\mathcal{Q}_1 - \mathcal{Q}_2| \leqslant \mathcal{Q}_{\text{add}} \leqslant \mathcal{Q}_1 + \mathcal{Q}_2, \tag{1.126}$$

$$\mathcal{Q}_{z,\text{add}} = -\mathcal{Q}_{\text{add}}, \mathcal{Q}_{\text{add}} + 1, \ldots, \mathcal{Q}_{\text{add}} - 1, \mathcal{Q}_{\text{add}}. \tag{1.127}$$

Equation (1.126) is also called the *triangle inequality*. Values of $\mathcal{Q}_{\text{add}}$ are quantized and therefore two possible values allowed by Equation (1.126) have to be spaced by integer values, at least by 1.

A special kind of angular momentum quantum number is the spin quantum number of a particle. It behaves as any other angular momentum quantum number and also obeys Equations (1.126) and (1.127). If it is a composite particle, the spin quantum number can be visualized intuitively as the sum of the angular momenta of its constituents, in analogy to classical mechanics. This is not possible for fundamental particles such as electrons, however, because these are pointlike particles in the standard model of particle physics and therefore the classical notion of angular momentum of a rotating body does not apply. Nevertheless, it is often spoken of "intrinsic spin" of a particle. For example, electrons, neutrons, and protons are Fermions, that means they have an intrinsic spin quantum number $s_q = 1/2$, which has to be added to their angular momentum to obtain the total spin.

The wavefunctions as solutions of a Schrödinger equation bear another property, which is an observable and thus a good quantum number, the *parity*. Mathematically, parity is related to the concept of even and odd functions and signals how a function changes when reflected at the origin. A (wave)function with $\psi(\vec{r}) = \psi(-\vec{r})$ is said to have even parity. Odd parity is assigned to the behavior $\psi(\vec{r}) = -\psi(-\vec{r})$. For example, $\sin(x)$ has odd parity whereas $\cos(x)$ has even parity. In polar coordinates the spatial reflection $\vec{r} \to -\vec{r}$ translates to $r \to r$, $\theta \to \pi - \theta$, $\phi \to \pi + \phi$.

In quantum physics, parity π_q is represented as eigenvalues obtained by application of the parity operator $\widehat{P}$. There are only two eigenvalues, $\pi_q = +1$ for even functions and $\pi_q = -1$ for odd functions. The parity of a composite system ψ_{add} is the *product* of the individual parities of the constituents,

$$\pi_{q,\text{add}} = \pi_{q,1}\pi_{q,2}. \tag{1.128}$$

The equation of motion in classical physics is parity invariant. Likewise, parity is not changed by gravitational or electromagnetic interactions. Parity is also conserved in nuclear reactions conveyed by the strong nuclear force (see Chapter 3). On the other hand, the weak interaction, responsible for nuclear β-decay (see Sections 4.1.6 and 4.1.7), does not conserve parity.

Quantized electromagnetic radiation also obeys conservation laws for angular momentum and parity. Photons carry angular momentum according to the multipolarity $\mathcal{M}$ of the radiation. The multipolarity can assume the values $\mathcal{M} = 1, 2, 3, \ldots$ for dipole, quadrupole, octupole, ... radiation, respectively. Monopole radiation ($\mathcal{M} = 0$) does not exist. When added to another angular momentum quantum number (for example of the atomic or nuclear state from which the photon is emitted), Equation (1.126) applies as for any other angular momentum addition.

Regarding parity, two types of radiation have to be distinguished: electric multipoles (classically motivated by changing electric field because of a moving charge) and magnetic multipoles (changing magnetic field by varying current or magnetic moment). The parity assigned to each of these two types is

$$\pi_{\rm q} = (-1)^{\mathcal{M}} \quad \text{for electric radiation,} \tag{1.129}$$

$$\pi_{\rm q} = (-1)^{\mathcal{M}+1} \quad \text{for magnetic radiation.} \tag{1.130}$$

The notation for electromagnetic radiation shows the type of the radiation together with its multipolarity: E1 for electric dipole radiation, M1 for magnetic dipole radiation, E2 for electric quadrupole radiation, and so on. Because of the differing parities for electric- and magnetic-type radiation only one of the types can be emitted if angular momenta and spins of initial and final state (atomic or nuclear) are fixed. It further is to be noted that radiation emitted from states with zero angular momentum or between states with angular momentum 1/2 is pure. This means that only photons of a single multipolarity are emitted and only of either electric- or magnetic-type. Moreover, radiation between states with zero angular momentum and identical parity is forbidden. Examples for allowed photon transitions are given in Table 1.3.

The emission probability for a given radiation type (electric or magnetic) drops by several orders of magnitude with each increase in multipolarity. In consequence, for example, radiation will be E1 even when E3 would be allowed according to the angular momentum and parity summation rules. For a given multipolarity, on the other hand, an electric-type radiation is more probable than a magnetic-type. This may lead to the situation that, for example, E2 radiation may have comparable probability to M1 radiation if both are allowed. Allowed but suppressed transitions are put in brackets in Table 1.3.

Emission probabilities can be calculated for the emission of photons from electron shells in atoms very accurately by applying quantum electrodynamics (QED). Emission probabilities from processes in atomic nuclei, on the other hand,

Table 1.3. Allowed Radiation in Transitions from Given Initial State to Final State, Tabulated According to Difference $\Delta\mathcal{Q}$ in Angular Momentum Quantum Numbers and in Parity Quantum Numbers of Initial and Final States

$\lvert\Delta\mathcal{Q}\rvert$ Parity Change?	0 (no 0 → 0)	1	2	3	4	5
Yes	E1	E1	M2	E3	M4	E5
	(M2)	(M2)	E3	(M4)	E5	(M6)
No	M1	M1	E2	M3	E4	M5
	E2	E2	(M3)	E4	(M5)	E6

strongly depend on the type of process, the structure of the initial and final states, and further electromagnetic properties. Their calculation is difficult and beyond the scope of this book. Two approximate formulae for the emission probabilities of the two types of radiation for nuclear γ-radiation appearing in nuclear reactions and de-excitations (see Chapter 3) are given below. The emission rate for electric-type radiation can be approximated by

$$\lambda_{\mathrm{E}}(\mathcal{M}) = \frac{4.4 \times 10^{21}(\mathcal{M}+1)}{\mathcal{M}[(2\mathcal{M}+1)!!]^2}\left(\frac{3}{\mathcal{M}+3}\right)^2\left(\frac{E_\gamma}{197}\right)^{2\mathcal{M}+1} R_{\mathrm{nuc}}^{2\mathcal{M}} \tag{1.131}$$

and for magnetic-type radiation by

$$\lambda_{\mathrm{M}}(\mathcal{M}) = \frac{1.9 \times 10^{21}(\mathcal{M}+1)}{\mathcal{M}[(2\mathcal{M}+1)!!]^2}\left(\frac{3}{\mathcal{M}+3}\right)^2\left(\frac{E_\gamma}{197}\right)^{2\mathcal{M}+1} R_{\mathrm{nuc}}^{2\mathcal{M}-2}. \tag{1.132}$$

In the above equations, the energy of the γ-ray E_γ has to be entered in units of MeV and the nuclear radius R_{nuc} (see Section 3.2) in fm. Then the emission rate is obtained in units of s^{-1}. Equations (1.131) and (1.132) also define the *Weisskopf units*, abbreviated w.u. and often used in nuclear physics. Actual, measured γ-emission rates often are given as multiples of the values obtained from these equations. Further properties of nuclear γ-emission are discussed in Section 3.7.1.2 (see Equations (3.138) and (3.139)).

Further Reading

Carroll, B. W., & Ostlie, D. A. 2017, An Introduction to Modern Astrophysics (2nd ed.; Cambridge: Cambridge University Press)

Iliadis, C. 2015, Nuclear Physics of Stars (2nd ed.; New York: Wiley)

Kippenhahn, R., Weigert, A., & Weiss, A. 2012, Stellar Structure and Evolution (2nd ed.; Berlin: Springer)

Landau, L. D., & Lifshitz, E. M. 1958, Theoretical Physics V Statistical Mechanics (London: Pergamon)

Messiah, A. 2014, Quantum Mechanics (two volumes in one) (New York: Dover)

Ostlie, S. 1958, An Introduction to the Study of Stellar Structure (New York: Dover)

Phillips, A. C. 1999, The Physics of Stars (2nd ed.; Chichester: Wiley)

Reif, F. 1965, Fundamentals of Statistical and Thermal Physics (McGraw-Hill: New York)

Ryan, S. G., & Norton, A. J. 2015, Stellar Evolution and Nucleosynthesis (Cambridge: Cambridge University Press)

Thomas Rauscher

Chapter 2

Stellar Models

2.1 Introduction

Modeling the internal structure of stars and the changes it undergoes with time is a complex endeavor, the details of which are beyond the scope of this introductory book. Important for understanding stellar evolution from the first ignition of nuclear burning to the final states of stars in the form of compact objects is the connection of stellar plasma properties to the nuclear reactions releasing energy and changing the plasma composition. Nuclear reaction networks have to be coupled to the hydrostatic or hydrodynamic equations describing the properties of the stellar plasma. Reaction networks are discussed in Section 4.2 and the sequence of hydrostatic burning phases of stars with their respective key reactions are presented in Chapter 7. In this chapter we are concerned with understanding the basic features of hydrostatic burning and how it affects the properties of stars.

Only a simple introduction into the topic can be given here, constricted mainly to one-dimensional, spherical symmetric treatments of stellar structure and energy transport. Nevertheless, a basic understanding of many features can be conveyed through such a treatment and thus provides the basis for further expansion on the topics in other courses. The spherical symmetric approach may appear overly simplistic in the days of advanced computer simulations but it still describes early burning phases well. Convection clearly is a 3D effect but can be effectively included by using mixing length theory (see Section 2.7.4).

What is implied by speaking of a *stellar model* is a table of plasma properties (such as matter density ρ, temperature T, pressure P, composition Y) as a function of a coordinate covering the interior of the star. A straightforward choice for this coordinate would be the radius r, running from $r = 0$ at the center of the star to $r = R$ at the stellar surface. All structure equations could be cast in this *Eulerian coordinate system* but it has proven to be more practical to use another choice of coordinates

doi:10.1088/2514-3433/ab8737ch2

because $R = R(t)$ may strongly vary with time t during stellar evolution. The total mass $M = M(t)$ of a star varies much less, except for phases with strong stellar winds. The *Lagrangian coordinate system* uses the enclosed mass $m = m(r)$ as independent variable, which assumes values $0 \leqslant m \leqslant M$. Eulerian and Lagrangian coordinates can be converted by realizing that the general transformation rule between two coordinates r and m is

$$\frac{\partial}{\partial r} = \frac{\partial}{\partial m}\frac{\partial m}{\partial r}. \tag{2.1}$$

The partial derivative symbols ∂ indicate as usual that one independent variable (m or r) is held constant while differentiating with respect to the other. If there is a dependence on time, the general time derivative operator would be

$$\left(\frac{\partial}{\partial t}\right)_m = \frac{\partial}{\partial r}\left(\frac{\partial r}{\partial t}\right)_m + \left(\frac{\partial}{\partial t}\right)_r. \tag{2.2}$$

The subscripts indicate which of the variables is kept constant. This is the so-called *substantial time derivative of hydrodynamics*, describing the temporal change of a physical property when following a mass element. Using

$$m(r) = \frac{4}{3}\pi\rho r^3 \tag{2.3}$$

and applying Equation (2.1) leads to

$$\frac{\partial r}{\partial m} = \frac{1}{4\pi r^2 \rho}, \tag{2.4}$$

describing the connection between the Eulerian and Lagrangian differentials.

In computer-implemented stellar equations, Lagrangian coordinates are used for better numerical stability. In particular, Equation (2.2) provides a significant advantage for time-dependent solutions whereas using only a time derivative in local space $(\partial/\partial t)_r$ would lead to much more complicated terms, especially with respect to convective terms with a velocity $(\partial r/\partial t)_m$.

In the following, the more intuitive Eulerian coordinates are used for most derivations but the Lagrangian counterparts are given for the most important equations. The distinction between partial and full derivatives (∂ versus d) will not be written explicitly anymore but assumed implicitly.

2.2 Hydrostatic Equations of Stellar Structure

The first equation to consider is the one for mass conservation. It provides a relation between enclosed mass $m(r)$ below radius r and matter density ρ, which just follows from the definition of density as already used in Equation (2.3). Written in differential form, this yields for the spherically symmetric case

$$\frac{\mathrm{d}m(r)}{\mathrm{d}r} = \frac{\rho\,\mathrm{d}V}{\mathrm{d}r} = 4\pi r^2 \rho(r). \tag{2.5}$$

The second equation is derived from the equation of motion of a mass element inside the star. In one-dimensional spherical symmetry the mass element is a spherical shell with mass dm and thickness dr (see Figure 2.1). Two forces act on its center-of-mass, the gravitational force

$$dF_G = -G\frac{m(r)\,dm}{r^2} \tag{2.6}$$

and the resulting net force dF_P exerted by the surrounding shells (note the distinction between $m(r)$ and dm, with $m(r + dr) = m(r) + dm$). This force is given by the pressure difference at inner boundary r and outer boundary r + dr multiplied by the surface area dA of the shell

$$dF_P = [P(r) - P(r + dr)]\,dA = -dP\,dA = -dP\frac{dm}{dr}\rho. \tag{2.7}$$

The last equality arises because $dm = \rho\, dV = \rho\, dA\, dr$. The minus sign appears because this force acts against the gravitational force and thus its force vector points opposite to the one of dF_G. The equation of motion $F = m\ddot{r}$ of the mass shell is given by

$$dF_G + dF_P = dm\ddot{r} = dm\frac{d^2r}{dt^2}, \tag{2.8}$$

leading to

$$\frac{dP(r)}{dr} = -\rho\left(\frac{Gm(r)}{r^2} - \frac{d^2r}{dt^2}\right). \tag{2.9}$$

In hydrostatic equilibrium, dF_G and dF_P cancel each other, the mass shell is at rest and thus the acceleration term on the right-hand side vanishes. This yields the second basic stellar structure equation

$$\frac{dP(r)}{dr} = -\rho\frac{Gm(r)}{r^2}. \tag{2.10}$$

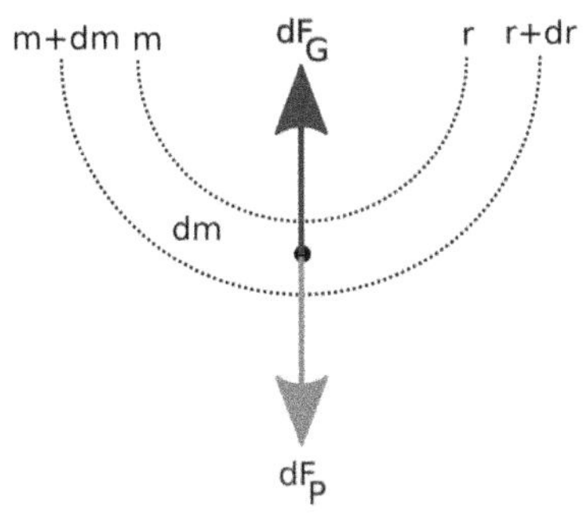

Figure 2.1. Forces on a spherically symmetrical shell.

Using Equation (2.4), it can easily be shown that the Lagrangian form of this second equation is

$$\frac{\mathrm{d}P}{\mathrm{d}m} = -\frac{Gm(r)}{4\pi r^4}. \tag{2.11}$$

The two Equations (2.5) and (2.10) are not sufficient for solving, as they contain the three unknowns ρ, m, P. Thus, an independent third equation is required to provide a further connection between at least two of the three properties. An equation such as $P = P(\rho,\ldots)$ provides such a connection and is called the equation-of-state (EOS). A general EOS $P = P(\rho, T, Y, Y_i,\ldots)$ does not only provide a connection between pressure and density but also depends on further thermodynamic variables such as plasma temperature T or composition Y (computed from the sum of individual abundances Y_i of i plasma components). Further variables also require further equations to fully determine the system of equations, as will be discussed later.

2.3 Barotropic and Polytropic EOS

The simplest EOS suited for supplementing Equations (2.5) and (2.10) is the *barotropic* EOS $P = P(\rho)$, which only depends on density and thus allows to construct a well-determined set of equations in combination with Equations (2.5) and (2.10). A special case of the barotropic EOS is the *polytropic* EOS, arising from thermodynamic considerations leading to the pressure being a polytrope of the density. It is written as

$$P = K\rho^{\tilde{\Gamma}} = K\rho^{1+\frac{1}{\tilde{n}}}, \tag{2.12}$$

where K is the polytropic constant, $\tilde{\Gamma}$ the polytropic exponent, and $\tilde{n}$ the polytropic index. (Note that the polytropic index is not necessarily an integer number but can be fractional.) Adiabatic processes are characterized by polytropes and that is why $\tilde{\Gamma}$ is sometimes called the *adiabatic exponent* (or, slightly misleading, adiabatic index). An adiabatic thermodynamic process is described by a polytrope with an adiabatic exponent

$$\tilde{\Gamma} = \frac{C_\mathrm{P}}{C_\mathrm{V}} \tag{2.13}$$

constructed from the ratio of the heat capacities at constant pressure C_P and at constant volume C_V (see also Section 1.4). The polytropic exponent is also a measure of the internal degrees of freedom g in a gas,

$$\tilde{\Gamma} = \frac{1 + g/2}{g/2} = 1 + \frac{2}{g}. \tag{2.14}$$

A classical mono-atomic gas has $g = 3$ (the three spatial directions of motion) and thus $\tilde{\Gamma} = 5/3$. In a molecular gas, additional degrees of freedom absorbing energy—such as rotation, vibration, ionization, dissociation—can be excited and this lowers $\tilde{\Gamma}$.

Various EOSs are discussed in Section 1.5. Among those, the Fermi-gas EOS (Section 1.5.4) has polytropic form, whereas the EOS for an ideal gas (Maxwell–Boltzmann gas, Section 1.5.2) depends on density and temperature and the photon-gas EOS only on temperature. The latter two, however, turn into polytropes when assuming constant entropy. As shown in Section 2.6, polytropic EOSs arise naturally in stars with convection because they show an adiabatic ρ–T structure and with entropy S = const it follows $T = T(\rho)$ and thus $P(T, \rho) = P(\rho)$, even for ideal and photon gases. Table 2.1 summarizes the polytropic EOSs that will be applied to discuss stellar structure and evolution in the following.

2.4 Lane–Emden Equations

Using a polytropic EOS with the hydrostatic equations of stellar structure, it is possible to find closed solutions for the set of equations describing the interior of the star. The dependence on m is eliminated by combining Equations (2.5) and (2.10) using the radial Laplace operator

$$\Delta_{\mathrm{L}} = \frac{1}{r^2}\frac{\mathrm{d}}{\mathrm{d}r}\left(r^2\frac{\mathrm{d}}{\mathrm{d}r}\right). \tag{2.15}$$

This is applied to P/ρ and by making use of Equation (2.10) we obtain

$$\Delta_{\mathrm{L}}\left(\frac{P}{\rho}\right) = \frac{1}{r^2}\frac{\mathrm{d}}{\mathrm{d}r}\left(\frac{r^2}{\rho}\frac{\mathrm{d}P}{\mathrm{d}r}\right) = \frac{1}{r^2}\frac{\mathrm{d}}{\mathrm{d}r}(-\mathrm{G}m(r)) = \frac{1}{r^2}\left(-\mathrm{G}\frac{\mathrm{d}m}{\mathrm{d}r}\right). \tag{2.16}$$

Inserting Equation (2.5) leads to

$$\frac{1}{r^2}\left(-\mathrm{G}\frac{\mathrm{d}m}{\mathrm{d}r}\right) = -\frac{1}{r^2}\mathrm{G}4\pi r^2\rho = -4\pi\mathrm{G}\rho. \tag{2.17}$$

Note that the final result

$$\Delta_{\mathrm{L}}\left(\frac{P}{\rho}\right) = -4\pi\mathrm{G}\rho \tag{2.18}$$

is a Poisson-type equation.

Table 2.1. Polytropic Exponents $\tilde{\Gamma}$ and Indices $\tilde{n}$ of Selected Equations-of-State (EOSs)

EOS	Comment	$\tilde{\Gamma}$	$\tilde{n}$
Fermi gas, non-relativistic, degenerate	e^- pressure	5/3	3/2
Fermi gas, relativistic, degenerate	e^- pressure at high density	4/3	3
Adiabatic photon gas (S_γ = const)	Radiation pressure (Section 2.6)	4/3	3
Adiabatic ideal gas (S_{gas} = const)	Particle pressure (Section 2.6)	5/3	3/2

Equation (2.18) involves the two unknowns $P(r)$ and $\rho(r)$. It can be solved by using a polytropic EOS of the form $P = K\rho^{1+1/\tilde{n}}$ as introduced in Section 2.3. Inserting such an EOS into Equation (2.18) yields a non-linear second-order differential equation for the density inside the star,

$$\frac{1}{r^2}\frac{\mathrm{d}}{\mathrm{d}r}\left(\frac{r^2}{\rho}\frac{\mathrm{d}}{\mathrm{d}r}K\rho^{1+1/\tilde{n}}\right) = -4\pi \mathrm{G}\rho. \tag{2.19}$$

Two boundary conditions are required to solve it. The first is taken to be a constant, the central density ρ_c at the center of the star at $r = 0$ and thus $\rho(0) = \rho_c$. The second boundary condition is to require $\mathrm{d}\rho/\mathrm{d}r = 0$ at $r = 0$. This follows from using a polytropic EOS in Equation (2.10) and realizing that $m(0) = 0$ by definition.

The Poisson Equation (2.19) can be recast when using a coordinate transformation $r \to a_{\mathrm{LE}}\xi$, $\mathrm{d}r \to a_{\mathrm{LE}}\,\mathrm{d}\xi$ and writing

$$\rho(r) = \rho_c \Phi^{\tilde{n}}(\xi). \tag{2.20}$$

The structure function $\Phi(\xi)$ fulfills the boundary conditions $\Phi(0) = 1$ (center of the star) and $\Phi(\xi_1) = 0$ (stellar surface, $\xi_1 = R/a_{\mathrm{LE}}$). It can also be shown that $\mathrm{d}\Phi/\mathrm{d}\xi = 0$ at $\xi = 0$. The normalization a_{LE} is given by

$$a_{\mathrm{LE}} = \left[\frac{(\tilde{n}+1)K\rho_c^{\frac{(1-\tilde{n})}{\tilde{n}}}}{4\pi \mathrm{G}}\right]^{1/2}. \tag{2.21}$$

Equation (2.19) then transforms to

$$\frac{1}{\xi^2}\frac{\mathrm{d}}{\mathrm{d}\xi}\left(\xi^2\frac{\mathrm{d}\Phi}{\mathrm{d}\xi}\right) = -\Phi^{\tilde{n}}, \tag{2.22}$$

with fixed polytropic index $\tilde{n}$. This is called the *Lane–Emden equation* of stellar structure. It allows to determine density and pressure throughout the star after having chosen a polytropic EOS and a central density ρ_c.

Solutions for such types of equations have been studied extensively by mathematicians and astrophysicists since the late 19th century. There are analytical solutions only for three cases:

- $\tilde{n} = 0$: $1 - \frac{\xi^2}{6}$
- $\tilde{n} = 1$: $\frac{\sin \xi}{\xi}$
- $\tilde{n} = 5$: $\frac{1}{\sqrt{1 + \frac{\xi^2}{3}}}$

Numerical solutions can be found for $0 \leqslant \tilde{n} \leqslant 5$, also covering the physically more relevant cases. Also the root ξ_1 of Φ, where Φ becomes zero, has to be determined in this manner. Several solutions are shown in Figure 2.2.

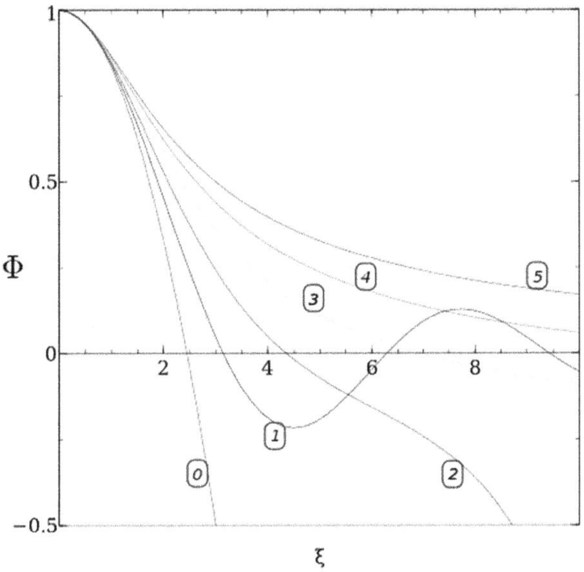

Figure 2.2. Solutions of the Lane–Emden equation, Equation (2.22), for integer polytropic index $\tilde{n} = 0, 1, 2, 3, 4, 5$. Note that the solutions for $\tilde{n} \geqslant 5$ only asymptotically approach the ξ-axis.

Setting ρ_c for a given choice of polytropic EOS also sets the radius R of the object. From $\Phi(\xi_1) = 0$ follows

$$R = a_{\mathrm{LE}}\xi_1 = \left[\frac{(\tilde{n}+1)K}{4\pi G}\right]^{1/2} \rho_c^{\frac{1-\tilde{n}}{2\tilde{n}}} \xi_1. \tag{2.23}$$

On the other hand, the total mass

$$M(R) = \int_0^R 4\pi r^2 \rho \, \mathrm{d}r \tag{2.24}$$

sustained by a central density ρ_c and with given radius R is

$$M(R) = -4\pi\left[\frac{(\tilde{n}+1)K}{4\pi G}\right]^{3/2} \rho_c^{\frac{3-\tilde{n}}{2\tilde{n}}} \left(\xi^2 \frac{\mathrm{d}\Phi}{\mathrm{d}\xi}\right)_{\xi=\xi_1}. \tag{2.25}$$

Often, the total mass M and the EOS (setting $\tilde{n}$) are known. Then the central density of the object can be calculated from Equation (2.25) and the resulting radius R by using this central density in Equation (2.23). Since the pressure throughout the object is determined by the polytropic EOS, also the central pressure P_c is given by

$$P_c = K\rho_c^{1+1/\tilde{n}}. \tag{2.26}$$

Note that the central temperature T_c remains undetermined because the EOS is independent of temperature. Introducing a temperature dependence will be discussed later. For real objects, a purely polytropic behavior can still be an adequate description but it may become necessary to switch to another polytropic EOS at a certain temperature. This will be discussed later.

Inspection of Equation (2.25) shows that the central density becomes undetermined for $\tilde{n} = 3$ ($\tilde{\Gamma} = 4/3$) polytropes because for them the term containing ρ_c becomes constant. This implies that there is no connection anymore between mass and radius, as ρ_c can be freely chosen in Equation (2.23), mathematically speaking. From a physics point-of-view, this describes an unstable equilibrium, a borderline case between bound and unbound, where the total energy (as sum of potential and kinetic energy) of the system is independent of its radius. Although initially stable, such an equilibrium is extremely sensitive to the smallest perturbations. Any tiny perturbation may cause the object to expand or shrink, as no internal forces counteract the perturbative, external forces. An intuitive example for such a fragile equilibrium is a pen perfectly balanced on its point. Although it would be stable in an ideal situation without any further forces acting, the smallest perturbation, such as the faintest flow of air, will topple it. As shown later in Section 2.6, stars have EOS polytropes close to $\tilde{n} = 3$ but sufficiently different to (barely) establish hydrostatic equilibrium and thus the possibility of restoring forces.

2.5 Properties of White Dwarfs

2.5.1 Mass–Radius Relation

As an example of a simple system that can be understood by studying a polytropic EOS is a white dwarf. A *white dwarf* is an endpoint in the evolution of stars with initial masses less than a few solar masses (see Chapter 6). It is basically the hot core of a star that has lost its envelope and does not further contract even when all energy-generating nuclear reactions have ceased and it is cooling. The reason why this object is stable is because the pressure generated by its EOS is sufficient to counteract gravitational contraction.

The main contribution to pressure is from the degenerate gas of electrons in the white dwarf. Other contributions, such as the ones by radiation or atomic nuclei, are negligible. The EOS of the electron gas is the one of a non-relativistic, degenerate Fermi-gas (see Section 1.5.4.1), which is a polytropic equation with $\tilde{\Gamma} = 5/3$, $\tilde{n} = 3/2$ and a polytropic constant

$$K = \frac{(2\pi\hbar)^2}{20m_e}\left(\frac{3}{\pi}\right)^{2/3}(N_A Y_e)^{5/3}. \tag{2.27}$$

For the $\tilde{n} = 3/2$ polytrope,

$$\begin{aligned} \xi_1 &= 3.65375, \\ \left(\xi^2\frac{d\Phi}{d\xi}\right)_{\xi=\xi_1} &= -2.71406. \end{aligned} \tag{2.28}$$

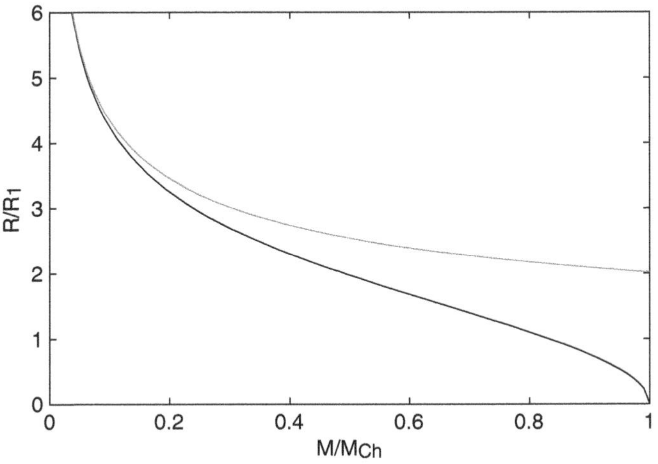

Figure 2.3. Mass–radius relation for a degenerate Fermi-gas ranging from non-relativistic to relativistic conditions (black line) and of a purely non-relativistic, degenerate Fermi-gas (red line). The mass is given in M_{Ch}, the radius in units of R_1 with $R_1 = 5.585 \times 10^{-3}(2Y_e)\ R_\odot$.

Using the above in Equations (2.23) and (2.25) with the K shown in Equation (2.27) yields

$$R = 1.122 \times 10^5 \rho_c^{-1/6}(2Y_e)^{5/6} \tag{2.29}$$

$$M = 4.964 \times 10^{-4} \rho_c^{1/2}(2Y_e)^{5/2}. \tag{2.30}$$

The numerical prefactors in the above equations have been chosen so that the radius appears in km and the mass in solar masses when ρ_c is input in units of g cm^{-3}.

Equations (2.29) and (2.30) can be combined to eliminate ρ_c, arriving at the *mass–radius relation for white dwarfs*,

$$M = 7.011 \times 10^{-1} R^{-3}(2Y_e)^5. \tag{2.31}$$

Here, the numerical prefactor is chosen so that the mass appears in solar masses $M_\odot$ when the radius R is input in units of 10^4 km. The mass–radius relation allows to study what radius is required for a given mass when requiring that the object is stably supported by a non-relativistic, degenerate Fermi-gas. Under the same assumption, it also allows to calculate the mass of such an object with given radius.

According to Table 2.1 and Section 1.5.4.2, the EOS of the degenerate, relativistic Fermi-gas has a polytropic index $\tilde{n} = 3$. In Section 2.4 it was shown that $\tilde{n} = 3$ polytropes do not exhibit restoring forces. Therefore a contraction initiated by the addition of mass will continue when the phase transition $\tilde{n} = 3/2 \rightarrow 3$ occurs. Thus, a white dwarf cannot exceed the mass limit for $\tilde{n} = 3$. Using Equation (2.25) with $\tilde{n} = 3$ leads to the mass limit

$$M_{\tilde{n}=3} = -2.01824\left(\frac{K}{\pi G}\right)^{3/2} 4\pi = 3.062\left(\frac{\hbar c}{G}\right)^{3/2} (N_A Y_e)^2 = 1.44(2Y_e)^2, \tag{2.32}$$

with the numerical prefactor chosen so that the resulting mass is in solar masses ($M_\odot$). The mass $M_{\tilde{n}=3} = M_{Ch}$ is also called the *Chandrasekhar mass.*

The behavior of the mass–radius relation of a white dwarf is shown in Figure 2.3. Increasing mass leads to the counterintuitive behavior of reducing the radius of the object. This is due to the fact of the different dependence of mass and radius on central density. Objects with larger mass also have a larger ρ_c but larger ρ_c implies smaller radius. This leads to the unphysical expectation that the radius would asymptotically shrink to zero for increasing mass. This notion does not consider that the EOS may change with increasing ρ_c. In fact, with increasing density the electrons become more energetic as the Fermi energy is increased (see Equation (1.79)). Electrons in the highest energy levels become relativistic and thus the EOS gradually changes from the one of a non-relativistic, degenerate Fermi-gas (Equation (1.84)) to the one of a relativistic, degenerate Fermi-gas (Equation (1.90)) when the mass is increased. Since this is a local change and the densities may be different at different distances from the center of the white dwarf, it is not possible anymore to describe the whole object with a single EOS. Rather, different shells within the white dwarf may be described by EOSs with different polytropic index, which have to be continued analytically at the shell boundaries. In Figure 2.3 the following approximate relation covering the regimes from non-relativistic to fully relativistic degeneracy is used:

$$\frac{R}{R_1} = 2.02\left[1 - \left(\frac{M}{M_{Ch}}\right)^{4/3}\right]^{1/2}\left(\frac{M}{M_{Ch}}\right)^{-1/3}. \tag{2.33}$$

The surface luminosity of a white dwarf can be related to its mass via Equation (2.31). The surface luminosity of a star depends on its effective surface temperature T_s and radius R (see Section 1.2),

$$\mathcal{L}^* = 4\pi R^2 \frac{\tilde{a}c}{4} T_s^4. \tag{2.34}$$

The radiation density constant is denoted by $\tilde{a}$ (see Section 1.5.3). Using Equation (2.31) to express the radius and inserting it in Equation (2.34), we find the surface luminosity of a white dwarf in terms of solar luminosity $L_\odot$ to be

$$\mathcal{L}^* \simeq \frac{1}{74^2} M^{2/3} T_s^4, \tag{2.35}$$

provided we insert the mass in units of solar masses $M_\odot$ and T_s in multiples of 6000 K. Since no energy is generated in a white dwarf anymore because all nuclear reactions have ceased, it cools down by emitting radiation from its surface (as it is implied by having a luminosity). Its luminosity is very rapidly decreasing with decreasing temperature, by the fourth power of T_s. The cooling track of a white dwarf in the Hertzsprung–Russell diagram is a straight line, which can be seen easily when taking the logarithm of both sides of Equation (2.34) or (2.35),

$$\log \mathcal{L}^* = 4\log(T_s) + \log(\pi\tilde{a}cR^2) \tag{2.36}$$

or

$$\log \mathcal{L}^* = 4\log(T_s) + \frac{2}{3}\log(M) - 2\log(74), \tag{2.37}$$

and remembering that the Hertzsprung–Russell diagram plots the logarithm of luminosity versus the logarithm of temperature. The second term on the right hand sides is an offset between the cooling curves of white dwarfs with different radii, i.e., different masses, since, e.g., $R \propto M^{-1/3}$ as long as a $\tilde{n} = 3/2$ polytrope can be assumed. Thus, there is a (theoretical) line for white dwarfs with $M_{\rm Ch}$. No white dwarf objects should be found beyond this distinct mass boundary at heavier masses and, so far, none have been found observationally. There is also a less precise lower limit for the mass since white dwarfs are final stages of stellar evolution and less massive stars evolve more slowly. In the limited age of the universe, only stars above a certain mass can have formed white dwarfs that are currently observable. This results in the fact that white dwarfs are located only in a rather narrow band in the Hertzsprung–Russell diagram.

The gravitational acceleration on the surface of a white dwarf is much larger than, e.g., the one on the solar surface. Using Equations (2.29) and (2.30), the gravitational acceleration at the surface of a white dwarf with 0.4 $M_\odot$ can be calculated to be approximately 4×10^5 m s^{-2}. Such an acceleration results in an observable gravitational redshift of a photon emitted from the surface of a white dwarf, which can be used to check this result.

In the weak field limit of General Relativity, a photon with frequency $\nu_{\rm ph}$ can be assigned an effective momentum $p_{\rm ph} = 2\pi\hbar\nu_{\rm ph}/{\rm c}$ and an effective mass $m_{\rm ph} = 2\pi\hbar\nu_{\rm ph}/{\rm c}^2$. Such a photon in the gravitational field at radius R has a total energy

$$E_{\rm ph,tot} = E_{\rm ph,kin} + E_{\rm ph,pot} = 2\pi\hbar\nu_{\rm ph} - {\rm G}\frac{m_{\rm ph}M}{R}. \tag{2.38}$$

As the photon is emitted upwards, its potential energy $E_{\rm ph,pot}$ increases and its frequency decreases to compensate as the total energy has to remain constant. Thus, the frequency change is

$$\Delta\nu_{\rm ph} = -{\rm G}\frac{m_{\rm ph}M}{Rh} = -{\rm G}\frac{\nu_{\rm ph}M}{R{\rm c}^2} \tag{2.39}$$

and the fractional change in wavelength is

$$\frac{\Delta\lambda_{\rm ph}}{\lambda_{\rm ph}} = \frac{\Delta\nu_{\rm ph}}{\nu_{\rm ph}} \approx {\rm G}\frac{M}{R{\rm c}^2}. \tag{2.40}$$

Using the mass–radius relation Equation (2.31), the radius cancels out and it is found that $\Delta\lambda_{ph}/\lambda_{ph} \propto M^{4/3}$ for white dwarfs. The calculation can be compared, for example, to the observed wavelength shift for the first discovered white dwarf *40 Eridani B*. Its mass is 0.57 $M_\odot$ and the resulting calculated shift is 7.55×10^{-5}. The observed shift agrees well with this estimate, it is $(7.97 \pm 0.43) \times 10^{-5}$.

2.5.2 More General Chandrasekhar Masses

An object stabilized by the pressure of electron gas cannot support a total mass larger than the numerical limit given by Equation (2.32). Since a white dwarf is nothing else than the core of a star after it has lost its envelope, it is obvious that the evolution of stellar cores and thus the ignition of additional burning phases is also governed by the properties of the electron EOS (as soon as it provides the largest pressure contribution, see Section 2.6). When mass is added to a core due to nuclear burning in surrounding shells after burning in the core has ceased, the core shrinks as required by the mass–radius relation of Equation (2.31). Exceeding M_{Ch}, the contraction would turn into a rapid collapse. Shrinkage or collapse trigger a phase of nuclear burning in the core once its temperature (increasing due to heating from outside shells and due to the release of gravitational binding energy) and density reach sufficiently large values to allow further nuclear reactions. Then another phase of hydrostatic equilibrium is established (see Chapters 6 and 7 for details).

For the cores of evolved massive stars (stellar mass $M > 8\ M_\odot$) further corrections have to be applied to Equation (2.32), which do not play a role in white dwarfs. The degenerate Fermi-gas pressure in the core (Equations (1.84), (1.90)) is reduced by the Coulomb repulsion between charged particles. Three interactions have to be considered: e^-–e^- repulsion, nucleus–nucleus repulsion, and e^-–nucleus attraction (see also Section 4.1.4). The combined effect of these interactions leads to a decrease in pressure by

$$\Delta P_{Coul} = -\frac{3}{10}\left(\frac{4\pi}{3}\right)^{1/3} Z^{2/3} e_C^2 (Y_e N_A \rho)^{4/3}. \tag{2.41}$$

Since the density dependence of this correction is the same as of the relativistic Fermi-gas (Equation (1.90)), this leads to a simple correction factor for the Chandrasekhar mass,

$$M_{Ch}^{Coul} = M_{Ch}\left[1 - 0.0226\left(\frac{Z}{6}\right)^{2/3}\right]. \tag{2.42}$$

The Coulomb correction reduces the critical mass. For example, $M_{Ch}^{Coul} = 0.947 M_{Ch}$ for pure ^{12}C and $0.915 M_{Ch}$ for pure ^{56}Fe.

Another correction to apply is the one for finite entropy when the electron gas is only partially degenerate. Using Equation (1.90) together with Equations (1.78) and (2.27), it can be shown that

$$M_{\mathrm{Ch}}^{*} \simeq M_{\mathrm{Ch}}^{\mathrm{Coul}}\left[1+\left(\frac{\pi \mathrm{k_B} T}{E_{\mathrm{F}}}\right)^{2}\right] \approx M_{\mathrm{Ch}}^{\mathrm{Coul}}\left[1+\left(\frac{S^{\mathrm{b}}}{\pi Y_{\mathrm{e}}}\right)^{2}+\cdots\right], \tag{2.43}$$

with S^{b} being the entropy per baryon as defined in Equation (1.64). This finite entropy correction is not important for white dwarfs because they are too cool. It is, however, very important for understanding the evolution of massive stars. (The relation given in Equation (2.33) and shown in Figure 2.3 is derived in a similar manner as the equation above.)

For other objects deriving their stability from the pressure of a degenerate Fermi-gas, not necessarily electrons, similar considerations apply. For instance, neutron stars are stabilized by a degenerate neutron-gas. A simple estimate of neutron star radii and limiting masses could be made by using the neutron mass instead of electron mass (and Y_{n} instead of Y_{e}) in Equation (2.27) to derive relations similar to Equations (2.31) and (2.32). A realistic numerical value of their mass limit, however, is much more difficult to calculate because of interactions between neutrons affecting the EOS and because of general relativistic effects.

2.6 Mixture of Gas and Radiation in Hydrostatic Equilibrium

2.6.1 Contributions of the Plasma Components

The plasma in the interior of a star is composed of radiation, free electrons, and atomic nuclei. Each of these components contributes to the total pressure,

$$P = P_{\gamma} + P_{\mathrm{e}} + \sum_{i} P_{i}, \tag{2.44}$$

where i enumerates the different types of nuclei present in the plasma. As explained in Section 1.5.4, depending on the density of the gas, electrons may have to be treated as an ideal gas or a degenerate Fermi-gas. Equations (1.51) and (1.57) can be used to select the appropriate treatment. As we have seen in Section 2.5, the electron gas can even become relativistic, further changing the EOS.

If electrons are not degenerate, the pressure is simply given by

$$P = \frac{\tilde{\mathrm{a}}}{3}T^{4} + \left(Y_{\mathrm{e}} + \sum_{i} Y_{i}\right)\mathrm{N_A}\rho \mathrm{k_B} T = \frac{\tilde{\mathrm{a}}}{3}T^{4} + Y\mathrm{N_A}\rho\mathrm{k_B}T. \tag{2.45}$$

This is fulfilled, e.g., in central hydrogen burning of stars with $M \geqslant \mathrm{M}_{\odot}$ and even for advanced burning phases of more massive stars due to lower central densities (see below for an explanation of the dependence of central density on stellar mass).

How does the electron abundance Y_{e} compare to the abundances Y_i of the nuclei? Because of charge neutrality, the electron abundance and total composition are given by Equations (1.16) and (1.17), respectively. The composition of a regular star is mainly hydrogen and helium, with much lower abundances of elements up to Fe and (negligible) traces of even heavier nuclides (e.g., solar metallicity is $\mathcal{Z}_{\odot} \approx 0.0196$). According to Equation (1.17)

$$Y \approx 2X_{\rm H} + \frac{3}{4}X_{\rm He} + \sum_{i>{\rm He}} \frac{Z_i + 1}{A_i} X_i. \tag{2.46}$$

The most abundant nuclei, with the exception of hydrogen, all have $Z_i/A_i \approx 1/2$ and thus

$$Y \approx \frac{1}{2} + \frac{3}{2}X_{\rm H} + \frac{1}{4}X_{\rm He}. \tag{2.47}$$

This shows that, as expected from the term $(Z + 1)/A$, the heavy nuclei mainly contribute to the total composition by adding electrons according to their charge number and not so much by the added nucleus. An expression for the electron abundance can be derived in a similar manner and this leads to

$$Y_{\rm e} \approx X_{\rm H} + \sum_{i \geqslant {\rm He}} \frac{X_i}{2} = \frac{X_{\rm H} + 1}{2}. \tag{2.48}$$

Therefore, we obtain for the sum of nuclide abundances appearing in Equation (2.45)

$$\sum_i Y_i = X_{\rm H} + \frac{1}{4}X_{\rm He} \tag{2.49}$$

by subtracting Equation (2.48) from Equation (2.47). Thus, in a pure hydrogen plasma, electrons and protons provide the same amount of pressure, whereas the relative contribution of electrons increases with increasing amount of heavier elements. At such high densities that the electron gas becomes degenerate, however, electrons provide the dominant pressure contribution in any case, also dominating over radiation pressure. A further discussion of which pressure component dominates at which conditions is provided in Section 2.6.2.

Since the EOS of the ideal gas has a dependence on ρ and T, it is not obvious how a gas mixture of photons and classically behaving particles could be described by a polytropic EOS as introduced in Section 2.3. The additional variable can be eliminated, however, by assuming adiabaticity and thus obtaining another connection between temperature and density through the entropy, see Equations (1.27) and (1.28). The adiabatic approximation is motivated by numerical studies showing that small oscillations in the plasma conserve entropy and that even large scale convection exhibits an adiabatic temperature and density behavior.

The entropies of the ideal gas and the photon gas are given in Equations (1.63) and (1.76), respectively. It is advantageous to parameterize the pressure contributions to the total pressure P of the particle pressure $P_{\rm gas}$ and the radiation pressure P_γ as $P = P_{\rm gas} + P_\gamma = \beta P + (1 - \beta)P$, with a mixing parameter $0 \leqslant \beta \leqslant 1$. For simplicity, we choose to relate the total pressure to the radiation pressure by $P = P_\gamma/(1 - \beta)$. With the help of Equations (1.74) and (1.76), this can be written as

$$P = \frac{1}{1 - \beta}\left(\frac{3}{\tilde{a}}\right)^{1/3}\left(\frac{S_\gamma}{4}\right)^{4/3} \rho^{4/3}, \tag{2.50}$$

which is a $\tilde{n} = 3$ polytropic EOS (see also Equation (2.12)). Realizing that $\beta P = Y \mathrm{N_A} \rho \mathrm{k_B} T$, Equation (2.50) can be recast as

$$\begin{aligned} P &= \frac{1}{1-\beta}\left(\frac{3}{\tilde{a}}\right)^{1/3}\left(\mathrm{N_A k_B} Y \frac{1-\beta}{\beta}\right)^{4/3} \rho^{4/3} \\ &= \left[\frac{3}{\tilde{a}}(\mathrm{N_A k_B} Y)^4 \frac{1-\beta}{\beta^4}\right]^{1/3} \rho^{4/3} = K(\beta)\rho^{4/3}. \end{aligned} \tag{2.51}$$

In Section 2.4 we found that $\tilde{n} = 3$ polytropes only provide a fragile equilibrium. In a realistic stellar model, including energy transport, it is found that the EOS is driven toward $\tilde{n} = 3$ but remains slightly below, thus (barely) allowing for hydrostatic equilibrium. Furthermore, β is not constant throughout the star and thus $K = K(\beta)$ may be different in different zones of the star. Nevertheless, Equation (2.51) is a good approximation to understand some basic stellar properties.

For example, relations between central temperature T_c, central pressure P_c, central density ρ_c, and mass M can be derived. We remember that the pressure component of the gas is the one for an ideal gas

$$P_{\mathrm{gas,c}} = \beta P_c = \beta K \rho_c^{4/3} = \mathrm{N_A} Y \mathrm{k_B} T_c. \tag{2.52}$$

Solving for T_c gives

$$T_c = \beta \frac{K}{\mathrm{k_B N_A} Y} \rho_c^{1/3}. \tag{2.53}$$

The polytropic constant in this equation can be determined from Equation (2.25), which yields

$$K \propto M^{2/3} \tag{2.54}$$

and thus

$$T_c \propto M^{2/3} \rho_c^{1/3}. \tag{2.55}$$

Equation (2.55) has two important implications:

1. For a given stellar burning phase, T_c is fixed within a narrow range defined by the main nuclear reactions keeping the core in hydrostatic equilibrium. This implies that more massive stars undergo a burning phase at lower ρ_c than stars with less mass.
2. For a sequence of burning phases within a star of fixed M,

$$T_c \propto \rho_c^{1/3}. \tag{2.56}$$

Using numerical factors, T_c can be calculated in units of K from

$$T_c \simeq 17.4 \times 10^4 \frac{\beta}{Y} M^{2/3} \rho_c^{1/3}, \tag{2.57}$$

when M is inserted in solar masses $M_\odot$ and ρ_c in units of $g\,cm^{-3}$. Figure 6.2 shows central temperature and density for various evolutionary stages of a star (see Chapter 7 for a detailed explanation of the burning phases), illustrating the law found in Equation (2.56).

Furthermore, combining Equations (2.25) and (2.51), a connection between mass and mixing parameter is found,

$$M(\beta) = -\frac{4}{G^{3/2}}\left(\frac{3}{\pi\tilde{a}}\right)^{1/2}(N_A k_B Y)^2 \frac{\sqrt{1-\beta}}{\beta^2}\left(\xi^2\frac{d\Phi}{d\xi}\right)_{\xi=\xi_1}. \tag{2.58}$$

Inserting the value for ξ_1 and choosing the numerical factors so that the resulting mass is given in units of $M_\odot$, we obtain

$$M \approx 18Y^2\frac{\sqrt{1-\beta}}{\beta^2}. \tag{2.59}$$

We recall that the total pressure contains the contributions from particles and photons as given by the mixing parameter β. The pressure of the ideal gas depends linearly both on ρ and T, see Equation (2.52). The pressure of the photon gas, on the other hand, is independent of ρ but depends on T^4, $P_\gamma = (\tilde{a}/3)T^4$. This means that the radiation pressure strongly increases with even only slightly rising temperature. This implies decreasing β but strongly increasing $\sqrt{1-\beta}/\beta^2$. Thus, stars with larger mass M are called *radiation dominated* because their hydrostatic equilibrium is sustained by radiation pressure. This can more than compensate for the diminishing pressure of the particle gas at lower density ρ.

The above result concerning the domination of radiation pressure in massive stars can also be found when realizing that Equation (2.55) implies that $T_c \propto M/R$ because $\rho_c = M/R^3$. Assuming the gas pressure can be described by the pressure of an ideal gas, depending linearly on density and temperature, we find $P_{gas,c} \propto M^2/R^4$. On the other hand, the radiation pressure depends on T^4 and is independent of the density and thus $P_\gamma \propto M^4/R^4$. Finally, the resulting ratio

$$\frac{P_\gamma}{P_{gas,c}} \propto M^2 \tag{2.60}$$

shows that the pressure of the photons must increase much faster with the stellar mass than the gas pressure.

Equation (2.51) strictly applies only for $S_\gamma = $ const. For the case of an adiabatic volume change, however, actually the total entropy $S_{tot} = S_\gamma + S_{gas}$ is constant (with S_{gas} containing the summed contributions from electrons and atomic nuclei). Such a volume change also affects β and thus also the adiabatic exponent $\tilde{\Gamma}$ (see the definition in Equation (2.12)). A more general expression for $\tilde{\Gamma}$ can be derived for the assumption $S_{tot} = $ const, which leads to

$$\tilde{\Gamma} = \frac{32 - 24\beta - 3\beta^2}{24 - 21\beta}. \tag{2.61}$$

It is easy to verify that this expression reduces to $\tilde{\Gamma} = 4/3$ for a pure photon gas ($\beta = 0$) and to $\tilde{\Gamma} = 5/3$ for a pure particle gas ($\beta = 1$). Radiation dominated stars are less bound than those in which the particle gas provides the dominating pressure contribution, as inferred from the value for $\tilde{\Gamma}$.

2.6.2 Pressure Domains in the ρ–T Diagram

To further understanding, it is convenient to derive the density and temperature ranges where a certain component of the stellar plasma dominates the total pressure. The boundary between the regions where the ideal gas and the photon gas, respectively, provide the dominant contributions is found by equating the expressions for their pressures and solving for T, yielding

$$T = \left(\frac{3\mathrm{N_A} Y}{\tilde{a}}\right)^{1/3} \rho^{1/3}. \tag{2.62}$$

For a gas mix of solar composition, the numerical relations are $T = 3.781 \times 10^7 \rho^{1/3}$ or $\ln T = 7.578 + 1/3 \ln \rho$, with density inserted in kg m^{-3} to obtain the temperature in K.

In a completely analogous manner the boundaries are obtained between the domains of an ideal gas and those of the non-relativistic and relativistic, degenerate Fermi-gases:

$$\begin{aligned} \text{Non-relativistic:} \quad & T = 1.207 \times 10^3 \frac{Y_\mathrm{e}^{5/3}}{Y} \rho^{2/3}, \\ \text{Ultra-relativistic:} \quad & T = 1.496 \times 10^6 \frac{Y_\mathrm{e}^{4/3}}{Y} \rho^{1/3}, \end{aligned} \tag{2.63}$$

again with density and temperature in SI units. For solar composition ($Y_\mathrm{e} = 1.17$) the numerical values are

$$\begin{aligned} \text{Non-relativistic:} \quad & T = 566.8\rho^{2/3} \quad \text{or} \quad \ln T = 2.753 + 2/3 \ln \rho, \\ \text{Ultra-relativistic:} \quad & T = 7.402 \times 10^5 \rho^{1/3} \quad \text{or} \quad \ln T = 5.869 + 1/3 \ln \rho, \end{aligned} \tag{2.64}$$

using the same units as before.

The boundary between non-relativistic and relativistic, degenerate Fermi-gas does not depend on temperature and is given in Equation (1.91). For solar composition, the numerical value is $\rho = 1.14 \times 10^9$ kg m^{-3} or $\ln \rho = 9.057$ ln(kg m^{-3}).

Since the pressure depends only on T and/or ρ, the regions of the dominant pressure contributions to the total pressure can be conveniently identified in a diagram plotting $\ln T$ versus $\ln \rho$. Figure 2.4 shows such a diagram. A star with $M \geqslant 0.08$ $\mathrm{M_\odot}$ starts its life in hydrogen burning in the region where the particle gas dominates the pressure in the stellar core. Equation (2.56) shows that during its

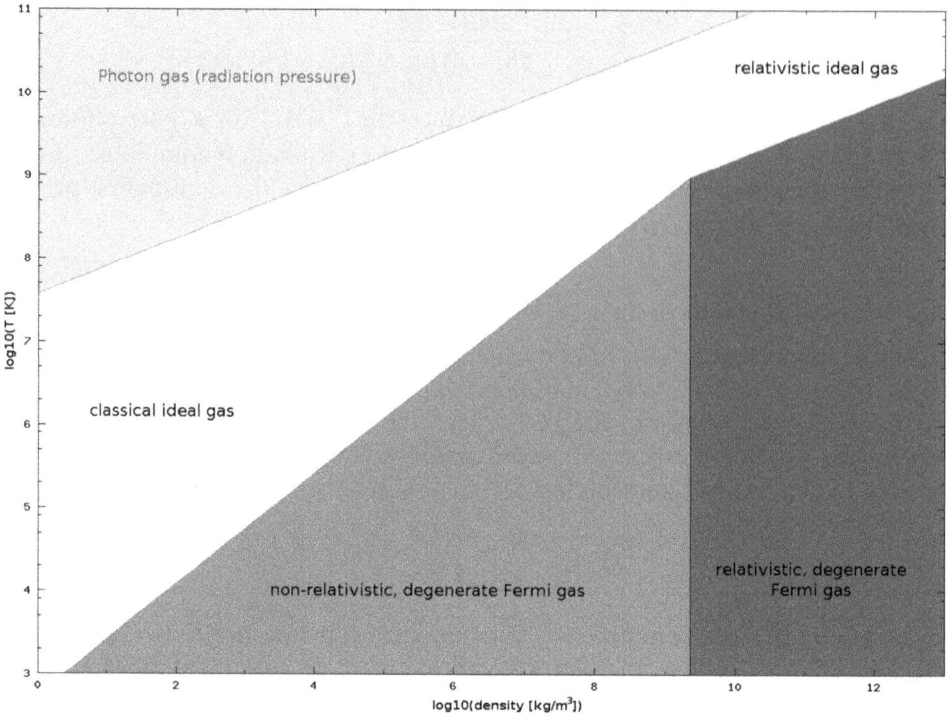

Figure 2.4. Dominant pressure contributions for solar composition shown in the ρ–T plane.

subsequent evolution (see Chapter 6 for more details on stellar burning phases) the core conditions evolve roughly along a track with slope 1/3 in the $\ln T$–$\ln \rho$ diagram. According to the first implication drawn from Equation (2.55), tracks for more massive stars must run parallel but shifted to lower densities, i.e., to the left in the shown diagram. Depending on their mass, their track will cross the boundary to the radiation dominated region at an earlier or later stage in their evolution. On the other hand, the tracks of stars with lower mass run closer to the boundary to the region of dominance of the degenerate Fermi-gas and may enter that region at some evolutionary stage because the slope of the boundary is steeper than 1/3. Hydrostatic equilibrium cannot be sustained by energy production under degenerate conditions because the pressure is independent of the temperature and thus the contracting star cannot self-regulate by increased heat production through nuclear reactions and this will result in a thermonuclear runaway. For further details, see Chapter 6.

2.7 Energy Transport in Stars

2.7.1 Energy Balance in a Mass Shell

All simple cases of stellar models discussed in the previous sections relied on the fact that they could be described by a polytropic EOS, only depending on matter density. This was true even for the case of a mixture of an ideal particle gas and a photon gas which, under the assumption of constant entropy, led to the polytropic EOS shown in Equation (2.51). In the more general case, the EOS is not polytropic anymore but rather additionally depends on further properties, such as temperature and composition. As already pointed out in Section 1.5, further variables entering through the EOS require further equations to solve the full set of hydrostatic equations. Such equations will describe the change of temperature due to energy production and energy transport in different layers of the star and thus will give the temperature gradient $\mathrm{d}T/\mathrm{d}r$ and $\mathrm{d}T/\mathrm{d}m$, respectively.

To start tackling the problem of deriving further relations, the change in the heat content within a stellar layer has to be described. In the spherical symmetric approach, the layer is a spherical mass shell as shown in Figure 2.5, with mass $\mathrm{d}m$, thickness $\mathrm{d}r$, volume $\mathrm{d}V = 4\pi r^2\,\mathrm{d}r$, and heat content Q. In the shell volume, energy can be released (or consumed) by nuclear reactions, with an energy generation rate ε per volume and time. Heat or energy can also flow into or out of the shell. In Section 1.2, the total power output integrated over an area was introduced as stellar luminosity. In that context, Equation (1.4) defined the stellar luminosity $\mathcal{L}^*$ as integrated energy flux through the surface of the star. More generally, in the spherically symmetric case a luminosity $L(r)$ is defined as an integrated energy flux through a sphere with radius r,

$$L(r) = 4\pi r^2 F_\gamma, \tag{2.65}$$

with (photon) energy flux F_γ. Similarly, $L(m)$ can be defined in Lagrange coordinates as flux through a sphere at mass coordinate m. Two boundary conditions (at $r = 0$ and $r = R$) are easily found. At the center of a star $L(0) = 0$. Obviously, the flow at the stellar radius R is just the previously defined stellar luminosity, $L(R) = \mathcal{L}^*$. Furthermore, $\mathcal{L}^*$ comprises the energy generated throughout the star,

$$\begin{aligned} \mathcal{L}^* &= L(R) = \int_0^R 4\pi r^2 \varepsilon(r)\,\mathrm{d}r, \\ \mathcal{L}^* &= L(M) = \int_0^M q(m)\,\mathrm{d}m, \end{aligned} \tag{2.66}$$

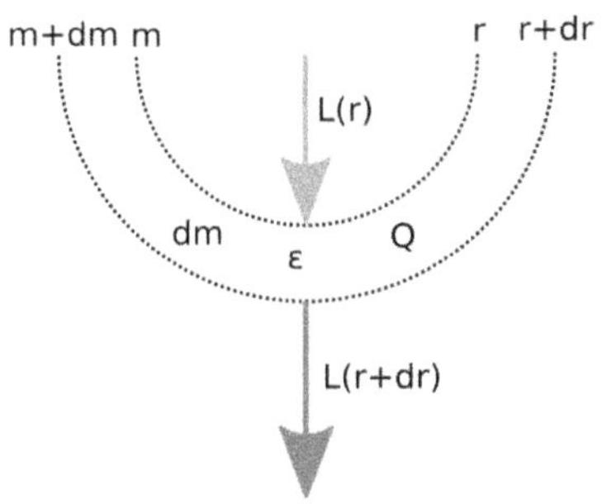

Figure 2.5. Energy balance in a mass shell.

where the second line uses Lagrange coordinates (see Section 2.1). The net energy generation per volume and time is denoted by ε, whereas the net energy generation per mass and time is denoted by q. Since $q\,\mathrm{d}m = \varepsilon\,\mathrm{d}V$ and in spherical symmetry $\mathrm{d}m = 4\pi r^2\rho\,\mathrm{d}r$, $\mathrm{d}V = 4\pi r^2\,\mathrm{d}r$ it is obvious that

$$q = \frac{\varepsilon}{\rho}. \tag{2.67}$$

A part of the energy released by nuclear reactions is emitted as neutrinos. Since these interact only very weakly with the plasma in regular stellar burning phases and thus escape freely without heating the plasma, their contribution is usually not included in the energy generation term. Using the full nuclear energy generation $\varepsilon_{\mathrm{nuc}}$ (q_{nuc}), the energy carried away by neutrinos would have to be subtracted, $\varepsilon = \varepsilon_{\mathrm{nuc}} - \varepsilon_\nu$ ($q = q_{\mathrm{nuc}} - q_\nu$). In a similar manner as above, a *neutrino luminosity* of a star,

$$\begin{aligned}\mathcal{L}_\nu^* &= \int_0^R 4\pi r^2\varepsilon_\nu(r)\,\mathrm{d}r,\\ \mathcal{L}_\nu^* &= \int_0^M q_\nu(m)\,\mathrm{d}m,\end{aligned} \tag{2.68}$$

can be defined, comprising all energy emitted by neutrinos.

The change in heat content within the mass shell is given by the energy produced in the shell (if any) and used to heat matter, and the energy flow into and out of the shell (in Lagrange coordinates),

$$\left.\frac{\mathrm{d}Q}{\mathrm{d}t}\right|_m = q - \frac{\mathrm{d}L}{\mathrm{d}m}. \tag{2.69}$$

According to the first law of thermodynamics, the change in heat has contributions from the change in internal energy and the volume work required to expand or contract the volume, $\mathrm{d}Q = \mathrm{d}U + P\,\mathrm{d}V$. Since a hydrostatic model is used here, only the stationary case without expansion or contraction of the mass shell is considered in the derived equations. In this stationary case, the heat flows into and out of the shell are balanced with the internal energy generation and therefore $\mathrm{d}Q/\mathrm{d}t = 0$. In this case, the luminosity gradient is obtained directly from Equation (2.69) as

$$\begin{aligned}\frac{\mathrm{d}L}{\mathrm{d}m} &= q,\\ \frac{\mathrm{d}L}{\mathrm{d}r} &= 4\pi r^2\rho q = 4\pi r^2\varepsilon.\end{aligned} \tag{2.70}$$

The second line above is obtained by applying Equations (2.1) and (2.4) for transforming Lagrange coordinates to Euler coordinates.

In order to derive the connection between this luminosity gradient and the temperature gradient required to fully solve the hydrostatic equations with a general

EOS, it is necessary to consider energy transport. In the following, the three well-known energy transport mechanisms—radiation (Section 2.7.2), conduction (Section 2.7.3), and convection (Section 2.7.4)—and their relevance in the astrophysical context are discussed in turn.

2.7.2 Radiation

2.7.2.1 Diffusion Treatment

The mean free path of a particle traveling in a medium and interacting with other particles was introduced in Section 1.6.2. Applying Equation (1.115) to photons in a stellar plasma, the interaction cross section σ_{int} includes all possible interactions of a photon with the constituents of the plasma, i.e., scattering on ions and electrons as well as atomic absorption and re-emission. It is convenient to introduce a mean absorption coefficient per unit mass κ and using the density ρ instead of the number density, leading to

$$l_\gamma = \frac{1}{\rho\kappa} \tag{2.71}$$

for the mean free path of a photon. Since l_γ decreases with increasing κ, this mean absorption coefficient is also called *opacity* of the plasma. It follows directly that $1/\kappa$ is the *transparency*. A typical value in stellar matter is $\kappa = 0.1\ \text{m}^2\,\text{kg}^{-1}$. A lower limit for κ is given by pure Thomson scattering on electrons in a fully ionized hydrogen gas (see Equation (2.91)). This yields $\kappa = 0.04\ \text{m}^2\,\text{kg}^{-1}$.

For a better insight into the meaning of the opacity κ it is helpful to note that the attenuation of the intensity I_γ of light traveling the distance x in an opaque medium with density ρ is given by

$$I(x) = I_0 \text{e}^{-\kappa\rho x}, \tag{2.72}$$

where I_0 is the intensity at the starting point $x = 0$.

A very crude estimate for l_γ in a star, e.g., the Sun, can be made by assuming that density and opacity remain constant throughout the star. For example, the mean density of the Sun as estimated from its mass and radius is

$$\overline{\rho}_\odot = \frac{3\ \text{M}_\odot}{4\pi R_\odot^3} \approx 1400\ \text{kg m}^{-3}. \tag{2.73}$$

Using this value with the typical $\kappa = 0.1\ \text{m}^2\,\text{kg}^{-1}$ in Equation (2.71) results in $l_\gamma \approx 0.007$ m. Of course, both density and opacity will vary strongly throughout a star, with considerably higher values in the stellar core than close to the surface. Nevertheless, stellar matter remains a very opaque medium from the center up to close to the surface. Thus, it is possible to describe the path of the photons in almost the whole object by a random walk with a very short mean free path. The distance traveled in a random walk is given by (see Section 1.6.2)

$$D \approx l\sqrt{N_{\text{int}}}, \tag{2.74}$$

with N_{int} being the number of scattering interactions. Since photons are covering the distance of their mean free paths at the speed of light c, the time taken to travel from the center of the Sun to its surface can be estimated to

$$t_{R_\odot} = \frac{R_\odot^2}{c l_\gamma}. \tag{2.75}$$

A time of $t_{R_\odot} \approx 7300$ yr is obtained with $l_\gamma = 0.007$ m. This is a crude estimate, of course, as density and opacity will vary strongly throughout the star. The mean free path is much shorter than the average value especially in the core. On the other hand, close to the surface the opacity drops rapidly. Most of the photons coming from the Sun are actually emitted in the zone of last scattering, called the *photosphere*. At the radius of the photosphere the mean free path of the photons becomes longer than the remaining distance to the surface and thus it becomes unlikely that another scattering occurs on the way out.

Another consideration concerns the question whether temperature changes considerably over the mean free path of a photon. Again, a crude estimate can be made to answer this by calculating the average temperature gradient in the Sun,

$$\frac{\Delta T}{\Delta r} = \frac{T_{core} - T_{surface}}{R_\odot} \approx \frac{1.6 \times 10^7 \text{ K}}{7 \times 10^8 \text{ m}} \approx 2 \times 10^{-2} \text{ K m}^{-1}. \tag{2.76}$$

Therefore the temperature change over the distance of the mean free path is approximately 2×10^{-4} K. This is a tiny difference relative to the actual temperatures encountered in the plasma. The local radiation field can be considered to be embedded in an isothermal layer and the change of T over the mean free path of the photons can be neglected, simplifying the mathematical treatment. Close to the surface, however, this will not be allowed anymore because of the increasing l_γ.

The photons carry energy and therefore photon motion provides an energy flow. Assuming $l_\gamma \ll R_\odot$ and local isothermal layers, it is possible to treat this flow as a diffusion. The general, time-independent diffusion equation for particles is Fick's first law

$$\vec{F}_{part} = -K_\mu \vec{\nabla} \mu_c = -D \vec{\nabla} \bar{n}, \tag{2.77}$$

with the particle flux $\vec{F}_{part}$ between particle densities and diffusion coefficients K_μ, D, which are related by $K_\mu = D\bar{n}/(N_A k_B T)$. The diffusion coefficient D can be defined in terms of the mean free path l of a moving particle as

$$D = \frac{\bar{v} l}{3}, \tag{2.78}$$

where $\bar{v}$ is the average particle speed and the factor 1/3 appears because of the three spatial directions in which a particle can travel. For photons, $\bar{v} = c$. Since the interesting quantity is the radiation flux (energy flux) F_γ, we consider photon flow between energy densities $\tilde{u}_\gamma = \tilde{a}T^4$ (see Equation (1.73)). In spherical symmetry the nabla operator changes into a simple derivative and Equation (2.77) can be written as

$$F_\gamma = -\frac{cl_\gamma}{3}\frac{d\tilde{u}_\gamma}{dr} = -\frac{cl_\gamma}{3}\frac{d\tilde{u}_\gamma}{dT}\left(\frac{dT}{dr}\right)_\gamma, \tag{2.79}$$

where we have introduced a temperature gradient. The derivative of the energy density with respect to temperature is actually the *heat capacity* of the photon gas at constant volume $C_V = d\tilde{u}_\gamma/dT$. Taking this derivative and inserting Equation (2.71) for l_γ, the energy flux

$$F_\gamma = -\frac{cl_\gamma}{3}4\tilde{a}T^3\left(\frac{dT}{dr}\right)_\gamma = -\frac{4\tilde{a}c}{3}\frac{T^3}{\kappa\rho}\left(\frac{dT}{dr}\right)_\gamma \tag{2.80}$$

is obtained. Note the analogy to the expression for thermal conduction

$$\vec{F} = -D_{\mathrm{cond}}\vec{\nabla}T. \tag{2.81}$$

From Equation (2.65) it is seen that the flux can be expressed in terms of luminosity,

$$F_\gamma(r) = \frac{L(r)}{4\pi r^2}. \tag{2.82}$$

Equating Equations (2.80) and (2.82) and solving for the temperature gradient leads to

$$\left(\frac{dT}{dr}\right)_\gamma = -\frac{3}{16\pi\tilde{a}c}\frac{\kappa\rho}{r^2T^3}L(r). \tag{2.83}$$

The same can be expressed in Lagrange coordinates (remember that $dm = 4\pi r^2\rho\, dr$) as

$$\left(\frac{dT}{dm}\right)_\gamma = -\frac{3}{64\pi^2\tilde{a}c}\frac{\kappa}{r^4T^3}L(m). \tag{2.84}$$

This is the temperature gradient caused by radiative energy transport. As already mentioned, the diffusion approximation used to obtain it is only valid for short mean free paths, i.e., large opacity. It does not apply close to the surface of a star where a more complicated treatment has to be applied. Furthermore, as temperature drops toward the surface partially ionized atoms are formed, which add absorption and emission modes for the photons to the ones available in a fully ionized plasma and this affects κ (see below).

2.7.2.2 Opacity and the Rosseland Mean

In the above derivation it is assumed that the opacity is independent of the wavelength or frequency of the photons. In reality, the interaction of photons with electrons, nuclei, and ions depends strongly on the photon energy. Therefore photons in one energy range may be affected by certain interaction processes while photons in another energy range may not. This implies a different mean free path l_γ and a different opacity coefficient κ for each frequency ν. To keep equations as simple as possible, it is desirable to still be able to use a single coefficient, which provides an opacity suitably averaged over frequency. Due to the non-linear dependence of opacity on frequency, however, the averaging has to be more sophisticated than a simple mean value.

Its derivation starts with the use of frequency-dependent versions of the quantities previously appearing in the derivation of Equations (2.83) and (2.84): $F_{\gamma,\nu}$, l_{γ_ν}, κ_ν, D_ν, $\tilde{u}_{\gamma,\nu}$. The subscript ν implies that the values are valid in the frequency range $[\nu, \nu + \mathrm{d}\nu]$. The frequency-dependent energy flux is

$$\vec{F}_{\gamma,\nu} = -D_\nu \vec{\nabla} \tilde{u}_{\gamma,\nu}, \tag{2.85}$$

with

$$D_\nu = \frac{\mathrm{c} l_{\gamma_\nu}}{3} = \frac{\mathrm{c}}{3\kappa_\nu \rho}. \tag{2.86}$$

The frequency distribution in the photon field is given by the Planck function $\tilde{u}_{\gamma,\nu} = B_\nu$ defined in Equation (1.69). To get the total energy flux, Equation (2.85) has to be integrated over all frequencies,

$$\vec{F}_\gamma = \int_0^\infty \vec{F}_{\gamma,\nu}\,\mathrm{d}\nu = -\int_0^\infty \frac{\mathrm{c}}{3\kappa_\nu \rho} \vec{\nabla} \tilde{u}_{\gamma,\nu}\,\mathrm{d}\nu = -\frac{\mathrm{c}}{3\rho}\left(\int_0^\infty \frac{1}{\kappa_\nu}\frac{\mathrm{d}\tilde{u}_{\gamma,\nu}}{\mathrm{d}T}\,\mathrm{d}\nu\right)\vec{\nabla} T. \tag{2.87}$$

The temperature gradient was obtained similarly as for Equation (2.80) and also this result bears similarity to one for conduction. Comparing Equation (2.87) for the spherically symmetric case to Equation (2.80), it can be seen that

$$\frac{1}{\kappa} = \frac{\pi}{\tilde{\mathrm{a}}\mathrm{c}T^3} \int_0^\infty \frac{1}{\kappa_\nu}\frac{\mathrm{d}\tilde{u}_{\gamma,\nu}}{\mathrm{d}T}\,\mathrm{d}\nu. \tag{2.88}$$

This harmonic mean of $1/\kappa_\nu$ over all frequencies ν (i.e., of *transparencies*) with a weight function $C_{V_\nu} = \mathrm{d}\tilde{u}_{\gamma,\nu}/\mathrm{d}T$ is called the *Rosseland average*. It puts more weight on contributions at frequencies ν where the energy flux is largest, as can be seen from Equation (2.87). This is around the maximum of the Planck function at $\nu = 2.8\mathrm{k_B}T/(2\pi\hbar)$. (Note that C_{V_ν} is the frequency-dependent heat capacity at constant volume.) Since the opacity is connected to the mean free path (Equation (2.71)), the Rosseland average for the mean free path is given by

$$l_\gamma = (4\tilde{\mathrm{a}}T^3)^{-1} \int_0^\infty l_{\gamma_\nu} \frac{\mathrm{d}\tilde{u}_{\gamma,\nu}}{\mathrm{d}T}\,\mathrm{d}\nu, \tag{2.89}$$

showing that the longest mean free paths are assigned the largest weight, i.e., frequencies at which the stellar plasma is more transparent. The opacity κ obtained by Equation (2.88) has to be used in Equations (2.83) and (2.84), respectively, to obtain the temperature gradient in the stellar gas caused by electromagnetic radiation. It depends on the composition of the stellar gas as well as on its temperature and density.

2.7.2.3 Opacity Sources

Mean free paths, opacities or transparencies of two mixed gases cannot be simply combined to obtain the opacity of the mixture, as there can be additional interactions between the gases also affecting these properties. Nevertheless, there are certain temperature ranges in which specific interaction processes, usually tied to one atomic species, dominate the opacity and other interactions can be neglected. The temperature of the stellar gas not only determines its degree of ionization but also sets the most probable frequency of the photons.

Principal interaction processes are the following:

- *Bound–bound transitions* excite or de-excite electrons in bound atomic systems. For example, an electron is lifted to a higher, still bound, energy level by absorption of a photon or an electron drops to a lower energy level by emission of a photon. Because of the given energy differences between the energy levels, photons can only be absorbed or emitted at specific frequencies. This results in a spectrum showing discrete lines. Bound–bound transitions are only important in the stellar atmosphere where they give rise to the line spectra of stars. Their correct treatment is a complicated atomic physics problem but essential in the construction of model atmospheres.
- *Bound–free transitions* ionize atoms fully or partially. In such a transition an electron receives enough energy from an absorbed photon to become unbound. The inverse process—capture of an electron into a bound state—is also possible but has a much smaller cross section at stellar temperatures due to phase space considerations. The energy spectrum is continuous but since the process can only occur when the electron receives sufficient energy to become unbound, there is an energy threshold below which the cross section is zero. This means that when the temperature is low, only a small number of photons will have sufficient energy to undergo such a transition. On the other hand, when the temperature is high, all atoms are fully ionized and no electron are in bound states, anymore, which also renders this kind of transitions unimportant.
- *Free–free transitions* occur when a free electron and a photon interact in the vicinity of a heavy atomic nucleus. The electron increases its kinetic energy by absorbing any photon energy. The inverse process of an electron slowing down by emission of electromagnetic radiation is called *Bremsstrahlung*. The energy transfer leads to a continuous spectrum.
- *Thomson scattering* of an isolated electron with a photon can occur at any energy and its cross section (and thus opacity) does not depend on the photon frequency. The cross section for Thomson scattering is given by

$$\sigma_{\mathrm{Thom}} = \frac{8\pi}{3}\left(\frac{e^2}{4\pi\varepsilon_0 m_e c^2}\right)^2 \tag{2.90}$$

and the resulting opacity is

$$\kappa_{\text{Thom}} = \frac{\bar{n}_{\text{e}}\sigma_{\text{Thom}}}{\rho}. \tag{2.91}$$

This can be approximated by $0.02(1 + X_{\text{H}})\ \text{m}^2\ \text{kg}^{-1}$. In a pure, fully ionized hydrogen gas, the Thomson opacity can be further approximated by

$$\kappa_{\text{Thom}} \approx \frac{\sigma_{\text{Thom}}}{m_{\text{p}}}. \tag{2.92}$$

This follows from the charge neutrality of the plasma and the realization that the electron mass is negligible compared to the proton mass.

Both bound–free (above the threshold) and free–free transitions lead to a temperature dependence of the opacity given by Kramer's rule,

$$\kappa \propto \frac{1}{T^{3.5}}, \tag{2.93}$$

implying a rapid drop in opacity toward higher temperature. Since the opacity contribution coming from Thomson scattering is independent of T, it will dominate at high temperature. For a standard (solar) mix of hot gas at densities as found in the interior of the Sun, Thomson scattering dominates the opacity for $T > 10^5$ K. This limit is shifted to higher temperature for higher density.

2.7.2.4 Eddington Limit

The *Eddington limit* or *Eddington luminosity* $\mathcal{L}^*_{\text{Ed}}$ is the maximal surface luminosity a star can achieve when there is a balance between radiation pressure acting outwards and gravitational force acting to contract the star or, in other words, when there is local hydrostatic equilibrium in a radiation-dominated surface zone. When the Eddington limit is exceeded, a radiation-driven *stellar wind* (particle flow from the stellar surface) sets in.

The Eddington luminosity is derived by starting with the equation of motion of a mass shell as given in Equation (2.8) and equating the forces exerted by radiation and by gravity, leading to zero acceleration of the shell. Radiation pressure is related to the photon flux by

$$P_\gamma = \frac{F_\gamma}{\text{c}} \tag{2.94}$$

and the force exerted by radiation is

$$F_\gamma^{\text{force}} = \frac{1}{\rho}\frac{\text{d}P_\gamma}{\text{d}r} = \frac{\kappa}{\text{c}}F_\gamma. \tag{2.95}$$

Inserting this in the second basic stellar structure equation (Equation (2.10)) leads to

$$\frac{\kappa}{\mathrm{c}}F_\gamma(r) = -\frac{\mathrm{G}m(r)}{r^2}. \tag{2.96}$$

Applying Equation (2.82) at the stellar surface ($r = R$) and determining the magnitude of the energy flux by the above equation yields

$$\mathcal{L}^*_{\mathrm{Ed}} = 4\pi R^2 |F_\gamma| = 4\pi R^2 \frac{\mathrm{c}}{\kappa}\frac{\mathrm{G}m(R)}{R^2} = \frac{4\pi \mathrm{cG}M}{\kappa}. \tag{2.97}$$

The opacity κ is determined accounting for the various opacity sources as described above. Originally, Sir Arthur Eddington only applied the opacity $\kappa = \kappa_{\mathrm{Thom}}$ arising from Thomson scattering on electrons in an ionized hydrogen gas. The resulting $\mathcal{L}^*_{\mathrm{Ed}}$ is called the *classical Eddington limit* and is given by

$$\mathcal{L}^*_{\mathrm{Ed}} = \frac{4\pi \mathrm{cG}Mm_{\mathrm{p}}}{\sigma_{\mathrm{Thom}}}, \tag{2.98}$$

making use of Equation (2.92). Expressed in solar luminosities $L_\odot$, this evaluates to

$$\mathcal{L}^*_{\mathrm{Ed}} \simeq 32{,}000\left(\frac{M}{\mathrm{M}_\odot}\right). \tag{2.99}$$

Although Thomson scattering only acts on electrons, it can result in a (proton) matter flow because of electric forces arising when the electrons are driven away.

The Eddington luminosity also limits the accretion onto an object. In an accretion process, hot matter falls into a gravitation well and further heats up through the release of gravitational binding energy (see Section 2.9.1). If the resulting radiation luminosity exceeds the Eddington limit, the radiation pressure acts against the infalling material and prevents further accretion. Since this cuts off the energy supply, radiation pressure may decrease again and this leads to pulsed accretion. From the equivalence of mass and energy $E = mc^2$ the limit for the accretion rate can be derived as

$$\left(\frac{\mathrm{d}m}{\mathrm{d}t}\right)_{\mathrm{accret}} = \frac{\mathcal{L}^*_{\mathrm{Ed}}}{\mu_\gamma \mathrm{c}^2}, \tag{2.100}$$

where $\mu_\gamma \leqslant 1$ describes the efficiency in the conversion to radiation during accretion.

Since many simplifying assumptions enter the derivation of $\mathcal{L}^*_{\mathrm{Ed}}$ it should not be viewed as an absolute limit. Many stars with luminosities below the Eddington luminosity show stellar winds. This indicates that further physical processes are contributing to lifting matter off the stellar surface. On the other hand, stars can exceed the limit for brief episodes without triggering stellar winds. Super-Eddington accretion has been found on black holes and neutron stars. The Eddington luminosity depends on the composition of the stellar matter (for example, in helium gas the radiation acts on two electrons that have to lift a ^{4}He nucleus having the mass of almost four protons; therefore twice the classical $\mathcal{L}^*_{\mathrm{Ed}}$ would be required) and the spectral distribution of the radiation (affecting the opacity). Moreover, the limit was derived considering only one spatial dimension. More complex processes in 2D or 3D include density fluctuations in the stellar atmosphere, turbulences, and the appearance of local photon bubbles, where the radiation pressure locally exceeds the gas pressure. In accretion disks, such effects may support luminosities as high as 10–100 times the Eddington limit without triggering accretion instabilities.

2.7.3 Conduction

When considering thermal motion of photons in the previous section, the resulting equation looked like the general equation for heat transfer by thermal conduction (Equation (2.81)), which is repeated here:

$$\vec{F} = -D_{\mathrm{cond}}\vec{\nabla}T. \tag{2.101}$$

Whether conduction can be an efficient energy transport mechanism depends on the cross sections for particle collisions. Heat transfer by electron–electron collisions has small cross sections and is not very effective because identical particles basically bounce off each other. It is usually neglected. Cross sections for charged-particle collisions increase with Z^2 and therefore are much larger for heavier ions. There are fewer ions than electrons in the stellar plasma, however, and they move much more slowly. Thus, the coefficient D_{cond} is even smaller than for electrons and thermal motion of ions is an even less effective means to transport energy.

For the above reasons, thermal conduction is negligible for energy transport under most stellar conditions. The notable exception is the case of a degenerate electron gas (see Section 1.5.4), which has a high thermal conductivity. This is due to two effects:

1. Increasing Fermi energies increase the typical electron speeds and thus decrease the heat capacity.
2. An electron can only be scattered when the final energy state after scattering is unoccupied. In a degenerate gas, most lower energy states are already filled and therefore the mean free path of an electron will become much longer than in a non-degenerate gas.

The combined action of the two effects leads to an extremely efficient transport of energy, which is important in degenerate stellar cores (see Chapter 6). It is also important in white dwarfs (see Section 2.5), which are also stabilized by the pressure of a degenerate electron gas. This energy transport is so efficient that degenerate cores and white dwarfs exhibit a uniform temperature throughout, with the exception of a thin outside layer where degeneracy breaks down. The same is true for neutron stars, which obtain an isothermal structure through conduction shortly after they have been born as hot proto-neutron stars (see Sections 9.3.1 and 9.4.5).

Compact objects such as white dwarfs and neutron stars do not cool efficiently by thermal radiation from their surface. Therefore neutrino cooling becomes important. This occurs via the Urca process, which is presented in more detail in Section 9.3.1.

2.7.4 Convection

2.7.4.1 Introduction

Convection is a phenomenon well known in daily life, from air rising above a hot surface to boiling water in a kettle. The surface of the Sun exhibits moving structures that are the top end of *convection cells*, similar to, but much larger than, those formed in a large pot filled with liquid and heated on a stove. Hot matter is rising to the surface where it cools. Cooler matter is moving downwards between the streams of upwardly flowing matter, giving rise to the typical pattern. On the solar surface this can be seen as a tight grid of thin, darker, cooler lines delimiting brighter patches of higher temperature.

It is apparent that the matter flow not only transports temperature and therefore energy but also matter of a given composition itself. Convection transports matter over large distances within a star and thereby mixes contents of different layers, thus changing the composition gradient within a star. Convective mixing has profound consequences for nuclear burning and nucleosynthesis. Nuclei can be transported into zones where they were previously depleted or were never present. This allows prolongation of certain nucleosynthesis processes or enabling processes that could not occur without mixing. This not only affects the nucleosynthetic signature of a star but also its evolution, which is determined by nuclear energy production. This will be discussed in Chapter 6. Here, the focus is on convection as an energy transport mechanism.

A comprehensive treatment of convection obviously requires 3D hydrodynamic models, solving Boltzmann transport equations. Full numerical modeling, however, still is limited by current computing power, as are 3D stellar models, and thus most simulations of stellar evolution and nucleosynthesis are still done in spherical symmetry, applying approximations of multi-D phenomena such as convection. Historically, a lot of effort has been put into developing effective theories to be applied in 1D models. These approaches are well suited to provide a basic understanding of the main features and effects of convection (and other multi-D phenomena), which may not be conveyed so easily through mathematically and computationally much more demanding models. The more complex a model, the more likely it is to be used as a black box and the more difficult it becomes to validate it. Providing details on the more involved treatment goes beyond the scope of this introduction. Here, only a simplified discussion is given to convey a basic understanding of the role of convection in a star, still resorting to a 1D approach.

2.7.4.2 Convection Criteria

The most important part in understanding convection is to study when it occurs. Already from daily experience it is evident that convection does not always appear. The heating of water in a pot starts without the immediate formation of convection cells. Although they appear quickly, and before the actual boiling, energy to heat the bulk of water is initially distributed within the pot and along the side walls by other means. Onset of convection requires that matter rises due to an increase in buoyancy. This, in turn, will occur for a pocket (or bubble) only if it is locally heated sufficiently to expand and thus reducing its density. The buoyancy increase has to be stable for an extended duration because if it is only a local fluctuation like a vibration or oscillation, the bubble will be restored to its initial position and will not be able to travel an extended distance.

This behavior can be modeled easily using two assumptions, which do not limit generality:

1. The gas bubble is always in pressure equilibrium with its environment, $P_b = P$;
2. While the bubble is in motion, its entropy S_b remains constant, i.e., it is an adiabatic process without thermal exchange between bubble and environment.

What is the condition for a long-distance motion of the gas bubble due to its buoyancy? If a displaced bubble, after adiabatic expansion, has a higher density than the surrounding gas it will sink back to its original position. Such a stratification will be stable against convection. It is unstable, on the other hand, if the rising bubble, again after adiabatic expansion due to decreasing pressure, has a lower density than the environment it will continue to rise. This stratification is unstable.

To quantify this, consider the situations before and after displacement of a bubble as depicted in Figure 2.6. A bubble is located at an initial position 1 at distance r with a pressure $P^* = P(r) = P_1$, temperature $T^* = T(r) = T_1$, and density $\rho^* = \rho(r) = \rho_1$. The values of these quantities in the surroundings of the bubble are the same as inside. After adiabatic movement of the bubble to a location 2 at a distance $r + \mathrm{d}r$, the values in the bubble and the environment change to the values shown in Table 2.2. Because of the pressure equilibrium, environment and bubble always keep the same pressure. Temperature and density, however, evolve differently inside and outside the bubble. As just mentioned, the densities inside and outside the bubble have to be compared to decide whether the gas bubble is buoyant. Therefore, a restoring force preventing further motion will be acting if $\mathrm{d}\rho^*/\mathrm{d}r > \mathrm{d}\rho/\mathrm{d}r$ (density in

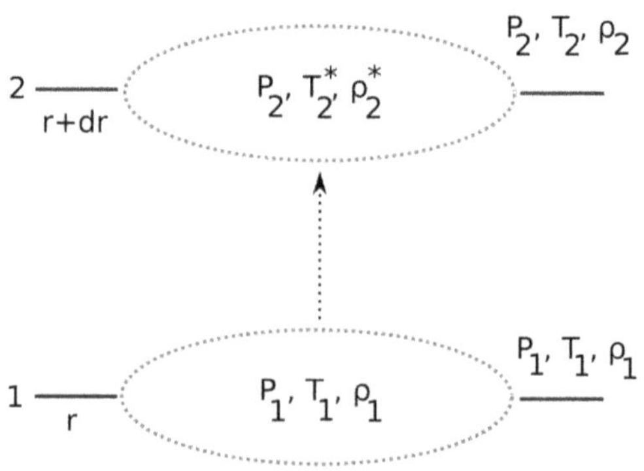

Figure 2.6. Displacement of a gas bubble; quantities within the bubble are identified by an asterisk, see also Table 2.2.

Table 2.2. Values of Pressure, Temperature, and Density in a Gas Bubble and its Environment Before and After Adiabatic Displacement from a Location 1 to a Location 2 in Spherical Symmetry

	Location 1		Location 2	
	Bubble	Environment	Bubble	Environment
Position	r	r	$r + \mathrm{d}r$	$r + \mathrm{d}r$
Pressure	$P_1 = P(r)$	$P_1 = P(r)$	$P_2 = P(r + \mathrm{d}r) = P_1 + \mathrm{d}P/\mathrm{d}r$	$P_2 = P(r + \mathrm{d}r) = P_1 + \mathrm{d}P/\mathrm{d}r$
Temperature	$T^* = T_1 = T(r)$	$T_1 = T(r)$	$T_2^* = T_1 + \mathrm{d}T^*/\mathrm{d}r$	$T_2 = T(r + \mathrm{d}r) = T_1 + \mathrm{d}T/\mathrm{d}r$
Density	$\rho^* = \rho_1 = \rho(r)$	$\rho_1 = \rho(r)$	$\rho_2^* = \rho_1 + \mathrm{d}\rho^*/\mathrm{d}r$	$\rho_2 = \rho(r + \mathrm{d}r) = \rho_1 + \mathrm{d}\rho/\mathrm{d}r$

Note. See also Figure 2.6.

bubble varies faster than density in environment). The stratification will be *unstable against convection* if

$$\frac{d\rho^*}{dr} < \frac{d\rho}{dr}, \tag{2.102}$$

because the density in the bubble is not changing rapidly enough to remove the buoyancy and to stabilize the bubble. This is the criterion for the onset of convection. It can also be written as

$$\frac{d\rho^*}{dr} - \frac{d\rho}{dr} < 0. \tag{2.103}$$

The relations for $d\rho/dr$ and $d\rho^*/dr$ remain to be determined. They can be expressed in terms of the thermodynamic variables (P, T, Y) or (P, S, Y). For the environment surrounding the bubble,

$$\frac{d\rho}{dr} = \left(\frac{d\rho}{dP}\right)_r \frac{dP}{dr} + \left(\frac{d\rho}{dT}\right)_r \frac{dT}{dr} + \left(\frac{d\rho}{dY}\right)_r \frac{dY}{dr}, \tag{2.104}$$

$$\frac{d\rho}{dr} = \left(\frac{d\rho}{dP}\right)_r \frac{dP}{dr} + \left(\frac{d\rho}{dS}\right)_r \frac{dS}{dr} + \left(\frac{d\rho}{dY}\right)_r \frac{dY}{dr}, \tag{2.105}$$

and for the bubble itself,

$$\frac{d\rho^*}{dr} = \left(\frac{d\rho}{dP}\right)_r \frac{dP^*}{dr} + \left(\frac{d\rho}{dT}\right)_r \frac{dT^*}{dr} + \left(\frac{d\rho}{dY}\right)_r \frac{dY^*}{dr}, \tag{2.106}$$

$$\frac{d\rho^*}{dr} = \left(\frac{d\rho}{dP}\right)_r \frac{dP^*}{dr} + \left(\frac{d\rho}{dS}\right)_r \frac{dS^*}{dr} + \left(\frac{d\rho}{dY}\right)_r \frac{dY^*}{dr}, \tag{2.107}$$

are obtained. The first terms in the products of gradients appearing in Equations (2.104)–(2.107) are always taken at radius r, which is the starting position of the bubble motion. At this location 1 they are equal in the bubble and its environment. Inside the bubble, $dS^*/dr = 0$ and $dY^*/dr = 0$ because of the assumption of adiabatic motion, which also implies that there is no particle exchange with the environment changing the composition. Taking the difference of the expressions in Equations (2.104)–(2.107) for use in Equation (2.103), it has to be noted further that $dP/dr = dP^*/dr$ due to the pressure equilibrium between bubble and environment. Thus, the difference in density gradients inside and outside the bubble becomes

$$\frac{d\rho^*}{dr} - \frac{d\rho}{dr} = \frac{d\rho}{dT}\left(\frac{dT^*}{dr} - \frac{dT}{dr}\right) - \frac{d\rho}{dr}\frac{dY}{dr}, \tag{2.108}$$

$$\frac{d\rho^*}{dr} - \frac{d\rho}{dr} = -\frac{d\rho}{dS}\frac{dS}{dr} - \frac{d\rho}{dY}\frac{dY}{dr}. \tag{2.109}$$

With the above, the criterion for onset of convection can be written as

$$\frac{\mathrm{d}T}{\mathrm{d}r} > \frac{\mathrm{d}T^*}{\mathrm{d}r} - \frac{\mathrm{d}\rho/\mathrm{d}Y}{\mathrm{d}\rho/\mathrm{d}T}\frac{\mathrm{d}Y}{\mathrm{d}r} \tag{2.110}$$

or equivalently

$$\frac{\mathrm{d}S}{\mathrm{d}r} > -\frac{\mathrm{d}\rho/\mathrm{d}Y}{\mathrm{d}\rho/\mathrm{d}S}\frac{\mathrm{d}Y}{\mathrm{d}r}. \tag{2.111}$$

This is called the *Ledoux criterion*. Verbally expressed, Equation (2.110) means that convection sets in as soon as the temperature gradient in the gas is larger than the critical value on the right-hand side. This temperature gradient becomes steep when other energy transport mechanisms (radiation, convection) are ineffective in carrying off the energy generated in a layer of the star.

In the standard situation of a centrally burning star, the terms including the composition gradient $\mathrm{d}Y/\mathrm{d}r$ on the right-hand side of Equations (2.110) and (2.111) hinder the onset of convection by increasing the critical value on the right-hand side. Note that several gradients appearing in these inequalities can be negative, such as $\mathrm{d}Y/\mathrm{d}r$ in a normally layered star. A gradient $\mathrm{d}Y/\mathrm{d}r < 0$ implies that heavier material is found toward the center of the star. The inhibiting action of this gradient on convection can be understood by the fact that it requires additional energy to lift heavier material against the gravitational force and thus a larger temperature gradient is required before convection can set in. There are astrophysical conditions for which $\mathrm{d}Y/\mathrm{d}r > 0$ (at least locally), e.g., off center ignition of burning phases or during a final Fe core collapse (see Chapter 6). In such cases, convection is even made easier by heavy matter falling downwards.

Assuming $\mathrm{d}Y/\mathrm{d}r = 0$, i.e., in an object exhibiting the same composition throughout or by just neglecting the composition terms, the following, simpler equations are obtained:

$$\frac{\mathrm{d}T}{\mathrm{d}r} > \frac{\mathrm{d}T^*}{\mathrm{d}r}, \tag{2.112}$$

$$\frac{\mathrm{d}S}{\mathrm{d}r} > 0. \tag{2.113}$$

These are two ways to express the *Schwarzschild criterion* for the onset of convection. Due to the missing composition terms, it allows convection for $\mathrm{d}Y/\mathrm{d}r < 0$ even when the Ledoux criterion does not. Numerical simulations compared to observational data have shown that it is more appropriate to apply the Schwarzschild criterion in deciding whether convection occurs in early burning phases. This can be explained through vibrations in the stellar plasma providing an additional push to enable convection. Furthermore, composition gradients are smoothed by overshooting moving bubbles mixing their contents with the environment (this is called *semi-convection*). In general, convection occurs when either of the

Table 2.3. Comparison of Convection Criteria and Actual Conditions as Obtained from Numerical Simulations; Case 3 Includes Semi-convection

Case	Ledoux	Schwarzschild	Convection?
1	No	No	No
2	Yes	Yes	Yes
3	No	Yes	Yes
4	Yes	No	Yes

criteria, Schwarzschild or Ledoux, allows it. Table 2.3 gives an overview about the possible situations.

The temperature gradient in the bubble, $\mathrm{d}T^*/\mathrm{d}r$ is identical to the *adiabatic gradient*, if the assumption of a purely adiabatic motion of the bubble holds (see further down for a generalization). The EOS of an adiabatic gas is a polytrope (see Section 2.3). Using a polytrope $P = K\rho^{\tilde{\Gamma}}$, the density gradient in the bubble becomes

$$\frac{\mathrm{d}\rho^*}{\mathrm{d}r} = \frac{\mathrm{d}\rho^*}{\mathrm{d}P}\frac{\mathrm{d}P}{\mathrm{d}r} = \frac{1}{K\tilde{\Gamma}\rho^{*\tilde{\Gamma}-1}}\frac{\mathrm{d}P}{\mathrm{d}r} = \frac{\rho^*}{\tilde{\Gamma}P}\frac{\mathrm{d}P}{\mathrm{d}r} \tag{2.114}$$

and the adiabatic gradient can be expressed through logarithmic derivatives as

$$\frac{\mathrm{d}T^*}{\mathrm{d}r} = \frac{\mathrm{d}\ln T^*}{\mathrm{d}\ln P}\frac{T^*}{P}\frac{\mathrm{d}P}{\mathrm{d}r}. \tag{2.115}$$

For a mixture of an ideal gas and radiation with mixing parameter β (see Section 2.6.1), this is given by

$$\frac{\mathrm{d}T^*}{\mathrm{d}r} = \left(1 - \frac{1}{\tilde{\Gamma}'}\right)\frac{T^*}{P}\frac{\mathrm{d}P}{\mathrm{d}r}, \tag{2.116}$$

with $\tilde{\Gamma}'$ defined as

$$\tilde{\Gamma}' = \frac{32 - 24\beta - 3\beta^2}{24 - 18\beta - 3\beta^2}. \tag{2.117}$$

The *pressure scale height* H_P is used often in the context of adiabatic gradients and convection. It is defined as

$$H_\mathrm{P} = -P\frac{\mathrm{d}r}{\mathrm{d}P} = -\frac{\mathrm{d}r}{\ln P} \tag{2.118}$$

and is a characteristic length over which pressure changes are not negligible.

Once convection is invoked, it is such an efficient way to transport energy that any temperature gradients are driven toward the adiabatic gradient. This is the reason, on the other hand, why stellar matter can be approximated by a $\tilde{n} = 3$ polytrope (except close to the surface, see below). Only a degenerate electron gas is more efficient in transporting energy. This is the reason why convection never appears in degenerate conditions.

The simplest approach to treat convection in stellar models without a multi-dimensional, fully hydrodynamic treatment is to use the criterion given in Equation (2.112) or (2.113) to check whether convection occurs. This is done, e.g., by comparing the temperature gradient caused by radiation to the adiabatic gradient. When convection is acting, the adiabatic temperature gradient is applied for further calculations in this layer. The matter transport then also has to be treated in an approximative manner.

2.7.4.3 Mixing Length Theory

A more sophisticated approach is to use *mixing length theory*, which allows to estimate an effective energy flow even in the case of a spherical symmetric model. Therein it is assumed that a gas bubble rises adiabatically for a typical length $l_{\rm mix}$ before thermalizing with the local environment. This approach is inspired by particle heat transfer but with gas bubbles instead of particles and the *mixing length* $l_{\rm mix}$ instead of the mean free path. The mixing length is determined by the physical properties of the stellar layer and may be different in different parts of the star. In this model, the temperature difference between bubble and local environment at the final position of the bubble is given by

$$\Delta T = l_{\rm mix}\left(\left|\frac{{\rm d}T}{{\rm d}r}\right| - \left|\frac{{\rm d}T^*}{{\rm d}r}\right|\right) = l_{\rm mix}\Delta T_{\rm grad}. \tag{2.119}$$

Since the released heat per mass at constant pressure (see also Section 1.4) at the final location is $\Delta Q = C_{\rm P}\Delta T$, the resulting heat flow is

$$F_{\rm mix} = \rho v\Delta Q = \rho v C_{\rm P} l_{\rm mix}\Delta T_{\rm grad}. \tag{2.120}$$

The average speed v of a rising bubble can be roughly estimated by using the buoyant force causing the bubble to accelerate over the length $l_{\rm mix}$. According to Archimedes' principle, the buoyant force is caused by the density difference $\Delta\rho$, which is given by

$$\Delta\rho = l_{\rm mix}\left(\left|\frac{{\rm d}\rho}{{\rm d}r}\right| - \left|\frac{{\rm d}\rho^*}{{\rm d}r}\right|\right) = l_{\rm mix}\Delta\rho_{\rm grad}. \tag{2.121}$$

Bubbles originating from different depths, and thus with different $\Delta\rho_{\rm grad}$ and different velocities v, pass any given radius r. The mean density difference of these can be taken as $\overline{\Delta\rho} = l_{\rm mix}\Delta\rho_{\rm grad}/2$ and the resulting mean buoyant force per unit volume $F_{\rm Archi} = g\overline{\Delta\rho} = g l_{\rm mix}\Delta\rho_{\rm grad}/2$ causes a mean acceleration $\bar{a} = F_{\rm Archi}/\rho = g l_{\rm mix}\Delta\rho_{\rm grad}/(2\rho)$. The average speed v of the bubbles is approximated by using half the terminal speed obtained when accelerating them by $\bar{a}$ for a distance $l_{\rm mix}$,

$$v = \frac{l_{\rm mix}}{2}\sqrt{\frac{g}{\rho}\Delta\rho_{\rm grad}} = \frac{l_{\rm mix}}{2}\sqrt{\frac{Gm(r)}{\rho r^2}\Delta\rho_{\rm grad}}. \tag{2.122}$$

Inserting this into Equation (2.120) leads to

$$F_{\mathrm{mix}} = \frac{C_{\mathrm{P}}\rho l_{\mathrm{mix}}^2}{2}\Delta T_{\mathrm{grad}}\sqrt{\frac{Gm(r)}{\rho r^2}\Delta\rho_{\mathrm{grad}}}\,. \tag{2.123}$$

For an ideal, mono-atomic gas it can be shown that

$$\Delta\rho_{\mathrm{grad}} = \frac{\rho}{T}\Delta T_{\mathrm{grad}} \tag{2.124}$$

and finally this yields

$$F_{\mathrm{mix}} = \frac{C_{\mathrm{P}}\rho l_{\mathrm{mix}}^2}{2}\Delta T_{\mathrm{grad}}^{3/2}\sqrt{\frac{Gm(r)}{T\rho r^2}}\,. \tag{2.125}$$

This shows that the energy flow is proportional to the difference in temperature gradients to the power of 3/2 and to the square of the mixing length. A reasonable value for the mixing length, which is a free parameter, is a distance for which pressure and density have considerably changed and thus the pressure scale height H_{P} (see Equation (2.118)) is a natural choice.

A more rigorous treatment also includes the energy loss from the bubble surface while it is moving in the stellar environment. This results in a temperature change inside the bubble, which is different from a purely adiabatic temperature gradient $|\mathrm{d}T/\mathrm{d}r|_{\mathrm{ad}}$ and therefore ΔT_{grad} should contain the actual temperature gradient in the bubble and not just the adiabatic gradient. Taking the energy loss into account,

$$\Delta T_{\mathrm{grad}}^{3/2} = \frac{8U_{\mathrm{mix}}}{9}\left(\left|\frac{\mathrm{d}T}{\mathrm{d}r}\right|_{\gamma} - \left|\frac{\mathrm{d}T}{\mathrm{d}r}\right|\right) \tag{2.126}$$

and the temperature gradient in a stellar layer $\mathrm{d}T/\mathrm{d}r$ resulting from convection is obtained by numerically solving the cubic equation

$$\left(\left|\frac{\mathrm{d}T}{\mathrm{d}r}\right| - \left|\frac{\mathrm{d}T}{\mathrm{d}r}\right|_{\mathrm{ad}} + U_{\mathrm{mix}}\right)^3 + \frac{8U_{\mathrm{mix}}}{9}\left(\left|\frac{\mathrm{d}T}{\mathrm{d}r}\right| - \left|\frac{\mathrm{d}T}{\mathrm{d}r}\right|_{\gamma}\right) = 0, \tag{2.127}$$

containing the adiabatic temperature gradient $(\mathrm{d}T/\mathrm{d}r)_{\mathrm{ad}}$, the temperature gradient resulting from the radiative energy flow $(\mathrm{d}T/\mathrm{d}r)_{\gamma}$ (see Equation (2.83)), and the dimensionless quantity U_{mix}, defined for convenience as

$$U_{\mathrm{mix}} = \frac{3\tilde{a}crT^3}{C_{\mathrm{P}}\rho^2\kappa l_{\mathrm{mix}}^2}\sqrt{\left(-\frac{\mathrm{d}\ln T}{\mathrm{d}\ln\rho}\right)\frac{8H_{\mathrm{P}}}{Gm(r)}}\,. \tag{2.128}$$

The opacity κ is determined as shown in Equation (2.88). For an ideal gas, $\mathrm{d}\ln T/\mathrm{d}\ln\rho = -1$.

It is interesting to note that there is a hierarchy of temperature gradients,

$$\left(\frac{\mathrm{d}T}{\mathrm{d}r}\right)_\gamma > \frac{\mathrm{d}T}{\mathrm{d}r} > \frac{\mathrm{d}T^*}{\mathrm{d}r} \geqslant \left(\frac{\mathrm{d}T}{\mathrm{d}r}\right)_{\mathrm{ad}}, \tag{2.129}$$

provided that convection is occurring. In this case, $\mathrm{d}T/\mathrm{d}r > \mathrm{d}T^*/\mathrm{d}r$ is the Schwarzschild criterion for convection (Equation (2.112)) and it is also obvious that the radiative gradient overestimates the actual temperature gradient, $\mathrm{d}T/\mathrm{d}r < (\mathrm{d}T/\mathrm{d}r)_\gamma$, because radiation alone is not sufficient to transfer all of the energy. The equality sign before the last term in Equation (2.129) applies for purely adiabatic bubble motion, otherwise part of the energy is lost through the bubble surface during motion, as described above. This additional energy loss is especially important close to the surface of a star.

There are two limiting cases simplifying the hierarchy shown in Equation (2.129). In the very dense central part of the star, where the opacity is high and the mean free path of the photons is short, $U_{\mathrm{mix}} \to 0$ and fully adiabatic convection dominates the energy transport, driving the temperature gradient toward the adiabatic gradient $\mathrm{d}T/\mathrm{d}r \approx \mathrm{d}T^*/\mathrm{d}r = (\mathrm{d}T/\mathrm{d}r)_{\mathrm{ad}}$. If $U_{\mathrm{mix}} \to \infty$, on the other hand, convection is inefficient and the temperature gradient will be given by the gradient for radiative transport $\mathrm{d}T/\mathrm{d}r = (\mathrm{d}T/\mathrm{d}r)_\gamma$. This applies close to the photosphere of a star. In both cases, the convective energy transport is independent of l_{mix} and this justifies the simplified approach described above, without application of mixing length theory. The full equations of mixing length theory have to be solved for cases in between the two extremes, with a resulting *superadiabatic* temperature gradient as shown in Equation (2.129). This is necessary, for example, in the upper part of an outer convective envelope.

As can be seen in the above derivation, this version of mixing length theory includes a number of assumptions and at least one free parameter, the mixing length. Although it is reasonable to assume that the mixing length is of the order of the pressure scale height, it is not guaranteed that it is exactly equal to it or that it is the same fraction of the scale height throughout the star, as often assumed. Furthermore, it was assumed above that bubbles of similar "average" volume and shape move at an average speed. In reality, there is a spectrum of smaller and larger turbulent eddies moving in the plasma. The advance in computational power has allowed to employ more and more sophisticated treatments of these eddies but a fully hydrodynamic treatment in 3D is still not possible for extended regions inside a star, not even to speak of following the full evolution of a star. Nevertheless, comparison of stellar models calculated with standard mixing length theory have been found to agree well with observed stellar properties and also predict well stellar distributions of globular clusters in the Hertzsprung–Russell diagram. Numerical treatments taking into account a distribution of turbulent eddies have confirmed that the effective mixing length is close to, although about 30% smaller than, the pressure scale height. All of this indicates that mixing length theory remains a good approximation for treating convective energy transport in the interior of a star.

2.8 Convective and Diffusive Matter Transport

In the preceding sections we were only concerned with the transport of energy. Since there is a composition gradient $\mathrm{d}Y/\mathrm{d}r$ in the star (see also Equation (2.110)), there will be physical processes tending to gradually flatten this composition gradient. The most effective way to change composition is by convection, not only transporting energy but also material from hotter to cooler regions. This matter transport can become extremely important for nucleosynthesis, as it not only moves heavier material upwards but also lighter nuclei, such as protons and α-particles, downwards. The mixing of the content of different layers enables nuclear reactions to occur that could not occur without convective mixing because a deeper layer is already depleted in lighter nuclei, for example, protons (see, for example, Section 8.2). Therefore the onset of convection and the extension of convective zones not only impacts the duration of certain evolutionary stages but is also crucial in the determination of the nucleosynthetic signature of a star.

Although a nuclear reaction between two nuclei proceeds very quickly (see Section 3.7), in hydrostatic burning the resulting change in abundances is slow even compared to the macroscopic motion of matter. Therefore, instantaneous mixing can be assumed in most cases. This implies that there is no abundance gradient within a convective region,

$$\left.\frac{\mathrm{d}X_i}{\mathrm{d}r}\right|_{\mathrm{zone}} = \left.\frac{1}{A_i}\frac{\mathrm{d}Y_i}{\mathrm{d}r}\right|_{\mathrm{zone}} = 0. \tag{2.130}$$

The composition at the boundaries of the convective zone, however, may be different from the one inside. This is especially true when the zone changes its extension and encompasses additional stellar layers. The resulting change in abundance for a zone extending between $r_1 < r < r_2$ is described by

$$\left.\frac{\mathrm{d}X_i}{\mathrm{d}t}\right|_{\mathrm{zone}} = \frac{1}{m(r_2) - m(r_1)}\left[\frac{\mathrm{d}r_2}{\mathrm{d}t}\left(X_i(r_2) - X_i|_{\mathrm{zone}}\right) - \frac{\mathrm{d}r_1}{\mathrm{d}t}\left(X_i(r_1) - X_i|_{\mathrm{zone}}\right)\right]. \tag{2.131}$$

The same equation in Lagrangian mass coordinates is

$$\left.\frac{\mathrm{d}X_i}{\mathrm{d}t}\right|_{\mathrm{zone}} = \frac{1}{m_2 - m_1}\left[\frac{\mathrm{d}m_2}{\mathrm{d}t}\left(X_i(m_2) - X_i|_{\mathrm{zone}}\right) - \frac{\mathrm{d}m_1}{\mathrm{d}t}\left(X_i(m_1) - X_i|_{\mathrm{zone}}\right)\right], \tag{2.132}$$

with the zone extending between $m_1 < m < m_2$. The boundary values $X_i(r_1)$, $X_i(r_2)$ and $X_i(m_1)$, $X_i(m_2)$, respectively, at the discontinuity between the convection zone and its neighboring layers are taken outside of the convection zone, in the layers into which the convection region is expanding.

Both convective energy transport as described in Section 2.7.4 and convective matter transport outlined in this section are strongly affected by *stellar rotation*. All equations given in this and previous sections consider a non-rotating, spherically symmetric case. Rotation diverts matter flows and gives rise to additional flows and mixing. The treatment of matter and energy flows in rotating stars is difficult and beyond the scope of this introduction. Many questions connected to rotation and its impact on stellar evolution and nucleosynthesis (by introducing additional mixing between different stellar layers) are subject to current research but yet not completely solved.

In the absence of convection, a composition gradient can also be changed by diffusion processes. Several types of diffusion can be distinguished. If there is a composition gradient in a layer, *concentration diffusion* tends to flatten it. *Temperature diffusion* drives heavier ions to regions of higher temperature even in initially chemically homogeneous layers without initial composition gradients. Pressure gradients also cause heavier particles to migrate toward higher pressure. This is called *pressure diffusion.*

The general shape of a diffusion equation was already shown in Equation (2.77). Particle diffusion in a stellar layer is not further considered in this book, as it is only important for certain special cases. The change in the positional distribution of abundances due to the three diffusion types can be expressed in one equation,

$$\left.\frac{\mathrm{d}X_i}{\mathrm{d}r}\right|_{\mathrm{zone,diffu}} = -\frac{1}{r\rho^2}\frac{\mathrm{d}}{\mathrm{d}r}\left[r^2 X_i T^{5/2}\left(A_{\mathrm{P},i}\frac{\mathrm{d}\ln P}{\mathrm{d}r} + A_{\mathrm{T},i}\frac{\mathrm{d}\ln T}{\mathrm{d}r} + \sum_{j\neq \mathrm{e},{}^4\mathrm{He}} A_{\mathrm{C},i,j}\frac{\mathrm{d}\ln C_i}{\mathrm{d}r}\right)\right]. \tag{2.133}$$

The diffusion parameters $A_{\mathrm{P},i}$, $A_{\mathrm{T},i}$, and $A_{\mathrm{C},i,j}$ for pressure diffusion, temperature diffusion, and concentration diffusion, respectively, have to be determined appropriately. The concentration diffusion (the third term in the bracket in the right-hand side of Equation (2.133)) includes a sum over all species j in the plasma because the concentration of one specific species i may also change due to the diffusion of all other species. Conservation of mass and charge reduce the number of linearly independent concentration diffusion coefficients $A_{\mathrm{C},i,j}$ and thus two of the species do not have to be included. Here, ^{4}He and electrons have been left out of the sum. The concentration $C_i = c_i/c_\mathrm{e}$ appearing in Equation (2.133) is defined as ratio of the usual particle concentration c_i and the electron concentration c_e.

Finally, the composition in a stellar layer can be changed by nuclear reactions, which not only generate (or absorb) energy but also transform nuclei. The composition changes due to nuclear reactions are followed in nuclear reaction networks (see Section 4.2) which have to be considered in the stellar equations as shown in Section 2.9. The resulting composition change within a layer is taken into account by adding another term to Equations (2.131) and (2.132), respectively, as shown in Equation (2.134).

2.9 Complete Equations for the Hydrostatic Model

This chapter commenced with the introduction of the basic equations of hydrostatic equilibrium (Section 2.2). As discussed, the use of more complex equations-of-state increases the number of variables. This increase requires additional equations to constrain them. Such equations are found in the physical modeling of energy transport. Finally, matter transport and abundance changes due to nuclear reactions also play a role. All these phenomena have to be included in the equations describing the properties of a star in hydrostatic equilibrium.

The full set of equations for the spherically symmetric, hydrostatic case can be summarized as

$$\begin{aligned}
\frac{\mathrm{d}m(r)}{\mathrm{d}r} &= 4\pi r^2 \rho(r) \\
\frac{\mathrm{d}P(r)}{\mathrm{d}r} &= -\rho(r)\mathrm{G}\frac{m(r)}{r^2} \\
\frac{\mathrm{d}L(r)}{\mathrm{d}r} &= 4\pi r^2(\varepsilon_{\mathrm{nuc}} - \varepsilon_\nu) \\
\left.\frac{\mathrm{d}T}{\mathrm{d}r}\right|_\gamma &= -\frac{3}{16\pi\tilde{\mathrm{a}}\mathrm{c}}\frac{\kappa\rho}{r^2T^3}L(r) \\
\left.\frac{\mathrm{d}T}{\mathrm{d}r}\right|_{\mathrm{conv}} &= -\frac{T\mathrm{G}m(r)}{Pr^2}\frac{\mathrm{d}\ln T}{\mathrm{d}\ln P} \\
\left.\frac{\mathrm{d}X_i}{\mathrm{d}t}\right|_{\mathrm{zone}} &= \frac{1}{m(r_2) - m(r_1)}\left[\frac{\mathrm{d}r_2}{\mathrm{d}t}\left(X_i(r_2) - X_i|_{\mathrm{zone}}\right)\right. \\
&\quad \left. - \frac{\mathrm{d}r_1}{\mathrm{d}t}(X_i(r_1) - X_i|_{\mathrm{zone}}) + \int_{r_1}^{r_2}\left.\frac{\mathrm{d}X_i}{\mathrm{d}t}\right|_{\mathrm{nuc}}\mathrm{d}r\right]
\end{aligned} \tag{2.134}$$

using Eulerian radial coordinates and

$$\begin{aligned}
\frac{\mathrm{d}r(m)}{\mathrm{d}m} &= \frac{1}{4\pi r^2\rho(m)} \\
\frac{\mathrm{d}P(m)}{\mathrm{d}m} &= -\mathrm{G}\frac{m}{4\pi r^4} \\
\frac{\mathrm{d}L(m)}{\mathrm{d}m} &= q_{\mathrm{nuc}} - q_\nu \\
\left.\frac{\mathrm{d}T}{\mathrm{d}m}\right|_\gamma &= -\frac{3}{64\pi^2\tilde{\mathrm{a}}\mathrm{c}}\frac{\kappa}{r^4T^3}L(m) \\
\left.\frac{\mathrm{d}T}{\mathrm{d}r}\right|_{\mathrm{conv}} &= -\frac{T\mathrm{G}m}{4P\pi r^4}\frac{\mathrm{d}\ln T}{\mathrm{d}\ln P} \\
\left.\frac{\mathrm{d}X_i}{\mathrm{d}t}\right|_{\mathrm{zone}} &= \frac{1}{m_2 - m_1}\left[\frac{\mathrm{d}m_2}{\mathrm{d}t}\left(X_i(m_2) - X_i|_{\mathrm{zone}}\right)\right. \\
&\quad \left. - \frac{\mathrm{d}m_1}{\mathrm{d}t}(X_i(m_1) - X_i|_{\mathrm{zone}}) + \int_{m_1}^{m_2}\left.\frac{\mathrm{d}X_i}{\mathrm{d}t}\right|_{\mathrm{nuc}}\mathrm{d}m\right]
\end{aligned} \tag{2.135}$$

using Lagrangian mass coordinates. Particle diffusion, as given by Equation (2.133), is neglected in the above sets of equations. Radiation determines the temperature gradient only in the absence of diffusion. The opacity κ used above is the one obtained from the Rosseland mean, see Section 2.7.2. The last lines in Equations (2.134) and (2.135) include an integral over $(\mathrm{d}X_i/\mathrm{d}t)_{\mathrm{nuc}}$, i.e., the mass fraction change in a zone caused by nuclear reactions. This change is calculated using a further set of coupled differential equations, called a nuclear reaction network. A full nuclear network includes reactions between all nuclides present in the stellar plasma and allows to track their abundance changes over time due to these reactions. At the same time, the energy generated or absorbed by nuclear transformations and by neutrinos is calculated. Thus, also the determination of $\varepsilon_{\mathrm{nuc}}$ and ε_ν require the use of these additional equations. Reaction networks are at the core of nucleosynthesis investigations and are presented in all details in Section 4.2.

2.9.1 Virial Theorem

The virial theorem is a generally applicable relation between averaged kinetic and averaged potential energy. Originally derived by Rudolf Clausius in 1870 for a system of classical particles bound by potential forces, it can be shown to also apply to relativistic and quantum mechanical systems. The significance of the theorem is in the fact that it allows to make statements on the distribution of energy between kinetic (including the internal energy of gases) and potential energy even for complex many-body systems which otherwise defy a straightforward, exact solution. More specifically, the average kinetic energy can easily be derived from the potential for such systems. Although not directly related to the nucleosynthesis questions this book focuses on, the importance of the virial theorem in a wide range of applications, from statistical physics to cosmology, warrants the inclusion of a more detailed discussion here, with emphasis on the relation to the stellar properties derived in the previous sections.

For the case that the force between two particles in the system can be described by a potential $V(r)$ following a power law $V(r) \propto r^k$, the virial theorem states

$$\langle E_{\mathrm{kin}} \rangle = \frac{k}{2} \langle E_{\mathrm{pot}} \rangle = \frac{k}{k+2} E_{\mathrm{tot}}, \tag{2.136}$$

for the relation between averaged total kinetic energy $\langle E_{\mathrm{kin}} \rangle$ and averaged total potential energy $\langle E_{\mathrm{pot}} \rangle$ including all particles in the system. The total energy of the system is given by $E_{\mathrm{tot}} = E_{\mathrm{kin}} + E_{\mathrm{pot}}$. The potential energy usually is defined in such a way that $E_{\mathrm{tot}} < 0$ for bound systems. The value of k depends on the potential, e.g., $k = -1$ for the gravitational and the Coulomb potential, and $k = 2$ for the harmonic oscillator.

The validity of Equation (2.136) can be easily seen from the simplest possible case, a two-body system with a mass m_1 in a circular orbit around another mass $m_2 \gg m_1$ at a distance r. The potential energy of the smaller mass is given by the gravitational potential V_G as $E_{\mathrm{pot},m_1} = m_1 V_G = -Gm_1m_2/r$. Its kinetic energy $m_1v^2/2$ is derived by equating the centripetal force m_1v^2/r and the gravitational one,

$$E_{\mathrm{kin},m_1} = \frac{Gm_1m_2}{2r}, \tag{2.137}$$

while the kinetic energy of the much larger mass m_2 is negligible. This result reproduces what is shown in Equation (2.136) with the averaged values replaced by the time-independent values for kinetic and potential energy because this is a stable, unchanging system. The virial theorem would, of course, also apply to generalized systems including many interacting bodies or elliptical orbits, and in fact to arbitrary gravitationally bound systems. In such cases the kinetic and potential energy change with time and time averages of the kinetic and potential energy have to be used.

A general proof of the theorem starts with the definition of the virial

$$G_{\mathrm{vir}} = \sum_i \vec{p}_i \cdot \vec{r}_i, \tag{2.138}$$

which is the sum over all the particles of the dot product of each particle's momentum with its position. (The virial is related to the moment of inertia of the particle system around the origin.) The change of the virial with time is

$$\begin{aligned}\frac{\mathrm{d}G_{\mathrm{vir}}}{\mathrm{d}t} &= \left(\sum_i \vec{p}_i \cdot \frac{\mathrm{d}\vec{r}_i}{\mathrm{d}t}\right) + \left(\sum_i \frac{\mathrm{d}\vec{p}_i}{\mathrm{d}t} \cdot \vec{r}_i\right)\\ &= \left(\sum_i m_i \frac{\mathrm{d}\vec{r}_i}{\mathrm{d}t} \cdot \frac{\mathrm{d}\vec{r}_i}{\mathrm{d}t}\right) + \left(\sum_i \vec{F}_i \cdot \vec{r}_i\right) = 2E_{\mathrm{kin}} + \sum_i \vec{F}_i \cdot \vec{r}_i,\end{aligned} \tag{2.139}$$

with m_i being the particle mass and $\vec{F}_i$ the total force on a particle. The asymptotic limit of the time average of the left-hand side of Equation (2.139) is

$$\lim_{\tau\to\infty}\left(\frac{1}{\tau}\int_0^\tau \frac{\mathrm{d}G_{\mathrm{vir}}}{\mathrm{d}t}\mathrm{d}t\right) = \lim_{\tau\to\infty}\frac{G_{\mathrm{vir}}(\tau) - G_{\mathrm{vir}}(0)}{\tau}. \tag{2.140}$$

The basic assumption entering the derivation is that the positions and velocities of all particles in the system are bounded for all times (e.g., stable, periodic orbits or otherwise gravitationally bound systems fulfill this requirement). Under this assumption Equation (2.140) evaluates to zero.

The time average also has to be applied to the right-hand side of Equation (2.139). In that equation, $\vec{F}_i$ is the total force on each particle i, which results from the sum of forces exerted by the other particles j,

$$\vec{F}_i = \sum_j \vec{F}_{ij}, \tag{2.141}$$

and applying Newton's third law of motion, $\vec{F}_{ij} = -\vec{F}_{ji}$, it can be shown that

$$\sum_i \vec{F}_i \cdot \vec{r}_i = \sum_i \sum_{j<i} \vec{F}_{ji} \cdot (\vec{r}_i - \vec{r}_j). \tag{2.142}$$

To further rewrite the sum over the forces on the particles it is helpful to remember the relation between force $\vec{F}$, momentum $\vec{p}$, and potential V,

$$\vec{F} = \frac{\mathrm{d}\vec{p}}{\mathrm{d}t} = -\vec{\nabla} V. \tag{2.143}$$

Although the virial theorem is applicable generally (provided the assumption of boundedness mentioned above is fulfilled), for simplicity a spherically symmetric potential depending only on the distance $r_{ij} = |\vec{r}_{ij}| = |\vec{r}_i - \vec{r}_j|$ between two particles i, j is taken here and thus

$$\vec{F}_{ji} = -\vec{\nabla}_{\vec{r}_i} V = -\frac{\mathrm{d}V}{\mathrm{d}r}\left(\frac{\vec{r}_i - \vec{r}_j}{r_{ji}}\right). \tag{2.144}$$

Then Equation (2.142) becomes

$$\sum_i \vec{F}_i \cdot \vec{r}_i = -\sum_i \sum_{j<i} \frac{\mathrm{d}V}{\mathrm{d}r} \frac{|\vec{r}_i - \vec{r}_j|^2}{r_{ji}} = -\sum_i \sum_{j<i} \frac{\mathrm{d}V}{\mathrm{d}r} r_{ji}. \tag{2.145}$$

Further assuming a power law for the potential, i.e., the potential energy depending on a power of the distance, $V(r_{ji}) = Cr_{ji}^k$, where k can also be non-integer and/or negative, Equation (2.145) becomes

$$-\sum_i \sum_{j<i} \frac{\mathrm{d}V}{\mathrm{d}r} r_{ji} = -\sum_i \sum_{j<i} kCr_{ji}^{k-1} r_{ji} = -\sum_i \sum_{j<i} kV(r_{ji}). \tag{2.146}$$

Realizing that $\sum_i \sum_{j<i} V(r_{ji})$ is just the total potential energy E_{pot} of the system, the full time-average of the left-hand and right-hand sides of Equation (2.139) finally is given by

$$0 = 2\langle E_{\mathrm{kin}}\rangle - k\langle E_{\mathrm{pot}}\rangle, \tag{2.147}$$

which is equivalent to Equation (2.136).

As mentioned above, the virial theorem has been generalized for a wide range of applications in physics and astronomy, including relativistic and quantum mechanical systems. It is even applicable to systems that are not in thermodynamic equilibrium (such as self-gravitating systems). The basic requirement is the definition of a virial that involves a bounded function, i.e., a function which does not yield values growing beyond any limit. Although the usual derivation includes an integration over infinite time, the virial theorem may still be valid for finite time. For example, a star is losing mass during its evolution, i.e., particles become unbound, but the virial theorem is still applicable as long as mass loss is negligible within the finite time interval considered. Similarly, the virial theorem is applied to galactic motion but also stars are ejected from galaxies and the time interval has to be chosen appropriately. On the other hand, in the presence of stable, periodic motion it is not necessary to integrate over infinite time, averaging over one period is sufficient.

The virial theorem also appears as a consequence of hydrodynamic equilibrium of a bound system, such as the one assumed for deriving the basic equations of stellar structure in Section 2.2. This can be seen by multiplying the second structure equation, Equation (2.10), by $4\pi r^3$ and integrating both sides over dr from the center of the star to the surface, yielding

$$\int_0^R 4\pi r^3 \frac{\mathrm{d}P}{\mathrm{d}r}\,\mathrm{d}r = -\int_0^R \frac{Gm(r)}{r} 4\pi r^2 \rho\,\mathrm{d}r. \tag{2.148}$$

(Using Equation (2.11) and integrating over dm would yield the same result in the discussion below.) The right-hand side of this equation is the *gravitational potential energy* E_G of the star, i.e., the sum of the potential energies of all mass elements in the star. This is easier to see when using the alternative expression $E_G = -\int_0^M (Gm(r)/r)\,\mathrm{d}m$. It is negative because a star is a bound system. The potential energy scale is chosen so that the potential energy of a mass element approaches zero for $r \to \infty$. Therefore, $-E_G$ has to be invested to overcome gravitational attraction and to move all mass zones to infinite distance. Because of hydrostatic equilibrium $E_G = \langle E_G \rangle$.

The left-hand side of Equation (2.148) can be solved by partial integration,

$$\begin{aligned}\int_0^R 4\pi r^3 \frac{\mathrm{d}P}{\mathrm{d}r}\,\mathrm{d}r &= [4\pi r^3 P(r)]_0^R - 3\int_0^R 4\pi r^2 P(r)\,\mathrm{d}r \\ &= -3\int_0^R 4\pi r^2 P(r)\,\mathrm{d}r = -3\langle P\rangle V,\end{aligned} \tag{2.149}$$

with the volume-integrated average pressure $\langle P \rangle$. Thus, the gravitational potential energy can be related to the average pressure,

$$\langle P\rangle = -\frac{\langle E_G\rangle}{3V}. \tag{2.150}$$

That this is just another version of the virial theorem can be seen when remembering that the (average) pressure in an ideal, non-relativistic gas is proportional to 2/3 of the average kinetic energy (see also Section 1.5.2),

$$P = \langle P\rangle = \frac{2\bar{n}}{3}\langle U\rangle = \frac{2\bar{n}}{3}\left\langle \frac{mv^2}{2}\right\rangle = \frac{2\bar{n}}{3}\langle E_{\mathrm{kin}}\rangle = \frac{2}{3V}\langle E_{\mathrm{kin}}\rangle. \tag{2.151}$$

Comparing this to Equation (2.150), it can be seen that $2\langle E_{\mathrm{kin}}\rangle + \langle E_G\rangle = 0$, which is equivalent to Equations (2.136) and (2.147) for $k = -1$.

The virial theorem allows to quickly estimate certain conditions and energy scales of bound systems. For example, for a gas cloud or star contracting under quasi-hydrostatic conditions the theorem requires that half of the released gravitational binding energy is used to heat the cloud or star and half is radiated away. This can be seen combining the definition of the total energy $E_{\mathrm{tot}} = E_{\mathrm{kin}} + E_{\mathrm{G}}$ with the definition of the stellar luminosity $\mathcal{L}^*$ as the total energy output of an object through its surface, see Equations (1.4) and (2.65). Conservation of energy requires that

$$\frac{\mathrm{d}E_{\mathrm{tot}}}{\mathrm{d}t} + \mathcal{L}^* = 0 \tag{2.152}$$

and thus

$$\mathcal{L}^* = -\frac{1}{2}\frac{\mathrm{d}E_{\mathrm{G}}}{\mathrm{d}t} = \frac{\mathrm{d}E_{\mathrm{kin}}}{\mathrm{d}t}. \tag{2.153}$$

This also shows that systems become hotter during gravitational contraction and therefore tightly bound systems are hot. In stars, the ignition of nuclear burning stops the contraction by establishing the hydrostatic equilibrium (see Chapter 7) and thus nuclear reactions prevent the star from becoming hotter, contrary to the intuitive notion of heating through nuclear reactions. This behavior, which also allows the hydrostatic equilibrium to be established, can also be understood by assigning a negative heat capacity to the star.

2.9.2 Typical Timescales

Several typical timescales connected to the behavior of the stellar plasma and the evolution of a star can be derived from the equations given in the preceding sections.

An important timescale is connected to the equilibrium between the gravitational force and a pressure force in a mass shell which was used in Section 2.2 to derive Equation (2.10), the second of the basic hydrostatic structure equations. The full equation of motion of a mass element is shown in Equation (2.8). To arrive at the solution for the hydrostatic case, the acceleration term in the resulting Equation (2.9) was assumed to evaluate to zero. Here, another solution is considered by assuming that gravitational force and pressure force are not canceling each other, i.e., the right-hand side of Equation (2.8) is not zero. Then the acceleration term in Equation (2.9) cannot be neglected. The pressure gradient appearing in Equation (2.9) is still acting against the pull of gravity, slowing the fall of the mass element toward the center of the star. The extreme, free-fall limit is obtained by assuming no pressure gradient, leading to

$$\frac{\mathrm{d}^2 r}{\mathrm{d}t^2} = \frac{\mathrm{G}m(r)}{r^2}, \tag{2.154}$$

which is just the gravitational acceleration of a falling mass element. A typical timescale (time for non-negligible motion of the mass element) is given by

$$\tau_{\mathrm{ff}} = \left|\frac{r}{\mathrm{d}r/\mathrm{d}t}\right| = \left|\frac{r}{v}\right|, \tag{2.155}$$

where v is the velocity of the falling mass. It can be determined, e.g., by equating kinetic energy and the potential energy released when starting the fall at a radius r_0,

$$\frac{\mathrm{d}m\, v^2}{2} = \frac{\mathrm{d}m}{2}\left(\frac{\mathrm{d}r}{\mathrm{d}t}\right)^2 = \frac{Gm(r_0)\,\mathrm{d}m}{r} - \frac{Gm(r_0)\,\mathrm{d}m}{r_0}, \tag{2.156}$$

leading to

$$|v| = \left|\frac{\mathrm{d}r}{\mathrm{d}t}\right| = \left|\sqrt{2Gm(r_0)\left(\frac{1}{r} - \frac{1}{r_0}\right)}\right|. \tag{2.157}$$

For $r_0 \to \infty$ this yields the escape velocity at radius r

$$|v_\infty(r)| = \left|\sqrt{\frac{2Gm(r)}{r}}\right|, \tag{2.158}$$

which is also the velocity at r of a mass element falling in from infinite distance (the sign of v would have to be chosen appropriately). Then the *free-fall timescale* evaluates to

$$\tau_{\mathrm{ff}} = \sqrt{\frac{r^3}{2Gm(r)}} = \sqrt{\frac{R^3}{2GM}}, \tag{2.159}$$

where the stellar radius R and total mass M were used. This is also called the *(hydro)dynamical timescale*. Stars react on this timescale to deviations from hydrostatic equilibrium. On the other hand, it is sufficient to assume hydrostatic equilibrium as long as this hydrodynamical timescale remains much shorter than the timescale of any evolutionary changes. For the Sun, $\tau_{\mathrm{ff}} \approx 1130$ s, for a red giant star with $R = 200\ R_\odot$, $\tau_{\mathrm{ff}} \approx 3.2 \times 10^6$ s, and for a white dwarf ($R = 0.014\ R_\odot$) $\tau_{\mathrm{ff}} \approx 1.9$ s.

The virial theorem as introduced in Section 2.9.1 can also be used to derive a timescale for a star settling into or dropping out of thermal equilibrium. It is also the time for sustaining a given luminosity by conversion of gravitational potential energy alone. According to Equation (2.153), the luminosity is of the order of $|\mathrm{d}E_{\mathrm{pot}}/\mathrm{d}t|$ and thus a timescale

$$\tau_{\mathrm{KH}} = \frac{|E_{\mathrm{pot}}|}{\mathcal{L}^*} \approx \frac{E_{\mathrm{kin}}}{\mathcal{L}^*} \tag{2.160}$$

can be defined. This is the *thermal timescale*, also called *Kelvin–Helmholtz timescale*. A rough estimate for the gravitational potential energy makes use of the average mass $\overline{m(r)}$ throughout the layers of a star at average radius $\bar{r}$, which in turn can be very roughly approximated by $\overline{m(r)} = M/2$, $\bar{r} = R/2$. This leads to

$$\tau_{\mathrm{KH}} \approx \frac{GM^2}{2R\mathcal{L}^*}. \tag{2.161}$$

For example, for the Sun with a luminosity of $\mathcal{L}^*_\odot = 3.848 \times 10^{33}$ erg s^{-1} (3.848×10^{26} W), a timescale $\tau_{\mathrm{KH}} \approx 1.7 \times 10^7$ yr is found. This is longer than the dynamical timescale τ_{ff} but since this still is much shorter than the typical solar (or stellar) lifetime, the Sun (and stars) establishes thermal equilibrium comparatively quickly. On the other hand, it shows that the Sun cannot sustain its luminosity from gravitational contraction alone for a time comparable to the age of the Earth. Only with an additional energy source—nuclear reactions—a constant power output can be sustained over billions of years. The fact that $\tau_{\mathrm{ff}} \ll \tau_{\mathrm{KH}}$ implies that there can be episodes in stellar evolution when a star is in hydrostatic equilibrium but not in thermal equilibrium (see Chapter 6).

When a star compensates its energy loss $\mathcal{L}^*$ by drawing energy from a nuclear energy reservoir E_{nuc} the corresponding *nuclear timescale* is given by

$$\tau_{\mathrm{nuc}} = \frac{E_{\mathrm{nuc}}}{\mathcal{L}^*}. \tag{2.162}$$

The actual amount of available energy E_{nuc} depends on the amount of matter undergoing nuclear reactions and the energy released per reaction. It will therefore depend on factors such as the volume in which nuclear burning takes place and the types of possible reactions. As discussed in Chapter 7, these will be different for each hydrostatic burning phase established during stellar evolution. To get an idea of the magnitude of the timescale, a simple estimate can be made by assuming that a solar mass of hydrogen is converted to ^{4}He (hydrogen burning, see Chapter 7). The net energy release in hydrogen burning is given by the net reaction Q–value Q_{nuc} (see Section 4.1.8), which is 26.73 MeV per created ^{4}He or 6.3×10^{18} erg g^{-1} (6.3×10^{14} J kg^{-1}). Therefore the maximal energy obtainable from 1 $M_\odot$ is 1.25×10^{52} erg and using this together with $\mathcal{L}^*_\odot$ in Equation (2.162) results in $\tau_{\mathrm{nuc}} \approx 10^{11}$ yr. Stars do not use up all their theoretically available nuclear fuel, only a small fraction of the stellar mass is converted to energy and thus this timescale is only a rough upper limit. Nevertheless, it shows that although most nuclear reactions occur on extremely short timescales of 10^{-22}–10^{-16} s (see Section 3.7) the long stellar lifetimes are a result of the available nuclear energy reservoir and how fast the contained energy can be released. In the case of hydrogen burning, the long nuclear timescale is set by the very slow pp-reaction, though (see Section 7.2.1). The star adapts thermally to a local energy release on the Kelvin–Helmholtz timescale τ_{KH} and hydrodynamically on the dynamic timescale τ_{ff}. From above considerations it became obvious that

$$\tau_{\mathrm{ff}} \ll \tau_{\mathrm{KH}} \ll \tau_{\mathrm{nuc}}. \tag{2.163}$$

2.9.3 Numerical Solution of the Hydrostatic Equations

The stellar structure equations, Equations (2.134) and (2.135), comprise coupled differential equations that have to be solved numerically and simultaneously. The complete set of equations includes $4 + N_{\mathrm{nuclide}}$ equations for the quantities L, T, P,

X_i, and m or r as functions of r or m, respectively. The index i runs from 1 to N_{nuclide}, the number of nuclides included in the nuclear reaction network. As long as Equation (2.163) holds, the calculation of the X_i can be decoupled and the mass fractions can be assumed to be constant while solving the remaining structure equations for a given point in time t. For the next timestep $t + \Delta t$, new values for X_i are then calculated by the given equations and the other quantities are determined using these new mass fractions.

This is not possible, however, when the energy generation ε_{nuc} by the nuclear reactions significantly affects the hydrodynamic properties of the stellar plasma. For this reason, a common approach in stellar modeling is to fully couple only a reduced reaction network, accounting for the most important reactions heating the plasma, to the structure equations while the remaining reactions are only considered in a decoupled network. Moreover, when the energy generation in a shell leads to expansion or contraction from one time step to another, then the purely hydrostatic equations presented in Equations (2.134) and (2.135) are not sufficient anymore. In this case, the change in heat content due to volume work $P\,\mathrm{d}V$ has to be included in Equations (2.69) and (2.70). Formally, also the acceleration term in Equation (2.9) would have to be included but this second order time derivative is usually too small to be of significance unless a layer is almost in free fall and thus only when the changes occur on a timescale shorter than or comparably to τ_{ff}.

For the numerical solution at a specified time t, the interior of the star is subdivided into hundreds to thousands of different layers, into radial shells (Eulerian grid) or mass shells (Lagrangian grid). Using a Lagrangian grid leads to higher numerical stability and simplifies the choice of boundary conditions. The initial total mass M of the star and its composition $X_i(m)$ at time t are chosen. At the boundaries between layers, the obtained solutions have to match analytically. This leaves four more boundary conditions to be determined at the outer edges of the grid. In the center of the star $L(0) = 0$, $m(0) = 0$. Making a choice for boundary conditions at the outer edge is more difficult. The simplest approach would be $P(0) = 0, T(0) = 0$. Although this leads to a correct description of the interior, it clearly does not reproduce the observed surface conditions. A more realistic approach would be to define the stellar radius R by the layer in which the optical depth (including the overlying layers) has assumed the value of 2/3 (this defines the photosphere). The pressure at that radius is then given by the weight of the layers above and thus $P = 2GM/(3\kappa_{\text{atm}}R^2)$. This uses the mean opacity of the stellar atmosphere κ_{atm}. Then the temperature at this radius is the effective stellar temperature T_{eff} as defined by Equation (1.5). Since only two of the four variables are constrained at each outer boundary, matching solutions have to be found iteratively. The best approach, however, would be to fit the interior solution smoothly to an atmosphere model.

There are several ways to solve the coupled differential equations of stellar structure. Historically, the most widely used one is the *Henyey method*, based on a relaxation scheme improving an initial guess of a solution by iteration with a Newton–Raphson method. Further details are beyond the scope of this book.

While the use of a Lagrangian grid simplifies the computation, unfortunately it is not always feasible to use such a grid in 2D or 3D models because Lagrangian grids suffer severe distortion and tangling in the presence of discontinuities, singularities, or other non-linear phenomena. This is the reason why multi-D codes frequently use Eulerian grids whereas most 1D codes use the Lagrangian approach. Using Eulerian grids comes at a cost, however. Not only are they harder to implement, they also have severe limitations in applicability because they cannot handle expanding fluids properly as material may leave the computational domain, even when using dynamically adaptive grids. To circumvent such problems, grid-less methods have seen an increased use in recent astrophysical applications. Such a method, which does not use any kind of grid, is *smoothed-particle hydrodynamics* (SPH). It is closely related to many-body methods in which the cells of grid-based methods are replaced by effective "particles" with a definite extension. In SPH, a physical quantity of a fluid element at a given spatial position is obtained by interpolation of the properties of the neighboring particles. This is done by integrating over a weighting function (kernel) with a preselected shape and extension (smoothing length). For the implementation in a computer code, the integration is replaced by a summation, that is, the usual discretization of integrals is applied. Extending simpler many-body approaches, SPH also includes pressure terms required to fully describe astrophysical plasmas. Of course, SPH has to obey the same conservation laws as grid-based methods, that is, conservation of mass, energy, and momentum of electrically neutral plasmas. A main application of SPH in astrophysics is the description of merging compact objects—such as white dwarfs, neutron stars, or black holes—but it also has been used for simulating core-collapse and thermonuclear supernovae (see Chapter 9).

One long-standing question in the evolution of stellar models concerned whether there is a solution—and only one, unique solution—of the equilibrated structure equations for any choice of parameters. The ad hoc *Vogt–Russell theorem* from the 1920s postulated that this is true, even when numerical solutions, such as the Henyey method, may encounter convergence problems in some cases. Based on a simplified mathematical approach, however, the Vogt–Russell theorem lacks a sound mathematical justification. It had to be abandoned when multiple solutions co-existing for some test cases were found. Stellar evolution, nevertheless, proceeds deterministically as unique solutions are either imposed by the previous time evolution or by break-down of the approximations applied, e.g., of full thermal equilibrium. Therefore, a sequence of locally unique solutions is followed during stellar evolution and a star always "knows" how to evolve.

Further Reading

Böhm-Vitense, E. 1992, Introduction to Stellar Astrophysics (Vol. 3; Cambridge: Cambridge University Press)

Carroll, B. W., & Ostlie, D. A. 2017, An Introduction to Modern Astrophysics (2nd ed.; Cambridge: Cambridge University Press)

Chandrasekhar, S. 1958, An Introduction to the Study of Stellar Structure (New York: Dover)

Clayton, D. D. 1984, Principles of Stellar Evolution and Nucleosynthesis (Chicago: University of Chicago Press)

Hansen, C. J., Kawaler, S. D., & Trimble, V. 1995, Stellar Interiors: Physical Principles, Structure, and Evolution (2nd ed.; Berlin: Springer)

José, J. 2016, Stellar Explosions (Boca Raton, FL: CRC Press)

Kippenhahn, R., Weigert, A., & Weiss, A. 2012, Stellar Structure and Evolution (2nd ed.; Berlin: Springer)

Phillips, A. C. 1999, The Physics of Stars (2nd ed.; Chichester: Wiley)

Prialnik, D. 2009, An Introduction to the Theory of Stellar Structure and Evolution (2nd ed.; Cambridge: Cambridge University Press)

Rosswog, S., & Brüggen, M. 2011, Introduction to High-Energy Astrophysics (Cambridge: Cambridge University Press)

Ryan, S. G., & Norton, A. J. 2010, Stellar Evolution and Nucleosynthesis (Cambridge: Cambridge University Press)

AAS | IOP Astronomy

Essentials of Nucleosynthesis and Theoretical Nuclear Astrophysics

Thomas Rauscher

Chapter 3

Nuclear Physics for Astrophysics

3.1 Introduction

In this chapter, a brief introduction to nuclear physics and nuclear properties is given. This provides the foundation for understanding nuclear transformations in hot stellar plasmas through decays and nuclear reactions, which will be discussed in the next chapter. Due to the intended nuclear astrophysics focus of this textbook, not all of the nuclear physics and all derivations can be discussed in detail. In particular, a historic perspective is not given. For a more detailed presentation of nuclear physics, the reader is referred to a textbook on nuclear physics.

Basic properties of atomic nuclei are

- mass and charge;
- spin, parity, and isospin;
- size and deformation;
- decay half-life;
- electric and magnetic moments.

These are determined by the quantum mechanical interaction of the protons and neutrons—the *nucleons*—constituting an atomic nucleus. A single neutron is not stable (half-life $t_{1/2} \approx 880$ s) but a single proton can already act as an atomic nucleus: it is the nucleus of the hydrogen atom ^{1}H. Further elements are characterized by their proton number in the nucleus and, as explained below, certain combinations of proton number Z and neutron number N result in stable atomic nuclides whereas the majority of combinations yields unstable nuclides that "fall apart" with a certain average lifetime. Nuclides with the same Z but different N are called *isotopes*, nuclides with the same N but different Z are called *isotones*, and nuclides with different Z and N but the same mass number $A = Z + N$ are called *isobars*. In literature, the terms *nuclei* and *nuclides* often are used interchangeably but "nuclei"

doi:10.1088/2514-3433/ab8737ch3

actually refers to each single nucleus (of whatever type) in an ensemble whereas "nuclide" refers to a representative of a type of nucleus, regardless of how many there are. Also, "isotopes" often is used instead of "nuclei" or "nuclides" but strictly would be correct only in the context of a specific element.

Nucleons are not fundamental particles, they are *baryons* containing three fundamental particles called *quarks*. Protons are made of two up-quarks and one down-quark. Neutrons are made of two down-quarks and one up-quark. Table 3.1 summarizes properties of nucleons and fundamental particles. For each fundamental particle there is also an anti-particle, not shown in the table, with opposite electrical charge but otherwise identical properties. Quarks interact through a strong force due to their color charge, the *color force*. It can be described by an exchange of virtual particles, the *gluons*. Gluons are *mesons*, i.e., they are composed of quark–antiquark pairs. The underlying theory is called *quantum chromodynamics* (QCD). Quarks in baryons, such as the nucleons, are so strongly bound that the color force is almost completely shielded outside of the baryon. Only a residual interaction between two baryons remains, analogous to a van der Waals force in atoms and

Table 3.1. Properties of Nucleons and Fundamental Particles (as of 2018); Quarks and Leptons also have Antiparticles with Opposite Charge and Helicity, Respectively

Particle	Type	Charge (e)	Spin s_q ($\hbar$)	I_3 ($\hbar$)	Mass (MeV/c^2)
Nucleons					
Proton	Baryon	+1	1/2	+1/2	938.272
Neutron	Baryon	0	1/2	−1/2	939.565
Mesons (Examples)					
π^0	Meson	0	0	0	135.0
π^-	Meson	−1	0	−1	139.0
π^+	Meson	+1	0	+1	139.0
ρ^0	Meson	0	1	0	775.26
ω^0	Meson	0	1	0	782.65
Fundamental Particles					
Up	Quark, generation 1	+2/3	1/2	+1/2	2.2
Down	Quark, generation 1	−1/3	1/2	−1/2	4.7
Charm	Quark, generation 2	+2/3	1/2	0	1280
Strange	Quark, generation 2	−1/3	1/2	0	96
Top	Quark, generation 3	+2/3	1/2	0	173,000
Bottom	Quark, generation 3	−1/3	1/2	0	4180
Electron	Lepton, generation 1	−1	1/2	—	0.511
Electron–neutrino ν_e	Lepton, generation 1	0	1/2	—	$<2 \times 10^{-6}$
Muon	Lepton, generation 2	−1	1/2	—	105.7
Muon–neutrino ν_μ	Lepton, generation 2	−1	1/2	—	<0.19
Tau	Lepton, generation 3	−1	1/2	—	1777
Tau–neutrino ν_τ	Lepton, generation 3	−1	1/2	—	<18.2

molecules. This residual force of the color interaction between quarks in the nucleons is a short-range, strongly attractive force between the nucleons, the *strong nuclear force*. (In literature, often the color force between quarks is also called strong force. Throughout this book, however, strong force is used interchangeably with strong nuclear force.) The strong nuclear force was intensively studied even before the conception of the quark model. Due to the complexity of QCD it is still not possible to fully treat nuclear physics from a first-principle point of view, i.e., to derive all nuclear properties directly from QCD. Instead, an effective interaction between nucleons is assumed, mediated by virtual mesons. The short range of the force is due to the mass of the virtual mesons acting as force carriers. The lightest mesons, the π mesons (see Table 3.1), have masses of about 140 MeV/c^2. The lifespan Δt of virtual particles with energy ΔE is limited by the *Heisenberg uncertainty principle* for energy and time,

$$\Delta E \Delta t \leqslant \hbar. \tag{3.1}$$

Assuming that the virtual particles travel almost with the speed of light, they can cover at most a distance of

$$r = \mathrm{c}\Delta t = \frac{\mathrm{c}\hbar}{\Delta E} \approx 1.4 \text{ fm}, \tag{3.2}$$

which implies that the attractive force drops off quickly for distances larger than 1–1.4 fm. This is approximated by an effective nucleon–nucleon potential, the *Yukawa potential*,

$$V_{\mathrm{Yuk}}(r) = -g_{\mathrm{coupl}} \mathrm{e}^{-m_{\mathrm{ex}} c r/\hbar} \frac{1}{r}, \tag{3.3}$$

where g_{coupl} is the coupling constant determining the basic strength of the force and m_{ex} is the mass of the virtual force carriers. For $m_{\mathrm{ex}} = 0$ (as would be the case for photons) a Coulomb-type potential is recovered. That is why a potential as in Equation (3.3) is also called a *screened Coulomb potential*. At a distance of about 1 fm, the strong nuclear force is 137 times stronger than the Coulomb force and 10^{38} times stronger than gravitation. It is equal to the Coulomb force at a distance of about 2.5 fm and its strength is dropping much faster with increasing distance due to the exponential term in Equation (3.3). At distances smaller than 0.5 fm the strong force becomes repulsive. This is not due to the Pauli principle but rather to the spin–spin interaction of the quarks, which is felt at shorter distance and aligns the quark spins. This leads to an increase in potential energy.

The strong nuclear force is also spin-dependent. It is stronger for nucleons with parallel spins than for combinations with anti-parallel spin alignment. For example, this requires the proton and the neutron forming a deuteron to have parallel spins because otherwise the deuteron would not be stable. In another example, a putative di-proton (^{2}He) or di-neutron cannot exist as a stable nucleus because the Pauli principle requires its two constituents to have anti-parallel spins and the nuclear force then is too weak to bind the nucleons together.

Table 3.1 also lists the z-projection I_3 of the isospin. The concept of the *isospin* was introduced when it was realized that the strong nuclear force does not distinguish between neutrons or protons and therefore both could be seen as two states of one particle, the nucleon. Later, the concept was also incorporated into the quark model. Quarks—and therefore all particles made from quarks—carry an isospin. Isospin can be treated mathematically in exactly the same way as angular momentum quantum numbers (see Section 1.7).

3.2 Nuclear Radii and Masses

It was experimentally found (see Equation (1.110)) that nuclei exhibit a roughly constant central density regardless of size and that their radial size scales with the third root of the number of nucleons contained in the nucleus. This is explained by both the nucleon–nucleon force being short-ranged and becoming repulsive at small distances. This gives rise to a preferred nucleon separation at the force minimum at roughly 0.5 fm. The rapid drop-off in strength implies that only nearest neighbors are attracting each other while distant nucleons are not felt. Except for the lightest nuclei or exotic nuclei with pronounced surplus in neutrons or protons, the matter density distribution as function of radius r inside a (spherically symmetric) nucleus can be parameterized as

$$\rho_{\mathrm{nuc}} = \frac{\rho_0}{1 + \mathrm{e}^{(r-R_{\mathrm{nuc}})/a_{\mathrm{nuc}}}}, \tag{3.4}$$

with the radius R_{nuc} at which ρ_{nuc} is $\rho_0/2$ (by definition this is the nuclear radius because a_{nuc} is small) and the nuclear surface diffuseness parameter a_{nuc}. The surface diffuseness usually is small ($a_{\mathrm{nuc}} \approx 0.55$ fm), providing a well-defined, distinct surface of the nucleus. The central density ρ_0 is about 2.4×10^{14} g cm^{-3}. This value is often referred to as *(standard) nuclear density*. Since the density stays constant, adding more nucleons increases the volume of a nucleus roughly linearly with the number of nucleons. The radius of a sphere is related to its volume by $R \propto V^{1/3}$ and thus the observed dependence $R_{\mathrm{nuc}} \propto A^{1/3}$ is also understood from the properties of the strong nuclear force.

Because nucleons are bound by the strong force, the total mass m_{nuc} of a nucleus with Z protons and N neutrons is not just the sum of its nucleon masses but rather this sum reduced by the amount of energy in the binding $\mathcal{B}$,

$$m_{\mathrm{nuc}}(Z, N) = Zm_{\mathrm{p}} + Nm_{\mathrm{n}} - \frac{\mathcal{B}(Z, A)}{\mathrm{c}^2}, \tag{3.5}$$

according to the energy–mass equivalence $E = m\mathrm{c}^2$. The total binding energy $\mathcal{B}(Z, N)$ is required to remove all nucleons from a nucleus, that is to move them to infinite distance. This is equivalent to the definition of the gravitational binding energy used in Chapter 2. Often, the binding energy is given in energy per nucleon, $\mathcal{B}/A$, with $A = Z + N$ (see Section 1.3). Empirically, $\mathcal{B}/A$ increases up to $A \approx 60$, with ^{62}Ni being the nucleus with the largest value, and decreases for larger A. This shows that the binding energy is not only proportional to the number of nucleons in a nucleus but must have contributions from further effects. An indication for this is also the fact that an odd–even staggering is observed, with nuclei having an even number of neutrons and protons being stronger bound.

To be noted is the fact that mass tables in literature often quote the *atomic* masses, especially for experimental masses. The atomic mass m_{ato} is related to the nuclear mass m_{nuc} by

$$m_{\mathrm{ato}} = m_{\mathrm{nuc}} + Zm_{\mathrm{e}} - \frac{\mathcal{B}_{\mathrm{e}}}{\mathrm{c}^2}, \tag{3.6}$$

with m_{e} being the electron mass and $\mathcal{B}_{\mathrm{e}}$ the total binding energy of the electrons in the atom. When using masses from a table, it should be carefully checked whether these are atomic or nuclear masses.

Another often used convention is to quote the *mass excess*

$$\Delta_{\mathrm{ato}} = (m_{\mathrm{ato}} - Am_{\mathrm{u}})\mathrm{c}^2 \tag{3.7}$$

instead of the atomic mass itself. By definition, $\Delta_{\mathrm{ato}} = 0$ for ^{12}C. The same definition could be used involving nuclear masses instead of atomic ones but this is less common.

The liquid drop model by Bethe and Weizsäcker proved useful in describing nuclear masses. The fact that nuclides are well-defined collections of nucleons with a sharp surface suggests a treatment similar to a liquid drop with the particles making up the drop only feeling nearest-neighbor interactions. This gives rise to a surface tension as particles at the outside only have neighbors further inside, similar as in a drop of liquid. The leading term in the Bethe–Weizsäcker formula stems from the contribution of the nucleon bulk and is proportional to the number of nucleons A. This value overestimates the binding energy and has to be corrected by introducing additional terms. There are five terms in the traditional formula:

$$\begin{aligned}\mathcal{B}(Z, A) = a_{\mathrm{V}}A - a_{\mathrm{S}}A^{2/3} - a_{\mathrm{C}}Z(Z-1)A^{-1/3} - a_{\mathrm{asym}}(N-Z)^2A^{-1} \\ + a_{\mathrm{pair}}\Delta_{\mathrm{pair}}(Z, A)A^{-1/2}.\end{aligned} \tag{3.8}$$

The four constant parameters a_{V}, a_{S}, a_{C}, a_{asym} are fit to best reproduce experimental masses. This is why this formula sometimes is referred to as *semi-empirical mass formula*. The fourth and fifth terms in Equation (3.8) are quantum mechanical corrections that would not appear in a fully classical picture of a liquid drop (see below). The strength of the fifth term depends on the number of protons Z, number of neutrons N, and thus also on the total nucleon number (mass number) $A = Z + N$. The standard version of the formula shown here assumes spherical liquid drops but it could be adapted for deformed nuclides.

The meaning of the five terms in Equation (3.8) is as follows:

- $a_{\mathrm{V}}A$: *Volume energy*; since only the nearest neighbors contribute to the binding of a nucleon, the binding of a nucleon saturates to a constant value and thus the total binding energy scales with the number of nucleons A in the bulk;

- $a_S A^{2/3}$: *Surface energy*; the volume term overestimates the actual binding energy because nucleons at the surface of the nuclear droplet have fewer neighbors and thus the total binding energy is reduced by an amount proportional to the surface;
- $a_C Z(Z-1)A^{-1/3}$: *Coulomb energy*; protons repel each other and thus the binding energy is reduced by the number of proton pairs (for larger proton numbers $Z(Z-1) \approx Z^2$); the $A^{-1/3}$ dependence reflects the $1/r$ dependence of the Coulomb potential;
- $a_{\rm asym}(N-Z)^2 A^{-1}$: *Asymmetry energy* (sometimes called symmetry energy); nuclides with an equal number of neutrons and protons are more strongly bound (see Section 3.3) as a consequence of the Pauli exclusion principle, which restricts packing of nucleons into the same energy states;
- $a_{\rm pair}\Delta_{\rm pair}A^{-1/2}$: *Pairing energy*; this term can only be understood fully in the quantum mechanical treatment of the shell model of atomic nuclei (see Section 3.3); due to the spin dependence of the nucleon–nucleon force, spin-0 pairs are more strongly bound and thus the total nuclear binding is modified according to whether N, Z, or both are even:

$$\Delta_{\rm pair} = \begin{cases} +1 & \text{for even proton and neutron numbers (even–even nucleus),} \\ 0 & \text{for nuclides with odd mass number,} \\ -1 & \text{for odd proton and odd neutron numbers (odd–odd nucleus).} \end{cases} \tag{3.9}$$

A good fit to experimental mass data is found with the following set of parameter values (all values in MeV/c^2): $a_V = 16$, $a_S = 18.5$, $a_C = 0.72$, $a_{\rm asym} = 23.4$, $a_{\rm pair} = 12$. Figure 3.1 shows a comparison between the semi-empirical formula and experimental binding energies for stable nuclides. Larger deviations are seen for light nuclei but the agreement for nuclides with $A \geqslant 40$ is within a range of $\pm 10\%$.

We will see that the knowledge of masses and binding energies for very neutron- or proton-rich, unstable nuclei is essential for the modeling of nucleosynthesis. For a better prediction of these properties a more sophisticated theoretical treatment than the simple semi-empirical formula is required. Nevertheless, some basic features of nuclear masses can be inferred from the simple formula. Assuming constant mass number A (and $Z(Z-1) \approx Z^2$) yields a quadratic equation

$$\mathcal{B}(A = \text{const}, Z) = C - a_1 Z^2 - a_2(N-Z)^2, \tag{3.10}$$

with a constant C and two parameters a_1, a_2. Thus, a cut through the Z, N plane with $A = Z + N = \text{const}$ results in a *mass parabola*. The charge number Z_0 closest to the maximum of the binding energy (which is the minimum in mass, according to Equation (3.5)) within such a parabola is given by its vertex at

$$Z_0 = \frac{A}{2 + 0.0153A^{2/3}}. \tag{3.11}$$

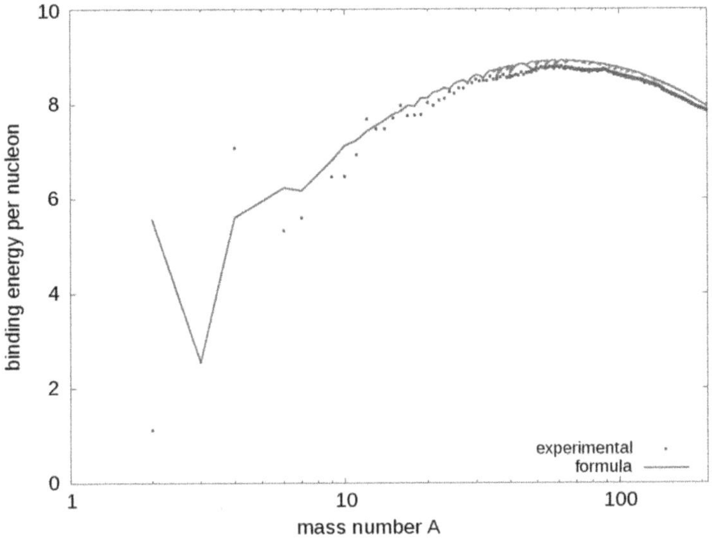

Figure 3.1. Comparison of the Bethe–Weizsäcker semi-empirical formula for binding energies with experimental values for stable nuclides. The binding energy per nucleon $\mathcal{B}/A$ is given in MeV/c^2.

As can be seen in Figure 3.2, the line of maximum binding energy follows the $N = Z$ line for light nuclides and gradually bends off toward the neutron-rich side for $A > 20$. This shows the action of the Coulomb term in Equation (3.8), which decreases the binding when more protons are contained in a nucleus. Along this line stable nuclei are found whereas isotopes with larger or smaller neutron number will spontaneously decay, that is, transform into another nuclide (see Section 3.9). The average time it takes such a nucleus to decay is given as lifetime or half-life (for a further definition of these, see Section 4.1.7).

The binding energy curve for stable nuclei, shown in Equation (3.8), has a maximum in the region of Fe and Ni isotopes. This limits the possibility to release energy by fusing light nuclei, which is essential for stellar evolution because it determines how a star can generate energy to stay in hydrostatic equilibrium (see Section 2.2 and Chapter 6). The solar system abundances exhibit a peak structure at ^{56}Fe, see Figure 5.1. This is often mistaken for an indication that ^{56}Fe has the highest $\mathcal{B}/A$. The tightest bound nuclei, however, are ^{62}Ni (8.795 MeV/A) and ^{58}Fe (8.792 MeV/A), followed by ^{56}Fe (8.790 MeV/A) and ^{60}Ni (8.781 MeV/A). The reason for the high ^{56}Fe abundance is found in the way it is created in stars and stellar explosions, which is discussed in Sections 7.7 and 9.4.3.

Equation (3.5) shows that the mass of a nucleus is less than the sum of the masses of its constituents. Although this is a general principle for any bound system being a consequence of the theory of relativity, the effect is especially pronounced in atomic nuclei due to the strength of the binding. The *separation energy* required to remove a nucleon or a number of nucleons from a nucleus is equal to the binding energy released when adding the same to a nucleus. Table 3.2 shows the relations for

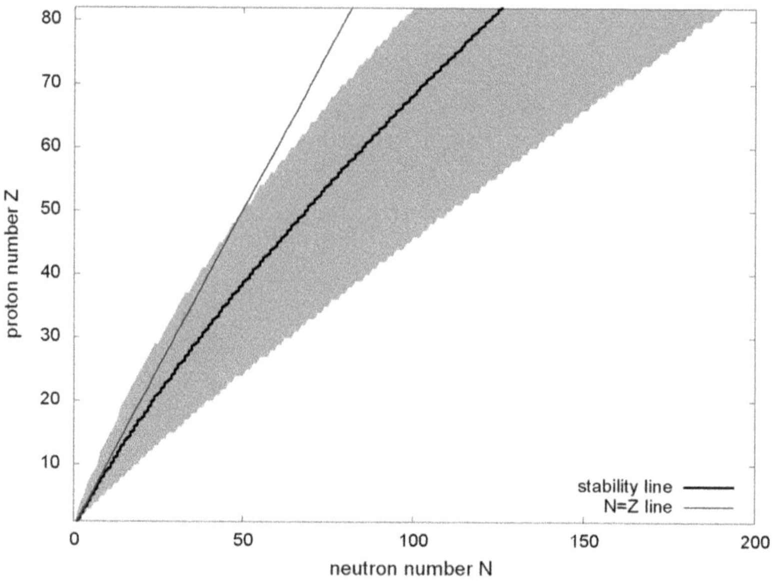

Figure 3.2. Line of stability according to Equation (3.11) compared to the $N = Z$ line. Also shown are approximate neutron and proton driplines obtained with the semi-empirical formula in Equation (3.8) as the borders of the gray area marking bound nuclei.

Table 3.2. Equivalent Formulae for Calculating Selected Separation Energies

Neutron-separation Energy S_n
$S_n = m_{nuc}(Z, N-1) + m_n - m_{nuc}(Z, N)$
$S_n = m_{ato}(Z, N-1) + m_n - m_{ato}(Z, N)$
$S_n = \Delta_{ato}(Z, N-1) + \Delta_{ato}(0, 1) - \Delta_{ato}(Z, N)$
Proton-separation Energy
$S_p = m_{nuc}(Z-1, N) + m_p - m_{nuc}(Z, N)$
$S_p = m_{ato}(Z-1, N) + m_{ato}(1, 0) - m_{ato}(Z, N)$
$S_p = \Delta_{ato}(Z-1, N) + \Delta_{ato}(1, 0) - \Delta_{ato}(Z, N)$
α-Particle Separation Energy
$S_\alpha = m_{nuc}(Z-2, N-2) + m_\alpha - m_{nuc}(Z, N)$
$S_\alpha = m_{ato}(Z-2, N-2) + m_{ato}(2, 2) - m_{ato}(Z, N)$
$S_\alpha = \Delta_{ato}(Z-2, N-2) + \Delta_{ato}(2, 2) - \Delta_{ato}(Z, N)$

Note. The mass excess of the neutron is $\Delta_{ato}(0, 1) = \Delta_{ato}(n) = 8.071\ \mathrm{MeV}/c^2$, the one of the proton is $\Delta_{ato}(1, 0) = \Delta_{ato}(^1\mathrm{H}) = 7.289\ \mathrm{MeV}/c^2$, and the one of the α-particle is $\Delta_{ato}(2, 2) = \Delta_{ato}(^4\mathrm{He}) = 2.425\ \mathrm{MeV}/c^2$.

neutron-, proton-, and α-separation energy. Similar relations can be derived for other nucleon groups. Instead of masses, always also the mass excesses as defined in Equation (3.7) can be used because the Am_u term cancels in the subtraction. For processes conserving the proton number Z (no decays mediated by the weak force, as discussed in Section 3.9), nuclear masses and atomic masses can be used interchangeably because the electron masses and binding energies cancel.

Because the neutron separation energy is decreasing with increasing number of neutrons in a nucleus with given Z, it will become zero or even negative at a specific neutron number N_{drip}. The location of this *neutron dripline* is defined as the last nucleus before $S_{\mathrm{n}} \leqslant 0$. A *proton dripline* can be defined in a completely analogous manner to define the border to $S_{\mathrm{p}} \leqslant 0$. A separation energy equal to or smaller than zero implies that the nucleon in question is not bound anymore. No additional energy is needed to remove it from the nucleus. With a negative value, energy actually is gained when removing the nucleon. The nucleus may still have a finite lifetime due to a Coulomb barrier to overcome (for protons or α-particles) or because of the Heisenberg uncertainty relation. Approximate neutron and proton driplines are shown in Figure 3.2. It should also be noted that the driplines are not sharply defined boundaries. Due to, e.g., the pairing term in Equation (3.8) a nucleus with $S_{\mathrm{n}}(Z, N) \leqslant 0$ can be followed by a nucleus with $S_{\mathrm{n}}(Z, N+1) > 0$ (and similarly for protons). The locations of these driplines vary between different theories for predicting masses. Since their location determines nucleosynthesis at extreme conditions, it is essential to predict them correctly and currently this remains a challenge for nuclear theory.

Table 3.2 also shows relations for calculating the energy required to remove α-particles (the nucleus of a ^{4}He atom made from two protons and two neutrons) from a nucleus. In nuclides with $N > 50$ negative S_α is found also close to the line of stability, giving rise to a region of spontaneous α-emitters. Despite the fact that it is energetically advantageous for such nuclides to emit an α-particle, they can have very long half-lives because of the combined probabilities that neutrons and protons within the nucleus form an α-particle and that this α-particle overcomes the Coulomb barrier inside the nucleus (see Section 3.8).

3.3 The Independent-particle Model and the Nuclear Shell Model

3.3.1 Independent-particle Model

The formula shown in Equation (3.8) was derived assuming that the atomic nucleus behaves like a droplet of an incompressible fluid. In addition, the symmetry term and the pairing term already considered the special behavior of the strong nuclear force. Although this semi-empirical formula was quite successful in reproducing the general behavior of binding energies with increasing A, a more fundamental, quantum mechanical treatment is needed to improve the accuracy of mass predictions and to explain features such as the increased binding at specific neutron and proton numbers. It was experimentally found that nuclides with neutron numbers 2, 8, 20, 28, 50, 82, 126 or proton numbers 2, 8, 20, 28, 50, 82 are more strongly bound than expected from the droplet model. These are called the *magic numbers*. (Proton number 126 is also expected to be "magic" but has not been reached experimentally, yet.) Nuclides with magic neutron number *and* magic proton number are called *double-magic nuclides*. An important example is ^{56}Ni. The liquid drop mass model can be extended to include these features by introducing an additional term in the formula, the *shell correction* Δ_{shell}. The actual physical meaning of this term can only be understood in the context of the quantum mechanical treatment outlined below.

Equation (3.4) implied that nucleons in an atomic nucleus are confined to a sharply bounded spatial region. Due to the short range of the strong nuclear force it can be assumed that a nucleon moves in a potential generated by the average forces by the nearest neighbors. Due to the incompressibility caused by the repulsive part of the strong force, this potential is approximately the same everywhere within the nucleus and rapidly goes to zero at the nuclear surface. Thus, the nucleons can be assumed to behave like independent particles moving in an averaged potential, enclosed in a well-defined space. Then their statistical properties can be described by a Fermi gas as introduced in Section 1.5.1.2 and specifically in Section 1.5.4. In the ground state (see Section 3.4 for other configurations) of an atomic nucleus all nucleons are in their lowest possible energy states and thus only occupy energies up the Fermi energy as given in Equation (1.78). Since $\bar{n}$ is roughly the same for all nuclei, also E_F must be the same. By the definition of the number density in Equation (1.7) it is inversely proportional to the volume. The volume of a nucleus, in turn, is proportional to A, as discussed above. Using $g = 2$ for nucleons, ρ_{nuc}, the nucleon mass, and assuming separate Fermi-gases for protons and neutrons, one arrives at $E_F \approx 29$ MeV. The average energy of a particle in a Fermi gas is $3/5E_F$. Neutron and proton-separation energies along the line of stability are of the order of 8–12 MeV and this energy will unbind the nucleon with the highest energy in the potential well, at E_F. Therefore we can conclude that the depth V_{nuc} of the average potential between nucleons must be around $E_F + S_n$ or $E_F + S_p$, respectively, which amounts to about 40 MeV. Moreover, the required energy to unbind all nucleons of the nucleus (having an average energy of $3/5E_F$) is $A(V_{nuc} - 3/5E_F)$. Thus, we can conclude that the total binding energies of nuclides must scale with the number of nucleons. This is consistent with assuming the volume term in Equation (3.8) for the starting point of the semi-empirical formula. Likewise, the asymmetry term can be derived from the Fermi gas picture.

A full quantum mechanical treatment of a nucleus containing A nucleons would require to find the A-body wave function

$$\Psi(\vec{r}_1, \vec{r}_2, \vec{r}_3, \ldots, \vec{r}_{A-1}, \vec{r}_A) = \hat{A} \prod_{i=1}^{A} \phi_i(\vec{r}_i). \tag{3.12}$$

The ϕ_i are the wave functions of the individual nucleons and the indices are assumed to also contain all quantum numbers characterizing the system. As nucleons are indistinguishable fermions, the individual wave functions have to obey the Pauli exclusion principle, implying $\phi_i(\vec{r}_i)\phi_j(\vec{r}_j) - \phi_j(\vec{r}_i)\phi_i(\vec{r}_j) = \tilde{\delta}_{ij}$. This is achieved by the *antisymmetrization operator*

$$\hat{A} = \frac{1}{\sqrt{A!}} \sum_{k=1}^{A!} (-1)^k \hat{P}, \tag{3.13}$$

where the *permutation operator* $\hat{P}$ exchanges the particle coordinates. For example, for a system of two particles $\hat{P}(\phi_1(\vec{r}_1)\phi_2(\vec{r}_2)) = \phi_1(\vec{r}_2)\phi_2(\vec{r}_1)$. As usual, the total wave function is normalized to

$$\int \mathrm{d}^3 r_1 \int \mathrm{d}^3 r_2 \cdots \int \mathrm{d}^3 r_A \, |\Psi|^2 = 1. \tag{3.14}$$

The state energies $\mathcal{E}$ of Ψ are eigenvalues of the $3A$-dimensional *Schrödinger equation*,

$$\mathcal{H}\Psi = \mathcal{E}\Psi. \tag{3.15}$$

The *Hamiltonian* of the system is

$$\mathcal{H} = \sum_{i=1}^{A} {}^{i}E_{\mathrm{kin}} + \sum_{i<j=1}^{A} V_{\mathrm{Yuk},ij}(|\vec{r}_i - \vec{r}_j|), \tag{3.16}$$

where ${}^{i}E_{\mathrm{kin}}$ is the kinetic energy of the ith nucleon and $V_{\mathrm{Yuk},ij}$ is the potential between two nucleons i, j describing the nucleon–nucleon interaction caused by the strong nuclear force. For such Hamiltonians including interactions between all particles, and even when interactions between three or more particles are ignored (as was done here), neither classical nor quantum systems can be solved analytically and exactly for $A \geqslant 3$. There has been large progress in numerical solutions based on such many-body Hamiltonians for larger systems but they are computationally expensive and currently only feasible up to $A \approx 20$. Therefore various ways have been devised to circumvent the problem by modifying the Hamiltonian to obtain equations that are easier to handle. The independent-particle model is such an approach. As mentioned above, the sum of interactions between two nucleons is replaced by an effective potential felt by each nucleon. Its strength (or depth) V_0 must be close to the 40 MeV as inferred from the Fermi gas considerations above. The geometric shape has to look similar to the inverse of the density distribution shown in Equation (3.4) (which, in turn, is similar to a Fermi-function as shown in Figure 1.3) because in the independent particle picture a nucleon may move freely within the potential inside the nucleus but cannot escape beyond the nuclear radius when it is bound because it does not have sufficient energy. This led to the introduction of the *Woods–Saxon potential*,

$$V_{\mathrm{SW}}(r) = \frac{V_0}{1 + \mathrm{e}^{(r - R_{\mathrm{nuc}})/a_{\mathrm{nuc}}}}. \tag{3.17}$$

Such a potential can be used to solve a simplified version of Equations (3.15), (3.16). We can assume $\mathcal{H} = \mathcal{H}_0 + V_{\mathrm{res}}$ with

$$\mathcal{H}_0 = \sum_{i=j}^{A} ({}^{i}E_{\mathrm{kin}} - {}^{i}V_{\mathrm{nuc}}) = \sum_{i} \underline{h}_i, \tag{3.18}$$

where $\underline{h}_i = {}^iE_{\text{kin}} - {}^iV_{\text{nuc}}$ and V_{res} is the residual interaction

$$V_{\text{res}} = \left(\sum_{i<j=1}^{A} V_{\text{Yuk},ij}(|\vec{r}_i - \vec{r}_j|) \right) - \sum_i {}^iV_{\text{nuc}}. \tag{3.19}$$

Not knowing the appropriate mean potential ${}^iV_{\text{nuc}}$, it would be possible to minimize the residual in an interactive procedure. This is called the *Hartree–Fock method,* which was introduced for describing wavefunctions for atomic electrons. The application of this method is not straightforward for nucleons because of the short-range repulsive part of the strong force but it can be adapted. It yields a potential similar to the one introduced in Equation (3.17) (strictly only for neutrons because a Coulomb potential has to be added for protons) and previously deduced from experimental scattering data. Once such an appropriate potential is found, the V_{res} term is negligible and the Schrödinger equation of Equation (3.15) reduces to $\mathcal{H}_0\Psi = \varepsilon\Psi$. One solution is the product of all individual wave functions ψ_i fulfilling the equation

$$\underline{h}_i\psi_i = E_i\psi_i. \tag{3.20}$$

Thus, the resulting total wave function Ψ is the antisymmetrized product of the A single-particle wave functions ψ_i with the total energy eigenvalue $\varepsilon = E_1 + E_2 + E_3 + \cdots + E_{A-1} + E_A$.

3.3.2 Time-independent Schrödinger Equation

Looking at Equation (3.20) for a single nucleon (dropping the index i) and writing the Hamiltonian explicitly, we arrive at a version of the *time-independent Schrödinger equation*

$$\underline{h}\psi(\vec{r}) = \left(-\frac{\hbar^2}{2m}\vec{\nabla} - V_{\text{nuc}} \right)\psi(\vec{r}) = E\psi(\vec{r}). \tag{3.21}$$

Note that $V_{\text{nuc}} = V_{\text{SW}} + V_{\text{C}}$. The mass m appearing in the equation is the one of the particle affected by the potential. In the context of the independent particle model, this is the mass of a nucleon m_{nucleon}. But the equation is completely general and can be applied to any particle and to bound as well as scattering states (see Section 3.5). Therefore in this chapter the solution of this equation in spherical symmetry is discussed in a more general way before returning to its application to bound states in the independent-particle model in Section 3.3.3.

Since both the effective strong interaction potential V_{SW} from Equation (3.17) and the Coulomb potential V_{C} are *central potentials* depending only on the distance r from the center of the nucleus, also the effective, averaged potential V_{nuc} is a central potential and the best choice for solving Equation (3.21) is to use spherical coordinates (r, θ, ϕ). (Note that Cartesian coordinates were used for the Schrödinger equation in the discussion of the Fermi gas in Section 1.5.1.2.) With this assumption the Schrödinger equation becomes separable into an equation for the radial motion and one for the angular motion. The angular motion is force-free and thus does not depend on the central potential. Then the full solution for the wavefunction ψ can be written as the product of the solutions of two separate equations,

$$\psi(r, \theta, \phi) = \sum_{\ell,\ell_z} \underline{R}_\ell(r) c_{\ell,\ell_z} \mathcal{Y}_{\ell,\ell_z}(\theta, \phi). \tag{3.22}$$

The *spherical harmonic functions* $\mathcal{Y}_{\ell,\ell_z}(\theta, \phi)$ do not depend on the central potentials and fully describe the angular dependence of the wavefunction ψ. The separation introduced an explicit dependence on the *orbital angular momentum quantum number* ℓ and the *magnetic quantum number* ℓ_z (which is the projection of the angular momentum onto the z-axis; see also Section 1.7). Equation (3.22) is also called the *partial wave expansion* of the wave function $\psi(r, \theta, \phi)$. The individual terms in the sum for a given ℓ are called *partial waves*. The constants c_{ℓ,ℓ_z} are normalization coefficients for the spherical harmonic functions.

The spherical harmonics have the property that $\mathcal{Y}_{\ell,\ell_z}(\pi - \theta, \pi + \phi) = (-1)^\ell \mathcal{Y}_{\ell,\ell_z}(\theta, \phi)$. Therefore they have even parity for even ℓ and odd parity when ℓ is odd (see Section 1.7). They determine the parity of the wave solution ψ of Equation (3.21) as the radial part is symmetric in r, see below.

The *radial, time-independent Schrödinger equation* has to be solved to obtain $\underline{R}_\ell(r)$,

$$-\frac{\hbar^2}{2m}\left(\frac{\mathrm{d}^2\underline{R}_\ell}{\mathrm{d}r^2} + \frac{2}{r}\frac{\mathrm{d}\underline{R}_\ell}{\mathrm{d}r}\right) + \left[V_{\mathrm{nuc}}(r) + \frac{\ell(\ell+1)\hbar^2}{2\,mr^2}\right]\underline{R}_\ell = E\underline{R}_\ell. \tag{3.23}$$

Substituting $u_\ell(r) = r\underline{R}(r)$, the equation is transformed to the more familiar form

$$\frac{\mathrm{d}^2 u_\ell}{\mathrm{d}r^2} + \frac{2m}{\hbar^2}\left[E - V_{\mathrm{nuc}}(r) - \frac{\ell(\ell+1)\hbar^2}{2\,mr^2}\right]u_\ell = 0, \tag{3.24}$$

which sometimes is also written as

$$\frac{\mathrm{d}^2 u_\ell}{\mathrm{d}r^2} + \left[\mathcal{K}^2 + E_{\mathrm{pot},\ell}\right]u_\ell = 0 \tag{3.25}$$

when using $E = p^2/(2m) = \hbar^2\mathcal{K}^2/(2m)$ and $\mathcal{K}$ being the wave number (see also Equation (1.112) for comparison). Note that the energy E or $\mathcal{K}^2$ can be negative and that the function values of ψ, $\underline{R}_\ell$, and u_ℓ will be complex numbers in the most general case.

The potential energy term is

$$E_{\mathrm{pot},\ell} = -\frac{2m}{\hbar^2}V_{\mathrm{nuc}}(r) - \frac{\ell(\ell+1)}{r^2}. \tag{3.26}$$

The term $\ell(\ell+1)/r^2$ is often called the *centripetal potential*. For $\ell > 0$ it acts like a repulsive potential at small radii. This shows the utility of the partial wave expansion introduced in Equation (3.22). The wave function of a bound state (particle with $E < 0$ or trapped between infinitely high potential walls, see below) is comprised of only one partial wave. The wave functions of scattering states ($E > 0$ in finite potential) mathematically involve an infinite sum. In practicality, however, only a few partial waves have to be considered. This is because of the short range of the effective nuclear interaction V_{SW} and the centripetal potential preventing particles with higher ℓ to reach the interaction region (that means the amplitude of the wave function approaches zero at small radii). The free Schrödinger equation without potential (or only with a Coulomb potential) can be solved analytically.

For a charged particle in a potential, i.e., a proton, the actual potential V_{nuc} not only contains the effective, averaged nuclear potential V_{SW} but also a *Coulomb potential* V_C. Assuming that the charges of the protons within the nucleus generating the potential are "smeared" out spherical symmetrically inside the nuclear radius, the Coulomb potential will be repulsive but constant inside the nucleus and only falling off with the well-known $1/r$ dependence outside the nuclear radius R_{nuc},

$$V_C(r) = \begin{cases} \dfrac{Z_1 Z_2 e_C^2}{R_{nuc}} & r < R_{nuc}, \\ \dfrac{Z_1 Z_2 e_C^2}{r} & r \geqslant R_{nuc}, \end{cases} \tag{3.27}$$

where e_C is the elementary charge. Here, $Z_1 = 1$ for a proton and $Z_2 = Z - 1$ for its repulsion by the other protons in the nucleus. For bound states, the constant Coulomb potential within the nucleus shifts the energy eigenvalues to higher energies and thereby reduces the binding energy, that is the difference between the energy eigenvalue of the bound state and the top of the potential wall. This is phenomenologically included in the Coulomb term of the semi-empirical mass formula given in Equation (3.8). A useful approximate for the maximum height of the Coulomb potential (the *Coulomb barrier* or *Coulomb wall*) in practical units is $1.44 Z_1 Z_2 / A^{1/3}$ MeV, with Z_1, Z_2 being the charge numbers of the interacting nuclei or particles, and A being their summed mass number.

Analytic solutions of Equation (3.25) only exist for special shapes of the potential term given in Equation (3.26). For the more realistic case of the radial potential with the shape of Equation (3.17) as well as for $\ell > 0$ and charged particles, Equation (3.25) has to be solved numerically. In principle, any numerical method for solving differential equations of the form

$$\frac{d^2 f(r)}{dr^2} + w(r) f(r) = 0 \tag{3.28}$$

could be used, where $w(r)$ is an arbitrary function independent of u. For the radial Schrödinger equation, $f(r) = u_\ell$ and

$$w(r) = w_\ell(r) = \frac{2m}{\hbar^2}\left[E - V_{nuc}(r) - \frac{\ell(\ell+1)\hbar^2}{2\,mr^2}\right] = \mathcal{K}^2 + E_{pot,\ell}, \tag{3.29}$$

which can easily be seen by comparison to Equations (3.24) and (3.25). A quickly converging, computationally efficient method is known as the *Numerov method*,

Cowell's method, or the *Fox–Goodwin method*. This recursive approach is one order more accurate than using the fourth-order Runge–Kutta method, another well known approach to numerically solve differential equations. It is based on approximating the second-order derivative by a three-point difference formula and cleverly canceling higher order correction terms. In the resulting three-point recursion the value of the wavefunction at the radius $r + s$ is given by

$$u_\ell(r+s) = \frac{2u_\ell(r) - u_\ell(r-s) - [10w_\ell(r)u_\ell(r) + w_\ell(r-s)u_\ell(r-s)]s^2/12}{1 + s^2 w_\ell(r+s)/12}. \quad (3.30)$$

This evaluates the wavefunction on an equidistant radial mesh with stepsize s. The method is not very sensitive to the actual value of s as long as it does not skip over roots of u_ℓ. For nuclear wavefunctions, a stepsize $s \leqslant 0.025$ fm usually is sufficient. Since Equation (3.28) describes a second-order differential equation, two boundary conditions are required to completely determine the solution. This is also seen in the recursion relation. Since the values of u_ℓ have to be known at $r - s$ and r in order to compute its value at radius $r + s$ two such values have to be chosen to start the recursion. They can be obtained from boundary conditions and normalization requirements. The first condition is that the amplitude of the radial wavefunction has to be zero at the origin, $u_\ell(0) = 0$, because otherwise the total wavefunction ψ is not normalizable (see the behavior of ψ at $r = 0$ in Equation (3.22), remembering that $\underline{R}_\ell = u_\ell/r$). For the same reason, the product $w_\ell(0)u_\ell(0) = 0$ and therefore $w_\ell(0)$ must assume a finite value despite of the fact that Equation (3.29) diverges for $r \to 0$. A simple choice is $w_\ell(0) = 0$ for implementation in a computer code.

For the second condition, the first derivative $\mathrm{d}u_\ell/\mathrm{d}r$ close to the starting point of the iteration can be used. It is not known initially but can be obtained from normalization requirements. In the practical application of Equation (3.30) this means that an arbitrary value can be chosen for $u_\ell(s)$ initially and the solutions at radii further out are calculated recursively. For scattering solutions ($E > 0$ at $r \to \infty$ and finite potential wells), the obtained wavefunction can be normalized by comparing it to its asymptotic form for the free solution far outside the nucleus, where $V_{\mathrm{SW}} = 0$. Without Coulomb potential, the free solution is

$$u_\ell(r) \to N \sin\left(\mathcal{K}r - \frac{\ell\pi}{2} + \delta_\ell\right), \quad (3.31)$$

with $N = 1$. The *scattering phase* is denoted by δ_ℓ. It accounts for a phase shift in the wave caused by the interaction potential inside the nucleus (see Section 3.5). In the absence of any nuclear interaction $\delta_\ell = 0$. The Coulomb potential for charged particles is only falling off with $1/r$ and therefore is a long-range potential also felt far outside the nucleus. The asymptotic solution for the wavefunction of a charged particle is given by

$$u_\ell(r) \to N \sin\left(\mathcal{K}r - \frac{\ell\pi}{2} + \delta_\ell - \eta_{\mathrm{S}} \ln(2\mathcal{K}r) + \delta_\ell^{\mathrm{C}}\right). \quad (3.32)$$

This expression includes the Sommerfeld parameter η_S (see Equation (4.24)) and the *Coulomb phase shift* δ_ℓ^C, which accounts for the phase shift caused by the Coulomb interaction alone. The Coulomb phase shift is given by the real part of the complex Bernoulli–Euler gamma-function Γ_{B-E},

$$\delta_\ell^C = \arg(\Gamma_{B-E}(\ell + 1 + i\eta_S)). \tag{3.33}$$

An arbitrary choice of $u_\ell(s)$ will lead to an arbitrary value of N in Equations (3.31) and (3.32), respectively. Dividing u_ℓ by N at all radii yields a wavefunction with its amplitude asymptotically normalized to unity.

The scattering wavefunctions obtained with the above procedure will be needed to calculate reaction cross sections (Sections 3.5 and 3.6) which, in turn, are essential to determine astrophysical reaction rates (Section 4.1). On the other hand, bound state wavefunctions are required for the determination of shell model states (Section 3.3.3) and in the application of some reaction mechanisms to cross section calculations. A slightly different normalization is applied for bound states to serve as the second condition required to solve the Schrödinger equation. Instead of normalizing to the asymptotic wavefunction, the fact can be used that a bound particle is confined to a finite volume and the probability to find the particle in the volume is unity. Therefore the value of the integration over the probability density, which is the squared absolute value of the amplitude, is known. Thus the second condition applied for the computation of bound state wavefunctions is

$$\int_0^{R_{\text{large}}} |u_\ell(r)|^2 \, dr = 1. \tag{3.34}$$

While R_{large} would define the sphere in which the particle is localized, for practical application R_{large} is any large distance from the center of the nucleus but a few times the nuclear radius R_{nuc} usually is sufficient. As before, calculating the normalization with the trial wavefunctions obtained with a trial value of $u_\ell(s)$ will lead to a value $M \neq 1$. In this case, the properly normalized wavefunction is obtained by taking $u_\ell/\sqrt{M}$ at every radius.

Within a potential barrier, that means for $\mathcal{K}^2 + E_{\text{pot},\ell} < 0$, the possible solutions of Equation (3.25) are a mixture of exponentially decaying and exponentially increasing functions. Which solution is physically meaningful depends on the boundary conditions. The Fox–Goodwin method, however, is always numerically more stable when applied to an integration out of such a barrier. When integrating into a barrier, the admixture of the exponentially increasing solution, even if ruled out by boundary conditions, may make the solution unstable. Therefore for bound states the method described above should be turned around and applied to a starting point inside the barrier and then proceeding toward smaller radii. Having barriers on two sides or for tunneling through a potential barrier (see also Figure 3.12) solutions obtained by integrating in two directions may be matched together by requiring equal amplitudes and first derivatives at an intermediate point where both solutions are assumed to be accurate. This considerably increases numerical stability and accuracy.

Whereas scattering solutions can be found for arbitrary energies, bound state wavefunctions can only be found for a discrete, albeit infinite, set of energies, corresponding to the eigenvalues of the differential equation. For a given potential and angular momentum quantum number ℓ, the solutions at different energies are enumerated by assigning them an additional quantum number n_{q} to allow distinguishing them. For example, for $\ell = 0$ without Coulomb potential and taking an infinite well potential (with constant potential for $0 < r < L$ and $V(r) = \infty$ at $r = 0$ and $r = L$) the solution wavefunctions can be found analytically. They are $u_{n_{\mathrm{q}}} = \sqrt{2/L}\,\sin(n_{\mathrm{q}}\pi r/L)$, which fulfill the requirement of being zero at $r = 0$ and $r = L$ where the potential is infinite. Due to the cyclic nature of the trigonometric functions, an infinite number of solutions can be found. The solutions are numbered by the main (sometimes called nodal) quantum number n_{q} (see Section 3.3.3). The energy eigenvalues of these solutions are $E_{n_{\mathrm{q}}} = n_{\mathrm{q}}^2\pi^2\hbar^2/(2mL^2)$. For realistic potentials, for which the solutions have to be found by applying the Fox–Goodwin method, the obtained energy eigenvalues for bound states will not be equidistant in energy and also will not follow the simple quadratic spacing as for the infinite well potential.

To find numerically the bound state energies and their wave functions, the Fox–Goodwin method can be applied iteratively, for example by integrating from large radii to the origin and checking whether the wave function fulfills the boundary condition $u_\ell(0) = 0$. If it does not, the energy has to be varied in small steps until it does. If the energy eigenvalue is known, an appropriate potential can be found by a similar approach, this time varying, for example, the potential depth V_0 in Equation (3.17).

3.3.3 Nuclear Shell Model

In general, there is an infinite number of mathematical solutions to the radial Schrödinger equation, even with fixed boundary conditions and chosen values of s_{q}, ℓ. As seen above, the solutions can be labeled by an additional quantum number n_{q}, the *main* or *nodal quantum number*. There are different conventions used in literature for the numbering scheme, either starting with zero, $n_{\mathrm{q}} = 0, 1, 2,\ldots$, or with one, $n_{\mathrm{q}} = 1, 2, 3,\ldots$ Starting the count with zero, the value of n_{q} also specifies the number of nodes (or roots) of the respective wavefunction, i.e., the number of times it crosses zero between the boundaries. Starting the count with one, n_{q} is the number of energy quanta, which obviously is the number of nodes plus 1 or alternatively the number of halfwaves of the wavefunction. The latter interpretation was also given in Section 1.5.1.2. Both interpretations are mostly useful in bound systems but may lose their meaning in unbound systems with complicated potential shapes and wavefunctions extending without finite borders.

The *Pauli principle* states that no two fermions can be identical in all quantum numbers. (In the Schrödinger picture using wave mechanics this is realized in that the total wave function ψ would vanish.) As previously mentioned, in principle the solutions for the energy may depend also on the chosen quantum number of orbital angular momentum $\ell = 0, 1, 2, \ldots$ and the quantum number ℓ_z for its projection onto the z-axis. This magnetic quantum number ℓ_z can assume the values

$\ell_z = -\ell, -\ell + 1, \ldots, \ell - 1, \ell$ (these are $2\ell + 1$ values), see also Section 1.7. Therefore the radial energy eigenvalues are degenerate in ℓ_z unless the nucleus is in a very strong magnetic field with a strength sufficient to lift this degeneracy (required to produce measurable effects would be fields exceeding 10^{13} T; such high magnetic fields are not even found in the strongest magnetars, i.e., neutron stars with extremely strong magnetic fields up to 10^{12} T). Furthermore, there is the particle spin quantum number s_q. Its z-projection s_z again can assume the values $s_z = -s_q$, $-s_q + 1, \ldots, s_q - 1, s_q$ (again $2s_q + 1$ values). The quantum number for particle spin s_q does not appear in Equation (3.25) and thus the obtained energy eigenstates are *degenerate* in both ℓ_z and s_z, i.e., several sets of quantum numbers $n_q, \ell, \ell_z, s_q, s_z$ result in a state with the same energy. How many sets are these? A fermion with $s_q = 1/2$ has two possible spin orientations $s_z = -1/2, +1/2$ and therefore two fermions can be put in each state characterized by a wavefunction with a specific set of quantum numbers n_q, ℓ, ℓ_z. The degeneracy in s_z is said to be two. As shown above, the degeneracy in ℓ_z is $2\ell + 1$. Therefore $2(2\ell + 1)$ fermions can have the same energy in Equations (3.24) and (3.25). Note that the general connection between quantum numbers and angular momenta in quantum physics is discussed in Section 1.7.

Due to the degeneracy of the solutions of the Schrödinger equation as described above, $2(2\ell + 1)$ neutrons or protons can be at the same energy. Filling the energy states with fermions starting at the lowest energy, there will be jumps in binding energy only when all positions for a given set (n_q, ℓ) are full. The next fermion has to be put in a state with a higher energy and this results in a sudden change in binding energy as this fermion will be less bound than all the others before. In analogy to the electron shells of an atom, the degenerate eigenenergies are called *nuclear shells*. The discontinuities in binding energies at the magic numbers (see Section 3.3.1), which could not be accounted for in the continuous evolution of binding energies in the semi-empirical droplet model, find their explanation within the framework of the nuclear shell model. It should not be forgotten that neutrons and protons have separate shells. The energies of the proton shells are shifted to higher energy with respect to the neutron shells due to the Coulomb repulsion.

The Schrödinger equation as given in Equations (3.24), (3.25), however, is not sufficient to obtain the observed values of the magic numbers. It proved necessary to add an additional term, which lifts the degeneracy in s_z for given ℓ. The effective potential is extended with a *spin–orbit (LS) term* as

$$V_{\rm nuc}(r) = V_{\rm SW}(r) + V_{\rm C}(r) + \frac{V_{0,\rm LS}}{r}\frac{\mathrm{d}V_{\rm SW}}{\mathrm{d}r} f_{\rm LS}, \tag{3.35}$$

where

$$f_{\rm LS} = \begin{cases} \dfrac{\ell}{2} & \text{for } j_{\rm nuc} = \ell + 1/2, \\ -\dfrac{\ell + 1}{2} & \text{for } j_{\rm nuc} = \ell - 1/2. \end{cases} \tag{3.36}$$

The additional term causes the energy eigenvalues for a given ℓ to have different values depending on the relative orientation of (orbital) angular momentum and intrinsic spin of the nucleon. Because $V_{0,\mathrm{LS}}/r\, \mathrm{d}V_{\mathrm{SW}}/\mathrm{d}r < 0$, the state with $j_{\mathrm{nuc}} = \ell - 1/2$ (angular momentum and spin anti-parallel) is located at higher energy than the corresponding state $j_{\mathrm{nuc}} = \ell + 1/2$ (angular momentum and spin parallel) for the same ℓ. Therein j_{nuc} denotes the total angular momentum quantum number of the nucleon, calculated by adding angular momentum and intrinsic spin according to the rule specified in Equation (1.126). Also note that the strength of the splitting is proportional to ℓ. Because of the state splitting induced by the spin–orbit interaction, a nuclear shell only holds $2j_{\mathrm{nuc}} + 1$ nucleons.

A spin–orbit interaction is also found in electron shells of atoms. It is responsible for the fine structure in atomic line spectra. Its cause and strength, however, are different from the spin–orbit effect in nuclear physics. In atomic physics, the spin–orbit interaction is caused by the Coulomb force whereas in nuclear physics it is a result of the spin–orbit dependence of the strong nuclear force. The atomic spin–orbit interaction is also much weaker than the nuclear one. The nuclear spin–orbit interaction leads to a reordering of energy states especially for larger ℓ. This is in accordance with experimental evidence and places larger energy gaps between the nuclear shells at the positions required to reproduce the magic neutron and proton numbers.

The negative energy (for bound states) solutions of the radial Schrödinger equation with the full potential from Equation (3.35) are called *single-particle states.* Their energy eigenvalues (*single-particle energies*) can be enumerated according to sets of quantum numbers (n_{q}, ℓ, s_z) or alternatively by (n_{q}, ℓ, j_{nuc}). Again in analogy to the notation for atomic line spectra originating from electron shells, nuclear states with orbital angular momentum $\ell = 0, 1, 2, 3, 4, \ldots$ are called s-, p-, d-, f-, g-,··· states (for $\ell > 3$ the naming continues alphabetically). The commonly used full notation for single-particle states specifies n_{q}, angular momentum ℓ in the spectral notation, and the total angular momentum j_{nuc}, for example 1s1/2 ($n_{\mathrm{q}} = 1$, $\ell = 0$, $j_{\mathrm{nuc}} = 1/2$), 3f5/2 ($n_{\mathrm{q}} = 3$, $\ell = 3$, $j_{\mathrm{nuc}} = 5/2$), 3f7/2 ($n_{\mathrm{q}} = 3$, $\ell = 3$, $j_{\mathrm{nuc}} = 7/2$), and so on. (As pointed out before, it has to be made sure whether the enumeration of n_{q} starts with 0 or 1.) Table 3.3 lists the single-particle states in the order they are obtained from the shell model for a spherical nucleus.

So far we have used a spherically symmetric model for a nucleus. Many nuclides are deformed, though. The inclusion of deformation leads to more complicated forms of Schrödinger equations than discussed here and also require to choose a different set of appropriate quantum numbers to characterize a state because ℓ is not a good quantum number anymore. The classification according to j_{nuc} is not appropriate for deformed nuclides. Deformation splits the shells existing in spherical symmetry and shifts their energy eigenvalues up and down according to the strength of the deformation. Moreover, it acts differently on different shells. This can cause a change in the sequence of shells and therefore also modify the spin of the nucleus from the one expected from a spherical shell model. Extreme deformation may also affect magic numbers by removing them or changing

Table 3.3. Order of Nuclear Shell Model States for Spherical Nuclides Close to Stability

State	#	tot #
4s1/2	2	168
3d3/2	4	166
2g7/2	8	162
3d5/2	6	154
1i11/2	12	148
2g9/2	10	136
3p1/2	2	126
2f5/2	6	124
1i13/2	14	118
3p3/2	4	104
2f7/2	8	100
1h9/2	10	92
3s1/2	2	82
1h11/2	12	80
2d3/2	4	68
2d5/2	6	64
1g7/2	8	58
1g9/2	10	50
2p1/2	2	40
2p3/2	4	38
1f5/2	6	34
1f7/2	8	28
2s1/2	2	20
1d3/2	4	18
1d5/2	6	14
1p1/2	2	8
1p3/2	4	6
1s1/2	2	2

Notes. Larger energy gaps between two neighboring states are marked by a horizontal line. The second column gives the number of nucleons in the shell, the third column the sum of nucleons in this shell and all shells at lower energy. Energy increases from bottom of the table to the top.

them to different values. Deformation not only changes the ground state properties but also the structure of the excited states (Section 3.4) and thus also affects nuclear reactions (Section 3.7).

Deviation from spherical symmetry of a charge distribution also gives rise to an electric moment, which can be determined experimentally. The measurement of electric moments, especially of the electric quadrupole moment, allows to obtain information on the deformation of a nucleus. For example, using a rotationally symmetric spheroid instead of a sphere, with semi-axis a_{sph} and b_{sph}, and assuming a homogeneous charge distribution, electric quadrupole moment is given by

$$\mathcal{Q}_4 = \frac{2Z}{5}\left(b_{\text{sph}}^2 - a_{\text{sph}}^2\right) = \frac{4}{5}Z\bar{R}^2\delta_{\text{def}} \tag{3.37}$$

with

$$\delta_{\text{def}} = \frac{b_{\text{sph}} - a_{\text{sph}}}{\bar{R}}, \tag{3.38}$$

the average radius $\bar{R} = (a_{\text{sph}} + b_{\text{sph}})/2$, and b_{sph} being oriented parallel to the rotation axis. The deformation parameter δ_{def} contains not only information on how strongly deformed a nucleus is but also whether it is a prolate deformation (cigar shape for $\delta_{\text{def}} > 0$) or an oblate deformation (pancake shape for $\delta_{\text{def}} < 0$). The assumption of symmetric spheroids is another approximation. Some nuclei seem to have triaxial deformation which will not be further discussed here.

Another way to describe the shape of a deformed object is to specify the surface as an expansion into spherical harmonics $\mathcal{Y}_{n,m}$,

$$R_{\text{nuc}}(\theta, \phi) = \bar{R}\left[1 + \sum_{n=2}^{\infty}\sum_{m=-n}^{n} a_{n,m}\mathcal{Y}_{n,m}(\theta, \phi)\right], \tag{3.39}$$

where $n \geqslant 2$ is equivalent to the multipole order (in case of oscillations it is the oscillation mode) and $\bar{R}$ is the radius of a sphere with the same volume as the deformed object. For axially symmetric nuclei, one usually chooses to consider only $n = 2$, $m = 0$ and $n = 4$, $m = 0$. Therefore the expansion is reduced to

$$\begin{aligned} R_{\text{nuc}}(\theta, \phi) &= \bar{R}\left[1 + a_{2,0}\mathcal{Y}_{2,0}(\theta, \phi) + a_{4,0}\mathcal{Y}_{4,0}(\theta, \phi)\right] \\ &= \bar{R}\left[1 + \beta_2\mathcal{Y}_{2,0}(\theta, \phi) + \beta_4\mathcal{Y}_{4,0}(\theta, \phi)\right]. \end{aligned} \tag{3.40}$$

The deformation parameters β_2, β_4 are often used instead of δ_{def}. There is a relation between δ_{def} and β_2,

$$\delta_{\text{def}} = \beta_2\sqrt{\frac{45}{16\pi}} = 0.946\beta_2. \tag{3.41}$$

3.3.4 Residual Interaction

The prerequisite for the shell model approach to work is that the residual interaction defined in Equation (3.19) for the independent-particle model is negligible. Even when this residual is tiny and not impacting the shell energies too much, it still affects the coupling between single nucleons within a shell or across shells. There are two main effects requiring to invoke the residual interaction for an explanation: the ground-state spins of nuclides and pairing forces between nucleons.

When a nucleus is in its ground state all nucleons are in the lowest possible energy states, filling them from the lowest to the highest. The energy of the state containing the last nucleon is the Fermi energy. What is the total spin (angular momentum) J of the nucleus as sum of the angular momenta of its nucleons? Since in a filled shell all magnetic quantum numbers connected with the j_{nuc} of the shell are completely occupied, all nucleons in this shell have to couple to angular momentum zero. Another simple case is a single nucleon in a shell. Then the ground-state spin and parity of the nucleus is determined by the shell properties. The same is true for a single missing nucleon in an otherwise filled shell. For example, the ground state of ^{17}O has $J^{\pi_q} = 5/2^+$ because of a single neutron in the 1d5/2 shell whereas all other neutrons and all protons are in completely filled shells. On the other hand, for the ground state of ^{15}O $J^{\pi_q} = 1/2^-$ because a neutron is missing in the 1p1/2 shell. The situation becomes complicated with partially filled shells. Spin addition rules (see Section 1.7) allow several possibilities to couple (add up the spins of) nucleons in such shells. All these possibilities have the same energy in the independent-particle model, which prevents a prediction of a definitive value for the ground-state spin. This kind of degeneracy is lifted by the residual interaction. The actual calculation is complicated but a few simple rules can be derived. This is a summary of the rules to estimate ground-state spins (total angular momentum) of nuclides:

1. Even numbers of nucleons couple to total angular momentum zero. The ground-state spin and parity of nuclides with even Z and even N (even–even nuclides) always is $J^{\pi_q} = 0^+$.
2. If either N is odd and Z even or vice versa (even–odd or odd–even nuclides), nucleons in the j_{nuc}-shell with odd number of nucleons couple to $J = j_{\mathrm{nuc}}$; in rare cases they couple to $J = j_{\mathrm{nuc}} - 1$.
3. In nuclides with odd Z and odd N there is a tendency to couple to a large value of J. Looking at the unpaired nucleons, the so-called *Nordheim rules* state:
   (a) strong Nordheim rule: If angular momentum and spin of the unpaired proton are anti-parallel while angular momentum and spin of the unpaired neutron are parallel (see Equation (3.36)) or vice versa, they couple to the difference of the j_{nuc}-values of their shells, $J = \left|j^{\mathrm{n}}_{\mathrm{nuc}} - j^{\mathrm{p}}_{\mathrm{nuc}}\right|$.
   (b) weak Nordheim rule: If angular momentum and spin of the unpaired neutron and the unpaired proton are both either parallel or anti-parallel (see Equation (3.36)), then the j_{nuc}-values of their shells tend to add but not necessarily to the largest possible value $J = j^{\mathrm{n}}_{\mathrm{nuc}} + j^{\mathrm{p}}_{\mathrm{nuc}}$.

Again, these rules may not hold for strongly deformed nuclides.

The residual interaction is also responsible for the necessity to include the pairing term in Equation (3.8). This implies that it is causing stronger binding for pairs of nucleons. These pairs can include the same type of nucleons (proton–proton or neutron–neutron pairs) or proton–neutron pairs. The binding in the latter case is found not to be as strong as for same-type pairs. This explains the form of the pairing term in the semi-empirical mass formula as given in Equation (3.9).

The pairing force emerging from the residual interaction requires a sophisticated theoretical approach. Usually, its consequences are calculated within a framework of two nucleons scattering on each other within incompletely filled shell-model states. This implies that nucleons can be scattered to higher lying energy states, resulting in a softening of the Fermi edge. Recall the occupation probability as function of state energy in a Fermi gas as shown in Figure 1.3 and given by Equation (1.50). Instead of using $o(E)$, the occupation of a nuclear shell is usually denoted by $v^2_{j_{\rm nuc}}(E_{j_{\rm nuc}})$. This is equivalent to the notation used in Section 1.5.1.2 insofar as it gives the ratio of the number of nucleons in a $j_{\rm nuc}$-shell at energy $e_{j\rm nuc}$ to the maximal possible number of nucleons in that shell. This allows a direct comparison with Equation (1.50). Without pairing, there is a sharp edge at the Fermi energy $\mu_{\rm c} = E_{\rm F}$. This is equivalent to the curve for $T = 0$ in Figure 1.3 because at room temperature ($k_{\rm B}T \approx 25$ M eV) the thermal excitation of the nucleon gas is negligible. Switching on a pairing force scatters some nucleons from energies below $E_{\rm F}$ to states above $E_{\rm F}$, which results in an occupation distribution as for $T > 0$ in Figure 1.3. The slope of the distribution around $E_{\rm F}$ is determined by the so-called *pairing gap* $\Delta_{\rm break}$, which therefore plays the role of an effective T in Equation (1.50). This pairing gap causes the observed odd–even staggering in binding energies, which motivated the inclusion of the pairing term in Equation (3.8). Experimentally the actual value of the pairing gap can be determined approximately by taking appropriate mass differences between neighboring nuclides. For example, the odd and even neutron pairing gaps are

$$\begin{aligned}\Delta^{\rm n,odd}_{\rm break}(N) &= \frac{1}{2}[\mathcal{B}(Z, N+1) - 2\mathcal{B}(Z, N) + \mathcal{B}(Z, N-1)] \quad \text{for } N \text{ odd},\\ \Delta^{\rm n,even}_{\rm break}(N) &= -\frac{1}{2}[\mathcal{B}(Z, N+1) - 2\mathcal{B}(Z, N) + \mathcal{B}(Z, N-1)] \quad \text{for } N \text{ even}.\end{aligned} \tag{3.42}$$

The proton pairing gaps are obtained in a similar way. With the above definition the gaps are positive for normal pairing.

The occupation number $v^2_{j_{\rm nuc}}(E_{j_{\rm nuc}})$ of a $j_{\rm nuc}$-shell is given by

$$v^2_{j_{\rm nuc}}(E_{j_{\rm nuc}}) = \frac{1}{2}\left[1 - \frac{E_{j_{\rm nuc}} - \mu_{\rm c}}{\sqrt{(E_{j_{\rm nuc}} - \mu_{\rm c})^2 + \Delta^2_{\rm break}}}\right]. \tag{3.43}$$

The chemical potential $\mu_{\rm c}$ is determined by the conservation of particle number $n^{j_{\rm nuc}}_{\rm part}$,

$$n^{j_{\rm nuc}}_{\rm part} = \sum_{j_{\rm nuc}} (2j_{\rm nuc} + 1) v^2_{j_{\rm nuc}}. \tag{3.44}$$

Without pairing one obtains $\Delta_{\rm break} = 0$ and the independent-particle occupation number is recovered, with the usual single-particle state energies $E_{j\rm nuc}$ and $\mu_{\rm c} = E_{\rm F}$.

A nucleon scattered into a state above μ_c leaves a hole at lower energy with similar quantum numbers except for the magnetic quantum number for the given j_{nuc}. Due to the selection rules, this hole cannot be simply filled by another nucleon that stabilizes the particle–hole configuration. Therefore such a configuration can be viewed as a quasi-particle. Quasi-particles appear in various many-body systems and formalisms have been developed to predict quasi-particle spectra. For example, in the BCS theory superconductivity was explained by quasi-particles formed from electron pairs. A similar mathematical approach can be used for nucleons in atomic nuclei. The main difference is that in nuclei the nucleon correlations are in the angular correlations (magnetic quantum numbers) whereas in superconductivity there is a spatial correlation between electrons. The energy of such a quasi-particle in an atomic nucleus is given by

$$e_{j_{nuc}} = \sqrt{(E_{j_{nuc}} - \mu_c)^2 + \Delta_{break}^2}. \tag{3.45}$$

Analogous to Equation (3.43) "occupation" numbers $u_{j_{nuc}}^2$ can also be defined for holes,

$$u_{j_{nuc}}^2(E_{j_{nuc}}) = \frac{1}{2}\left[1 - \frac{E_{j_{nuc}} - \mu_c}{\sqrt{(E_{j_{nuc}} - \mu_c)^2 + \Delta_{break}^2}}\right]. \tag{3.46}$$

The $u_{j_{nuc}}^2$ have to fulfill the condition

$$n_{hole}^{j_{nuc}} = \sum_{j_{nuc}} (2j_{nuc} + 1)u_{j_{nuc}}^2, \tag{3.47}$$

with $n_{hole}^{j_{nuc}}$ being the number of hole states. Obviously, the occupation numbers for particles and holes are connected by

$$\sum_{j_{nuc}} \left(v_{j_{nuc}}^2 + u_{j_{nuc}}^2\right) = 1. \tag{3.48}$$

Quasi-particles cannot be created at arbitrary energies. The softening of the Fermi edge will be stronger when energy states are close together. An estimate of what energy differences can be bridged can be made by applying again the uncertainty relation of Equation (3.1). It specifies by how much energy conservation may be violated by the nucleon scattering for a short time. For a nucleon with 40 MeV the classical time to complete an "orbit" within the nucleus is $\Delta t \approx 4.3 \times 10^{-22}$ s. This leads to $\Delta E \leqslant 1.5$ MeV. This would not be enough to cross the gap at magic numbers but is larger than the difference of some other shell energy differences.

The quasi-particle view also explains the excitation spectra of nuclei (see also Section 3.4). To excite a nucleon in an even–even nucleus to a higher energy state, a nucleon pair has to be broken. This creates two particle–hole pairs. Therefore the minimal energy required is $2\Delta_{break}$. In consequence, a "gap" is found before the first excited state of such a nucleus because it cannot be below that energy.

Finally, it should be noted that mass formulae can also be constructed using the mean-field approach of the independent-particle model and including residual interactions. These so-called microscopic mass formulae yield the semi-empirical mass formula as an asymptotic solution. As an intermediate but easily applicable and quite successful approach, microscopic–macroscopic models are constructed by combining a (deformed) droplet model with a shell correction (or microscopic correction) term.

3.4 Excited States and the Nuclear Level Density

The previous sections discussed the properties of an atomic nucleus in its ground state, that means with all nucleons in their lowest energy state allowed by the Pauli principle. When sufficient energy is added to the nucleus, for example by a collision, to lift one or more nucleons into a higher energy state, the nucleus is said to be in an *excited state*. Having one or more nucleons at higher energies reduces the total binding energy of the nucleus by the amount of the absorbed energy. The new nucleon configuration may also result in a total nuclear spin and parity differing from the ones of the ground state. Nuclear energy schemes as shown in Figure 3.3 are used to depict the excitation possibilities, sorted by excitation energy. The energy levels shown in such a diagram are not single-particle states but configurations obtained from redistribution of nucleons among possible single-particle states.

As seen in Figure 3.3, an excited state of the nucleus is characterized by excitation energy (energy above the ground state of the nucleus), spin, and parity. Furthermore, excitations have a certain lifetime, depending on the emission probabilities of electromagnetic radiation (see Section 1.7). The nucleus will return to its ground state by radiating the surplus energy as photons (γ-quanta). This does not necessarily proceed in the same way as it was originally excited. If the excitation energy is larger than the energy required to remove a nucleon or a group of nucleons (such as an α-particle) from the nucleus (the particle separation energy, see Table 3.2), excess energy can also be removed from the excited state by emission of those particles. According to the uncertainty principle shown in Equation (3.1), in quantum physics a limited lifetime is always connected to an uncertainty in energy ΔE. The uncertainty in excitation energy of an excited state due to its lifetime τ_{life} is called *total width* and denoted by

$$\hat{\Gamma}_{\mathrm{tot}} = \Delta E = \frac{\hbar}{\tau_{\mathrm{life}}}. \tag{3.49}$$

A longer lifetime thus implies a smaller width. The total width can be decomposed into *partial widths* Γ for the individual de-excitation lifetimes, that is

$$\hat{\Gamma}_{\mathrm{tot}} = \Gamma_\gamma + \Gamma_{\mathrm{n}} + \Gamma_{\mathrm{p}} + \Gamma_\alpha + \cdots = \sum_i \Gamma_i, \tag{3.50}$$

with i enumerating the different de-excitation possibilities. Depending on the individual separation energies, one or most of the partial widths can be zero. The width Γ_γ is never zero but it can be extremely tiny when the emission of a photon to a

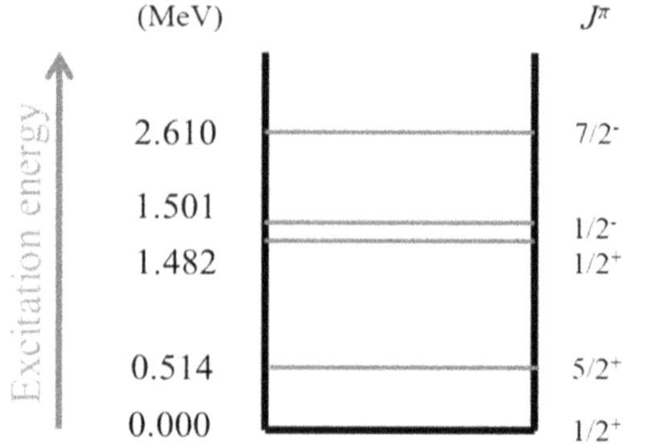

Figure 3.3. Energy scheme showing ground state and excited states of a fictitious nucleus, with excitation energy, spin, and parity shown for each state. As an additional property (not shown) each state has a certain energy width.

lower-lying state is forbidden by quantum mechanical selection rules (see Section 1.7). In such a case, when no other particle emission channels are open, the effective lifetime of the excited state may even exceed the age of the Galaxy or of the Universe although not being the stable ground state. Such excited states with long lifetimes larger than the ones encountered in usual nuclear decays or reactions are called meta-stable states or *isomers*. For some nuclei with unstable ground states, the isomer can be longer lived than the ground state. A striking example is ^{180}Ta. Its ground state ^{180g}Ta with spin and parity $J^\pi = 1^+$ is unstable against β^--decay and electron capture with a half-life of 8.15 h (for the connection between lifetime and half-life see Section 4.1.7). It has an excited state at 77.2 keV with $J^\pi = 9^-$ for which the emission of electromagnetic radiation is strongly suppressed by the spin selection rules. In consequence, this excited state has a half-life larger than 7×10^{15} yr. Therefore all ^{180}Ta naturally occurring on Earth is in the form of this isomer (denoted by ^{180m}Ta). (Note that decay half-lives and lifetimes of isomers can be altered in stellar environments, see Section 4.1.3.)

The spin and parity of an excited configuration is only easily predictable in the case of excitation of a single nucleon outside a closed shell or a hole in an otherwise filled shell. Then the rules as given in Section 3.3.4 still apply. The vast majority of configurations, however, involve the redistribution of many nucleons originating from several shells. What is subsumed as a single "excited state" may also be composed of the overlap of many different nucleon configurations having the same energy eigenvalues. Their coupling is determined by the residual interaction and very complicated to tract theoretically, not just because of the sheer number of possible nucleon configurations for a given energy. This also affects the total and partial widths assigned to an excited state.

Somewhat easier to treat are collective motions of nucleons. Just as the surface of a drop of liquid may vibrate and oscillate, also nucleons outside closed shells may undergo such collective motions. Since these are surface effects, the effective strong nuclear force suffices to model the motion and the extremely short-ranged effect of the residual interaction will not play a large role. The surface motions of the nucleus only contain little energy, insufficient to eject nucleons from the nucleus. They can, however, radiate electromagnetic waves and therefore their appearance can be studied in γ-ray spectra emitted from an excited nucleus.

Vibrational modes give rise to equidistant excitation energies

$$E_n(\lambda_{\text{vib}}) = \frac{\hbar\omega(\lambda_{\text{vib}})}{2}(2n + 2\lambda_{\text{vib}} + 1), \tag{3.51}$$

where n is the number of vibration quanta and ω the vibration frequency. The vibration frequencies are different for each nucleus and depend on the actual nucleon surface distribution. They can be estimated using the droplet model but strong shell effects render the approximation useless for practical application. Furthermore, the above relation is derived assuming a spherical symmetric nucleon distribution. It can be generalized to a deformed nucleus leading to a splitting of the degenerate excitation spectrum ($2\lambda_{\text{vib}} + 1$ states per mode λ_{vib}) arising from rotational states. Regardless of the details, such patterns of equidistant excitation energies have been identified in many nuclei.

Rotational energy is also quantized and thus gives rise to additional excited states in the spectrum of a nucleus. The energy difference between two rotational states is increasing with increasing rotational angular momentum. Furthermore, some nuclei show several sequences of rotational excitations, called *rotational bands*.

Symmetric objects do not show rotation around the symmetry axis in quantum mechanics because there is no angular dependence of the wave function around such an axis. Therefore spherically symmetric nuclei do not have rotational degrees of freedom. A detailed treatment of rotation is a complicated problem which is beyond the scope of this book but a few simple assumptions already allow to study the general behavior expected for deformed nuclei. Considering deformation along a single axis, the symmetry axis, we can distinguish between an elongation (prolate deformation) and a compression (oblate deformation) along this axis, see also the discussion around Equation (3.38). The deformed shape can be envisioned to be caused by additional nucleons, the *valence nucleons* around an undeformed, spherical nucleus. The spin $\vec{J}_{\text{val}}$ of these valence nucleons couples to the angular momentum of the collective rotation $\vec{R}_{\text{coll}}$ to the total angular momentum $\vec{I} = \vec{J}_{\text{val}} + \vec{R}_{\text{coll}}$, with its total angular momentum quantum number I. This gives rise to the band structure.

For a deformed object rotating in a plane perpendicular to its symmetry axis, the rotational energy is given by

$$E_{\text{rot}} = \frac{\hbar^2}{2\Theta}\left[I(I + 1) - K_{\text{coupl}}^2\right], \tag{3.52}$$

where K_{coupl} is the quantum number of the projection of $\vec{I}$ (and $\vec{J}_{\text{val}}$) onto the symmetry axis. The quantity $\sqrt{I(I + 1) - K_{\text{coupl}}^2}$ is the quantum number of the projection of $\vec{R}$ onto the rotation axis. Due to the quantum mechanical vector addition (see also Section 1.7), I_{coll} can assume the values $I_{\text{coll}} = K_{\text{coupl}}, K_{\text{coupl}} + 1,$

$K_{\text{coupl}} + 2, \ldots$ because $K_{\text{coupl}} \leqslant I_{\text{coll}}$. Different values of I_{coll} for a fixed value of K_{coupl} result in rotational bands. Even–even nuclei are a special case. Their valence nucleons are paired in the ground state and thus their spins couple to $\vec{J} = \vec{0}$. Therefore also $J = K = 0$ and Equation (3.52) reduces to

$$E_{\text{rot}} = \frac{\hbar^2}{2\bar{\Theta}}[I(I+1)]. \tag{3.53}$$

Here, the quantum number I of the rotating even–even nucleus can only assume even values $I = 0, 2, 4, \ldots$ due to the mirror symmetry of the nucleus with respect to the plane perpendicular to the symmetry axis.

The moment of inertia $\bar{\Theta}$, to be used in Equations (3.52) and (3.53), may vary not only from nucleus to nucleus but can also be a function of excitation energy. Two classical extremes can be considered for estimating the moment of inertia: rotation of the full body and rotation only of the outer shell. (This can be compared to the different momenta of inertia of cooked and raw eggs.) Assuming the nucleus to be a rigid, rotating sphere, its moment of inertia is

$$\bar{\Theta}_{\text{sph}} = \frac{2}{5} m_{\text{nuc}} R_{\text{nuc}}. \tag{3.54}$$

If rotation is considered to occur only on the surface of the sphere while the inner part remains stationary, the motion can be viewed as a wave (without vertices) propagating on the surface. This is called irrotational flow. The moment of inertia for this case is

$$\bar{\Theta}_{\text{irrot}} = \frac{45}{16\pi} \bar{\Theta}_{\text{sph}} \delta_{\text{def}}^2, \tag{3.55}$$

with δ_{def} defined by Equation (3.38). A realistic nuclear moment of inertia will lie between these two extremes.

Now we return to the excited states emerging from the configurations of nucleon excitations, as mentioned above. Rotational and vibrational states of a nucleus appear against this background of further excitations caused by a reordering of the nucleons in a nucleus. Energy added to a nucleus can be used to move nucleons from a lower energy state to an excited state. This may involve breaking of a nucleon pair or forming a new one. The full excitation spectrum is obtained from the wave functions of all energetically possible nucleon configurations. Monte Carlo methods have been introduced recently to tackle this complicated combinatorial problem. The number of states per energy interval is the *nuclear state density*. For practical purposes, the observable *nuclear level density* is more often used. Each observable nuclear level with spin J is degenerate in the magnetic quantum number Q_z (see Section 1.7) and there are $2J + 1$ states for each J, see Equation (3.61) below.

Without having to resort to a full solution of coupled wave equations for all participating nucleons, a simple expression for the level density can be derived in the independent-particle model, assuming the nucleons form a non-interacting Fermi gas. The number of states of such a Fermi gas has been derived in Section 1.5.4. The ground state of a nucleus has the lowest energy because all nucleons are in their lowest possible energy states. This implies that the single particle states are filled up to the Fermi energy E_F (see Equation (1.49)). Assuming the protons and neutrons form two Fermi-gases, one would expect two different Fermi energies.

In Equations (1.44)–(1.46) the state density was derived from the number of states in phase space volume. From these equations it can be seen that the number of states is expected to increase proportional to $\sqrt{E}$ (with E being the state energy corresponding to the momentum p). This is the number of single-particle states available for population with two fermions each. Assume that a specific amount of energy is added to the nucleus which is distributed among all its nucleons. Then the calculation of all possible configurations of nucleons resulting in the same total energy U_{NLD} of the excited nucleus is a combinatorial problem. It can be shown that for the simplified case of equidistant energy levels with energy spacing D_0^{lev}, the number of such excitations per energy interval (called the nuclear *level density*) is given by

$$\tilde{\rho}_{\mathrm{nuc}}(U_{\mathrm{NLD}}) \propto \frac{1}{U_{\mathrm{NLD}}} e^{2\sqrt{a_{\mathrm{NLD}} U_{\mathrm{NLD}}}} \tag{3.56}$$

with the *level density parameter*

$$a_{\mathrm{NLD}} = \frac{\pi^2}{6 D_0^{\mathrm{lev}}}. \tag{3.57}$$

This formula can be improved by using the single-particle level spacings from the Fermi gas model (see Section 1.5.4), leading to a modified level density parameter (see also below). A further improvement is to realize that the breaking of a nucleon-pair requires extra energy. Therefore there is an energy threshold for excitation because part of the energy supplied to the nucleus goes into breaking the (first) nucleon pair(s). Phenomenologically this can be treated by shifting the effective excitation energy for nuclei with an even number of nucleons,

$$U_{\mathrm{NLD}} = E^{\mathrm{x}} - E_{\mathrm{shift}}. \tag{3.58}$$

This is called the *back-shifted Fermi gas model* of the nuclear level density. Taking the values of the pairing correlations from the liquid drop model of nuclear masses (Equation (3.9)) the energy shift becomes

$$E_{\mathrm{shift}} = \begin{cases} \frac{12}{\sqrt{A}}\ \mathrm{MeV} & \text{for even–even nuclei} \\ 0\ \mathrm{MeV} & \text{for odd–A nuclei} \\ -\frac{12}{\sqrt{A}}\ \mathrm{MeV} & \text{for odd–odd nuclei} \end{cases}. \tag{3.59}$$

Alternatively, the pairing energies can be taken from local mass differences, see Equation (3.42).

The full expression for the total nuclear level density in the back-shifted Fermi gas model is

$$\tilde{\rho}_{\mathrm{nuc}}(U_{\mathrm{NLD}}) = \frac{1}{\sqrt{2\pi\sigma_{\mathrm{cut}}^2}} \frac{\sqrt{\pi}}{12a_{\mathrm{NLD}}^{1/4}} \frac{1}{U_{\mathrm{NLD}}^{5/4}} \mathrm{e}^{2\sqrt{a_{\mathrm{NLD}}U_{\mathrm{NLD}}}}. \tag{3.60}$$

Note that the above is the observable *level* density, not to be confused with the total *state* density. In the same approach, the nuclear state density (see also Equation (1.46)) would be

$$\tilde{\rho}(U_{\mathrm{NLD}}) = \sum_{J\pi_{\mathrm{q}}} {}^{J\pi_{\mathrm{q}}}\tilde{\rho}_{\mathrm{nuc}}(U_{\mathrm{NLD}}, J, \pi_{\mathrm{q}}) = \frac{\sqrt{\pi}}{12a_{\mathrm{NLD}}^{1/4}} \frac{1}{U_{\mathrm{NLD}}^{5/4}} \mathrm{e}^{2\sqrt{a_{\mathrm{NLD}}U_{\mathrm{NLD}}}}. \tag{3.61}$$

With Equation (3.60), the density of levels with a specific spin and parity at a given excitation energy of the nucleus is given by

$${}^{J\pi_{\mathrm{q}}}\tilde{\rho}_{\mathrm{nuc}}(U_{\mathrm{NLD}}, J, \pi_{\mathrm{q}}) = \mathcal{P}_{\pi_{\mathrm{q}}}(U_{\mathrm{NLD}}, \pi_{\mathrm{q}})\mathcal{P}_{\mathrm{J}}(U_{\mathrm{NLD}}, J)\tilde{\rho}_{\mathrm{nuc}}(U_{\mathrm{NLD}}). \tag{3.62}$$

Here, it is assumed that the spin projection $\mathcal{P}_{\mathrm{J}}$ and parity projection $\mathcal{P}_{\pi_{\mathrm{q}}}$ factorize. This is an acceptable approximation for most applications. In addition, often equidistributed parities are assumed, i.e., $\mathcal{P}_{\pi_{\mathrm{q}}} = 1/2$ independent of excitation energy. At high excitation energies, the parities are found to be distributed equally but this assumption obviously does not hold at low excitations with few levels. For practical applications, experimentally known discrete levels are explicitly counted or the projections determined from microscopic calculations. In the Fermi gas model the spin projection is derived as

$$\mathcal{P}_{\mathrm{J}}(U_{\mathrm{NLD}}, J) = \frac{2J+1}{2\sigma_{\mathrm{cut}}^2} \exp\left(-\frac{J(J+1)}{2\sigma_{\mathrm{cut}}^2}\right). \tag{3.63}$$

The spin cut-off parameter σ_{cut} limits the contribution of higher spins. It can be taken as open parameter and fitted to an experimentally determined spin distribution, or approximated with the help of the rotational energy of the nucleus (see also Equations (3.54) and (3.55)),

$$\sigma_{\mathrm{cut}}(U_{\mathrm{NLD}}) = \frac{\bar{\Theta}}{\hbar^2}\sqrt{\frac{U_{\mathrm{NLD}}}{a_{\mathrm{NLD}}}}. \tag{3.64}$$

In this case, the back-shifted Fermi gas model only has one free parameter if the level density parameter a_{NLD} is considered as fitting parameter (and the back-shift E_{shift} determined from nuclear pairing).

The level density parameter is larger for nuclei with a larger number of nucleons because of the larger number of possible nucleon configurations. Moreover, at magic nucleon numbers it should drop due to the extra energy required to excite nucleons across a shell energy gap (see Section 3.3.1). The level density parameter as shown in Equation (3.57) does not include the shell corrections added on top of the liquid drop model, accounting for these shell effects. A number of parameterizations have been suggested for predicting the level density parameter. A simple empirical relation is

$$a_{\mathrm{NLD}} = A(c_0 + c_1 C_{\mathrm{corr}}(Z, N)), \tag{3.65}$$

with c_0, c_1 being fitted constants and $C_{\mathrm{corr}}(Z, N)$ being the shell correction or microscopic correction (including everything beyond the simple spherical liquid drop model). It is known that shell effects are reduced with increasing excitation energy. Therefore an energy-dependence of a_{NLD} may be introduced at the expense of additional open parameters. For example,

$$a_{\mathrm{NLD}}(U_{\mathrm{NLD}}, Z, N) = (c_0 A + c_1 A^{2/3})\left[1 + \frac{C_{\mathrm{corr}}(Z, N)}{U_{\mathrm{NLD}}}(1 - \mathrm{e}^{-c_2 U_{\mathrm{NLD}}})\right] \tag{3.66}$$

with its three parameters c_0, c_1, c_2 has been successfully applied for large-scale predictions of the nuclear level density across the nuclear chart.

The back-shifted Fermi gas formula of Equation (3.60) is not applicable at low U_{NLD}. It is often assumed incorrectly that this is because the formula diverges for $U_{\mathrm{NLD}} \to 0$. The actual reason, however, is a breakdown of the assumptions made in the derivation. This breakdown already occurs at energies above those at which the divergence would become noticeable. In fact, the back-shifted Fermi gas formula predicts a level density which is too low because it only accounts for levels from single-particle excitations and neglects collective effects. At low excitation energies, it can be replaced by a parameterization including the thermodynamic temperature T_{nuc} of a nucleus (not to be confused with the plasma temperature used later and in the other sections of this book). This nuclear temperature is a measure of the effective temperature of the Fermi gas of nucleons. Thermodynamically, there are several ways to define it. For example,

$$\frac{1}{T_{\mathrm{nuc}}} = \frac{\mathrm{d}}{\mathrm{d}U_{\mathrm{NLD}}} \log \tilde{\rho}$$

or

$$\frac{1}{T_{\mathrm{nuc}}} = \frac{\mathrm{d}}{\mathrm{d}U_{\mathrm{NLD}}} \log \tilde{\rho}_{\mathrm{nuc}}.$$

At high excitation energies, these tend toward $\sqrt{a_{\mathrm{NLD}}/U_{\mathrm{NLD}}}$. This motivates a functional dependence

$$\tilde{\rho}_{\mathrm{nuc}}(E^{\mathrm{x}}) = \frac{1}{T_{\mathrm{nuc}}} \mathrm{e}^{(E^{\mathrm{x}} - E_0)/T}. \tag{3.67}$$

The two quantities T_{nuc} and E_0 can either be fitted to the number of experimentally observed levels or matched with Equation (3.60) at an appropriate excitation energy, giving a hybrid formula when considering $T_{\mathrm{nuc}} \simeq \sqrt{U_{\mathrm{NLD}}/a_{\mathrm{NLD}}}$ at the matching energy. In both cases, T_{nuc} is considered to be constant across the experimentally observed energy range or below the matching point, respectively. This explains the name *constant temperature formula*.

As mentioned above, collective effects are not included in the Fermi gas approach. They can be partially included by considering an energy-dependence in the level density parameter $a_{\rm NLD}$ (compare also Equation (3.66)) or at low excitation energy by using the constant temperature formula. An explicit inclusion can be performed by providing additional multiplicative factors in Equation (3.60), for vibrational and rotational excitations. Among these, rotational excitations provide the strongest enhancement but become weaker at higher excitation energy. In any case, it is not trivial to avoid double- or multiple-counting of excitations when using such enhancement factors and, for example, an energy-dependent level density parameter fitted to experimental data.

Finally, nuclear state densities and level densities can be obtained (in tabular form) from microscopic nuclear structure calculations going beyond the simple Fermi gas model.

3.5 Introduction to Nuclear Scattering

The basics of scattering theory are explained in Section 1.6. Here, the focus is on scattering of nucleons and α-particles on nuclei and the introduction of quantities mainly used in nuclear physics. The solutions of the time-independent radial Schrödinger equation (see Section 3.3.2) with a nuclear interaction potential are used to determine the probability of atomic nuclei to scatter on each other or undergo nuclear reactions which result in a rearrangement of the nucleons involved. This probability is measured by the cross section defined in Equations (1.99) and (1.100). A calculation of scattering and reaction processes is performed most conveniently in the center-of-mass system, so that it is not important whether both interaction partners are moving or whether one of them is at rest. It also reduces the number of geometrical parameters because only one angle relative to the symmetry axis connecting the two reaction partners (which is the direction of the projectile beam in laboratory experiments) is required to describe the motion after the interaction. It is important to note that in the following center-of-mass energies and reduced masses are used as defined in Equations (1.104) and (1.105), respectively, in all equations, even if not indicated by any label. For the conversion between center-of-mass energy and laboratory energy, see Equation (1.106).

Using the setup as shown in Figure 1.5, the scattering amplitude $\mathcal{F}_{\mathcal{K}}$ in Equation (1.113), which also determines the scattering cross section (Equation (1.114)), has to be determined from solutions of the time-independent Schrödinger equation given in Equation (3.21). The short-range interaction between projectile and target is described by an appropriate nuclear interaction potential, for example the one shown in Equation (3.17), whereas a Coulomb potential may have to be considered for the long-range interaction of charged scattering partners. Finding the correct nuclear interaction potential is one of the challenges in the theoretical treatment of nuclear scattering. Deriving such a potential from experimental scattering data is called the *inverse scattering problem*.

As before, the asymptotic solutions for the wavefunctions as given by Equations (3.31) and (3.32) are used for the total wavefunction and the outgoing wave. Using the trigonometric relation

$$\sin\phi = \frac{\mathrm{i}}{2}[\mathrm{e}^{-\mathrm{i}\phi} - \mathrm{e}^{\mathrm{i}\phi}] \tag{3.68}$$

and the partial wave expansion (Equation (3.22)), the incoming plane wave can be written as

$$\begin{aligned}\psi_{\mathcal{K}}^{-}(r, \theta) &= \mathrm{e}^{\mathrm{i}kr\cos\theta} \\ &= \frac{1}{2\mathcal{K}r}\sum_{\ell}(2\ell + 1)\mathrm{i}^{\ell+1}\,[\mathrm{e}^{-\mathrm{i}(\mathcal{K}r-\mathcal{K}\pi/2)} - \mathrm{e}^{\mathrm{i}(\mathcal{K}r-\mathcal{K}\pi/2)}]\,\mathcal{L}_{\ell}(\cos\theta).\end{aligned} \tag{3.69}$$

This uses the *Legendre polynomials* $\mathcal{L}_{\ell}(\cos\theta)$, which are also related to the spherical harmonic functions $\mathcal{Y}$ (compare Equation (3.22)). Choosing the symmetry axis connecting projectile and target (i.e., the beam direction in the laboratory) as z-axis simplifies the spherical harmonics $\mathcal{Y}$ to the harmonic functions $\underline{Y}$, which are connected to the Legendre polynomials by

$$\underline{Y}_{\ell,\ell_z}(\theta) = \sqrt{\frac{2\ell + 1}{4\pi}}\,\mathcal{L}_{\ell}(\cos\theta). \tag{3.70}$$

A dependence on ϕ does not appear anymore because of the choice of the symmetry axis as spin projection axis. The solutions are, of course, degenerate in ℓ_z.

In Equation (3.69) the plane wave was constructed from two spherical waves, one wave incoming onto the scattering center, the other wave outgoing from the scatterer. The total wave function of the stationary (time-independent) scattering problem is also a superposition of an incoming and an outgoing wave. Therefore we expect it to have the same form. In the absence of a scattering potential it is identical to the wave given in Equation (3.69) as there will be no additional scattering process. When there is scattering, the outgoing wave is modified. Therefore, the ansatz for the total wave function with scattering is

$$\psi_{\mathcal{K}}(r, \theta) = \frac{1}{2\mathcal{K}r}\sum_{\ell}(2\ell + 1)\mathrm{i}^{\ell+1}\,[\mathrm{e}^{-\mathrm{i}(\mathcal{K}r-\mathcal{K}\pi/2)} - S_{\ell}^{\alpha\alpha}\mathrm{e}^{\mathrm{i}(\mathcal{K}r-\mathcal{K}\pi/2)}]\,\mathcal{L}_{\ell}(\cos\theta). \tag{3.71}$$

There is only one wave number $\mathcal{K}$ because in elastic scattering the energy is not affected and therefore $\mathcal{K}$ must be the same before and after elastic scattering. The modification factor of the outgoing wave is the diagonal element of the *S-matrix (scattering matrix)* $S_{\ell}^{\alpha\alpha}$. The label α denotes the reaction channels, the double label implies that the incoming particle is the same as the outgoing particle, with the same quantum numbers and energy. The most general case, which includes inelastic scattering and reactions, also includes further channels with their respective wave functions, leading to a matrix containing the modification factors for all channels. In most practical applications, only one factor for a specific combination of incoming and outgoing channel is required, see Section 3.6.

Since the total wave function $\psi_{\mathcal{K}}$ is the sum of incoming and outgoing wave, the outgoing wave must be

$$\begin{aligned}\psi_{\mathcal{K}}^{+}(r,\theta) &= \mathcal{F}_{\mathcal{K}}(\theta)\frac{e^{ikr}}{r}\\ &= \frac{1}{2\mathcal{K}r}\sum_{\ell}(2\ell+1)i^{\ell+1}\left[1 - S_{\ell}^{\alpha\alpha}e^{i(\mathcal{K}r-\mathcal{K}\pi/2)}\right]\mathcal{L}_{\ell}(\cos\theta).\end{aligned} \tag{3.72}$$

Using $i^{\ell} = e^{i\pi\ell/2}$ yields

$$\mathcal{F}_{\mathcal{K}}(\theta) = \frac{i}{2\mathcal{K}}\sum_{\ell}(2\ell+1)\left[1 - S_{\ell}^{\alpha\alpha}\right]\mathcal{L}_{\ell}(\cos\theta). \tag{3.73}$$

Therefore the differential scattering cross section is related to the S-matrix by

$$\frac{d\sigma_{\text{el}}}{d\Omega}(\theta) = \frac{1}{4\pi}\left|\sum_{\ell}(2\ell+1)\left[1 - S_{\ell}^{\alpha\alpha}\right]\mathcal{L}_{\ell}(\cos\theta)\right|^{2}. \tag{3.74}$$

The nuclear interaction potential in the scattering center determines the value of $S_{\ell}^{\alpha\alpha}$, which may depend on the energy of the projectile. On the other hand, the $S_{\ell}^{\alpha\alpha}$ can be determined by measuring the differential cross section over a range of angles θ. The inverse scattering problem then is constituted by finding the potential yielding the appropriate $S_{\ell}^{\alpha\alpha}$ in a range of projectile energies.

Following Equation (1.100) and considering that the Legendre polynomials are orthogonally normalized to

$$\int \mathcal{L}_{\ell}(\cos\theta)\mathcal{L}_{\ell'}(\cos\theta)\, d\Omega = \frac{4\pi}{2\ell+1}\delta_{\ell\ell'} \tag{3.75}$$

(with the Kronecker symbol $\tilde{\delta}_{\ell\ell'}$), then the cross section for elastic scattering is obtained as a sum of cross sections for each partial wave,

$$\sigma_{\text{el}} = \frac{\pi}{\mathcal{K}^2}\sum_{\ell}(2\ell+1)|1 - S_{\ell}^{\alpha\alpha}|^2. \tag{3.76}$$

In the most general case $S_{\ell}^{\alpha\alpha}$ may be a complex number, see Section 3.6. Taking the square of the absolute value of $S_{\ell}^{\alpha\alpha}$ guarantees that the cross section will always be a real number. The elastic cross section and the S-matrix are directly connected to the reaction cross section which will be discussed in Section 3.6.1.

3.6 Reaction Cross Sections

In addition to elastic scattering, which leaves projectile and scattering center unchanged, interacting nuclei can also undergo changes by transferring energy between the interaction partners and/or by reordering the nucleons in the system, leading to the appearance of different nuclei in the outgoing channel, also called *exit channel*, as compared to the nuclei in the incoming channel, the *entrance channel*. For example, the interacting nuclei can become excited due to transfer of kinetic energy to excitation energy. This means that in at least one of the interacting nuclei some or all of the nucleons are moved to excited states (see Section 3.4) with higher energy. The transferred energy then is missing from the kinetic energy in the exit channel. This is often called *inelastic scattering*, although sometimes this term is used for any process that is not elastic scattering. If the nucleons available in the entrance channel are rearranged so that different nuclei appear in the exit channel, this is usually called a *nuclear reaction*. Again, this term is not consistently used in the scientific literature, sometimes including spontaneous nuclear decay as well, sometimes excluding it. Throughout this book, the term non-elastic is used to subsume inelastic and reaction processes whereas the term inelastic scattering is only used for nuclear excitations without change in the composition of the participating nuclei. In nature, all energetically possible processes in an interaction that are not forbidden by spin selection rules (see Section 1.7) will occur simultaneously but with different probabilities, that is with different cross sections. In many cases, one or a few processes will dominate. Nuclear reaction theory provides simplified approaches neglecting the unimportant processes and focusing only on the dominating ones. This implies that depending on interaction energy and number of involved nucleons, different reaction models have to be applied for a realistic description (see Section 3.7).

3.6.1 Total Reaction Cross Section from Nuclear Scattering

The cross section for elastic scattering was derived from elementary scattering theory in Section 3.5. It was shown that it is related to the diagonal element $S_\ell^{\alpha\alpha}$ of the S-matrix (sometimes also called scattering matrix or collision matrix). If energy and/or nuclei in entrance and exit channel are not the same, we have to use the appropriate entry of the S-matrix and denote it by $S_\ell^{\alpha\beta}$. The label β is used to show that the exit channel is not identical to the entrance channel α. It should not be confused with β-decay. The S-matrix is a unitary matrix because of the conservation of the probability current in quantum physics. In other words, and in the context of nuclear scattering: flux cannot vanish. It also cannot be generated which would imply that the outgoing wave function would have a larger amplitude than the incoming wave function. This restricts the elements of the scattering matrix to $|S_\ell^{\alpha\alpha}|^2 \leqslant 1$ and $|S_\ell^{\alpha\beta}|^2 \leqslant 1$. Moreover, all channels are in competition and therefore any reaction or inelastic process will remove flux from the elastic scattering channel. Due to flux conservation, however, the sum over all channels must remain constant. This argument leads to the unitarity property of the S-matrix,

$$|S_\ell^{\alpha\alpha}|^2 + \sum_{\beta\neq\alpha} |S_\ell^{\alpha\beta}|^2 = 1. \tag{3.77}$$

This has the important consequence that a knowledge of $S_\ell^{\alpha\alpha}$ also allows to determine the *total reaction cross section* σ_{reac}, accounting for all non-elastic processes. The element $S_\ell^{\alpha\beta}$ is connected to a wave function after a non-elastic process. The wave function for the non-elastic channel only contains outgoing spherical waves but no incoming ones and the outgoing waves ${}^{\alpha\beta}\psi_{K_\beta}^{+}$ are described by an expression similar to Equation (3.72) but replacing $S_\ell^{\alpha\alpha}$ by $S_\ell^{\alpha\beta}$. To find the relation between the elastic and non-elastic channels, the incoming flux (plane waves) j^- has to be compared to the total outgoing flux. The latter is obtained by an integration of the outgoing flux $j^+(\Omega)$ in the elastic channel over the full solid angle Ω as shown in Equation (1.112) but using Equation (3.72) with $S_\ell^{\alpha\beta}$. For purely elastic scattering (i.e., $|S_\ell^{\alpha\beta}|^2 = 0$ for all $\beta \neq \alpha$) both fluxes are the same because as many particles flow in as are flowing out. When including reaction channels this is not true because some particles from the elastic channel are scattered into the reaction channels and thus there is a net inflow of particles when the outgoing flux of the reaction channel is not included in the integration. The resulting reaction cross section for the channel $\alpha\beta$ is given by

$$\sigma_{\text{reac}}^{\alpha\beta} = \frac{r^2}{j^-}\int j^+(\Omega)\,\mathrm{d}\Omega = \frac{\pi}{\mathcal{K}^2}\sum_\ell (2\ell + 1)|S_\ell^{\alpha\beta}|^2. \tag{3.78}$$

Finally, the total reaction cross section is obtained by summing over all non-elastic channels β,

$$\sigma_{\text{reac}} = \sum_{\beta\neq\alpha}\sigma_{\text{reac}}^{\alpha\beta} = \frac{\pi}{\mathcal{K}^2}\sum_\ell (2\ell + 1)\sum_{\beta\neq\alpha}|S_\ell^{\alpha\beta}|^2. \tag{3.79}$$

Using the unitarity property of the S-matrix shown in Equation (3.77), the above equation can be recast as

$$\sigma_{\text{reac}} = \frac{\pi}{\mathcal{K}^2}(2\ell + 1)(1 - |S_\ell^{\alpha\alpha}|^2). \tag{3.80}$$

Therefore the knowledge of $S_\ell^{\alpha\alpha}$ suffices to determine also the total reaction cross section in addition to the elastic scattering cross section. Note that σ_{reac} includes all non-elastic processes but does not specify which processes are involved and how the incoming flux is distributed among the non-elastic channels. Thus, the total reaction cross section is sometimes also called *absorption cross section*, see Section 3.7.1.2 for a further discussion.

The labels α and β used in the identification of the reaction channel are to be understood to include the spin and magnetic quantum numbers of all incoming and outgoing nuclei, particles or photons. In most experimental measurements the projectiles are unpolarized and also the target nuclei are not aligned. Moreover, the final spin orientations of the final nuclei and ejected particles or photons are not measured, either. The experimentally measured cross section is therefore an average over the initial angular momentum quantum numbers of the nuclei in the entrance channel, and a sum over all final spin orientations. For example, for a reaction a + A → B + b (projectile a, target nucleus A, final nucleus B, ejectile b; with spin quantum numbers J_a, J_A, J_B, J_b, respectively, and their projections s_z^a, s_z^A, s_z^B, s_z^b) the relation between experimental cross section and the total reaction cross section derived above is given by

$$\sigma_{\mathrm{reac}}^{\mathrm{exp}} = \frac{1}{g_{\mathrm{a}} g_{\mathrm{A}}} \sum_{s_z^{\mathrm{a}} s_z^{\mathrm{A}} s_z^{\mathrm{B}} s_z^{\mathrm{b}}} \sigma_{\mathrm{reac}}^{\alpha\beta}, \tag{3.81}$$

using the spin factor notation

$$g_{\mathrm{x}} = 2J_{\mathrm{x}} + 1, \tag{3.82}$$

which is in this case, $g_{\mathrm{a}} = 2J_{\mathrm{a}} + 1$ and $g_{\mathrm{A}} = 2J_{\mathrm{A}} + 1$.

It is to be noted that an experimental determination of $S_\ell^{\alpha\alpha}$ by elastic scattering or only measuring the summed reaction products may not provide the actual values for $S_\ell^{\alpha\alpha}$ or the total reaction cross section in the strict definition of the S-matrix. This is because a reaction, for example a compound reaction (see Section 3.7.1), can release particles into various channels, including the elastic channel. This would lead to an underestimation of the actual reaction cross section, which is supposed to specify the probability that the target nucleus reacted with the projectile. Measuring the actual spin orientations and using polarized beams may help to distinguish elastically scattered projectiles from particles emitted from the compound nucleus.

The diagonal element of the S-matrix $S_\ell^{\alpha\alpha}$ can be calculated from the solutions of the time-independent Schrödinger equation. As before, an interaction potential V_{nuc} has to be assumed. In the past, the *equivalent square well potential* was used to save on computing time. It is a generalization of the infinite square well potential introduced in Section 3.3.2 and assumes a potential well with a sharp border and finite but constant depth, described by

$$V_{\mathrm{SW}}(r) = \begin{cases} V_0 + \mathrm{i}W_0 & \text{for } r \leqslant R_{\mathrm{nuc}} \\ 0 & \text{for } r > R_{\mathrm{nuc}} \end{cases}. \tag{3.83}$$

In the absence of a Coulomb potential, analytical solutions can be found for wave functions and the scattering matrix for such a potential.

More commonly, nowadays a Saxon–Woods type potential is used for the effective nuclear interaction V_{SW}, as shown in Equation (3.17) but with an imaginary part added. The imaginary part can again assume the same general shape (but with different depth W_0 instead of V_0 and a different diffuseness parameter $a_{\mathrm{nuc}}^{\mathrm{imag}}$ instead of a_{nuc}),

$$V_{\mathrm{SW}}(r) = V(r) + \mathrm{i}W(r) = \frac{V_0}{1 + \mathrm{e}^{(r-R_{\mathrm{nuc}})/a_{\mathrm{nuc}}}} + \mathrm{i}\frac{W_0}{1 + \mathrm{e}^{\left(r-R_{\mathrm{nuc}}^{\mathrm{imag}}\right)/a_{\mathrm{nuc}}^{\mathrm{imag}}}}. \tag{3.84}$$

Note that for an attractive potential V_0 and W_0 are negative. Often, $R_{\mathrm{nuc}}^{\mathrm{imag}} = R_{\mathrm{nuc}}$ is chosen. Another commonly used form for the imaginary part is one obtained when taking the derivative $\mathrm{d}/\mathrm{d}r V_{\mathrm{nuc}}$ for the shape of the potential, leading to

$$W_{\mathrm{surf}}(r) = 4\frac{W_{\mathrm{D}} \mathrm{e}^{\left(r-R_{\mathrm{nuc}}^{\mathrm{D}}\right)/a_{\mathrm{nuc}}^{\mathrm{D}}}}{\left[1 + \mathrm{e}^{\left(r-R_{\mathrm{nuc}}^{\mathrm{D}}\right)/a_{\mathrm{nuc}}^{\mathrm{D}}}\right]^2}. \tag{3.85}$$

This is called a surface potential and is combined with a real part of the potential. The factor 4 before the imaginary part is a frequently used convention. Sometimes the radius parameter $R_{\mathrm{nuc}}^{\mathrm{D}}$ and diffuseness $a_{\mathrm{nuc}}^{\mathrm{D}}$ are chosen differently from their values in the real part of the potential. A combination of volume and surface imaginary parts has also been tried in the past to define an optical potential as

$$V_{\mathrm{SW}} = V(r) + \mathrm{i}\underline{W}(r) = V(r) + \mathrm{i}(W(r) + W_{\mathrm{surf}}(r)), \tag{3.86}$$

usually when fitting experimental data.

The Woods–Saxon potential and the surface Woods–Saxon potential shown above are spherically symmetric. Many scattering and reaction calculations are performed assuming such spherical symmetry of the interacting nuclei. In Section 3.3.3 it was mentioned that nuclei can be deformed. With deformation, the potentials become dependent on the orientation of the nuclei and the potential parameters and the radius coordinate become vectors in space (see also Equation (3.88)). Another possibility to approximate the average effect of scattering on randomly oriented, deformed nuclei is to adapt the radius coordinate to obtain an effective spherical potential accounting for deformation. Using the expansion shown in Equation (3.40) for obtaining $\bar{R}$ and integrating over the angles θ and ϕ, the effective radius coordinate r_{def} is obtained which is replacing r in the formulas for the potentials (for example, the Saxon–Woods potentials shown above):

$$\begin{aligned} r_{\mathrm{def}} = \frac{r}{\pi}\int_0^{\pi/2} \Bigg[& 1 + \beta_2(3\cos^2(\theta) - 1)\sqrt{\frac{5}{16\pi}} \\ & + \beta_4\big(35\cos^4(\theta) - 30\cos^2(\theta) + 3\big)\sqrt{\frac{9}{256\pi}}\Bigg]^{-1} \mathrm{d}\theta. \end{aligned} \tag{3.87}$$

This results in potentials that are shallower than the unmodified spherical potentials and describe scattering on deformed nuclei well.

Instead of the parameterized shapes of the effective nuclear potential, also numerical tables specifying the potential depth at a given radius can be plugged into a numerical determination of $S_\ell^{\alpha\alpha}$. Such tables are derived with microscopic and semi-microscopic approaches based on nucleon–nucleon interactions. One such approach is the *folding potential*, which can be used as the real part of the effective nuclear potential. It determines the potential by folding the nucleon density distributions $\rho_{\mathrm{nuc}}^{\mathrm{a}}$, $\rho_{\mathrm{nuc}}^{\mathrm{A}}$ of projectile and target nucleus, respectively, with an effective nucleon–nucleon interaction $\mathcal{V}_{\mathrm{fold}}$,

$$V(\vec{r}) = \lambda_{\mathrm{fold}} \int\int \rho_{\mathrm{nuc}}^{\mathrm{a}}(\vec{r}_{\mathrm{a}})\rho_{\mathrm{nuc}}^{\mathrm{A}}(\vec{r}_{\mathrm{A}})\mathcal{V}_{\mathrm{fold}}(|\vec{r} + \vec{r}_{\mathrm{a}} - \vec{r}_{\mathrm{A}}|)\, \mathrm{d}\vec{r}_{\mathrm{a}}\mathrm{d}\vec{r}_{\mathrm{A}}. \tag{3.88}$$

The renormalization factor λ_{fold} is close to unity and accounts for the effect of the Pauli exclusion and antisymmetrization of the actual nucleon wave functions, which otherwise are not accounted for in Equation (3.88). Assuming spherical symmetry can be used to simplify the integration. Another simplification can be made when the projectile can be taken as being point-like. Then it can be described by a delta functional, changing the *double-folding potential* of Equation (3.88) into a *single-folding potential.* As the folding procedure only yields the real part of the potential, an imaginary part has to be added. Its shape is often taken to be the one of the Saxon–Woods potential or its derivative. The potential parameters are fitted to scattering data.

The concept of simultaneously describing elastic and non-elastic scattering by a single, effective interaction potential is called *optical model of nuclear reactions.* The non-elastic channels give rise to a loss of flux from the elastic channel. This is often compared to the wave scattering of light on an opaque, partly absorbing body, thus motivating the name of the model. The loss of flux is then related to a complex refraction index. The loss of flux from the elastic channels requires the introduction of an imaginary part in the scattering potential, as shown above. This is because the Schrödinger equation with real potentials conserves flux. Including an imaginary part implies that $S_\ell^{\alpha\alpha}$ becomes a complex number with an absolute value $|S_\ell^{\alpha\alpha}|^2 < 1$. This is related to a complex refraction index. Therefore such complex potentials, simultaneously describing elastic and non-elastic scattering, are categorized as *optical potentials.* The optical model of nuclear reactions makes use of such potentials to calculate reaction cross sections.

The optical potentials shown above were assumed to be energy independent. In reality, the potential parameters (especially the depths of the real and imaginary parts) should be energy-dependent, as the number of non-elastic channels depends on the interaction energy. For example, the imaginary part should vanish below the threshold of the non-elastic channel at lowest energy. Another reason for the appearance of an energy dependence is the fact that formal scattering theory shows that optical potentials should be non-local (that means, should not only depend on one radius value r). Non-local potentials can be approximated well by local potentials but only by introducing an additional energy dependence. However, often the energy dependence, regardless of its origin, is assumed to be weak and thus is neglected.

In a further simplification, real and imaginary parts of optical potentials are usually treated as being independent from each other. Such a view lends itself easily to fitting potential parameters to reproduce experimental elastic scattering data. In the majority of applications, the parameters of real and imaginary part are independently fitted until the best agreement with the data is found. There is, however, a physical connection between the real and the imaginary part, which is expressed in the *dispersion relation.* The energy-dependence of real part (its depth denoted by V) and imaginary part (with depth $\underline{W}$) is related by

$$V(r, E) = V_{\mathrm{SW}}(r, E) + \frac{E - E_{\mathrm{F}}}{\pi} \int_{E_{\mathrm{F}}}^{\infty} \frac{\underline{W}(r, E') - \underline{W}(r, E)}{(E' - E_{\mathrm{F}})^2 - (E - E_{\mathrm{F}})^2} \mathrm{d}E'. \tag{3.89}$$

This assumes that the energy dependence of $\underline{W}(r, E)$ is symmetric around the Fermi energy. The geometrical shape and a general energy-dependence still have to be chosen but using the dispersion relation there is less freedom in the parameter choice for the real and imaginary parts of the potential.

The diagonal element $S_\ell^{\alpha\alpha}$ of the scattering matrix can be found by numerically solving the Schrödinger equation with a given optical potential as described in Section 3.3.2 and matching the obtained wave function to the asymptotic solutions given by Equations (3.31) and (3.32). Using the phase shifts defined in Equations (3.31) and (3.33), the diagonal element can be written as

$$S_\ell^{\alpha\alpha} = \mathrm{e}^{2\mathrm{i}(\delta_\ell + \delta_\ell^{\mathrm{C}})}. \tag{3.90}$$

The Fox–Goodwin method to obtain the wave functions and phase shifts was given for real potentials in Section 3.3.2, resulting in real wave functions. The method can easily be adapted to optical potentials with a complex part. This also leads to complex valued wave functions and also complex asymptotic function values of the wave functions, replacing the real valued functions in Equations (3.31) and (3.32). Therefore also the phase shift will be complex with a real valued part δ_ℓ^{r} and an imaginary valued part δ_ℓ^{i}, $\delta_\ell = \delta_\ell^{\mathrm{r}} + \mathrm{i}\delta_\ell^{\mathrm{i}}$. Nevertheless, the mathematical framework remains the same otherwise.

Figure 3.4 shows the relation between the elastic scattering cross section σ_{el} and the reaction cross section σ_{reac} for all possible values of $S_\ell^{\alpha\alpha}$. Only values in the shaded region and at its boundary are possible. At the outer boundary, $S_\ell^{\alpha\alpha}$ is real valued. Note that this does not necessarily imply that reactions are not happening. The phase shifts are determined by the combined action of the real and imaginary parts of the wave function. Without imaginary part there is no absorption into non-elastic channels, leading to $|S_\ell^{\alpha\alpha}| = 1$ and thus $\sigma_{\mathrm{reac}} = 0$. Including an imaginary part, both the imaginary and the real part of the optical potential contribute to the value of the phase shift δ_ℓ and to $S_\ell^{\alpha\alpha}$ but absorption into non-elastic channels will always occur, even for real valued $S_\ell^{\alpha\alpha} \neq \pm 1$. On the other hand, there is always also elastic scattering when $\sigma_{\mathrm{reac}} > 0$. The maximum of σ_{reac} is obtained for $S_\ell^{\alpha\alpha} = 0$, with $\sigma_{\mathrm{reac}} = \sigma_{\mathrm{el}}$. The maximum value of the scattering cross section is obtained for $S_\ell^{\alpha\alpha} = -1$ and it is four times as large as the largest reaction cross section. This is because incoming and outgoing waves may coherently interfere in elastic scattering. When they interfere constructively, the wave amplitudes are doubled, which results in a cross section that is quadrupled.

As mentioned before, the cross section defined by Equation (3.80) is not only called reaction cross section but also absorption cross section because flux from the elastic channel is absorbed into otherwise unspecified non-elastic channels. Since the absorption is mathematically described by the imaginary part $\underline{W}$ of the optical potential, the absorption cross section in Equation (3.80) can also be written as

$$\sigma_{\mathrm{reac}} = \frac{8m\pi}{\hbar^2 K^3} \sum_\ell (2\ell + 1) \int_0^\infty [-\underline{W}(r)]\, |u_\ell(r)|^2 \,\mathrm{d}r. \tag{3.91}$$

This nicely shows how the strength of the imaginary potential at radius r is weighted by the probability to find the particle at that location. Since the amplitude of the radial wave function u_ℓ is suppressed in the interior of the nucleus in the presence of a potential barrier, this may result in the main contribution to the absorption cross section coming from radii further out than the location of the maximal depth of the

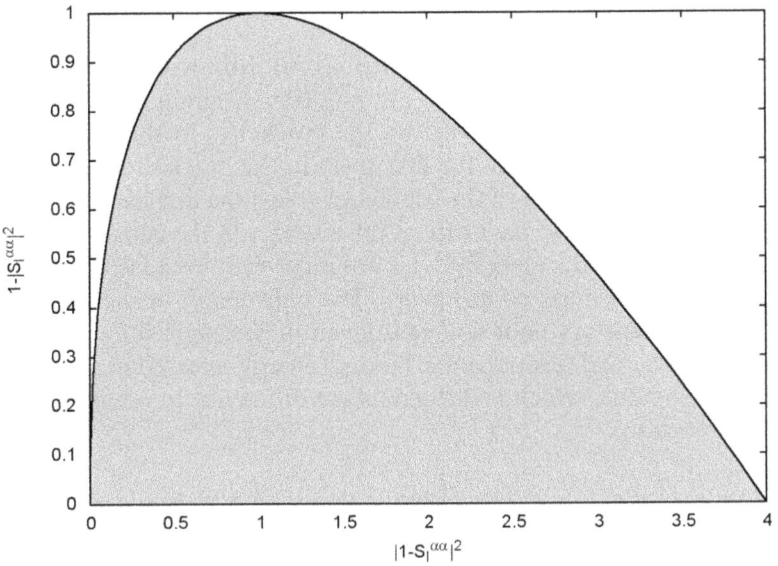

Figure 3.4. Comparison of the factors including the diagonal element of the S-matrix for the elastic scattering cross section (horizontal axis) and the reaction cross section (vertical axis).

imaginary potential. This is of particular interest in reactions with charged particles at low energies, as frequently encountered in nuclear astrophysics.

As can be seen in Equation (3.71), the prerequisite for $\sigma_{\rm reac} > 0$, which is $|S_\ell^{\alpha\alpha}|^2 < 1$, leads to a suppression in the amplitude of the outgoing wave. On the other hand, $|S_\ell^{\alpha\alpha}|^2 = 1$ implies perfect reflection without loss of flux into non-elastic channels. Therefore, again in analogy with wave optics, $|S_\ell^{\alpha\alpha}|^2$ can be viewed as a reflection coefficient and consequently

$$T_\ell = 1 - |S_\ell^{\alpha\alpha}|^2 \tag{3.92}$$

as a transmission coefficient. It can also be viewed as measuring the coupling described by the imaginary potentials between external scattering and all internal non-elastic processes not explicitly considered in a reaction model. Since there are no specific reactions considered so far, this includes all reactions. In this case, the absorption cross section and the reaction cross section are identical. If there were specific reaction channels modeled by additional imaginary potentials, the absorption cross section would be the remaining difference between the total reaction cross section given by Equation (3.80) and the cross sections for the explicitly modeled reaction channels.

The transmission coefficient T_ℓ should not be confused with penetrabilities, penetration coefficients, or similar barrier tunneling coefficients. Contrary to such penetration coefficients, $T_\ell = 0$ for real valued potentials without imaginary part and $T_\ell \simeq 1$ for strongly absorbed partial waves. (Penetrabilities are further discussed in Section 3.7.1.1 and transmission coefficients will be used in Section 3.7.1.2.)

3.6.2 Reaction Energetics

The reaction a + A → b + B is usually written as A(a,b)B, with projectile a, target A, ejectile b, and residual nucleus B. This notation originates from experimental nuclear physics where the nuclide before the bracket is in the target, bombarded by a beam of projectiles given by the first entry in the bracket. The second entry in the bracket then is the residual of the beam and at last the final nuclide created in the reaction is given. For theory, the order of the nuclides in the entrance channel a + A and the exit channel b + B, respectively, is not important because all calculations are performed in center-of-mass coordinates. (The conversion between center-of-mass coordinates and laboratory coordinates is given in Section 1.6.1.)

In analogy to chemical reactions, the binding energy released in a reaction is given by the reaction Q-value, which is defined as the difference in binding energy before and after the reaction,

$$Q_{\mathrm{nuc}} = \mathcal{B}_{\mathrm{before}} - \mathcal{B}_{\mathrm{after}} = \left[\left(\sum_{\mathrm{initial}} m_{\mathrm{nuc}}^{\mathrm{initial}}\right) - \left(\sum_{\mathrm{final}} m_{\mathrm{nuc}}^{\mathrm{final}}\right)\right] c^2, \tag{3.93}$$

which includes the sum of the masses of all reaction partners before the reaction (in the entrance channel) and the sum of masses of final nuclei (in the exit channel). For example, for a reaction a + A → b + B, the reaction Q-value is given by $Q_{\mathrm{nuc}} = (m_{\mathrm{nuc}}^{\mathrm{a}} + m_{\mathrm{nuc}}^{\mathrm{A}} - m_{\mathrm{nuc}}^{\mathrm{b}} - m_{\mathrm{nuc}}^{\mathrm{B}})c^2$. A positive Q-value means that there is a net energy release in the form of kinetic energy and/or emission of electromagnetic radiation. Such a reaction is called *exothermic*. An *endothermic* reaction, on the other hand, has a negative Q-value and can only proceed when additional energy is supplied. This results in an energy threshold below which the reaction cannot occur. Again, the required energy can be supplied either by kinetic energy between projectile and target or by electromagnetic radiation, depending on the type of reaction. Note that the Q-value definition in Equation (3.93) is completely general and does not depend on how a reaction actually proceeds. It also does not depend on whether any intermediate reactions occur but only depends on the initial and final masses. This is simply because we are looking at a difference in potential energies defined in a conservative force field. The Q-value, of course, depends on the direction of a reaction. The Q-value of a reverse reaction b + B → a + A will be the same as the one of the forward reaction a + A → b + B but with opposite sign. The reverse of an exothermic reaction is an endothermic one and vice versa.

Instead of nuclear masses m_{nuc}, also mass excesses Δ_{ato} as defined in Equation (3.7) can be used in Equation (3.93) to compute the reaction Q-value. Atomic masses, on the other hand, can only be used straightforwardly without the creation of positrons through the weak interaction. Otherwise, due to the definition of the atomic mass, $2m_{\mathrm{e}}c^2 = 1.022$ MeV per emitted positron have to be subtracted from the Q-value obtained with atomic masses.

Obviously, there is a relation between the reaction Q-value and the particle separation energies as given in Table 3.2. For capture reactions a + A → C + γ (i.e., the ejectile is a photon) the Q-value simply is the separation energy of the lighter nuclide a in the final nucleus C because this is the gained binding energy. For example, the energy required to remove a proton from ^{12}C is 15.957 MeV and therefore the Q-value of the reaction ^{11}B + p → ^{12}C + γ is $Q_{\rm nuc} = S_{\rm p}(^{12}{\rm C}) = 15.957$ MeV. When the ejectile is a particle, as in the reaction A(a,b)B, the Q-value is the difference of the separation energies of the projectile and the ejectile in the final nucleus B, $Q_{\rm nuc} = S_{\rm a}({\rm B}) - S_{\rm b}({\rm B})$.

To visualize the energetics of a reaction it is helpful to use energy schemes of excited states, as introduced in Figure 3.3, for the heavier nuclides appearing in the reaction, that is nuclides A and B. The energy schemes are drawn next to each other, vertically offset by the separation energies of the projectile or ejectile. A reaction may also proceed via intermediate steps, for example by forming an intermediate nucleus C (a compound nucleus, see Section 3.7.1.2). An example for such a combination of energy schemes for the involved nuclei is shown in Figure 3.5 for a reaction a + A → C → B + b. Note that this sketch automatically also includes a view of the energetics of the reactions A(a,γ)C and C(γ,b)B. Transitions between nuclear levels are shown by thick arrows and the relative energies of the transitions can be read off the sketch directly. For a more detailed discussion of the reaction mechanisms, that is the ways in which a reaction can physically proceed, is provided in Section 3.7.

The total reaction cross section $\sigma_{\rm reac}$ as obtained using Equation (3.80) includes all reactions to all energetically accessible final states. The final states comprise different final nuclei as well as different spin and parity configurations (depending on the available excited states in a final nucleus). Usually, one is interested in a specific *reaction channel*, such as A(a,b)B, with a final nucleus B and ejectile b. In the following, the reaction cross section for a reaction a + A and starting from excited state μ in nucleus A and ending in excited state ν in nucleus B is denoted by $\sigma_{\rm Aa}^{\mu\nu}$. This channel is part of the total reaction cross section $\sigma_{\rm reac} \geqslant \sigma_{\rm Aa}^{\mu\nu}$ for a + A. In laboratory experiments, the target nucleus is in its ground state, $\mu = 0$, and therefore the *laboratory reaction cross section* for such a process is given by

$$\sigma^{\rm lab} = \sigma^{\mu=0} = \sigma^0 = \sum_{\nu} \sigma_{\rm Aa}^{0\nu}. \tag{3.94}$$

It includes transitions from the ground state of nucleus A and a sum over all energetically possible transitions to final states ν in B. Each possible transition may have a different strength (that is, probability), which has to be calculated by solving a Schrödinger equation. Depending on the reaction mechanism (see Sections 3.6.4 and 3.7), different potentials and wave functions are used.

As will become clear in Section 4.1.3, astrophysical reactions also have to consider reactions proceeding on the excited states of target nuclei because of the elevated temperatures in stellar plasmas. In the laboratory, this cannot be directly studied for the majority of nuclei. An exception are isomers with sufficiently long half-lives. A special phenomenon can be observed for such reactions on isomers, called *super-elastic scattering*. This is the appearance of scattered projectiles with higher energy than the initial beam energy. Inspecting Figure 3.5 it becomes obvious that a projectile being scattered from an excited state in the target and forming a

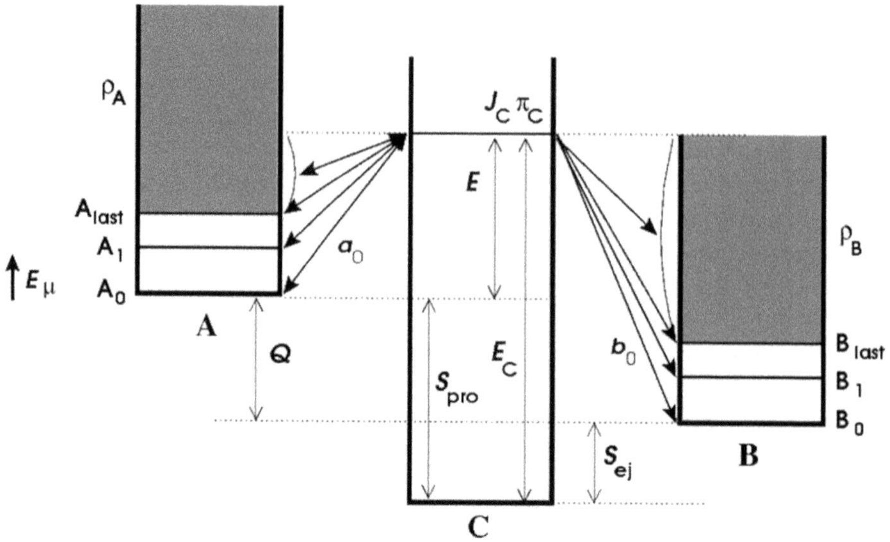

Figure 3.5. Schematic view of the transitions (thick arrows between nuclei denote particle transitions, γ transitions within a nucleus are not shown) in the reactions A(a,γ)C and A(a,b)B, proceeding via a compound state (horizontal dashed line) with spin J_C and parity π_C in the compound nucleus C. The reaction Q-value for the capture reaction ($Q_{\rm nuc}^{\rm cap} = S_{\rm pro}(C) = S_a(C)$) and the reaction A(a,b)B ($Q = Q_{\rm nuc}^{Aa} = S_{\rm pro}(C) - S_{\rm ej}(C) = S_a(C) - S_b(C)$) are given by the mass differences of the involved nuclei. Above the last explicitly included discrete states $A_{\rm last}$ and $B_{\rm last}$, respectively, transitions can be computed by integrating over nuclear level densities (shaded areas), see Section 3.7.1.2.

compound state (see Section 3.7.1.2) can be re-emitted with higher energy, resulting in a nucleus identical to the initial target nucleus but at a lower excitation (or in the ground state). Reaction model codes adapted to predict astrophysical reaction rates automatically include this effect (see Section 4.1.3) because they make use of the effective cross section $\sigma^{\rm eff}$ defined by Equation (4.44) but in some codes mainly aimed at providing cross sections for experiments it has to be included manually.

3.6.3 Reciprocity Theorem

There is a well-known reciprocity relation between the reaction cross section $\sigma_{\rm Aa}^{\mu\nu}$ for a forward reaction a + A connecting initial state μ in nucleus A and final state ν in nucleus B and the reaction cross section $\sigma_{\rm Bb}^{\nu\mu}$ of the reverse reaction connecting state ν in B to state μ in A. It is given by

$$\sigma_{\rm Bb}^{\nu\mu}(E_{\rm Bb}) = \frac{1+\tilde{\delta}_{\rm Bb}}{1+\tilde{\delta}_{\rm Aa}} \frac{g_\mu^{\rm A} g_{\rm a}}{g_\nu^{\rm B} g_{\rm b}} \frac{\mu_{\rm red}^{\rm Aa} E_{\rm Aa}}{\mu_{\rm red}^{\rm Bb} E_{\rm Bb}} \sigma_{\rm Aa}^{\mu\nu}(E_{\rm Aa}) \tag{3.95}$$

where $E_{\rm Aa}$ and $E_{\rm Bb}$ are the relative energies in the channels a + A and b + B, respectively. The spin factors g_x are defined in Equation (3.82). The Kronecker factors $1 + \tilde{\delta}_{\rm Aa}$ and $1 + \tilde{\delta}_{\rm Bb}$ are included to account for the quantum effect of increased cross section when two particles of the same kind are interacting. The *Kronecker symbol* $\delta_{\rm Aa}$ assumes a value of 1 when A and a are identical and is zero otherwise.

Since this relation applies to the transitions between a single initial and a single final state, it is not directly applicable to the regular laboratory reaction cross section defined in Equation (3.94), which connects one initial state (usually the ground state $\mu = 0$) of the target nucleus A with a number of possible final states in nucleus B. The laboratory cross section of the reverse reaction again only connects the ground state of nucleus B with excited states in nucleus A. Obviously, different transitions are involved and therefore the reciprocity relation cannot be directly used to connect two laboratory cross sections (unless both reaction directions only involve transitions from the ground states of the involved nuclei; this can be the case for reactions with very light nuclei). Of course, reciprocity applies to each of the transitions in the sum appearing in the laboratory cross section.

It is worth noting that the reciprocity relation is not just an expression of time reversal symmetry of the nuclear and Coulomb interactions but is a more general principle also applying to complex optical potentials not conserving time symmetry. Equation (3.95) can be proven in several different ways. One way is to use a phase space argument. The number of states in the momentum interval $[p, p + {\rm d}p]$ is proportional to p^2, as shown in Section 1.5.1.2 (see Equation (1.45)). Therefore $\sigma^{\mu\nu}_{\rm Aa} \propto p^2_{\rm Bb}$ and $\sigma^{\nu\mu}_{\rm Bb} \propto p^2_{\rm Aa}$. This has to be weighted by the statistical weight for the $2J + 1$ spin orientations and also identical particles have to be accounted for, leading to $g_{\rm a} g^{\rm A}_{\mu} \sigma^{\mu\nu}_{\rm Aa} \propto (1 + \tilde{\delta}_{\rm Aa}) p^2_{\rm Bb}$ and $g_{\rm b} g^{\rm B}_{\nu} \sigma^{\nu\mu}_{\rm Bb} \propto (1 + \tilde{\delta}_{\rm Bb}) p^2_{\rm Aa}$. Using $p^2 = \hbar^2 k^2 = 2\mu_{\rm red} E$ leads to

$$\frac{g_{\rm a} g^{\rm A}_{\mu}}{1 + \tilde{\delta}_{\rm Aa}} 2\mu^{\rm Aa}_{\rm red} E_{\rm Aa} \sigma^{\mu\nu}_{\rm Aa} = \frac{g_{\rm b} g^{\rm B}_{\nu}}{1 + \tilde{\delta}_{\rm Bb}} 2\mu^{\rm Bb}_{\rm red} E_{\rm Bb} \sigma^{\nu\mu}_{\rm Bb}, \tag{3.96}$$

which is identical to Equation (3.95). Note that the relative energies in the entrance and exit channel are related by

$$E_{\rm Bb} = E_{\rm Aa} + Q^{\mu\nu}_{\rm nuc}. \tag{3.97}$$

More commonly, the Q-value of a reaction is defined relative to the ground states of the involved nuclei, denoted by $Q^{\rm Aa}_{\rm nuc}$ for a reaction A(a,b)B. Then the channel energies can be expressed as

$$E_{\rm Bb} = E_{\rm Aa} + Q^{\rm Aa}_{\rm nuc} + E^{\rm x}_{\mu} - E^{\rm x}_{\nu}, \tag{3.98}$$

with the excitation energies $E^{\rm x}_{\mu}$, $E^{\rm x}_{\nu}$ of the initial and final level, respectively.

The formulae for reaction cross sections in any given reaction model (see later) also obey the reciprocity relation as long as specific initial and final states are specified. This is true even for reactions proceeding via an intermediate step. Let us use the compound nucleus reaction cross section to illustrate this point. Compound nucleus reactions are studied in more detail in Sections 3.7.1 and 3.7.1.2. The cross section for a reaction between specific initial and final states via an intermediate state x in the compound nucleus C is given by

$$\sigma_{\mathrm{Aa}}^{\mu\nu} = \frac{\pi}{k_{\mathrm{Aa}}^2} \frac{1 + \tilde{\delta}_{\mathrm{Aa}} g_x^{\mathrm{C}}}{g_{\mathrm{a}} g_\mu^{\mathrm{Aa}}} \frac{\hat{\mathcal{T}}^{\mathrm{Aa}} \hat{\mathcal{T}}^{\mathrm{Bb}}}{\mathcal{T}^{\mathrm{tot}}} \tag{3.99}$$

and for the reverse reaction via the same intermediate state it is

$$\sigma_{\mathrm{Bb}}^{\nu\mu} = \frac{\pi}{k_{\mathrm{Bb}}^2} \frac{1 + \tilde{\delta}_{\mathrm{Bb}} g_x^{\mathrm{C}}}{g_{\mathrm{b}} g_\nu^{\mathrm{Bb}}} \frac{\hat{\mathcal{T}}^{\mathrm{Bb}} \hat{\mathcal{T}}^{\mathrm{Aa}}}{\mathcal{T}^{\mathrm{tot}}}. \tag{3.100}$$

Here the particle transmission coefficients $\hat{\mathcal{T}}^{\mathrm{Aa}}$, $\hat{\mathcal{T}}^{\mathrm{Bb}}$ describe the transitions connecting entrance and exit channel, respectively, with the intermediate level in the compound nucleus. For more details on the definition of $\hat{\mathcal{T}}$ and $\mathcal{T}$, see Section 3.7.1.2. Using the channel energies E_{Aa}, E_{Bb} instead of wave numbers k_{Aa}, k_{Bb} and considering the relation between the energies in the two channels as given in Equations (3.97) and (3.98), the reciprocity relation shown in Equation (3.95) can again be recovered from the above equations because the factor $\hat{\mathcal{T}}^{\mathrm{Aa}}\hat{\mathcal{T}}^{\mathrm{Bb}}/\mathcal{T}^{\mathrm{tot}}$ cancels out. Therefore this relation does not depend on the properties of the intermediate level. Thus it is also applicable to a more general cross section including a sum over many intermediate states as assumed in the statistical model of nuclear reactions (Section 3.7.1.2).

In the reciprocity relation shown in Equation (3.95) the projectile and ejectile are particles with rest mass larger than zero. If the reaction includes a photon, such as in $\mathrm{A} + \mathrm{a} \rightarrow \mathrm{C} + \gamma$, then the reciprocity relation reads

$$\sigma_{\mathrm{C}\gamma}^{\nu\mu}(E_\gamma) = \frac{\mathrm{c}^2}{1 + \tilde{\delta}_{\mathrm{Aa}}} \frac{g_\mu^{\mathrm{A}} g_{\mathrm{a}}}{g_\nu^{\mathrm{B}}} \frac{\mu_{\mathrm{red}}^{\mathrm{Aa}} E_{\mathrm{Aa}}}{E_\gamma^2} \sigma_{\mathrm{Aa}}^{\mu\nu}(E_{\mathrm{Aa}}). \tag{3.101}$$

3.6.4 Shape-elastic Resonances

For the most part, the nuclear phase shifts δ_ℓ appearing in Equation (3.31) are a smoothly and slowly varying function of projectile energy and therefore also S_ℓ^{aa} is slowly varying with energy. This reflects the fact that the Schrödinger wave equation has a solution for each asymptotically unbound energy $E > 0$. Nevertheless, when scanning through energy it is found that the scattering cross section (or the reaction cross section if using complex optical potentials) exhibits large increases ("peaks") around certain projectile energies. These are called *resonances*. They are accompanied by a rapid increase in phase shift in a specific partial wave, which will then make an important contribution. The reason for this is in the appearance of maximal amplitudes of the wave function in the "interior" of the nucleus where the nuclear potential is strongest. This implies a strong interaction of projectile and target nucleus and thus leads to strong scattering or reactions, respectively. Mathematically, the optimal matching between the part of the wave function inside the nucleus and the part outside will occur when the wave amplitude is an extremum at the border between the two regions. In other words, this will occur when the *spatial derivative of the wave function* D_ℓ^{u} is zero at the edge of the nucleus,

$$D_\ell^{\mathrm{u}}(R_{\mathrm{nuc}}) = \left.\frac{\mathrm{d}u_\ell(r)}{\mathrm{d}r}\right|_{r=R_{\mathrm{nuc}}} = 0. \tag{3.102}$$

Note that for a smooth optical potential, such as the Saxon-Woods potential, there is no sharp edge of the nucleus. The basic idea, however, remains the same. The dependence of wave amplitudes inside the nucleus on the derivative can immediately be understood, for example, when a projectile has to tunnel through a barrier to get inside the nucleus. Inside a potential barrier ($\mathcal{K}^2 + E_{\mathrm{pot},\ell} < 0$, see Section 3.3.2) the solution of Equation (3.25) is not an oscillating wave but an exponentially decaying function $u_\ell \propto \exp(-\mathcal{K}r)$. Matching such a decaying function to an extremal amplitude at the border of the barrier guarantees that the amplitudes inside the barrier also achieve their maximally possible values.

The full derivation can be found in any nuclear physics textbook, showing that around the resonance energy there are two contributions to the asymptotic wave function, a resonance contribution stemming from the above described fact of comparatively large amplitudes inside the nucleus and a contribution from scattering at the edge of the nucleus. This can also be expressed as two contributing phase shifts,

$$\delta_\ell(E_{\mathrm{Aa}}) = \delta_\ell^{\mathrm{res}}(E_{\mathrm{Aa}}) + \delta_\ell^{\mathrm{edge}}(E_{\mathrm{Aa}}). \tag{3.103}$$

The resonant phase shift $\delta_\ell^{\mathrm{res}}$ will be varying quickly with energy, depending on how quickly the derivative varies when changing the energy, i.e., on $\mathrm{d}f_{\mathrm{deriv}}/\mathrm{d}E(E_{\mathrm{res}})$ around the resonance energy E_{res}. The phase shift $\delta_\ell^{\mathrm{edge}}$, on the other hand, will change more slowly when changing the energy. With the widths introduced in Equation (3.49), the resonant phase shift can be expressed as

$$\delta_\ell^{\mathrm{res}}(E_{\mathrm{Aa}}) = \arctan\left(\frac{\Gamma_{\mathrm{Aa}}}{2(E_{\mathrm{res}} - E_{\mathrm{Aa}})}\right) + n(E_{\mathrm{Aa}})\pi, \tag{3.104}$$

where Γ_{Aa} is the width for capturing or emitting the projectile from the compound nucleus C, formed from the combination of projectile a and target nucleus A with the given energy E_{Aa} in the a + A system. Note that $\hat{\Gamma}_{\mathrm{tot}} = \Gamma_{\mathrm{Aa}}$ for pure scattering in the absence of reactions. The integer $n(E_{\mathrm{Aa}})$ is optionally added to make $\delta_\ell^{\mathrm{res}}$ a continuous function in energy. Otherwise the phase shift jumps at $\pi/2$ (for $\delta_\ell^{\mathrm{edge}} = 0$) when crossing the resonance energy. (For $\delta_\ell^{\mathrm{edge}} \neq 0$ a resonance can occur without the phase shift passing through $\pi/2$.) The width of the resonance is taken to be the full width of the resonance peak (for $\delta_\ell^{\mathrm{edge}} = 0$) at half the value of the maximum and is usually given in MeV.

The two contributions also show up in $S_\ell^{\alpha\alpha}$, which depends on the phase shift, and therefore in the scattering and reaction cross sections. For $\delta_\ell^{\mathrm{edge}} = 0$ it can be shown that the energy dependence of the cross section assumes the shape of a *Lorentz curve*,

$$\sigma_{\mathrm{el}}^{\mathrm{Aa}}(E_{\mathrm{Aa}}) = \frac{4\pi}{\mathcal{K}_{\mathrm{Aa}}^2}(2\ell + 1)\frac{\Gamma_{\mathrm{Aa}}^2/4}{(E_{\mathrm{Aa}} - E_{\mathrm{res}})^2 + \hat{\Gamma}_{\mathrm{tot}}^2/4}. \tag{3.105}$$

This assumes that the cross section around the resonance is dominated by the partial wave ℓ giving rise to the resonance and therefore the sum over partial waves can be approximated by a single partial wave. A cross section with such a shape is called a *pure Breit–Wigner resonance*. For $\delta_\ell^{\mathrm{edge}} \neq 0$, the shape can also become asymmetric or even a dip instead of an increase. These interference effects can only appear for $\delta_\ell^{\mathrm{res}}$ and $\delta_\ell^{\mathrm{edge}}$ from the same partial wave ℓ. In the absence of reactions (for optical potentials without imaginary part), $\hat{\Gamma}_{\mathrm{tot}} = \Gamma_{\mathrm{Aa}}$ (see Equation (3.50)) and thus $\hat{\Gamma}_{\mathrm{tot}}^2$ instead of Γ_{Aa}^2 would appear in the numerator of Equation (3.105). With reactions, also the reaction cross section shows a resonance,

$$\sigma_{\mathrm{reac}}^{\mathrm{Aa}}(E_{\mathrm{Aa}}) = \frac{4\pi}{\mathcal{K}_{\mathrm{Aa}}^2}(2\ell + 1)\frac{\Gamma_{\mathrm{Aa}}\Gamma_{\mathrm{reac}}/4}{(E_{\mathrm{Aa}} - E_{\mathrm{res}})^2 + \hat{\Gamma}_{\mathrm{tot}}^2/4}, \tag{3.106}$$

where Γ_{reac} is the reaction width, comprising partial widths of all open reaction channels ($\Gamma_{\mathrm{reac}} = \hat{\Gamma}_{\mathrm{tot}} - \Gamma_{\mathrm{Aa}}$).

Comparing a level scheme such as the one shown in Figure 3.3 for a real nucleus with the resonances obtained from solutions of the Schrödinger equation with assumed optical potentials, it quickly becomes apparent that there is a larger number of resonances in the real nucleus and also resonance widths are different. Resonances described up to here are called *shape-elastic* resonances because they only depend on the “shape” (depth and radial extension) of the scattering potential. They do not account for the actual interaction processes between projectile and the nucleons inside the target nucleus. To obtain realistic resonance widths, additional modeling is required. This will be outlined in Section 3.7.1. Shape-elastic resonances are broader than resonances arising from the nucleon–nucleon interaction inside a nucleus. Therefore they are comparatively short-lived (see Equation (3.49)).

3.7 Introduction to Reaction Theory

Although reaction theory dates back as far as the early to middle 20th century, the special requirements of astrophysics and the need for cross section predictions of nuclei far from stability provide an interesting and stimulating environment for the application and further developments of different approaches. The challenges are manifold. On one hand, astrophysical energies are very low and various reaction mechanisms may contribute or interfere. On the other hand, even if the reaction mechanism is unique and well understood, nuclear properties entering the reaction model have to be predicted for nuclei far off stability. This proves challenging even for modern nuclear structure calculations. Although a fully microscopic treatment is preferable, good parameterizations and averaged quantities are still necessary in many cases due to the sheer number of reactions and involved nuclei, especially for intermediate and heavy nuclei consisting of more than 30–40 nucleons and thus not allowing the application of few-body models. Finally, the interpretation of experiments has to be supported by theory. This latter case may involve different methods than the one dealing with the prediction of astrophysical rates because experiments may be conducted at higher energies and theory is needed to extract the information to be included in the rates. For example, the properties of excited states and their spectroscopic factors can be studied by (d,p) reactions at comparatively high energy which otherwise are not directly relevant in astrophysics. In the following the focus is on theory frequently used for the prediction of reaction cross sections in studies related to astrophysics. How these are incorporated in the determination of astrophysical reaction rates and the special cases arising in the context of astrophysics are explained in detail in Chapter 4. Full details of the theoretical approaches and their derivation can be found in textbooks devoted to nuclear reaction theory which, however, usually require a knowledge in advanced quantum mechanics.

A fully microscopic theory of nuclear reactions would have to solve the full Hamiltonian of the interacting system (Equation (3.16)), including all nucleons and their interactions, yielding wave functions with exact antisymmetrization. This is only possible for few-nucleon systems and even then an approximation of the strong nuclear interaction has to be made, similar to the effective interaction used in shell model calculations. To simplify the mathematics, further approximations have to be made. Depending on the approximation, we speak of different *reaction mechanisms* and reaction models for describing these. It should be kept in mind that a reaction mechanism is a simplified picture. All energetically possible "mechanisms" always occur simultaneously in nature but may have vastly different contributions to the total reaction cross sections. In many instances, a single mechanism dominates and this allows to apply a special mathematical treatment, a specific reaction model, and justifies to speak of a mechanism. The interaction energy and the nuclear structure of the interacting nuclei determine which reaction mechanism dominates. To complicate things further, there may be energy ranges where different reaction mechanisms compete (or even coherently interfere).

There are different theoretical approaches available for each reaction mechanism. Theoretical models can be roughly sorted into three very general categories:

1. Models involving adjustable parameters, such as the pure Breit–Wigner resonance model or the R-matrix method (see Section 3.7.1.1). Parameters are fitted to the available experimental data and the cross sections are extrapolated down to astrophysical energies. These fitting procedures, of course, require the knowledge of data, which are sometimes too scarce for a reliable extrapolation.
2. "Ab initio" models, where the cross sections are determined from the wave functions of the system. The potential model and microscopic models are, in principle, independent of experimental reaction data. More realistically, however, these models depend on some physical parameters, such as a nucleus–nucleus or a nucleon–nucleon interaction, which can be reasonably

determined from experiment only. Most microscopic models are only tractable for light nuclei including few nucleons or nucleon clusters.
3. The above models can be used for systems with a few nuclear states involved and thus are limited to nuclei with comparatively low level-density. This condition is fulfilled in most of the reactions involving light nuclei ($A \leqslant 20$). However when the level density near the reaction threshold in the compound nucleus is large (i.e., more than a few levels per MeV), statistical models, using averaged optical transmission coefficients, are more suitable (see Section 3.7.1.2).

The two reaction mechanisms at the top of a hierarchy of mechanisms are direct reactions and compound nucleus reactions. They can experimentally be distinguished by yielding different energy patterns and angular distributions of reaction products. They also differ vastly in the reaction timescales. While direct reactions proceed on a timescale of the order of 10^{-22} s, the formation and subsequent de-excitation of a compound nucleus takes on the order of 10^{-16} s. A reaction proceeding via the formation of a compound nucleus at excitation energy (compound formation energy) E_{C} is shown in Figure 3.5. In a compound reaction, the combined nucleons of projectile and target nucleus form an intermediate nucleus. The energy in the system is shared by many or all nucleons. The compound nucleus can de-excite by emission of γ-radiation at various energies and/or particles (nucleons or nucleon groups) as allowed by quantum mechanical spin selection rules (see Section 1.7).

In direct reactions (Section 3.7.2), on the other hand, only few nucleons from the projectile (in direct capture and stripping reactions) or target nucleus (in pick-up reactions adding a nucleon to the passing projectile) are directly brought into their final quantum states, without the participation of other nucleons (termed "spectator nucleons"). The reaction Q-value, if positive, is released in single-energy photons or taken up by an ejectile.

The optical model introduced in Section 3.6.1 is well suited for describing transitions between states in intermediate and heavy nuclei and is used in treating compound nucleus reactions. It has been and is still used also to treat reactions with light nuclei although other methods exist for these.

3.7.1 Compound Reactions

Compound reactions form the compound nucleus at the *compound formation energy*

$$E_{\mathrm{C}} = E_{\mathrm{Aa}} + E_{\mu}^{\mathrm{x}} + S_{\mathrm{pro}}(\mathrm{C}), \tag{3.107}$$

where the energy convention of Equation (3.98) is used. This is the excitation energy of the compound nucleus and is computed from the projectile energy E and the separation energy of the projectile in the compound nucleus $S_{\mathrm{pro}}(\mathrm{C})$ (see also Figure 3.5). When the energy separation between the excited states at the compound formation energy is larger than the energy distribution of the incident projectiles, individual resonances will appear in the reaction cross section. These resonances are narrower than the shape-elastic resonances discussed in Section 3.6.4 and therefore exhibit longer life-times (contributing to the longer timescale of the compound nucleus reactions as compared to direct reactions). They correspond to excited states in the compound nucleus arising from the excitation of nucleons or nucleon groups by the deposition of the projectile's energy (see Section 3.4). This is also why the cross section as function of energy is often called *excitation function*. For the calculation of the reaction cross section, similar methods as for inclusion of shape-elastic resonances can be used, leading to a Breit–Wigner form but with adapted resonance widths to account for the internal structure of the nucleus and its excitations. This is outlined in Section 3.7.1.1.

The nuclear level density strongly increases with excitation energy (see, e.g., Equation (3.60)) and therefore compound resonances will move closer together in the excitation function with increasing E_C. When the distance between resonances is too small to be experimentally resolved, a smooth cross section is obtained despite of the fact that it is composed of a large number of resonances that also may constructively and destructively interfere. This allows a further approximation, using average resonance properties around E_C in a *statistical model of compound reactions*. This is discussed in Section 3.7.1.2.

It is to be noted that the superposition of amplitudes from a very large number of states with random phases does not necessarily result in completely featureless excitation functions. It has been shown that the so-called *Ericsson fluctuations* can appear in cross sections and lead to small-scale variations when scanning through the energy in the system. The usual statistical model approach does not account for these fluctuations. Therefore, a meaningful comparison to experimental data is only possible after averaging the data over a sufficiently wide energy range, comparable to the average resonance widths. In the computation of astrophysical reaction rates an integration over an energy range much wider than the average resonance width is performed (see Equation (4.2)). Therefore any such naturally occurring cross section fluctuations do not play a role. When using laboratory beams with a very narrow energy spread, however, the results may not be directly comparable to calculations.

It is worthwhile to point out that the reaction rate in general is rather "forgiving" to fluctuations around an average cross section value, provided the fluctuations cancel within the energy window mainly contributing to the integral in Equation (4.2). Therefore the statistical model approach may even be applicable to the calculation of astrophysical reaction rates in the presence of small but isolated resonances as long as their average contribution is correctly accounted for. Thus, the prediction of astrophysical reaction rates is different from, and perhaps less challenging than, a close reproduction of experimental cross sections. In simple numerical tests, ten levels within the relevant energy window have proven to be sufficient for the application of the statistical model to calculate reaction rates. It has to be emphasized that this is a rough estimate, to be used as an initial guide-line to assess the suitability of the statistical model. Sometimes even one or a few resonances, if broad enough to almost fill the energy window, may be sufficient. On the other hand, it is always desirable to get as much experimental information as possible for any nucleus but especially for those with a low nuclear level density at compound formation energies within the relevant energy window.

Applicability maps depicting for each target nucleus the lowest stellar temperature at which the statistical model is applicable to predict stellar reaction rates were derived using the criterion of minimally 10 resonances. These are shown in Figures 3.6–3.8 for rates dominated by the neutron-, proton-, and α-width, respectively. The importance of the reaction width determining a cross section or rate is discussed in Sections 3.7.1.2 and 4.1.2.4. The features seen in the figures can be understood by remembering that the nuclear level density $\tilde{\rho}_{\mathrm{nuc}}(E_{\mathrm{C}})$ at the compound nucleus excitation energy E_{C} determines the applicability. Around stability, S_{pro} usually is high (8–12 MeV, depending on the projectile) and dominates E_{C} because the projectile energies E are much lower in astrophysical sites. Statistical models will then be applicable for intermediate and heavy nuclei with sufficiently high $\tilde{\rho}_{\mathrm{nuc}}$. Even at stability, however, the nuclear level density may not be high enough in nuclei with closed shells. This is clearly seen in Figure 3.6. Figures 3.6 and 3.7 also clearly show another effect: Approaching the driplines, the neutron- or proton-separation energies strongly decrease, resulting in low E_{C} for neutron- or proton-induced reactions, respectively. This leads to low $\tilde{\rho}_{\mathrm{nuc}}(E_{\mathrm{C}})$ even for intermediate and heavy nuclei. The reason why the impact of reduced level density at magic nucleon numbers is not seen in Figure 3.7 is that the charge of the protons shifts the astrophysically relevant energy window to higher energies (see Section 4.1.2.2) and thus also increases the relevant E_{C}. At higher excitation energy, however, the level density is also higher, allowing the application of the statistical model. This shift is even more pronounced for α-particles. Furthermore, clearly identifiable in Figure 3.8 is the region of spontaneous α-emitters with negative α-separation energy S_{α}.

Below the estimated temperature limits shown in Figures 3.6–3.8 (these are nuclei with low $\tilde{\rho}_{\mathrm{nuc}}(E_{\mathrm{C}})$) isolated resonances (Section 3.7.1.1) or direct reactions (Section 3.7.2) have to be taken into account when calculating stellar reaction rates.

3.7.1.1 Compound Reactions with Isolated Resonances

Isolated resonances in the low and intermediate level density regimes can be treated by applying simple single-level or multi-level Breit–Wigner formulae or in the more involved *R*-matrix approach. The resonance properties (resonance energy, spin, partial and total widths) have to be known. Often, the resonance properties are derived by fits to experimental data. Where this is impossible, nuclear theory has to be invoked to predict the required quantities. This remains problematic, however, because the astrophysical reaction rates can be very sensitive to the resonance properties, especially the resonance energy (see Section 4.1.2.3), and high predictive power would be required. This is why there are large uncertainties in reaction rates off stability due to the unknown resonance contributions in nuclei with low $\tilde{\rho}_{\mathrm{nuc}}(E_{\mathrm{C}})$.

Cluster models (Section 3.7.3) have been successful in describing resonant cross sections in light nuclei but cannot be easily applied to nuclei at intermediate and heavy mass.

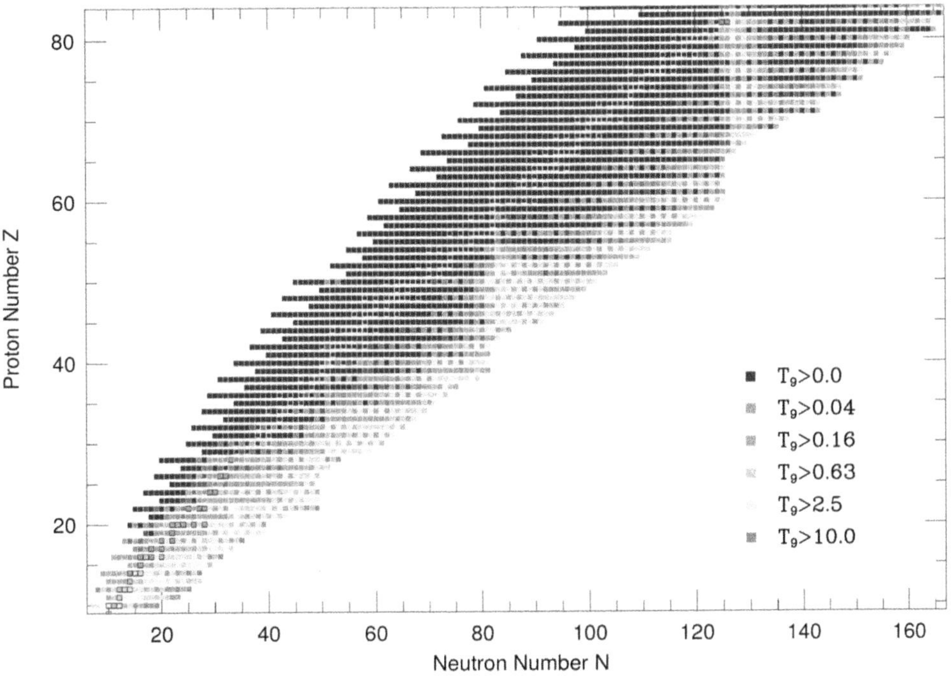

Figure 3.6. Estimated applicability of the statistical model for neutron-induced reactions. For each nucleus in the chart the color shade labels the temperature above which the statistical model becomes applicable to calculate the astrophysical reaction rate (T_9 is in GK). (Reprinted with permission from Rauscher et al. 1997. Copyright (1997) by the American Physical Society.)

Due to angular momentum conservation, the added spins of projectile $\vec{s}$, target $\vec{I}$, and orbital angular momentum $\vec{\ell}$ yield the spin(s) of accessible resonances $\vec{J} = \vec{s} + \vec{I} + \vec{\ell}$. The resulting value of the spin quantum number J has to be computed using the quantum mechanical spin addition rule given in Equation (1.126). Although resonances with the same spin quantum number J interfere and single resonances may also show interference with a direct reaction (Section 3.7.2), the Breit–Wigner formula is often used without interference terms:

$$\sigma_{\mathrm{BW}}^{\mathrm{Aa}\to\mathrm{Bb}}(E_{\mathrm{Aa}}) = \frac{\pi}{\mathcal{K}_{\mathrm{Aa}}^2}\sum_{i=1}^{n} {}^{i}\omega_{\mathrm{BW}} \frac{\Gamma_i^{\mathrm{Aa}}(E_{\mathrm{Aa}})\hat{\Gamma}_i^{\mathrm{Bb}}(E_{\mathrm{Bb}})}{\left(E_{\mathrm{Aa}} - E_{\mathrm{res},i}^{\mathrm{Aa}}\right)^2 + (\hat{\Gamma}_i^{\mathrm{tot}}(E_{\mathrm{Aa}})/2)^2}. \tag{3.108}$$

This uses the combined spin factor

$${}^{i}\omega_{\mathrm{BW}} = \frac{g_i}{g_A g_a}(1 + \tilde{\delta}_{\mathrm{Aa}}) \tag{3.109}$$

with $g_i = 2J_i + 1$ accounting for the spin of the ith resonance (see also Equation (3.82) for the definition of the statistical weights g). Also remember that the channel energies E_{Aa} and E_{Bb} are related by Equation (3.98). This provides the cross section for n non-interfering resonances, generalizing Equation (3.106). The total width

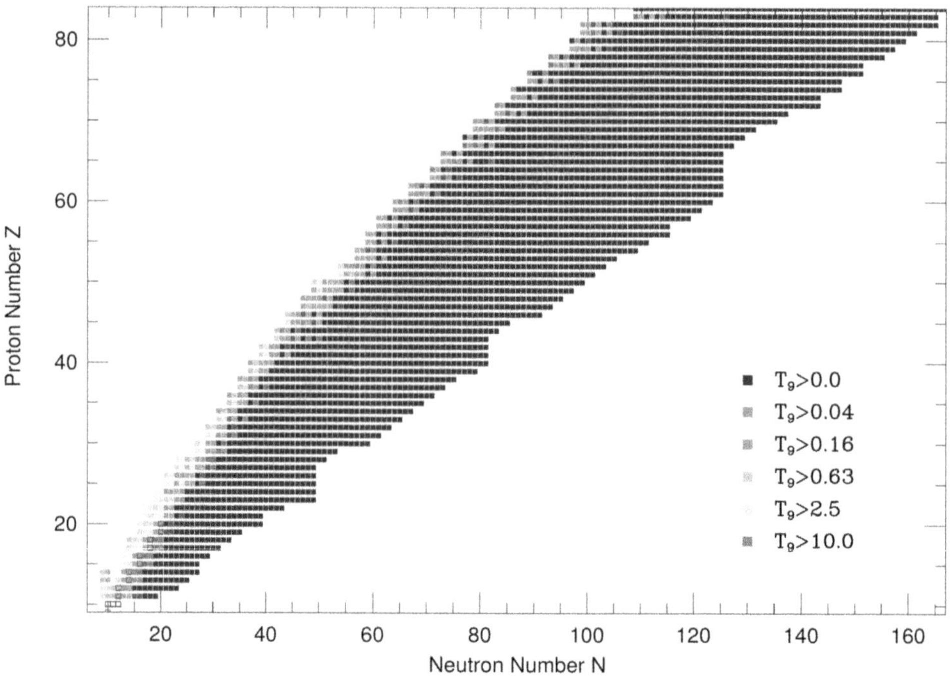

Figure 3.7. Estimated applicability of the statistical model for proton-induced reactions. For each nucleus in the chart the color shade labels the temperature above which the statistical model becomes applicable to calculate the astrophysical reaction rate (T_9 is in GK).

$\hat{\Gamma}_i^{\mathrm{tot}} = \hat{\Gamma}_{\mathrm{tot}}$ of a resonant state i in the compound nucleus is the sum over the widths of the individual decay channels

$$\hat{\Gamma}_{\mathrm{tot}} = \hat{\Gamma}^{\mathrm{Aa}} + \hat{\Gamma}^{\mathrm{Bb}} + \cdots, \tag{3.110}$$

also including transitions to other reaction channels beyond the exit channel B + b. The widths of the individual decay channels $\hat{\Gamma}^z$ (with z being Aa, Bb,…) are summed over transitions to all possible final states in the channel,

$$\hat{\Gamma}^z = \sum_{\xi} \Gamma_{\xi}, \tag{3.111}$$

where ξ labels a state in the final nucleus of the respective channel. Figure 3.5 shows the energy scheme and the contributing transitions in each channel.

Note that the width of the entrance channel in the numerator of Equation (3.108) is a partial width, whereas the total width includes a sum over all possible transitions to ground state and excited states also of nucleus A. Except for nuclei with long-lived isomeric states, for laboratory cross sections $\Gamma^{\mathrm{Aa}} = \Gamma^{\xi=0}$, with $\xi = 0$ denoting the ground state of nucleus A. When calculating astrophysical reaction rates, frequently also reactions on excited states of the target nucleus have to be considered. This is explained in detail in Section 4.1.3.

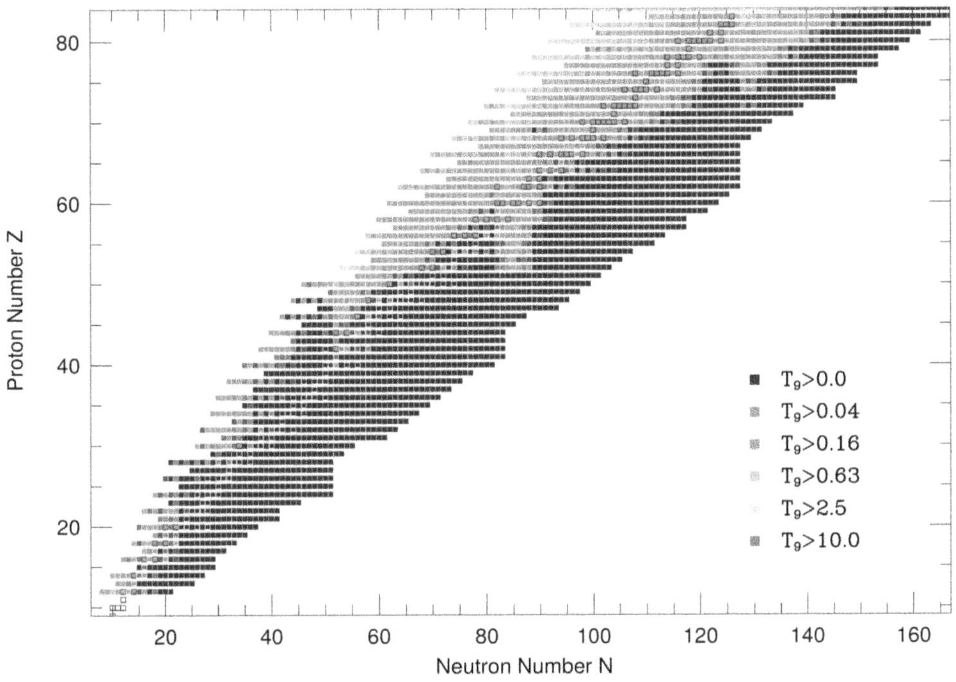

Figure 3.8. Estimated applicability of the statistical model for α-particle induced reactions. For each nucleus in the chart the color shade labels the temperature above which the statistical model becomes applicable to calculate the astrophysical reaction rate (T_9 is in GK). (Reprinted with permission from Rauscher et al. 1997. Copyright (1997) by the American Physical Society.)

The partial widths are related to the phase shift by the nuclear and Coulomb potentials as indicated in Equation (3.104). The phase shifts, in turn, are determined in the optical model by matching the solutions of the Schrödinger equation at the edge of the nucleus and far away from the nucleus, where the nuclear potential does not contribute to the phase shift anymore. Since the Coulomb potential for charged particles also has to be considered far away from the nucleus and the wave function at $r \to \infty$ will be a superposition of the *regular Coulomb wave function* $\mathcal{J}_\ell$ and *irregular Coulomb wave function* $\hat{\mathcal{J}}_\ell$, which are the solutions of a wave equation only including the Coulomb potential. Outgoing and incoming spherical waves are then given by

$$\begin{aligned} u_\ell^+(r) &= \mathrm{e}^{-\mathrm{i}\delta_\ell^C}\,[\hat{\mathcal{J}}_\ell(\eta_S, \mathcal{K}r) + \mathrm{i}\mathcal{J}_\ell(\eta_S, \mathcal{K}r)], \\ u_\ell^-(r) &= \mathrm{e}^{\mathrm{i}\delta_\ell^C}\,[\hat{\mathcal{J}}_\ell(\eta_S, \mathcal{K}r) - \mathrm{i}\mathcal{J}_\ell(\eta_S, \mathcal{K}r)], \end{aligned} \tag{3.112}$$

which asymptotically behave as shown in Equation (3.32). Using the Coulomb wave functions a *penetration factor* $\mathcal{O}_\ell$, a *shift factor* $\mathcal{I}_\ell$, and a *matching factor* f_ℓ^{match} can be defined:

$$\mathcal{O}_\ell = \left.\frac{\mathcal{K}r}{\mathcal{J}_\ell^2 + \hat{\mathcal{J}}_\ell^2}\right|_{r=R_{\text{nuc}}},$$

$$\mathcal{I}_\ell = -b_\ell + r\left.\frac{\mathcal{J}_\ell \mathrm{d}\mathcal{J}_\ell/\mathrm{d}r + \hat{\mathcal{J}}_\ell \mathrm{d}\hat{\mathcal{J}}_\ell/\mathrm{d}r}{\mathcal{J}_\ell^2 + \hat{\mathcal{J}}_\ell^2}\right|_{r=R_{\text{nuc}}}, \tag{3.113}$$

$$f_\ell^{\text{match}} = \left[\left.\frac{r}{u_\ell^+}\frac{\mathrm{d}u_\ell^+}{\mathrm{d}r}\right|_{r=R_{\text{nuc}}} - b_\ell\right]^{-1} = \mathcal{I}_\ell + \mathrm{i}\mathcal{O}_\ell.$$

All these quantities are determined by the conditions outside the nucleus. The matching coefficient b_ℓ can be set to any convenient value. A frequently used choice is $b_\ell = 0$. Another possibility is to choose b_ℓ so that $\mathcal{I}_\ell(R_{\text{nuc}}) = 0$.

Making use of the above quantities the transmission coefficient (see also Equation (3.92)) is given by

$$T_\ell = 4\mathcal{O}_\ell \text{Imag}\left(f_\ell^{\text{match}}\right)\Big\{\left[1 - \mathcal{I}_\ell \text{Real}\left(f_\ell^{\text{match}}\right) + \mathcal{O}_\ell \text{Imag}\left(f_\ell^{\text{match}}\right)\right]^2 + \left[\mathcal{O}_\ell \text{Real}\left(f_\ell^{\text{match}}\right) + \mathcal{I}_\ell \text{Imag}\left(f_\ell^{\text{match}}\right)\right]^2\Big\}^{-1}. \tag{3.114}$$

Real and imaginary parts of f_ℓ^{match} are appearing in the above equation and it becomes obvious that $T_\ell > 0$ can only be achieved with $\text{Imag}(f_\ell^{\text{match}}) \neq 0$. This implies complex wave functions generated by a complex (optical) potential for the nuclear interaction. If the real part is negligible T_ℓ can be approximated by

$$T_\ell = \frac{\hat{\tau}_\ell}{[1 + \hat{\tau}_\ell/4]^2} \approx 1 - \mathrm{e}^{-\hat{\tau}_\ell} \tag{3.115}$$

with $\hat{\tau}_\ell = 4\pi\mathcal{O}_\ell\hat{s}_\ell < 1$. This implies $T_\ell \approx \hat{\tau}_\ell$ for very small $\hat{\tau}_\ell$. The quantity $\hat{s}_\ell$ is called *nuclear strength function*.

Also partial widths can be expressed by the quantities defined in Equation (3.113),

$$\Gamma_\ell(E) = 2\mathcal{O}_\ell(E)\left[\frac{\mathrm{d}f_\ell^{\text{match}}}{\mathrm{d}E}\right]^{-1} = 2\mathcal{O}_\ell(E)\hat{\gamma}_\ell^2(E) \tag{3.116}$$

with

$$\Gamma = \sum_\ell \Gamma_\ell, \tag{3.117}$$

where the sum runs over all quantum mechanically allowed orbital angular momentum quantum numbers (see Section 1.7). This introduces the *reduced width* $\hat{\gamma}_\ell$. Note that for charged particles the penetration factor $\mathcal{O}_\ell$ is strongly energy dependent and quickly approaches zero with decreasing energy. The reduced width, on the other hand, depends on the potential in the interior of the nucleus and is assumed to have a weak dependence on projectile energy. Also the *observed resonance energy* E_{res} depends on $\hat{\gamma}_\ell$ and on the shift factor $\mathcal{I}_\ell$ because it is shifted with respect to the "true" resonance energy $E^{\text{s.p.}}$ (energy of the single particle state as generated by the potential in the interior of the nucleus) by

$$E_{\text{res}} = E^{\text{s.p.}} - [\mathcal{I}_\ell(E) - b_\ell]\hat{\gamma}_\ell^2. \tag{3.118}$$

With only one open channel, the matching coefficient b_ℓ can be chosen so that $E_{\text{res}} = E^{\text{s.p.}}$.

Multiple resonances with many possible reaction channels are usually treated in the *R-matrix approach*. The *R-matrix* is defined as

$$\mathcal{R}_{c'c}(E) = \sum_i \frac{\hat{\gamma}_{c,i}\hat{\gamma}_{c',i}}{E_i^{\text{s.p.}} - E}, \tag{3.119}$$

with c and c' being two different reaction channels into which the resonances can decay. The channel notation is to be understood as including all spins and partial waves allowed by the selection rules. Defined in this way, the R-matrix relates the wave function in the interior of the nucleus to its derivative at each channel entrance.

Within the R-matrix approach, the standard Lorentzian (Breit–Wigner) form without resonance energy shift, as given in Equation (3.108), can be recovered when redefining the widths to be inserted in Equation (3.108). The widths appearing in Equation (3.108) have to be replaced by *observed widths*. Their parameters (energy dependence, width) can be obtained from fits to experimental data. Calculations performed in R-matrix theory, on the other hand, will provide widths Γ related to the observed parameters as follows:

$$\Gamma_{i,c}^{\text{obs}} = \frac{\Gamma_{i,c}}{1 + \sum_{c''}\hat{\gamma}_{c''}^2 \mathrm{d}\mathcal{I}_{c''}/\mathrm{d}E}. \tag{3.120}$$

The summed partial widths $\hat{\Gamma}$ and the total width $\hat{\Gamma}_{\text{tot}}$ are then computed from these observed widths. In order to keep the connection between partial widths and reduced widths as shown in Equation (3.116), observed reduced widths $\hat{\gamma}^{\text{obs}}$ have to be introduced, which are related to the calculated reduced widths by

$$\hat{\gamma}_{i,c}^{\text{obs}} = \frac{\hat{\gamma}_{i,c}}{1 + \sum_{c''}\hat{\gamma}_{c''}^2 \mathrm{d}\mathcal{I}_{c''}/\mathrm{d}E}. \tag{3.121}$$

The $\hat{\gamma}^{\text{obs}}$ would then be used in Equation (3.116) to obtain the observed partial widths via the penetration factor. Observed widths, derived from experiment or converted from calculated widths, also have to be used in the calculation of resonant reaction rates (Section 4.1.2.3).

So far, the treatment accounted for shape resonances caused by the optical potential. As mentioned before, most of the observed resonances are narrower than anticipated for shape-elastic resonances. They are due to excitations of several nucleons or nucleon groups, as already mentioned at the beginning of Section 3.7.1. Modeling of such resonances is difficult as the individual forces between nucleons have to be accounted for and Equation (3.15) for a multi-particle quantum system has to be solved to obtain the energies of such excited states and their resonance properties. In a phenomenological, optical model approach the notion of partial and reduced widths is kept but they are modified by taking into account the probability that a certain nucleon configuration appears. A partial width for a compound resonance can be expressed as

$$\Gamma_\ell = \frac{2\hbar^2}{mR_{\rm nuc}^2}\mathcal{O}_\ell\left(\mathcal{X}^{I^{\rm a}_{\rm iso}I^{\rm a}_3 I^{\rm A}_{\rm iso}I^{\rm A}_3}_{I^{\rm C}_{\rm iso}I^{\rm C}_3}\hat{\gamma}_\ell^{\rm s.p.}\right)^2\mathcal{S}_\ell, \tag{3.122}$$

using the penetration factor $\mathcal{O}_\ell$, and a product of a squared isospin Clebsch–Gordan coefficient $\mathcal{X}$ (see Equation (1.122)) and a quantity called *spectroscopic factor* $\mathcal{S}_\ell$. The Clebsch–Gordan coefficient describes isospin transitions allowed in the reaction $a + A \leftrightarrow C$. The spectroscopic factor is a measure to what extent the resonance state can be described as a single-particle state, composed of the product wave functions of the initial nucleus and the projectile (or final nucleus and ejectile). Sometimes also the product $\mathcal{X}^2\mathcal{S}$ is called spectroscopic factor.

Comparison to Equation (3.116) shows that the reduced width was replaced by

$$\hat{\gamma}_\ell^2 = \frac{\hbar^2}{mR_{\rm nuc}^2}\left(I^{\rm C}_{\rm iso}\hat{\gamma}_\ell^{\rm s.p.}\right)^2\mathcal{S}_\ell, \tag{3.123}$$

where the *dimensionless single-particle reduced width* is calculated from

$$\hat{\gamma}_\ell^{\rm s.p.2} = \frac{R_{\rm nuc}}{2}|u_\ell(R_{\rm nuc})|^2. \tag{3.124}$$

Thus, for the case of a particle emission from a resonance state the expression for the partial width shown in Equation (3.122) can be interpreted physically as the product of the probability that the emitted nucleon configuration is from a single-particle product state (spectroscopic factor), the probability that the emitted nucleon configuration is located at the edge of the nucleus (single-particle reduced width), and the probability that the nucleon configuration escapes the nucleus (penetration factor).

Spectroscopic factor and single-particle reduced width can either be calculated or experimentally determined. The single-particle reduced width depends on radial wave functions u_ℓ obtained with an optical potential and these can be adapted to reproduce experimentally determined scattering phase shifts. The spectroscopic factor $\mathcal{S}$ is usually obtained by a simple comparison of the calculated magnitude of the differential cross section to measurements. Direct reactions populating an excited state also probe the spectroscopic factor, see Section 3.7.2.1. For reaction channels involving neutrons or protons, direct (d,p) or (d,n) reactions at energies above the astrophysically relevant ones are used, thereby avoiding the problem of low reaction cross sections at astrophysical energies.

In the absence of experimental data, the spectroscopic factor can be calculated microscopically from the overlap between initial and final state wave functions, for example, in the shell model. There is some ambiguity, however, because this overlap is not consistently defined in different microscopic approaches. It has been shown that the spectroscopic factors for depositing or picking up a single nucleon are closely related to the occupation factors of the participating quasi-particle states (see Equation (3.43)).

3.7.1.2 Resonant Reactions at Large Compound Level Density

As discussed at the beginning of Section 3.7.1, the number of excited states increases drastically with increasing excitation energy and this leads to regions of unresolved resonances in the excitation functions. When the average level spacing $D^{\mathrm{lev}} = 1/\tilde{\rho}_{\mathrm{nuc}}$ is larger than the average resonance width at E_{C}, then a statistical model can be applied to calculate the reaction cross section, making use of average level properties. In fact, the majority of reactions included in astrophysical reaction networks can be treated with this approach, as shown in Figures 3.6–3.8.

A cross section for elastic scattering (Equation (3.76)) and a reaction cross section (Equation (3.80)) can be defined in the optical model. The particular type of reaction was not specified in the definition of the reaction cross section. It is worthwhile, nevertheless, to further consider the possible reaction channels. In a compound reaction, a particle can be re-emitted into the scattering channel. This is why the total width in Equation (3.108) also includes transitions back into the entrance channel. When the particle is re-emitted with the same energy as the elastically scattered projectile, this is called *compound-elastic scattering*. Elastic scattering on the optical potential itself without formation of a compound nucleus is called *shape-elastic scattering*. The elastic scattering cross section σ_{el}, the one for reactions σ_{reac}, and the total cross section σ_{tot} in the optical model actually are related by

$$\begin{aligned}
\sigma_{\mathrm{el}} &= \sigma_{\mathrm{el}}^{\mathrm{pot}} + \sigma_{\mathrm{el}}^{\mathrm{comp}}, \\
\sigma_{\mathrm{reac}} &= \sum_x {}^x\sigma_{\mathrm{reac}} - \sigma_{\mathrm{el}}^{\mathrm{comp}} = \sigma_{\mathrm{abs}} - \sigma_{\mathrm{el}}^{\mathrm{comp}}, \\
\sigma_{\mathrm{tot}} &= \sigma_{\mathrm{el}}^{\mathrm{pot}} + \sum_x {}^x\sigma_{\mathrm{reac}} = \sigma_{\mathrm{el}}^{\mathrm{pot}} + \sigma_{\mathrm{abs}},
\end{aligned} \tag{3.125}$$

where the cross sections for shape-elastic scattering $\sigma_{\mathrm{el}}^{\mathrm{pot}}$ on the optical potential and compound-elastic scattering $\sigma_{\mathrm{el}}^{\mathrm{comp}}$ are introduced. The sums are to be understood as *coherent* sums, that means contributions for a given ℓ are combined before the sum over ℓ is performed. The true *absorption cross section* $\sigma_{\mathrm{abs}} = \sum_x {}^x\sigma_{\mathrm{reac}}$ includes all non-elastic processes, that is all reaction types x regardless of whether they re-introduce flux into the elastic channel or not. Note that it is different from σ_{reac} whenever a reaction process x populates the elastic channel. Figure 3.9 provides an overview of the connection of the different cross sections discussed above. Note that

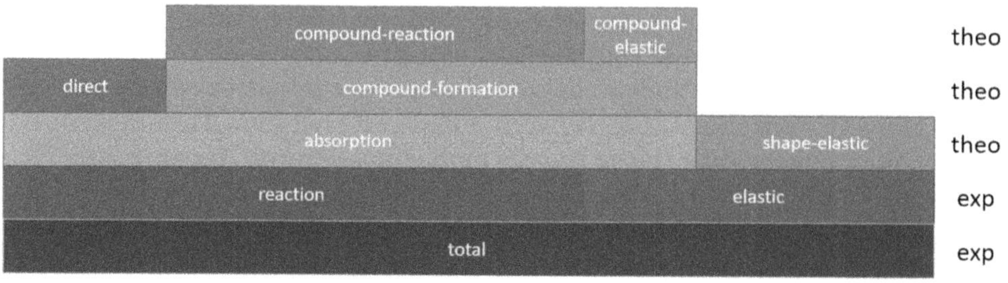

Figure 3.9. Hierarchy of cross sections: The bottom two lines include cross sections that can be measured whereas the distribution of the total cross section among different reaction mechanisms is only modeled with reaction theory. A measurement of angular distributions may help to distinguish direct and compound reactions.

only the total, elastic, and reaction cross section are measurable whereas the other cross sections appear in a theoretical modeling of reaction mechanisms.

Often, one reaction mechanism is dominating by far and σ_{abs} can be identified with the cross section for that mechanism, for example the compound cross section σ_{BW} for resonant processes as discussed in Section 3.7.1.1 or the Hauser–Feshbach or direct reaction cross sections discussed further below. But this is not always the case as—depending on the nucleus and the projectile energy—there may be several mechanisms contributing to σ_{abs}. In order to treat such a case self-consistently, a *coupled-channels approach* has to be applied, accounting for the interplay between different reaction channels and mechanisms. This is mathematically challenging and introduces additional parameters. Therefore it is not always feasible and most of low-energy reactions have been treated by assuming a single reaction mechanism or by incoherently adding cross sections of different reaction mechanisms. In many cases, this is completely justified.

Nevertheless, using an optical potential derived from scattering for reaction processes requires a cautious approach. First of all, it has to be realized that a measured σ_{el} may contain a compound reaction contribution as shown in the first line of Equation (3.125). Second, if only the compound mechanism contributes, the probability for this channel, that is of forming a compound nucleus, is not related to σ_{reac} as obtained from Equation (3.80) but to σ_{abs} as defined above. Only if the compound-elastic contribution is small, $\sigma_{\mathrm{abs}} \approx \sigma_{\mathrm{reac}}$. This is the case for highly charged particles, as their emission is suppressed by the Coulomb barrier, but not always for neutrons. Finally, it is important to note another impact of the fact that the absorption cross section may include contributions from several reaction mechanisms. While the experimental elastic cross section may be reproduced well with a given imaginary part of the optical potential, using the same imaginary part for a given reaction mechanism, for example, the statistical model or DWBA, may not be advisable because it would assume that all reactions proceed only through that mechanism. Instead, different imaginary parts or a coupled-channels approach should be used. An experimental determination of differential cross sections may help to distinguish the contributing reaction mechanisms, as these predict different angular distributions of the reaction products. Direct reactions have larger differential cross sections in the forward direction whereas the differential cross section for compound processes is symmetric with respect to the beam direction.

The compound-elastic contribution appears twice in the relations shown in Equation (3.125). How to determine the absorption cross section in the case of pure compound reactions? In the regime of narrow, overlapping, and experimentally unresolved resonances, it is sufficient to introduce an averaged cross section. Starting from Equation (3.76) such an averaged elastic cross section is given by

$$\begin{aligned}\langle\sigma_{\mathrm{el}}\rangle &= \frac{\pi}{\mathcal{K}^2}\sum_\ell(2\ell+1)\langle|1-S_\ell^{\alpha\alpha}|^2\rangle \\ &= \frac{\pi}{\mathcal{K}^2}\sum_\ell(2\ell+1)\left[1-2\mathrm{Real}(\langle S_\ell^{\alpha\alpha}\rangle)+\langle|S_\ell^{\alpha\alpha}|^2\rangle\right] \\ &= \frac{\pi}{\mathcal{K}^2}\sum_\ell(2\ell+1)\left[|1-\langle S_\ell^{\alpha\alpha}\rangle|^2+\langle|S_\ell^{\alpha\alpha}|^2\rangle-|\langle S_\ell^{\alpha\alpha}\rangle|^2\right],\end{aligned} \tag{3.126}$$

and therefore

$$\begin{aligned}\langle\sigma_{\mathrm{el}}^{\mathrm{pot}}\rangle &= \frac{\pi}{\mathcal{K}^2}\sum_\ell(2\ell+1)|1-\langle S_\ell^{\alpha\alpha}\rangle|^2, \\ \langle\sigma_{\mathrm{el}}^{\mathrm{comp}}\rangle &= \frac{\pi}{\mathcal{K}^2}\sum_\ell(2\ell+1)\left[\langle|S_\ell^{\alpha\alpha}|^2\rangle-|\langle S_\ell^{\alpha\alpha}\rangle|^2\right].\end{aligned} \tag{3.127}$$

The angle brackets denote an average over energy intervals smaller than the widths of the shape-elastic resonances but larger than the spacing of the compound resonances. Likewise, the average reaction cross section is written as

$$\begin{aligned}\langle\sigma_{\mathrm{reac}}\rangle &= \frac{\pi}{\mathcal{K}^2}\sum_\ell(2\ell+1)\langle1-|S_\ell^{\alpha\alpha}|^2\rangle \\ &= \frac{\pi}{\mathcal{K}^2}\sum_\ell(2\ell+1)\left[1-\langle|S_\ell^{\alpha\alpha}|^2\rangle\right].\end{aligned} \tag{3.128}$$

This yields the average absorption cross section as

$$\begin{aligned}\langle\sigma_{\mathrm{abs}}\rangle &= \langle\sigma_{\mathrm{el}}^{\mathrm{comp}}\rangle+\langle\sigma_{\mathrm{reac}}\rangle \\ &= \frac{\pi}{\mathcal{K}^2}\sum_\ell(2\ell+1)\left[\langle|S_\ell^{\alpha\alpha}|^2\rangle-|\langle S_\ell^{\alpha\alpha}\rangle|^2+1-\langle|S_\ell^{\alpha\alpha}|^2\rangle\right] \\ &= \frac{\pi}{\mathcal{K}^2}\sum_\ell(2\ell+1)\left[1-|\langle S_\ell^{\alpha\alpha}\rangle|^2\right] \\ &= \frac{\pi}{\mathcal{K}^2}\sum_\ell(2\ell+1)T_\ell.\end{aligned} \tag{3.129}$$

Comparison with Equation (3.92) leads to the last line, where the transmission coefficient is meant as the transmission arising from a mean (averaged) interaction potential.

The *statistical model of compound reactions* was initially developed by Bohr, who conceived the *independence hypothesis*. It states that the projectile forms a compound system with the target, shares its energy among all of the nucleons, and finally the compound nucleus decays by emitting photons or particles independently of the formation process. This implicitly requires long reaction timescales as the compound nucleus has to live long enough to establish complete statistical equilibrium among the nucleons. In the independence hypothesis, the (laboratory) cross section can be factorized into two terms,

$$\sigma_{\mathrm{Bohr}}^{\mathrm{Aa}\to\mathrm{Bb}} = \sigma_{\mathrm{abs}}^{\mathrm{Aa}} b_{\mathrm{dec}} = \sigma_{\mathrm{abs}}^{\mathrm{Aa}} \frac{\hat{T}^{\mathrm{Bb}}}{\mathcal{T}} = \sigma_{\mathrm{abs}}^{\mathrm{Aa}} \frac{\langle \hat{\Gamma}^{\mathrm{Bb}} \rangle}{\langle \hat{\Gamma}_{\mathrm{tot}} \rangle}, \tag{3.130}$$

the formation cross section $\sigma_{\mathrm{abs}}^{\mathrm{Aa}}$ and a branching ratio b_{dec} describing the probability for decay to the observed channel b + B. Note that this assumes the formation cross section being identical to the absorption cross section (see Figure 3.9), implying that there are no direct reactions. Since the absorption cross section is related to transmission coefficients as shown in Equation (3.129), also the transmission coefficients have to be related to an average over all (narrow) compound widths contributing to the cross section. Since the number of resonances per energy is given by the nuclear level density $\tilde{\rho}_{\mathrm{nuc}}$, the relation is

$$\begin{aligned} T_\ell &= 2\pi \tilde{\rho}_{\mathrm{nuc}} \langle \Gamma_\ell \rangle, \\ \hat{T} &= 2\pi \tilde{\rho}_{\mathrm{nuc}} \langle \hat{\Gamma} \rangle, \\ \mathcal{T} &= 2\pi \tilde{\rho}_{\mathrm{nuc}} \langle \hat{\Gamma}_{\mathrm{tot}} \rangle. \end{aligned} \tag{3.131}$$

The above transmission coefficients are defined in complete analogy to the definition of total and partial widths given in Equations (3.110) and (3.111), summing over individual transitions and partial waves, respectively. This is further elaborated below.

An early implementation of Bohr's independence hypothesis was the Weisskopf–Ewing theory. Since then, the Hauser–Feshbach approach has been widely used, which also incorporates conservation of angular momentum partially lifting the independence assumption but thus being more realistic. Nevertheless, although too simplified, Equation (3.130) is sometimes useful when estimating the relative feeding of different reaction channels.

The reaction cross section of the reaction a + A → C → B + b (proceeding via compound nucleus C) in the *Hauser–Feshbach model* can be written as

$$\sigma_{\mathrm{HF}}^{\mathrm{Aa}\to\mathrm{Bb}}(E_{\mathrm{Aa}}) = \frac{\pi}{\mathcal{K}_{\mathrm{Aa}}^2} \frac{1 + \tilde{\delta}_{\mathrm{Aa}}}{g_{\mathrm{A}} g_{\mathrm{a}}} \sum_{J\pi_{\mathrm{q}} \ell j j' \ell'} g_J \frac{T_{J\pi_{\mathrm{q}} \ell j}^{\mathrm{Aa}} \hat{T}_{J\pi_{\mathrm{q}} \ell' j'}^{\mathrm{Bb}}}{\mathcal{T}_{J\pi_{\mathrm{q}}}} W^{\mathrm{Aa}\to\mathrm{Bb}}. \tag{3.132}$$

The total transmission coefficient $\mathcal{T}$ in the denominator includes transitions to all channels accessible by particle or photon emission from the same compound nucleus C, not only (but including) a + A and B + b. The sums over channel spins j and partial waves ℓ are explicitly written to emphasize that the transmission coefficients must include these quantum numbers and also account for the quantum mechanical spin/parity selection rules. Each transmission coefficient includes transitions from a compound state at the compound energy E_C in nucleus C, with spin J and parity π_q. The statistical model assumes that many resonance states with all spins J and parities π_q are available at E_C, therefore the sum over all values. The transmission coefficients represent the average widths of the states with given J^{π_q} at E_C. Note that the nuclear level density at E_C does not enter the calculation. It is implicitly assumed that it is sufficiently high to allow application of this model and to use the sum over $J\pi_q$. While T^{Aa} only includes those transitions to the state in the target nucleus with which the projectile is interacting (in laboratory experiments this is usually the ground state or a long-lived isomeric state), the $\hat{T}$ include all transitions allowed by energetics and quantum selection rules. Note that this changes when dealing with reactions in an astrophysical plasma, where also transitions from and to excited states of nucleus A have to be considered (see Section 4.1.3).

It can be shown that the Hauser–Feshbach model results from averaging over a large number of Breit–Wigner resonances. Starting with the resonant cross section given in Equation (3.108), the sum of individual resonances can be replaced by an average over an energy interval ΔE using the mathematical relation

$$\begin{aligned}\left\langle \frac{\Gamma_i^{Aa}\hat{\Gamma}_i^{Bb}}{\left(E_{Aa} - E_{res,i}^{Aa}\right)^2 + \left(\hat{\Gamma}_i^{tot}/2\right)^2} \right\rangle &= \frac{1}{\Delta E}\int \frac{\Gamma_i^{Aa}\hat{\Gamma}_i^{Bb}}{\left(E_{Aa} - E_{res,i}^{Aa}\right)^2 + \left(\hat{\Gamma}_i^{tot}/2\right)^2}\, dE_{Aa} \\ &\approx \frac{2\pi}{\Delta E}\frac{\Gamma_i^{Aa}\hat{\Gamma}_i^{Bb}}{\hat{\Gamma}_i^{tot}}.\end{aligned} \tag{3.133}$$

Here, the angle brackets denote the average as defined by the above equation. Note that an approximation for narrow resonances, as shown in Equation (4.33) and also used in Equations (4.72) and (4.74), was applied to arrive at the last line of Equation (3.133). This is obviously allowed because of the assumption of a large number of narrowly spaced resonances. Then the sum over resonances in Equation (3.108) can be rewritten as

$$\begin{aligned}&\left\langle \sum_i g_i \frac{\Gamma_i^{Aa}\hat{\Gamma}_i^{Bb}}{\left(E_{Aa} - E_{res,i}^{Aa}\right)^2 + \left(\hat{\Gamma}_i^{tot}/2\right)^2} \right\rangle \\ &= \sum_{J\pi_q} g_J 2\pi \frac{\Delta n_{res}(J\pi_q)}{\Delta E} \left\langle \frac{\Gamma_{J\pi_q}^{Aa}\hat{\Gamma}_{J\pi_q}^{Bb}}{\hat{\Gamma}_{J\pi_q}^{tot}} \right\rangle \\ &= \sum_{J\pi_q} g_J 2\pi \tilde{\rho}_{nuc}(J, \pi_q) \frac{\left\langle \Gamma_{J\pi_q}^{Aa}\right\rangle \left\langle \hat{\Gamma}_{J\pi_q}^{Bb}\right\rangle}{\left\langle \hat{\Gamma}_{J\pi_q}^{tot}\right\rangle} \tilde{W}(J, \pi_q).\end{aligned} \tag{3.134}$$

The number of resonances per energy interval $\Delta n_{\mathrm{res}}/\Delta E$ was replaced by $\tilde{\rho}_{\mathrm{nuc}}$ in the last line. The averaged widths, the nuclear level density, and the $\tilde{W}$ are energy-dependent, of course. The *width fluctuation coefficients* $\tilde{W}$ account for the different averaging in the last line,

$$\tilde{W}(E, J, \pi_{\mathrm{q}}) = \left\langle \frac{\Gamma^{\mathrm{Aa}}_{J\pi_{\mathrm{q}}}(E)\hat{\Gamma}^{\mathrm{Bb}}_{J\pi_{\mathrm{q}}}(E)}{\hat{\Gamma}^{\mathrm{tot}}_{J\pi_{\mathrm{q}}}(E)} \right\rangle \frac{\left\langle \hat{\Gamma}^{\mathrm{tot}}_{J\pi_{\mathrm{q}}}(E) \right\rangle}{\left\langle \Gamma^{\mathrm{Aa}}_{J\pi_{\mathrm{q}}}(E) \right\rangle \left\langle \hat{\Gamma}^{\mathrm{Bb}}_{J\pi_{\mathrm{q}}}(E) \right\rangle}. \tag{3.135}$$

In terms of physics, they describe non-statistical correlations between the widths in the channels a + A and b + B. In practice, they differ from unity only close to channel openings. Noting the connection between averaged widths and transmission coefficients shown in Equation (3.131), a comparison of Equation (3.132) with Equations (3.108) and (4.33) readily shows that the statistical model cross section is an averaged Breit–Wigner cross section for narrow resonances, when $\tilde{W} = 1$.

Returning to Hauser–Feshbach cross section given in Equation (3.132), the total transmission coefficients $\hat{T}^z$ in each channel z = Aa, Bb,… include a sum over final states ξ in that channel. Due to the large number of final states at excitation energies approaching $E_{\mathrm{C}} - S^z_{\mathrm{sep}}$, with S^z_{sep} being the separation energy for channel z, the sum over discrete states has to be extended by an integration over a level density above the energy $E^{\mathrm{X}}_{\xi_{\mathrm{last}}}$ of the last discrete state included,

$$\begin{aligned}\hat{T}^z(E, J, \pi_{\mathrm{q}}) = &\left[\sum_{\xi} T^{\xi}\left(E, J, \pi_{\mathrm{q}}, E^{\xi}, J^{\xi}, \pi^{\xi}_{\mathrm{q}}\right)\right] \\ &+ \int_{E^{\mathrm{X}}_{\xi_{\mathrm{last}}}}^{E_{\mathrm{C}}} \sum_{J^z\pi^z_{\mathrm{q}}} T^z\left(E, J, \pi_{\mathrm{q}}, E^z, J^z, \pi^z_{\mathrm{q}}\right)\tilde{\rho}^z_{\mathrm{nuc}}\left(E^z, J^z, \pi^z_{\mathrm{q}}\right) \mathrm{d}E^z.\end{aligned} \tag{3.136}$$

The integration includes the nuclear level density $\tilde{\rho}^z_{\mathrm{nuc}}$ in the channel z, that is in the target nucleus A for channel a + A, in the final nucleus B for channel b + B, and so on. The transmission T^z is computed in the same manner as T^{ξ}, only that it is a transition to an artificial state with given $(E^z, J^z, \pi^z_{\mathrm{q}})$. In the integral these transitions are weighted by the nuclear level density $\tilde{\rho}^z_{\mathrm{nuc}}(E^z, J^z, \pi^z_{\mathrm{q}})$. The relative transition energy in channel z is $E^c_{\xi} = E_{\mathrm{C}} - S^z_{\mathrm{sep}} - E^{\mathrm{X}}_{\xi}$. The reader is advised to consult Figure 3.5 to get an overview of the included transitions and their relative energies.

Particle transmission coefficients have to obey spin selection rules (Section 1.7) and thus

$$T^{\xi}\left(E, J, \pi_{\mathrm{q}}, E^{\xi}, J^{\xi}, \pi^{\xi}_{\mathrm{q}}\right) = \sum_{\ell=|j-s|}^{j+s} \sum_{j=|J^{\xi}-J|}^{J^{\xi}+J} T_{j\ell}\left(E^c_{\xi}\right). \tag{3.137}$$

Again, the sums over the spin quantum numbers are specified explicitly. As before, the angular momentum $\vec{\ell}$ and the channel spin $\vec{j} = \vec{J} + \vec{J}^{\xi}$ are connected by

$\vec{j} = \vec{\ell} + \vec{s}$, including the particle spin s. Each $T_{j\ell}$ can be directly obtained from the solution of the (time-independent, radial) Schrödinger equation at the energy E_ξ^c with an appropriate optical potential, as shown in previous sections.

The calculation of radiative transmission coefficients proceeds equivalently to Equation (3.137) but electromagnetic selection rules (see Table 1.3) have to be obeyed. The photon transmission coefficients to be used on the right-hand side of Equation (3.137) are given by

$$T^{\gamma}_{X\ell}(E_\gamma) = 2\pi \vec{f}^{\gamma}_{X\ell}(E_\gamma) E_\gamma^{2\ell+1}. \tag{3.138}$$

The parities π_{q}, π_{q}^{ξ} and the angular momentum ℓ in the sums select the type of allowed electromagnetic transition ($X\ell$ = E1, E2, M1, M2,..., where X is E or M). To phenomenologically account for pre-equilibrium particle emission at higher compound excitation energy, the integration in Equation (3.136) can be carried out up to an energy taken as the minimum of E_{C} and an appropriately chosen E_{cut} (for example, the energy at which the γ transmission exceeds a certain fraction of the total transmission) for the γ transmission coefficient appearing in the numerator of Equation (3.132). The total transmission coefficient $\mathcal{T}_{J\pi}$ in the denominator, however, always has to include the full integration up to E_{C}.

At least the dominant γ-transitions (E1 and M1) have to be included in the calculation of the photon transmission coefficient for astrophysics. Among the collective modes of nuclei the electric dipole (E1) excitation has the special property that most of its strength is concentrated in the isovector giant dipole resonance (GDR). Macroscopically, this strong resonance is described as a vibration of the charged (proton) matter in the nucleus against the neutral matter (neutrons). The E1 GDR transmission coefficient can be parameterized as a split Lorentzian shape,

$$T^{\gamma}_{\mathrm{E1}}(E_\gamma) = \frac{8}{3}\frac{NZ}{A}\frac{e_{\mathrm{C}}^2}{\hbar c}\frac{1+\chi}{m_{\mathrm{p}}c^2}\sum_{i=1}^{2}\frac{i}{3}\frac{\Gamma_{\mathrm{GDR},i}E_\gamma^4}{\left(E_\gamma^2 - E_{\mathrm{GDR},i}^2\right)^2 + \Gamma_{\mathrm{GDR},i}^2 E_\gamma^2}. \tag{3.139}$$

The neutron–proton exchange contribution is accounted for by the term $\chi \approx 0.2$ and the summation over i includes two terms which correspond to the split of the GDR in statically deformed nuclei, with oscillations along ($i = 1$) and perpendicular ($i = 2$) to the axis of rotational symmetry. In this deformed case, the two resonance energies are related to the mean value calculated by the relations

$$\begin{aligned} E_{\mathrm{GDR},1} + 2E_{\mathrm{GDR},2} &= 3E_{\mathrm{GDR}}, \\ E_{\mathrm{GDR},2}/E_{\mathrm{GDR},1} &= 0.911\frac{D_{\parallel}}{D_{\perp}} + 0.089, \end{aligned} \tag{3.140}$$

using the ratio of the diameter along the nuclear symmetry axis $D_{\parallel}$ to the diameter perpendicular to it $D_{\perp}$, which is obtained from the experimentally known deformation or from mass model predictions (see Section 3.3.3). Many microscopic and macroscopic models have been devoted to the calculation of the GDR energies E_{GDR} and widths Γ_{GDR}. Different parameterizations are also available in literature.

Special attention is paid to the low-energy tail of the resonance, which has been found to have an energy-dependence different from a Lorentzian form. For example, a simple approach to improve the reproduction of low energy data is to introduce an energy-dependent width

$$\Gamma_{\mathrm{GDR}}(E_\gamma) = \Gamma^0_{\mathrm{GDR}}\sqrt{\frac{E_\gamma}{E_{\mathrm{GDR}}}}, \tag{3.141}$$

where $\Gamma^0_{\mathrm{GDR}} = \Gamma_{\mathrm{GDR}}(E_{\mathrm{GDR}})$. Low E_γ are also relevant for the calculation of cross sections at astrophysically relevant energies.

Photon M1 transitions strengths $\tilde{f}^{\gamma}_{\mathrm{M1}}$ usually are estimated by either assuming a constant (energy-independent) strength or a single Lorentzian with constant width. As for the E1 strength, microscopic calculations are also available, providing tabulated values for the strengths.

As can be inferred from Figure 3.5, the energies of emitted γ-rays are in the range $0 \leqslant E_\gamma \leqslant E_{\mathrm{C}}$. Since the strength of the γ-transition scales with some power of E_γ (see Equation (3.138)), γ-transitions with higher energies are favored. On the other hand, the number of available endpoints of the transitions increases with increasing excitation energy of the nucleus because the nuclear level density increases rapidly. This competition between transition strength and nuclear level density gives rise to a peak in the γ-emission energies as shown in Figure 3.10. This peak is fragmented when certain transitions to discrete excited states are dominating at low $\tilde{\rho}_{\mathrm{nuc}}$ (not shown in the figure). This is mainly the case far off stability for captures with low capture Q-value. It has been found that for astrophysically relevant projectile energies, the γ-energies with the strongest impact on the reaction cross sections and rates are between 2 and 4 MeV (unless the level density is so low that only few transitions are allowed). The γ-emission peak defines the range of γ-energies at which changes in the strength function have largest impact as well as the excitation energies at which the knowledge of $\tilde{\rho}_{\mathrm{nuc}}$ is most important for capture. This also holds for the reverse reaction (photodisintegration) under stellar conditions because of the weighting of transitions from excited states as given by Equation (4.102). In consequence, changes in the strength function around this energy have the largest impact. Unfortunately, such low energies cannot be probed directly by photodisintegration experiments because they are below the particle separation energies, at least close to stability. Such experiments would allow to study strength functions in the most direct way (but they cannot test the astrophysically relevant rates, either, see Section 4.1.5). Other types of experiments are complicated by the fact that the observables are generated by a convolution of different nuclear properties (such as the dependence on the nuclear level density or different spin selectivities of transitions) which have to be known and disentangled.

3.7.2 Direct Reactions

In its most general definition, the term "direct reaction" includes all processes directly connecting the initial and final states of a nuclear reaction without formation of an intermediate compound system. This includes elastic scattering as

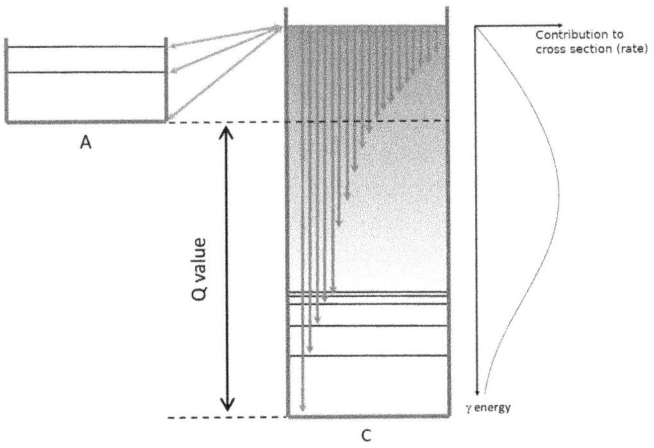

Figure 3.10. Sketch of particle transitions (brown diagonal arrows) and γ-transitions (green vertical arrows) contributing to the cross section or reaction rate of a reaction $a + A \rightarrow C + \gamma$, proceeding via a compound state. The increasing nuclear level density with increasing excitation energy of C is visualized by the color shade. Although γ-transitions with higher energy are stronger, there is a γ-emission peak at intermediate excitation energies because of the larger number of excited states at higher excitation.

described in the optical model, and inelastic scattering which predominantly excites collective states. The latter includes Coulomb excitation which has been found to be important in heavy ion collisions due to the high Coulomb barriers involved. In astrophysically relevant reactions, especially with α-particles, energies may also be close to or below the Coulomb barrier and Coulomb excitation may also become important, depending on the structure of the target nucleus.

Here, the focus is on direct reactions in which some (if it is a stripping reaction) or all (if it is a capture or charge-exchange reaction) nucleons of the projectile are incorporated in the target nucleus. In a pick-up reaction, one or more nucleons from the target nucleus are added to the projectile to form the ejectile, again in a direct manner. Pick-up and stripping reactions are subsumed under the term "transfer reactions." In contrast to the statistical model of compound reactions, direct reactions excite only few degrees of freedom because most of the nucleons included in the system of target nucleus plus projectile remain spectators. A nucleon of the projectile reaches its final state without sharing any energy with any of the other nucleons present and the excess energy is emitted as a discrete photon carrying the energy difference between initial and final state. Direct reactions can be identified experimentally because of their angular dependence of the differential cross sections, being peaked in forward direction. Direct processes are also faster by at least five orders of magnitude than compound reactions, with reaction timescales of the order of 10^{-22} s. This is comparable to the time the projectile requires to cover a distance of the size of a nucleus. Therefore, direct reactions are important at high projectile energies when compound formation is disfavored.

Although astrophysical energies are low by nuclear physics standards, direct reactions at higher energies can be used to extract certain properties, such as spin assignments and spectroscopic factors (see Equation (3.122) and below), of resonances in stable and unstable nuclei which are required for the calculation of cross sections and astrophysical reaction rates. For example, also low-lying neutron- or proton-states can be probed by direct (d,p) or (d,n) reactions, respectively.

Moreover, direct processes are also important at very low energy. In resonant reactions at lower energy, it is sometimes necessary to include a non-resonant background (which may show interference with resonances) but experimentally it is often difficult to distinguish between a direct component and contributions from tails of broad resonances. However, in systems with low nuclear level density, and thus widely spaced resonances, direct reactions become important even at astrophysically low interaction energies because compound formation is suppressed. This is most pronounced in reactions with light nuclei but also becomes important for intermediate and heavy nuclei far off stability, for example for neutron capture in nuclei with low neutron separation energy close to the neutron dripline (see also Figures 3.6–3.8). For such cases, for example, the direct capture cross section can become considerably larger than the compound cross section. Figure 3.11 compares (n,γ) cross sections at 30 keV calculated in a direct capture model and in the Hauser–Feshbach model for a sequence of Ti isotopes. The neutron separation energy is decreasing with increasing neutron number. With decreasing neutron separation energy, the direct component plays an increasingly important role because the compound nucleus is formed at lower excitation energy and thus also at lower level density, that means with fewer accessible resonances and final states. Similar considerations may apply to proton-induced reactions on proton-rich nuclei.

3.7.2.1 DWBA

Direct transfer reactions can be treated by solving the time-independent Schrödinger equation with optical potentials in the entrance and exit channels. A simple implementation of this is the Distorted Wave Born Approximation (DWBA). The differential cross section for the transfer of one nucleon or a cluster (denoted by "x") in the reaction a + A → b + B with a − x = b, A + x = B is given in zero-range DWBA by

$$\frac{d\sigma_{\mathrm{DWBA}}}{d\Omega} = \frac{\mu_{\mathrm{red}}^{\mathrm{Aa}}\mu_{\mathrm{red}}^{\mathrm{Bb}}}{(2\pi\hbar^2)^2}\frac{\mathcal{K}_{\mathrm{Bb}}g_{\mathrm{B}}}{\mathcal{K}_{\mathrm{Aa}}g_{\mathrm{A}}}\sum_{\ell sj}\mathcal{X}^2 S_{\ell j}N_0\frac{\sigma_{\ell sj}(\theta)}{2s+1}, \qquad (3.142)$$

with the zero-range normalization constant N_0. This is valid for projectiles with $A \leqslant 4$ and clusters with $A = 1$ or 3. The reduced cross section appearing in Equation (3.142) is given (without spin–orbit coupling) by

$$\sigma_{\ell sj}(\theta) = \sum_m \left| t_{\ell sj}^m \right|^2, \qquad (3.143)$$

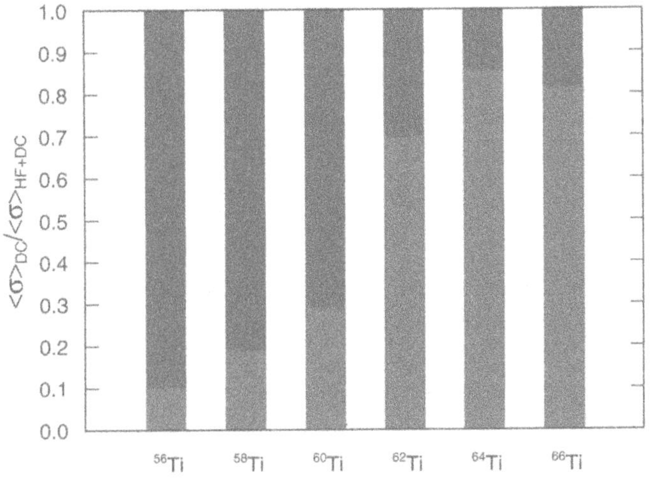

Figure 3.11. Relation between direct neutron capture and compound capture in the Hauser–Feshbach model for Ti isotopes.

with the reduced transition amplitude

$$t^{m}_{\ell sj} = \frac{1}{2\ell + 1} \int \chi^{(-)*}_{\rm Bb}\left(\mathcal{K}_{\rm Bb}, \frac{A_{\rm A}}{A_{\rm B}} r\right) u_{\ell j}(r)[i^{\ell}\mathcal{Y}_{\ell m}(\vec{r})]^{*} \chi^{(+)}_{\rm Aa}(\mathcal{K}_{\rm Aa}, r)\, {\rm d}r. \tag{3.144}$$

The orbital angular momentum quantum number ℓ, the spin quantum number s, and the total angular momentum quantum number j refer to the nucleon or cluster x bound in the residual nucleus B. The spectroscopic factor and the isospin Clebsch–Gordan coefficient for the partition B = A + x are given by $\mathcal{S}_{\ell j}$ and $\mathcal{X}$, respectively. The optical wavefunctions in the entrance and exit channels are given by $\chi^{(+)}$ and the time-reversed solution $\chi^{(-)}$. The bound state wave function is denoted by $u_{\ell j}$ and the $\mathcal{Y}_{\ell m}$ are the usual spherical harmonics. The zero-range approximation assumes a pointlike neutron–proton interaction. This simplifies the integrals to obtain the transition amplitude. Nevertheless, expressions similar to the above are obtained when a finite range interaction is taken into account.

The spectroscopic factor provides a measure of the probability to find the final nucleus B in a configuration given by the initial nucleus A + x. The wave functions are obtained, for example, from shell model wave functions, similar to the total wave function for a nucleus as given in Equation (3.12). Let $\Psi_{\rm B}(\vec{r}_1, \vec{r}_2, \vec{r}_3, \ldots, \vec{r}_{A-1}, \vec{r}_A)$ be the wave function for nucleus B with A nucleons and $\Psi_{\rm B-x}(\vec{r}_1, \vec{r}_2, \vec{r}_3, \ldots, \vec{r}_{A-1}, \vec{r}_{A-A_{\rm x}})$ the wave function for nucleus A = B − x. Likewise there is the wave function $\Psi_{\rm x}(\vec{r}_{A-A_{\rm x}}, \vec{r}_{A-A_{\rm x}+1}, \ldots, \vec{r}_A)$ for the nucleon cluster x. For a nucleus B′ composed of nucleus A and cluster x, the wave function can be separated into

$$\begin{aligned}\tilde{\Psi}_{\rm B'}(\vec{r}_1, \vec{r}_2, \vec{r}_3, \ldots, \vec{r}_{A-1}, \vec{r}_A) &= \hat{A}\,[U(\vec{R}_{\rm x})\Phi_{\rm x}(\vec{x}_1, \vec{x}_2, \ldots, \vec{x}_{A_{\rm x}}) \\ &\quad \Psi_{\rm B-x}(\vec{r}_1, \vec{r}_2, \vec{r}_3, \ldots, \vec{r}_{A-1}, \vec{r}_{A-A_{\rm x}})].\end{aligned} \tag{3.145}$$

The mass number of cluster x is denoted by A_x. The cluster wave function Ψ_x has been separated into a wave function Φ_x expressed with internal coordinates $\vec{x}_1, \vec{x}_2, \ldots$ of the cluster and a radial function $U(\vec{R}_x)$ describing the cluster motion inside the nucleus. The radial function is given in center-of-mass coordinates

$$\vec{R}_x = \sum_{i=A-A_x}^{A} \frac{\vec{r}_i}{A}. \tag{3.146}$$

The spectroscopic factor with respect to the cluster x then is given by the overlap wave function between nuclei B and B′,

$$\mathcal{S}_x = \left| \int \tilde{\Psi}^*_{B''} \Psi_B \, d^3r \right|^2. \tag{3.147}$$

Note that the asterisk is used in this chapter to denote complex conjugation whereas in Chapter 4 it is used to label quantities with modifications in a stellar environment.

Important for the successful application is to keep the number of open parameters as small as possible. For this reason, folding potentials (Equation (3.88)) were used in many astrophysical applications of the model, with the renormalization λ_{fold} either determined from scattering data or from global dependences.

Is it necessary to go beyond the DWBA? There are three fundamental assumptions contained in the DWBA treatment:

1. The reaction proceeds directly from initial to final state and all particles except the transferred one(s) remain unaffected spectators.
2. The wave function for the relative motion between the reactands is assumed to be correctly described by the optical potential.
3. The reaction is assumed to be sufficiently weak to be treated in lowest order.

To relax the first two assumptions, the *coupled-channel Born approximation* was introduced. Under rare circumstances the transfer amplitudes may be large and the third assumption has to be relaxed. This leads to a full *coupled-channels treatment* for the reaction.

One has to be aware of the fact that the relevant energies remain low for astrophysical reaction rates, also because transitions from excited states contribute considerably. This is contrary to what one is used to in the investigation of reactions proceeding at several tens of MeV. Due to the low energies involved, the reaction channel is weak (compared to, e.g., elastic scattering) and the third assumption is valid. The usual concern with the second assumption is that the optical potential also has to describe well the wave function even deep in the nuclear interior. The deep region, however, is crucial for reactions at higher energy whereas at the low astrophysical energies most contributions to the overlap integrals stem from regions close to the surface of the nucleus or even from outside of the nuclear radius. As long as these regions are described well by the optical potentials, the DWBA should work. The first assumption implies that either no indirect processes exist or that they can be treated separately (incoherently) as was suggested by the above, separate discussion of compound reactions and other mechanisms. Interference with isolated resonances can be treated explicitly by adding an interference term, for example, between the S-factor of the direct reaction $\tilde{S}_{\text{direct}}$ and the one of a Breit–Wigner resonance $\tilde{S}_{\text{BW}}$ (see Section 3.7.1.1)

$$\tilde{S} = \tilde{S}_{\text{direct}} + \tilde{S}_{\text{BW}} - 2\left(\sqrt{\tilde{S}_{\text{direct}}\tilde{S}_{\text{BW}}}\right)\cos\delta_{\text{inter}}, \tag{3.148}$$

where

$$\delta_{\text{inter}} = \arctan\left(\frac{2(E - E_{\text{res}})}{\hat{\Gamma}_{\text{tot}}}\right) \tag{3.149}$$

is the energy-dependent, relative phase shift. (The astrophysical S-factor is defined in Equation (4.23).)

3.7.2.2 Direct Capture

A potential model can also be used to calculate direct capture (DC). Although microscopic models are an alternative for light systems (see Section 3.7.3), a DC potential model has the advantage that it can be applied also to heavier nuclei. The DC cross section for a particular transition is determined by the overlap of the scattering wave function in the entrance channel, the bound-state wave function in the exit channel, and the electromagnetic multipole transition operator.

The DC differential cross section for the reaction a + A → B + γ is given by

$$\frac{\mathrm{d}\sigma_{\text{DC}}}{\mathrm{d}\Omega} = 2\left(\frac{e_{\text{C}}^2\mu_{\text{red}}^{\text{Aa}}c^2}{\hbar c}\right)\left(\frac{\mathcal{K}_\gamma}{\mathcal{K}_{\text{Aa}}}\right)^3\frac{1}{g_{\text{A}}g_{\text{a}}}\sum_{M_{\text{A}}M_{\text{a}}M_{\text{B}}\hat{\sigma}}|t_{M_{\text{A}}M_{\text{a}}M_{\text{B}},\hat{\sigma}}|^2. \tag{3.150}$$

The polarization $\hat{\sigma}$ of the electromagnetic radiation can be ±1. The multipole expansion of the transition matrices $t_{M_{\text{A}}M_{\text{a}}M_{\text{B}},\hat{\sigma}}$ including electric dipole (E1) and quadrupole (E2) transitions as well as magnetic dipole (M1) transitions is given by

$$t_{M_{\text{A}}M_{\text{a}}M_{\text{B}},\hat{\sigma}} = t^{\text{E1}}_{M_{\text{A}}M_{\text{a}}M_{\text{B}},\hat{\sigma}}d^1_{\delta\hat{\sigma}}(\theta) + t^{\text{E2}}_{M_{\text{A}}M_{\text{a}}M_{\text{B}},\hat{\sigma}}d^2_{\delta\hat{\sigma}}(\theta) + t^{\text{M1}}_{M_{\text{A}}M_{\text{a}}M_{\text{B}},\hat{\sigma}}d^1_{\delta\hat{\sigma}}(\theta). \tag{3.151}$$

The rotation matrices $d_{\delta\hat{\sigma}}$ depend on the angle between $\vec{\mathcal{K}}_{\text{Aa}}$ and $\vec{\mathcal{K}}_\gamma$ which is denoted by θ, where $\delta = M_{\text{A}} + M_{\text{a}} - M_{\text{B}}$.

Defining

$$\begin{aligned} C(\text{E1}) &= \mathrm{i}\mu_{\text{red}}^{\text{Aa}}\left(\frac{Z_{\text{a}}}{m_{\text{a}}} - \frac{Z_{\text{A}}}{m_{\text{A}}}\right), \\ C(\text{E2}) &= \frac{\mathcal{K}_\gamma}{\sqrt{12}}\mu_{\text{red}}^{\text{Aa}2}\left(\frac{Z_{\text{a}}}{m_{\text{a}}^2} + \frac{Z_{\text{A}}}{m_{\text{A}}^2}\right), \end{aligned} \tag{3.152}$$

the transition matrices for the electric dipole ($\mathrm{E}\mathcal{M} = \mathrm{E1}$) and quadrupole ($\mathrm{E}\mathcal{M} = \mathrm{E2}$) transitions can be written as

$$
\begin{aligned}
t^{\mathrm{E}\mathcal{M}}_{M_\mathrm{A}M_\mathrm{a}M_\mathrm{B},\hat{\sigma}} = \sum_{\ell_\mathrm{a}j_\mathrm{a}} & \mathrm{i}^{\ell_\mathrm{a}}\left\{\ell_\mathrm{a}0\; s_\mathrm{a}M_\mathrm{a}\middle|j_\mathrm{a}\,M_\mathrm{a}\right\}\left\{j_\mathrm{b}\,M_\mathrm{B}(-M_\mathrm{A})I_\mathrm{A}M_\mathrm{A}\middle|I_\mathrm{B}M_\mathrm{B}\right\} \\
& \times\left\{\mathcal{M}\delta\; j_\mathrm{b}\,M_\mathrm{B}(-M_\mathrm{A})\middle|j_\mathrm{a}\,M_\mathrm{a}\right\}C(\mathrm{E}\mathcal{M})\,\hat{\ell}_\mathrm{a}\hat{\ell}_\mathrm{b}\hat{j}_\mathrm{b} \\
& \times\{\ell_\mathrm{b}0\;\mathcal{M}0|\ell_\mathrm{a}0\}\mathcal{W}_\mathrm{R}(\mathcal{M}\ell_\mathrm{b}j_\mathrm{a}s_\mathrm{a};\,\ell_\mathrm{a}j_\mathrm{b})I^{\mathrm{E}\mathcal{M}}_{\ell_\mathrm{b}j_\mathrm{b}I_\mathrm{B};\ell_\mathrm{a}j_\mathrm{a}}.
\end{aligned}
\tag{3.153}
$$

In the above expressions the quantum numbers for the channel spin in the entrance channel and for the transferred angular momentum are denoted by j_a and j_b, respectively. The quantities I_A, I_B and s_a (M_A, M_B and M_a) are the spins (magnetic quantum numbers) of the target nucleus A, residual nucleus B and projectile a, respectively.

For magnetic dipole transitions ($\mathrm{M}\mathcal{M} = \mathrm{M1}$) we obtain

$$
\begin{aligned}
t^{\mathrm{M}\mathcal{M}}_{M_\mathrm{A}M_\mathrm{a}M_\mathrm{B},\,\hat{\sigma}} = \sum_{\ell_\mathrm{a}j_\mathrm{a}} & \mathrm{i}^{\ell_\mathrm{a}}\hat{\sigma}\left\{\left\{\ell_\mathrm{a}0\; s_\mathrm{a}\; M_\mathrm{a}\middle|j_\mathrm{a}\,M_\mathrm{a}\right\}\left\{j_\mathrm{b}\,M_\mathrm{B}(-M_\mathrm{A})I_\mathrm{A}M_\mathrm{A}\middle|I_\mathrm{B}M_\mathrm{B}\right\}\right. \\
& \times\left\{1\delta\, j_\mathrm{b}\,M_\mathrm{B}(-M_\mathrm{A})\middle|j_\mathrm{a}M_\mathrm{a}\right\} \\
& \times\left[\mu^{\mathrm{Aa}}_{\mathrm{red}}\left(\frac{Z_\mathrm{A}}{m_\mathrm{A}^2}+\frac{Z_\mathrm{a}}{m_\mathrm{a}^2}\right)\hat{\ell}_\mathrm{b}\,\hat{j}_\mathrm{b}\sqrt{\ell_\mathrm{a}(\ell_\mathrm{a}+1)}\;\mathcal{W}_\mathrm{R}(1\ell_\mathrm{a}j_\mathrm{a}s_\mathrm{a};\,\ell_\mathrm{a}j_\mathrm{b})\right. \\
& \left.+\,2\mu_\mathrm{a}^{\mathrm{mag}}(-1)^{j_\mathrm{b}-j_\mathrm{a}}\,\hat{s}_\mathrm{a}\hat{j}_\mathrm{b}\sqrt{s_\mathrm{a}(s_\mathrm{a}+1)}\;\mathcal{W}_\mathrm{R}(1s_\mathrm{a}j_\mathrm{a}\ell_\mathrm{a};\,s_\mathrm{a}j_\mathrm{b})\right] \\
& -\left\{\ell_\mathrm{a}0s_\mathrm{a}M_\mathrm{a}\middle|j_\mathrm{a}\,M_\mathrm{a}\right\}\left\{j_\mathrm{a}\,M_\mathrm{a}I_\mathrm{A}M_\mathrm{B}(-M_\mathrm{a})\middle|I_\mathrm{B}M_\mathrm{B}\right\} \\
& \times\{I_\mathrm{A}M_\mathrm{B}(-M_\mathrm{a})1\delta|I_\mathrm{A}M_\mathrm{A}\} \\
& \left.\times\,\mu_\mathrm{A}^{\mathrm{mag}}\delta_{j_\mathrm{a}j_\mathrm{b}}\sqrt{(I_\mathrm{A}+1)/I_\mathrm{A}}\right\}\left\{\frac{\hbar c}{2m_\mathrm{p}c^2}\right\}\delta_{\ell_\mathrm{a}\ell_\mathrm{b}}\,\hat{\ell}_\mathrm{a}I^{\mathrm{M1}}_{\ell_\mathrm{b}j_\mathrm{b}I_\mathrm{B};\ell_\mathrm{a}j_\mathrm{a}},
\end{aligned}
\tag{3.154}
$$

where the $\{\cdots|\cdots\}$ denote Clebsch–Gordan coefficients (Equation (1.122)), $\mathcal{W}_\mathrm{R}$ are Racah coefficients (Equation (1.125)), and the μ^{mag} are the magnetic moments.

The overlap integrals in Equations (3.153) and (3.154) are given by

$$
I^{\mathrm{E}\mathcal{M}}_{\ell_\mathrm{b}j_\mathrm{b}I_\mathrm{B};\ell_\mathrm{a}j_\mathrm{a}} = \int \mathrm{d}r\; u_{NLJ}(r)\;\tilde{\mathcal{O}}^{\mathrm{E}\mathcal{M}}(r)\;\chi_{\ell_\mathrm{a}j_\mathrm{a}}(r) \tag{3.155}
$$

for the electric dipole (E1) or quadrupole (E2) transition, and by

$$
I^{\mathrm{M1}}_{\ell_\mathrm{b}j_\mathrm{b}I_\mathrm{B};\ell_\mathrm{a}j_\mathrm{a}} = \int \mathrm{d}r\; u_{NLJ}(r)\;\tilde{\mathcal{O}}^{\mathrm{M1}}(r)\;\chi_{\ell_\mathrm{a}j_\mathrm{a}}(r) \tag{3.156}
$$

for the magnetic dipole transition (M1).

The radial part of the bound state wave function in the exit channel and the scattering wave function in the entrance channel are given by $u_{NLJ}(r)$ and $\chi_{\ell_a j_a}(r)$, respectively. The radial parts of the electromagnetic multipole operators are

$$\begin{aligned}
\tilde{\mathcal{O}}^{\mathrm{M1}}(r) &= \frac{1}{2\mathcal{K}_\gamma r}\left[\sin(\mathcal{K}_\gamma r) + \mathcal{K}_\gamma r\cos(\mathcal{K}_\gamma r)\right],\\
\tilde{\mathcal{O}}^{\mathrm{E1}}(r) &= \frac{3}{(\mathcal{K}_\gamma r)^3}\left[\left(\mathcal{K}_\gamma^2 r^2 - 2\right)\sin(\mathcal{K}_\gamma r) + 2\mathcal{K}_\gamma r\cos(\mathcal{K}_\gamma r)\right]r,\\
\tilde{\mathcal{O}}^{\mathrm{E2}}(r) &= \frac{15}{(\mathcal{K}_\gamma r)^5}\left[\left(5\mathcal{K}_\gamma^2 r^2 - 12\right)\sin(\mathcal{K}_\gamma r) + (12 - (\mathcal{K}_\gamma r)^2)\mathcal{K}_\gamma r\cos(\mathcal{K}_\gamma r)\right]r^2.
\end{aligned} \tag{3.157}$$

In the long wave-length approximation—applicable for $\mathcal{K}_\gamma r \ll 1$—these quantities reduce to

$$\begin{aligned}
\tilde{\mathcal{O}}^{\mathrm{M1}}(r) &\simeq 1,\\
\tilde{\mathcal{O}}^{\mathrm{E1}}(r) &\simeq r,\\
\tilde{\mathcal{O}}^{\mathrm{E2}}(r) &\simeq r^2.
\end{aligned} \tag{3.158}$$

Mostly, only the dominant E1 transitions have to be taken into account. Possible exceptions are captures far from stability with very low reaction Q-value because for these cases no final states may be energetically accessible through E1 transitions. However, because the astrophysical reaction rate involves summing over transitions originating from excited states, a larger spin range may be available and E1 (from excited target states) may again dominate in the calculation of a rate. For E1 transitions, the above expressions reduce to

$$\begin{aligned}
\sigma_{\mathrm{DC}}^{\mathrm{E1}} = &\frac{16\pi}{9}\left(\frac{E_\gamma \mu_{\mathrm{red}}^{\mathrm{Aa}}}{\mathcal{K}_{\mathrm{Aa}}\hbar c}\right)^3\left(\frac{e_{\mathrm{C}}}{\hbar}\right)^2\frac{3}{g_{\mathrm{a}} g_{\mathrm{A}}}\left(\frac{Z_{\mathrm{a}}}{m_{\mathrm{a}}} - \frac{Z_{\mathrm{A}}}{m_{\mathrm{A}}}\right)^2 \mathcal{X}^2 S_{\ell_\beta J_\beta}\\
&\times \sum_{\ell_\alpha J_\alpha}(2J_\beta + 1)(2J_\alpha + 1)\max(\ell_\alpha, \ell_\beta)\\
&\times \begin{Bmatrix} 1 & \ell_\beta & \ell_\alpha \\ I & J_\alpha & J_\beta \end{Bmatrix}_6^2 a_I^2 \left|\int u_\beta^*(r)\chi_\alpha(r) r\, \mathrm{d}r\right|^2.
\end{aligned} \tag{3.159}$$

The coefficients a_I^2 are calculated in LS coupling to

$$a_I^2 = g_{\mathrm{A}}(2I + 1)(2L_{\mathrm{B}} + 1)(2S_{\mathrm{B}} + 1)\begin{Bmatrix} I & L_{\mathrm{A}} & s_{\mathrm{B}} \\ L_{\mathrm{B}} & I_{\mathrm{B}} & \ell_\beta \end{Bmatrix}_6^2 \begin{Bmatrix} I & L_{\mathrm{A}} & S_{\mathrm{B}} \\ I_{\mathrm{A}} & I_{\mathrm{a}} & I_{\mathrm{A}} \end{Bmatrix}_6^2. \tag{3.160}$$

The orbital and total angular momentum quantum numbers of the nuclei in the entrance and exit channels are ℓ_α, J_α, ℓ_β and J_β, respectively. The spin quantum number, the orbital and total angular momentum quantum numbers are characterized by S, L and I, respectively, with indices a, A and B corresponding to the projectile, target and residual nucleus, respectively. The notation $\{\cdots\}_6$ stands for the

6j-symbol (Equation (1.124)), which is related to the Racah coefficient used further above (see Equation (1.125)). The radial wave functions in the entrance and exit channels are given by χ_α and u_β, respectively. The spectroscopic factor for the partition B = A + a and the isospin Clebsch–Gordan coefficient are given by $\mathcal{S}_{\ell_\beta J_\beta}$ and $\mathcal{X}$, respectively.

Some astrophysical calculations have not made use of the full DC equations but used simplifying assumptions. For example, astrophysical neutron capture on stable and neutron-rich nuclei was calculated in the *hard-sphere model*. This model assumes that the amplitude of the scattering wave function is zero at the edge of the target nucleus. The E1 neutron capture cross section can then be written as

$$\sigma^{\mathrm{E1}}_{\mathrm{DC,hard}} = \frac{8\pi}{3}\left(\frac{Z_{\mathrm{A}}}{A_{\mathrm{A}}}\right)^2 \frac{r_{\mathrm{hard}} e_{\mathrm{C}}^2}{\mathrm{c}^3\sqrt{2m^3_{\mathrm{A+n}}E}}\xi \frac{g_J}{2g_{J_{\mathrm{A}}}(2\ell_\beta + 1)}\left(\frac{\tilde{Y}+3}{\tilde{Y}+1}\right)^2 \left(Q^J_{\mathrm{nuc}} + E\right)\mathcal{X}^2\mathcal{S}_J, \quad (3.161)$$

where r_{hard} is the hard sphere radius and the multiplicity ξ is the number of incident channel spins which can lead to the same final state with spin J. It is $\xi = 1$ for $J_A = 0$, or for $J_A \neq 0$ and $J = J_A^\mu \pm 3/2$. The value $\xi = 2$ applies for $J_A \neq 0$ and $J = J_A \pm 1/2$. The quantity ℓ_β is the orbital angular momentum of the final bound state and $\mathcal{S}_J$ is the spectroscopic factor of this state. The dimensionless parameter $\tilde{Y}$ is given by

$$\tilde{Y} = \frac{r_{\mathrm{hard}}\sqrt{2m_{\mathrm{A+n}}\left(Q^J_{\mathrm{nuc}} + E\right)}}{\hbar}. \quad (3.162)$$

The reaction Q-value for populating the final state with spin quantum number J (which can also be an excited state in B = A + n) is given by Q^J_{nuc}. The advantage of this approach is that no explicit wave functions for scattering and bound states are required. On the other hand, the correct overlap of the wave functions may yield more accurate cross sections, especially for low projectile energies when considerable contributions to Equation (3.155) are coming from far outside the nuclear radius.

For the validity of the low-order potential model approach for DC while neglecting higher-order processes, similar arguments can be made as were presented for the DWBA toward the end of Section 3.7.2.1.

In astrophysical applications, the projectile energies are low by nuclear physics standards and often below the Coulomb barrier when the projectile is charged. Therefore the wave function χ in the entrance channel will be strongly suppressed inside the barrier, that is, inside the nucleus. In consequence, the radial overlap integrals with the bound state wave functions appearing in Equations (3.155) and (3.156), as well as in Equation (3.159), receive most contributions from larger radii, at the edge of the nucleus or even outside of it. This has the advantage that the cross sections become rather insensitive to the bound state wave function inside the nucleus and are mainly determined by the tail of the wave functions. Asymptotically the wave functions are given by the Coulomb wave functions and a nuclear phase shift (Equation (3.32)). Since the DWBA cross sections involve similar asymptotic wave functions, the asymptotic wave function determined through a DWBA analysis of the measurement of a transfer reaction (for example, (d,n)) can directly be used to also determine the capture cross sections (for example, (p,γ)). This is called the method of *Asymptotic Normalization Coefficients (ANC)*.

3.7.3 Cluster Models

Microscopic models are based on the treatment of all nucleons with their nucleon–nucleon interactions and exact antisymmetrization of the wave functions. An exact solution would require to solve coupled Schrödinger equations with the full Hamiltonian as given in Equation (3.16). As mentioned before, this is not yet possible for systems with more than a few nucleons and therefore certain simplifications have to be made. In cluster models, it is assumed that the nucleons are grouped in clusters and internal wave functions describing the relative cluster motions are generated. The main advantage of cluster models with respect to other microscopic theories is their ability to deal with reactions as well as with nuclear spectroscopy. Over the past years, much work has been devoted to the improvement of the internal wave functions through multicluster descriptions or monopolar distortion. The main limitation arises from the number of channels included in the wave function, which reduces the validity of the model at low energies. Also large nuclear level densities require many channels in the wave functions. Therefore also the application of cluster models is limited to light nuclei.

As an example for a cluster model, a system with two clusters is shown here. This provides a natural extension of the formalism of scattering states. The internal wave functions of the clusters are denoted as $\phi_i^{I_i\pi_q^i}(\vec{\xi}_i)$, where I_i and π_q^i are the spin and parity of cluster i, and $\vec{\xi}_i$ represents a set of its internal coordinates. One defines the channel function as

$$\varphi_{\ell I}^{JM\pi_q}(\Omega, \vec{\xi}_1, \vec{\xi}_2) = \sum_{\ell\ell_1 m m_1} \mathcal{X}_{JM}^{\ell\ell_1 m m_1} \mathcal{Y}_{\ell_1 m_1}(\Omega) \left(\sum_{I_1 I_2 m_1 m_2} \mathcal{X}_{\ell m}^{I_1 I_2 m_1 m_2} \phi_1^{I_1 m_1 \pi_1}(\vec{\xi}_1) \phi_2^{I_2 m_2 \pi_2}(\vec{\xi}_2) \right) \quad (3.163)$$

with the channel spin I, the relative angular momentum ℓ, the total spin J and the total parity $\pi_q = \pi_q^1 \pi_q^2 (-1)^\ell$.

The total wave function is written as

$$\Psi^{JM\pi_q} = \sum_{\alpha\ell I} \hat{A} R_{\alpha\ell I}^{J\pi}(r_x) \varphi_{\alpha\ell I}^{JM\pi_q}\left(\Omega, \vec{\xi}_1^{\,i}, \vec{\xi}_2^{\,j}\right), \quad (3.164)$$

Index α corresponds to different two-cluster arrangements, and $\hat{A}$ is the antisymmetrization operator (see Equation (3.13)). The relative wave function $R(r_x)$ (with r_x being the cluster separation) is also determined using the Schrödinger equation (see also Equation (3.147) for a similar approach). In most applications, the internal cluster wave functions are defined in the shell model. Accordingly, the nucleon–nucleon interaction must be appropriately chosen, which leads to the introduction of effective forces. Since such an effective force (once determined by comparison to experiment), however, is nearly the same for all light nuclei, the predictive power still is high for such nuclei. A straightforward application of Equation (3.164) is the

Resonating Group Method (RGM). In more recent applications, the relative wave function is expanded over Gaussian functions. This yields the *Generator Coordinate Method* (GCM). The GCM is equivalent to the RGM, but is better adapted to numerical calculations, as it makes uses of projected Slater Determinants.

3.8 Barrier Penetration: α-Decay and Nuclear Fission

The Coulomb barrier not only inhibits a projectile to enter a nucleus, it also is a barrier for nucleons or nucleon clusters to escape the range of the nuclear interaction even when they are unbound ($E > 0$), that is, not inside the potential well generated by the attractive nuclear force ($E < 0$). Above the magic neutron number $N = 50$ there are nuclei with $S_\alpha \leqslant 0$. Many such nuclei have very long half-lives of 10^{12}–10^{15} yr. For $N > 82$ there are even stable nuclei with negative α-particle separation energy S_α. This is only possible due to the action of the Coulomb interaction because an α-particle escaping the nucleus first has to tunnel through the Coulomb barrier. Since wave functions are exponentially suppressed inside a barrier (see Figure 3.12 and also Section 3.3.2), the emission probability is small when the particle energy inside the nucleus is well below the top of the Coulomb barrier.

The probability to emit an α-particle, and thus also the half-life with respect to α-emission, can be modeled very simply by assuming three factors: (1) the probability that an α-particle forms among the nucleon configuration of the nucleus, (2) the probability that it is present at the edge of the nucleus (at the inner radius of the Coulomb barrier), and (3) the probability that it tunnels through the barrier. The α-particle formation probability is given by the spectroscopic factor from Equation (3.147). The probability to be moving toward the Coulomb barrier is related to the kinetic energy of the α-particle E_α inside the nucleus and the nuclear radius $R_{\rm nuc}$,

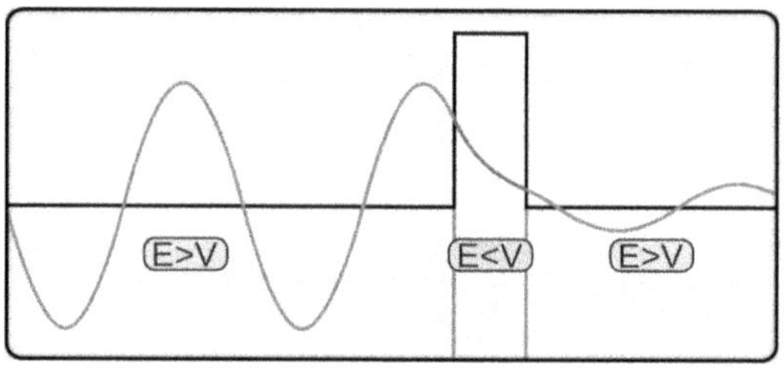

Figure 3.12. Sketch of the tunneling effect in quantum mechanics: shown is how the wave function of a particle moving from the left to the right can penetrate a potential barrier although its energy is less than the height of the barrier. The particle has an energy E larger than the constant potential V to the left and right of the barrier and therefore the wave can progress undisturbedly in these two regions. Within the potential barrier the wave amplitude is attenuated exponentially. When the wave is past the barrier it continues undisturbedly again but with smaller amplitude. Since the probability of finding a particle at a certain position is given by the square of the wave function, the probability to find the particle beyond the barrier can be very small, depending on the barrier width and height, but it is not zero. A barrier of arbitrary shape can be approximated by combining many narrow, rectangular barriers (see text). (This Tunnel effect image has been obtained by the author from the Wikimedia website where it was made available by F. Kling under a CC BY-SA 3.0 licence. It is included within this book on that basis. It is attributed to F. Kling.)

giving the time it takes for the particle to cross the nucleus. Therefore the "reduced probability" λ_0^α is given by

$$\lambda_0^\alpha = \mathcal{S}_\alpha \frac{1}{2R_\text{nuc}}\sqrt{\frac{2E_\alpha}{m_\alpha}}. \tag{3.165}$$

This is equivalent to a "reduced decay width" $\Gamma_0^\alpha = \hbar\lambda_0^\alpha$. The dominant factor, however, is the tunneling probability. This is given by a Coulomb barrier penetrability obtained by the ratio of the amplitudes of the squared α-particle wave function outside and inside of the nucleus $|\psi_\alpha(R_\text{out})|^2/|\psi_\alpha(R_\text{in})|^2$, that is, at the inner edge of the barrier R_in and the outside edge R_out. The actual value for an arbitrary nuclear potential plus Coulomb potential has to be obtained by the wave function which is a solution of the time-independent Schrödinger equation. As mentioned earlier, inside the barrier the wave amplitude is exponentially suppressed. Therefore the penetration probability P^α is expected to look like $P^\alpha \approx \exp(-b_\text{G})$. The calculation finds for a rectangular barrier and for E_α far below the barrier height H_barrier

$$P^\alpha \approx \mathrm{e}^{-\left(2(R_\text{out}-R_\text{in})/\hbar\right)\sqrt{2m_\alpha(H_\text{barrier}-E_\alpha)}}. \tag{3.166}$$

This expression can be generalized for a barrier with arbitrary shape by assuming that the shape can be approximated by a sum of rectangular strips of thickness dr. For the limit d$r \to 0$ the sum has to be replaced by an integration, leading to

$$P^\alpha \approx \mathrm{e}^{-(2/\hbar)\int_{R_\text{in}}^{R_\text{out}} \sqrt{2m_\alpha(H_\text{barrier}(r)-E_\alpha)}\,\mathrm{d}r}. \tag{3.167}$$

The same result is obtained when using the so-called WKB approximation in the solution of the Schrödinger equation. For the Coulomb potential, the integral can be given in closed form as

$$b_\text{G} = \frac{2}{\hbar}\sqrt{\frac{2m_\alpha}{E_\alpha}}Z_\text{a}Z_\text{A}e_\text{C}^2\left[\arccos\left(\sqrt{\frac{E_\alpha}{V_\text{C}}}\right) - \sqrt{\frac{E_\alpha}{V_\text{C}}\left(1-\frac{E_\alpha}{V_\text{C}}\right)}\right]. \tag{3.168}$$

The quantity b_G is called *Gamow factor*. It is also used in the definition of the astrophysical S-factor, see Equation (4.23). Using the above quantities, the lifetime τ_life^α of a nucleus against α-emission is given by

$$\frac{1}{\tau_\text{life}}^\alpha = \lambda_0^\alpha \mathrm{e}^{-b_\text{G}(E_\alpha)}. \tag{3.169}$$

The nuclear binding energy released is again given by the Q-value. It should be noted, however, that the kinetic energy of the emitted α-particle is dominated by the Coulomb repulsion at R_out.

Another type of barrier is encountered when deforming a nucleus. Treating the nucleus as a liquid drop, the surface tension (see, for example, Equation (3.8)) acts against any further deformation beyond the equilibrium deformation in the nuclear ground state. The Coulomb repulsion, on the other hand, acts in the opposite manner, trying to expand (deform) the nucleus. Assuming quadrupole deformation (Equation (3.38)) and symmetric fission (the nucleus splits into two equal parts), a critical ratio between charge number Z and mass number A of

$$\left(\frac{Z^2}{A}\right)_{\text{crit}} = \frac{10}{3}\frac{a_S R_{\text{nuc}}}{A^{1/3}} \approx 50 \tag{3.170}$$

is obtained. When this ratio is exceeded, the Coulomb force destabilizes the nucleus and small disturbances of the nuclear shape will lead to fission. The *fissionability parameter* is defined as

$$x_{\text{fiss}} = \frac{Z^2/A}{(Z^2/A)_{\text{crit}}}. \tag{3.171}$$

Coulomb energy and surface energy are a function of deformation. Plotting the sum of surface energy and Coulomb energy as a function of deformation it becomes apparent that the sum has a maximum at a given deformation and decreases again with stronger deformation. In classical terms this can be visualized by further and further pulling apart an already deformed liquid drop. At a certain point, the scission deformation, the drop develops a pinch in the middle which continues to narrow until it is split into two drops. Since the two drops are both positively charged, they will quickly accelerate and fly apart. Feynman liked to point out that the energy released in fission is mainly electric energy, not nuclear energy. Reaching the scission point requires energy and therefore one speaks of a (fission) barrier. Approximation of the barrier by a parabolic shape leads to the Hill–Wheeler barrier transmission

$$T^{\text{fiss}} = [1 + e^{2\pi(H_{\text{fiss}}-E)/(\hbar\omega)}]^{-1}, \tag{3.172}$$

where H_{fiss} is the height of the fission barrier and $\hbar\omega$ is the curvature energy of the barrier determining the barrier width.

The naive liquid drop picture cannot describe all the observed phenomena. In the hybrid model suggested by Strutinsky, a liquid drop description is combined with shell and pairing effects. The total energy of the drop is found as the droplet energy E_{FT}, which is the surface energy and the Coulomb energy, and shell corrections Δ_{shell} and pairing corrections Δ_{pair},

$$E = E_{\text{FT}} + \sum_{Z,N}\left(\Delta_{\text{shell}}(Z, N) + \Delta_{\text{pair}}(Z, N)\right). \tag{3.173}$$

This leads to a double-humped fission barrier. Between two energy maxima there is a secondary minimum in the barrier. This not only affects the total penetration probability through the barrier but also allows for additional phenomena. The nucleus can overcome both maxima in *prompt fission* or remain in the second minimum before propagating through the second hump, leading to *delayed fission*. Self-consistent microscopic treatments even show that there could be triple-humped barriers, underlining the complexity of the fission process.

Another complication is the description of the fission fragment distribution. The nucleus does not just split into two equal parts but due to the excitation of the nucleus and its fragments a distribution of fragment masses emerges. A realistic distribution shows two maxima, similar to the addition of two Gauss distributions around two preferred but unequal masses. The mean of the heavier fragment masses is determined by the magic numbers $Z = 50$ and $N = 82$. The mean of the lighter fragment masses increases linearly with the mass of the initial nucleus. This behavior of the averages is another indication of the importance of shell effects. Some microscopic theories predict more complicated fragment distributions, for example, a distribution with three maxima.

Apart from prompt and delayed fission, one distinguishes also between spontaneous and induced fission. Adding energy to a nucleus, for example, by a low-energy neutron, may "nudge" a nucleus close to the critical ratio into splitting apart. Another process playing a role in astrophysics is β-delayed fission. The nucleus produced in a β-decay may be in an excited state and this may enhance the probability that a fission follows the β-decay.

In summary, the fission process is a complicated process in a many-body system and the predictive power of macroscopic–microscopic and microscopic theories based on nucleon–nucleon interaction is still very limited in this respect. Fission barrier parameters are often only fitted to experimental data. This is a serious impediment for understanding the production of intermediate and heavy nuclei in the r-process that runs into the region of fissionable nuclei.

3.9 β-Decay and Electron Capture

In Section 3.2, Equation (3.11) was derived, defining the line of nuclear stability in the nuclear chart. For nuclei located off the line of stability, transformations changing neutron and proton number but leaving the total number of nucleons unaffected are energetically allowed. A spontaneous transition of such a kind is also called decay but is substantially different from the α-decay discussed in Section 3.8. Transformations of a neutron into a proton or vice versa are mediated by the weak interaction, not by the strong (nuclear) interaction, and thus have to be treated in their own mathematical framework allowing, for example, for different selection rules (parity is not conserved by the weak interaction) and different transition matrix elements. They involve the interaction between leptons (see Table 3.1) which are not affected by the nuclear interaction. Nevertheless, nuclear structure as given by the nuclear shell model affects the decays because it defines the initial and final states potentially available before and after the transformation of a nucleon.

The basic process is that of a transformation of a neutron into a proton or a proton into a neutron,

$$\begin{aligned} \mathrm{n} &\rightarrow \mathrm{p} + \mathrm{e}^- + \bar{\nu}_\mathrm{e}, \\ \mathrm{p} &\rightarrow \mathrm{n} + \mathrm{e}^+ + \nu_\mathrm{e}. \end{aligned} \tag{3.174}$$

The electron antineutrino is denoted as $\bar{\nu}_\mathrm{e}$. Free neutrons are unstable and decay into protons because their mass is slightly larger than the proton mass. For the same reason, on the other hand, free protons cannot spontaneously transform into neutrons.

Inside of nuclei, the situation is different. Neutrons cannot decay when there is no energy state available for the resulting proton, taking into account energy, spin selection, and the Pauli principle. A proton, on the other hand, may be transformed into a neutron when the neutron–proton mass difference can be supplied by some kind of excitation and when energy states are available for the resulting neutron. This already shows that the decays are sensitive to the nuclear structure determining the energy states and their occupation.

According to the above, a number of decay modes mediated by the weak interaction are found:

- β^- decay: In nuclei off the line of stability on the neutron-rich side, a neutron can change into a proton under emission of an electron and an electron antineutrino. The final nucleus may be in an excited state and emit further γ quanta during de-excitation.
- β^+ decay: In nuclei off the line of stability toward the proton-rich side, the proton excess allows for a proton to change into a neutron under emission of a positron and an electron neutrino. Again, the final nucleus may be in an excited state and emit further γ quanta during de-excitation.
- Electron capture (EC): In some unstable, proton-rich nuclei the energy difference between initial and final nucleus does not allow β^+ decay. Nevertheless, a proton can be changed into a neutron by capturing an electron in the reaction $\mathrm{p} + \mathrm{e}^- \rightarrow \mathrm{n} + \nu_\mathrm{e}$. Under standard laboratory conditions, the electron is captured from the lowest electron shell of the atom, the K-shell. That is why this type of "decay" is sometimes called K-capture.

The energy released in these processes is again given by the reaction Q-value taking into account the excitation energies of initial and final states of the nucleons. In β^-- and β^+-decays the released energy is distributed between the two emitted leptons because the kinetic energy and momentum of the final nucleus are practically zero due to its comparatively large mass (compared to the masses of the leptons). This results in a smooth energy distribution of the emitted leptons when observing many events (see also Figure 7.3). The possible energy range is bounded by zero energy at the lower end (the other particle is carrying the total released energy) and the reaction Q-value at the high end (the particle in question received the full energy). In EC only a neutrino leaves the nucleus, carrying the released energy. Therefore there is no smooth spectrum in EC, the neutrinos are emitted at a discrete energy.

These decay modes are affected by the environment of the nucleus. Obviously, for K-capture to occur the nucleus has to be in an atom. In a hot, fully ionized plasma, in which the nucleus is embedded in a sea of free electrons, it can still capture an electron but the probability is largely different from the one for K-capture. Moreover, nuclei are thermally excited in a stellar plasma and this changes also the probabilities, and thus the half-lives, for all three decay modes because different initial and final states are available. These changes can go both ways, increasing or decreasing half-lives. Finally, electron emission in β^- decay may be suppressed in dense plasmas due to the fact that the free plasma electrons already occupy the majority of energy states that could be populated by the emitted electron. All these effects are also addressed in Sections 4.1.3 and 4.1.4.

Free protons are known to be stable (or at least to have a lifetime several times exceeding the age of the Universe). They can be transformed into neutrons, nevertheless, by electron captures or by electron antineutrino-induced reactions. The required conditions occur in some astrophysical sites at extremely high temperature and/or density (see, for example, Section 9.3.3). The processes transforming neutrons and protons into each other can be summarized as

$$\begin{aligned} \mathrm{n} + \nu_\mathrm{e} &\leftrightarrows \mathrm{p} + \mathrm{e}^-, \\ \mathrm{p} + \bar{\nu}_\mathrm{e} &\leftrightarrows \mathrm{n} + \mathrm{e}^+. \end{aligned} \tag{3.175}$$

At least the neutron–proton mass difference has to be provided by the electron or the electron antineutrino, respectively, to allow a transformation of a proton into a neutron.

From the point of view of fundamental particle physics, in a β^- decay one of the two down quarks in the neutron (see Section 3.1) is transformed into an up quark by interaction with a W^- boson. The boson then decays into an electron and an electron antineutrino. Likewise, in a β^+ decay one of the up quarks in a proton is turned into a down quark by interaction with a W^+ boson subsequently decaying into a positron and an electron neutrino. A detailed, rigorous treatment of the weak interaction and its connected reactions is beyond the scope of this book. Here only a few definitions and basic concepts are introduced, which are of relevance to the astrophysical applications presented in the following chapters.

The transition probability for $\beta^{+,-}$-decay is proportional to a matrix element M_w. This is obtained from the overlap of the initial and final wave function of the system (including nucleus wave functions $\psi_{\mathrm{i,f}}$ and lepton wave functions $\psi_{\mathrm{e},\nu}$) before and after decay, respectively, with the Hamiltonian $\hat{H}_\mathrm{w}$ for the weak interaction,

$$M_\mathrm{w} = \int \psi_\mathrm{f}^* \psi_\mathrm{e}^* \psi_\nu^* \hat{H}_\mathrm{w} \psi_\mathrm{i} \, \mathrm{d}^3 r. \tag{3.176}$$

At the energies appearing in β-decays, the lepton wave functions are larger than the dimension of the nucleus. Therefore they can be approximated by a constant given by their value at the center of the nucleus and also the Hamiltonian can be simplified accordingly. Then it can be shown that the matrix element M_w is mainly determined by a combination of two nuclear matrix elements M_F and M_GT,

$$|M_\mathrm{w}|^2 = g_\mathrm{V}^2 M_\mathrm{F}^2 + g_\mathrm{A}^2 M_\mathrm{GT}^2 \tag{3.177}$$

where g_V and g_A are the coupling constants determining the strength of the interaction. The nuclear matrix elements only contain nuclear wave functions and depend on the structure of the decaying nucleus. The coupling constants are given by properties of the weak interaction and should not vary from nucleus to nucleus. The vector coupling of the weak interaction is given by g_V and the axial vector coupling by g_A. The existence of only vector and axial vector couplings implies that the helicity of the produced leptons is not equal. Scalar and tensor couplings, allowing for leptons with the same helicity, are permitted by theory but not observed. The ratio between the constant is experimentally determined as $g_A/g_V = -1.2670 \pm 0.0035$.

Decays in which M_w is dominated by M_F are called *Fermi decays* and those in which M_{GT} is dominating are called *Gamow–Teller decays.* Mixed decays are also possible, for example the decay of the free neutron. Spin selection rules are different for the two types of decay. Fermi decays preserve the total spin, parity, and isospin of the nucleus ($\Delta J = 0$, $\Delta\pi_q = 0$, $\Delta I_{iso} = 0$) whereas for Gamow–Teller transitions $\Delta\pi_q = 0$ and $\Delta J = 0$ or 1 but no $0 \to 0$ transitions are allowed.

Decays for which both nuclear matrix elements M_F and M_{GT} vanish according to the above derivation are called *forbidden decays.* Nevertheless, forbidden decays occur in nature, although with longer half-lives than *allowed decays.* They are only forbidden in the above model because of the two approximations made in deriving Equation (3.177): (1) constant lepton wave functions within the nucleus; (2) non-relativistic treatment. The first assumption resulted in approximating the lepton wave function by its value at the nuclear center. This is only possible when $\ell = 0$ for the lepton because wave functions with $\ell > 0$ have a node at the origin. When the lepton wave function is expanded as a plane wave,

$$\psi_\nu(r) = e^{i\mathcal{K}r} = 1 + \mathcal{K}r + \dots , \tag{3.178}$$

the constancy assumption is equivalent to taking only the first term of the expansion and neglecting all other terms, starting with the term linear in r. Taking into account also the linear term, lepton emission with $\ell > 0$ leads to a non-zero matrix element, yielding a half-life about 1000 times longer than for its allowed equivalent. Moreover, parity is changed in such a *first forbidden transition.* Similar values result when relaxing the second assumption and accounting for relativistic wave functions. Sometimes only taking the first two terms in the expansion still leads to vanishing matrix elements. Then higher order terms have to be considered, leading to *double forbidden, triple forbidden, ... transitions.*

It has become common to classify β-decays according to their $\log_{10}\overline{ft}$ values. The quantity $\overline{ft}$ is derived from multiplying the half-life with an integral over lepton momenta and a Coulomb correction. It is a convenient definition as $\overline{ft}$ only contains known constants and the nuclear matrix elements:

$$\overline{ft} = \frac{2\pi^3\hbar^7}{m_e^5c^4}\frac{\ln 2}{g_V^2M_F^2 + g_A^2M_{GT}^2}. \tag{3.179}$$

Allowed transitions have $\log_{10}\overline{ft}$ values of 3–6, first forbidden transitions have $\approx$6–9, second forbidden transitions have $\approx$10–13, and so on. Not all forbidden transitions are clearly separated by distinct $\log_{10}\overline{ft}$ ranges, there is some ambiguity.

More on decays and (anti)neutrino-induced reactions in astrophysics can be found in Sections 4.1.6, 7.2, 9.3.3, and 9.3.4. Neutrino oscillations are discussed in Section 7.2.3.

Further Reading

Bertulani, C. A., & Danielewicz, P. 2004, Introduction to Nuclear Reactions (Bristol: IOP Publishing)
Blatt, J. M., & Weisskopf, V. F. 1979, Theoretical Nuclear Physics (New York: Springer)
Descouvemont, P. 2003, Theoretical Models for Nuclear Astrophysics (New York: Nova Science Publishers)
Descouvemont, P., & Rauscher, T. 2006, NuPhA, 777, 137
Fröbrich, P., & Lipperheide, R. 1996, Theory of Nuclear Reactions (Oxford: Clarendon)
Glendenning, N. K. 1983, Direct Nuclear Reactions (New York: Academic)
Iliadis, C. 2015, Nuclear Physics of Stars (2nd ed.; New York: Wiley)
Koonin, S. E. 1986, Computational Physics (Menlo Park: The Benjamin/Cummings Publishing Company)
Lane, A. M., & Thomas, R. G. 1958, RvMP, 30, 257
Machner, H. 2005, Einführung in die Kern- und Elementarteilchenphysik (Weinheim: Wiley)
Mayer-Kuckuk, T. 1994, Kernphysik (Stuttgart: Teubner)
Rauscher, T. 2011, IJMPE, 20, 1071
Rauscher, T., Thielemann, F.-K., & Kratz, K.-L. 1997, PhRvC, 56, 1613
Satchler, G. R. 1983, Direct Nuclear Reactions (Oxford: Clarendon)
Schmid, E. W., Spitz, G., & Lösch, W. 1987, Theoretische Physik mit dem Personal-Computer (Berlin: Springer)
Thompson, I. J., & Nunes, F. M. 2009, Nuclear Reactions for Astrophysics (Cambridge: Cambridge University Press)

Thomas Rauscher

Chapter 4

Abundance Changes in Astrophysical Plasmas

4.1 Astrophysical Reaction Rates

4.1.1 Introduction

The central problem in nuclear astrophysics is the determination of the *astrophysical reaction rate*, either experimentally or—in the majority of cases—by theoretical prediction. Such an astrophysical reaction rate quantifies the number of reactions per time in a given volume of hot stellar plasma. The abundance of each atomic nucleus in the stellar plasma can be affected by a large number of reaction types with each, in turn, assigned a reaction rate. A complete set of differential equations incorporating all reactions occurring in a plasma and allowing to follow the abundance changes of all nuclei is called a *nuclear reaction network*. Before discussing such networks in Section 4.2, the definition of reaction rates for various types of reactions is presented below.

4.1.2 Two-body Rates for Interacting Nuclei

Two-body rates describe reactions between two interacting particles or nuclei. Assuming a unidirectional motion of particles hitting stationary particles with a velocity v, this would be similar to a laboratory scattering experiment in which a particle beam is shot on a stationary target as described in Section 1.6. With the definition of the cross section as given in Equation (1.100), the number of interactions per time and per nucleus simply is the product of cross section σ and the number of incoming particles per time and per area,

$$\lambda_{\mathrm{trans}} = \sigma \tilde{n} v = \sigma F_{\mathrm{inc}}, \tag{4.1}$$

where F_{inc} is the particle number current density as introduced in Section 1.6.1.

In a more general approach, target A and projectile a follow specific thermal momentum distributions dn_{A} and dn_{a} in an astrophysical plasma. With the relative

doi:10.1088/2514-3433/ab8737ch4

velocities $\vec{v}_A - \vec{v}_a$ between the reaction partners, the number of reactions per volume and time is given by

$$r_{Aa} = \frac{1}{1 + \tilde{\delta}_{Aa}} \int \int \sigma^*(|\vec{v}_A - \vec{v}_a|, T)|\vec{v}_A - \vec{v}_a| d\bar{n}_A \, d\bar{n}_a, \tag{4.2}$$

and involves the relative velocity in 3D space $\vec{v}_A - \vec{v}_a$ and the thermodynamic distributions of target and projectile $d\bar{n}_A$ and $d\bar{n}_a$. The factor before the integral is included to avoid double counting when two particles of the same kind are interacting. The *Kronecker symbol* $\tilde{\delta}_{Aa}$ assumes a value of 1 when A and a are identical and is zero otherwise.

Here, the reaction cross section could be either a total reaction cross section including all possible reactions between projectile and target (also leading to different final nuclei and ejectiles) or describe a specific process, e.g., only neutron capture. Furthermore, different reactions may be possible in a hot plasma than in laboratory scattering. This is made more visible by using the stellar cross section σ^* in place of the earlier used reaction cross section σ. It should be noted that, contrary to the usual laboratory cross section, the *stellar* cross section may not only depend on the relative velocities of the interacting particles but also explicitly on the plasma temperature T. The implications of using such a cross section modified in the stellar plasma, instead of the usual laboratory one, will be discussed in Section 4.1.3.

The thermodynamic distributions have to be chosen according to the type of particles (fermions, bosons) and temperatures involved. Note that Equation (4.2) can be generalized to three and more interacting nuclear species by integrating over the appropriate number of distributions (see Section 4.2 for further information). In the following, the focus is on the case of two interacting nuclei or nucleons. Reactions in which the projectile is a photon or a lepton are discussed in Sections 4.1.5 and 4.1.6, respectively. Although nucleons are fermions and atomic nuclei are either fermions or bosons, depending on their total spin, it was pointed out in Section 1.5.4 that the ion component of a stellar plasma fulfills Equation (1.51) and thus can be described as an ideal gas following the Maxwell–Boltzmann statistics (see Section 1.5.2). This only changes when a plasma approaches nuclear density and quantum effects in the nucleon gas cannot be neglected anymore. This would be the case, e.g., in the collapse to a neutron star at the end of the life of a massive star but regular and even explosive phases of nuclear burning never reach such densities.

It can be shown that Maxwell–Boltzmann statistics also apply when changing the coordinates used in Equation (4.2) in such a manner that only the one-dimensional motion of particles along the axis connecting two particles are considered, that is motion parallel to the vector $\vec{v} = \vec{v}_A - \vec{v}_a$, with $v = |\vec{v}|$. Using Equation (1.58) for $d\bar{n}_A$ and $d\bar{n}_a$, Equation (4.2) can be rewritten as

$$r_{Aa} = \frac{1}{1 + \tilde{\delta}_{Aa}} \int_0^\infty \int_0^\infty \sigma^*(v, T) v w(v_A) w(v_a) \, dv_A \, dv_a, \tag{4.3}$$

with v_A and v_a being the components of the particle velocity parallel to $\vec{v}$. The integration limits have been made explicit, velocities away from the interaction point (i.e., negative velocities) do not contribute. The Maxwell–Boltzmann distribution in velocity $w(v)$ is obtained from Equation (1.58) using $p = mv$,

$$w(v) = 4\pi\bar{n}\left(\frac{m}{2\pi k_B T}\right)^{3/2} v^2 e^{-mv^2/(2k_B T)}. \tag{4.4}$$

The product $w(v_A)w(v_a)$ expressed in center-of-mass coordinates (see Section 1.6.1) with the help of Equations (1.104) and (1.105) again leads to a product of two Maxwell–Boltzmann-like velocity distributions,

$$\begin{aligned}
\frac{w(v_A)w(v_a)}{\bar{n}_A\bar{n}_a} &= \mathcal{U}(v_A)\mathcal{U}(v_a) \\
&= 4\pi\left(\frac{m_A}{2\pi k_B T}\right)^{3/2} v_A^2 e^{-m_A v_A^2/(2k_B T)} \\
&\quad \times 4\pi\left(\frac{m_a}{2\pi k_B T}\right)^{3/2} v_a^2 e^{-m_a v_a^2/(2k_B T)} \\
&= 4\pi\left(\frac{m_A + m_a}{2\pi k_B T}\right)^{3/2} v_{\text{center}}^2 e^{-(m_A+m_a)v_{\text{center}}^2/(2k_B T)} \\
&\quad \times 4\pi\left(\frac{\mu_{\text{red}}}{2\pi k_B T}\right)^{3/2} v^2 e^{-\mu_{\text{red}} v^2/(2k_B T)} \\
&= \mathcal{U}(v_{\text{center}})\mathcal{U}(v),
\end{aligned} \tag{4.5}$$

with v_{center} being the velocity of the center-of-mass and μ_{red} being the reduced mass as introduced in Equation (1.105). The resulting double integral

$$r_{Aa} = \bar{n}_A\bar{n}_a \int_0^\infty \int_0^\infty \sigma^*(v, T) v \mathcal{U}(v)\mathcal{U}(v_{\text{center}})\, dv\, dv_{\text{center}} \tag{4.6}$$

can be simplified realizing that the cross section $\sigma^*(v, T)$ does not depend on the motion of the center of mass v_{center} and that the velocity distribution $\mathcal{U}$ is normalized,

$$\int_0^\infty \mathcal{U}(v_{\text{center}})\, dv_{\text{center}} = 1. \tag{4.7}$$

This leads to

$$\begin{aligned}
r_{Aa} &= \frac{1}{1 + \tilde{\delta}_{Aa}} \int_0^\infty \sigma^*(v, T) v w(v)\, dv \\
&= \frac{\bar{n}_A\bar{n}_a}{1 + \tilde{\delta}_{Aa}} \left(\frac{\mu_{\text{red}}}{2\pi k_B T}\right)^{3/2} \int_0^\infty \sigma^*(v, T) v^3 e^{-\mu_{\text{red}} v^2/(2k_B T)}\, dv \\
&= \frac{\bar{n}_A\bar{n}_a}{1 + \tilde{\delta}_{Aa}} \langle \sigma^* v \rangle,
\end{aligned} \tag{4.8}$$

which also can be expressed in terms of center-of-mass energy $E = \mu_{\mathrm{red}} v^2/2$,

$$\begin{aligned} r_{\mathrm{Aa}} &= \frac{\bar{n}_{\mathrm{A}} \bar{n}_{\mathrm{a}}}{1 + \tilde{\delta}_{\mathrm{Aa}}} \sqrt{\frac{8}{\pi \mu_{\mathrm{red}}}} \left(\frac{1}{\mathrm{k_B} T} \right)^{3/2} \int_0^\infty \sigma^*(E,T) E \mathrm{e}^{-E/(\mathrm{k_B}T)} \, \mathrm{d}E \\ &= \frac{\bar{n}_{\mathrm{A}} \bar{n}_{\mathrm{a}}}{1 + \tilde{\delta}_{\mathrm{Aa}}} \langle \sigma^* v \rangle, \end{aligned} \tag{4.9}$$

Note the necessary transformation of the differential, $\mathrm{d}E/\mathrm{d}v = \mu_{\mathrm{red}} v$, when recasting Equation (4.8) as Equation (4.9). The quantity $\langle \sigma^* v \rangle$ is the *reaction rate per particle pair* or *reactivity* under stellar conditions (see Section 4.1.3.1 for a detailed discussion of the impact of thermal plasma effects). The angle brackets stand for the appropriate averaging, i.e., integration, over the velocity and energy distribution, respectively. The notation $\langle \sigma v \rangle^*$ is used for the same quantity sometimes.

The two-particle reaction rate, giving *number of reactions per time and per volume*, also can be written using the abundances defined in Equation (1.14),

$$r_{\mathrm{Aa}} = \frac{Y_{\mathrm{a}} Y_{\mathrm{A}}}{1 + \tilde{\delta}_{\mathrm{Aa}}} \rho^2 \mathrm{N}_{\mathrm{A}}^2 \langle \sigma^* v \rangle. \tag{4.10}$$

This contains the plasma density explicitly because the abundances themselves are defined to be independent of the density. When only the destruction or creation of a particular nuclide is of interest, the *number of reactions per time and per target nucleus* is used

$$\lambda_{\mathrm{trans}}^{*,\mathrm{A}} = Y_{\mathrm{A}} \rho \mathrm{N}_{\mathrm{A}} \langle \sigma^* v \rangle. \tag{4.11}$$

This can be used in simple estimates of nuclear lifetimes or burning timescales, see Section 4.1.7 and in particular Equations (4.106) and (4.110). Contrary to the reaction rate as defined in Equations (4.8) and (4.9) it does not depend on the number density or abundance of the projectile and it has the dimension of 1/time.

Note that $\mathrm{N_A} \langle \sigma^* v \rangle$ has the dimension volume per time per amount of material. It is usually given as $\mathrm{cm^3\ s^{-1}\ mole^{-1}}$. In order to have consistent dimensions on the left-hand side and the right-hand side of Equations (4.10) and (4.11), respectively, the right-hand sides have to be multiplied by the molar mass constant $\mathrm{M_u}$, see also Equation (1.9) for further details. Because $\mathrm{M_u} \simeq 1$ in the cgs-system of units, it has become the de facto standard to not explicitly specify $\mathrm{M_u}$ in rate definitions and to implicitly assume an appropriate unit conversion. The molar mass conversion has to be explicitly included when using other mass units than grams.

4.1.2.1 Relevant Energies for Non-resonant Reactions with Neutrons

The energy distribution $E \exp(-E/(\mathrm{k_B}T))$ in Equation (4.9) has its peak at $E = \mathrm{k_B}T$. In practical units this is $E = \mathrm{k_B}T = T_9/11.6045$ MeV when T_9 is in units of 10^9 K. Although limits of the integral in Equation (4.9) are at zero and at infinity, in practice it is not necessary to integrate over the whole range due to this peak structure. Depending on the energy dependence of the cross section, most contributions to the value of the integral come from a range around the peak or around a peak shifted due to the folding of the energy distribution with the energy dependence of the cross section. In the following, a number of specific cases is discussed, distinguished by different types of reactions and their typical cross section behavior.

For constant cross section $\sigma^* = C = \text{const}$, the reactivity

$$\begin{aligned}\langle\sigma^* v\rangle_{\text{const}} &= C\sqrt{\frac{8}{\pi\mu_{\text{red}}}}\left(\frac{1}{\mathrm{k_B}T}\right)^{3/2}\int_0^\infty \mathrm{e}^{-E/(\mathrm{k_B}T)}\,\mathrm{d}E \\ &= C\sqrt{\frac{8}{\pi\mu_{\text{red}}}}\left(\frac{1}{\mathrm{k_B}T}\right)^{3/2}(\mathrm{k_B}T)^2 = C\sqrt{\frac{8\mathrm{k_B}T}{\pi\mu_{\text{red}}}}\end{aligned} \tag{4.12}$$

is obtained. Obviously, in this case the cross section would have to be known only at a single energy to determine the factor C and the reactivity (as well as the rate) would be proportional to $\sqrt{T}$. This is only of mathematical interest. A more realistic case are non-resonant neutron capture cross sections which show an energy dependence according to the dominating partial waves. It was shown in Section 3.6 that the cross section for s-wave direct neutron capture is proportional to $1/v = 1/\sqrt{E}$. Inserting $\sigma^* = C/v = C/\sqrt{E}$ into Equation (4.9), the resulting integral still can be solved analytically,

$$\begin{aligned}\langle\sigma^* v\rangle_{\text{s}} &= C\sqrt{\frac{8}{\pi\mu_{\text{red}}}}\left(\frac{1}{\mathrm{k_B}T}\right)^{3/2}\int_0^\infty \sqrt{E}\mathrm{e}^{-E/(\mathrm{k_B}T)}\,\mathrm{d}E \\ &= C\sqrt{\frac{8}{\pi\mu_{\text{red}}}}\left(\frac{1}{\mathrm{k_B}T}\right)^{3/2}\sqrt{\frac{\pi}{4}}(\mathrm{k_B}T)^{3/2} = C\sqrt{\frac{2}{\mu_{\text{red}}}} = \text{const.}\end{aligned} \tag{4.13}$$

This is a remarkable result, as the reactivity turns out to be independent of the plasma temperature. Also, it suffices to evaluate the cross section at one energy to know the reactivity at all temperatures. Because s-waves often dominate the neutron capture cross sections for the *s* process (here the *s* stands for "slow neutron capture," the nucleosynthesis process further explained in Section 8.2), this is an important result since this relation is applied when experimentally determining cross sections for this process.

For p- and d-wave direct neutron capture ($\sigma^* \propto \sqrt{E}$ and $\sigma^* \propto E^{3/2}$, respectively), we obtain the reactivities

$$\langle\sigma^* v\rangle_{\text{p}} = C\sqrt{\frac{8}{\pi\mu_{\text{red}}}}\left(\frac{1}{\mathrm{k_B}T}\right)^{3/2}\frac{3}{4}\sqrt{\pi}(\mathrm{k_B}T)^{5/2} = \frac{3C}{\sqrt{2\mu_{\text{red}}}}\mathrm{k_B}T \tag{4.14}$$

and

$$\langle\sigma^* v\rangle_{\text{d}} = C\sqrt{\frac{8}{\pi\mu_{\text{red}}}}\left(\frac{1}{\mathrm{k_B}T}\right)^{3/2}\frac{15}{8}\sqrt{\pi}(\mathrm{k_B}T)^{7/2} = 30C\sqrt{\frac{2}{\mu_{\text{red}}}}(\mathrm{k_B}T)^2, \tag{4.15}$$

respectively. They are not independent of plasma temperature.

Although the reactivity for s-wave neutron capture is independent of temperature, the energy range within which the cross section has to be dominated by s-waves in order to simplify the integral accordingly is not. For neutron captures, the main contribution to the integral comes from the energy region around the peak of the Maxwell–Boltzmann distribution. Its location is only weakly affected by the energy dependence of the cross sections for neutron capture as long as there are no resonances (see below and Section 3.6.4). A rough approximation of the relevant energy window $\mathcal{G} \pm \Delta_G/2$ in practical units is

$$\mathcal{G} = 0.172 T_9 \left(\ell + \frac{1}{2}\right) \text{MeV}, \tag{4.16}$$

$$\Delta_G = 0.194 T_9 \sqrt{\ell + \frac{1}{2}} \ \text{MeV}, \tag{4.17}$$

where $\ell = 0, 1, 2,\ldots$ is the dominating partial wave and T_9 the temperature in GK. Within the energy region significantly contributing to the integral, the cross section has to display the energy dependence of a single partial wave, otherwise the simplification of the integral as shown in Equations (4.13)–(4.15) cannot be applied. With two or more contributing partial waves, the integration has to be performed numerically.

With increasing energy, higher partial waves become more important but s-waves will always remain non-negligible (if permitted by spin selection rules). Thus, the product $\sigma^* v$ (i.e., σv in the absence of stellar effects) becomes a slowly varying function of neutron velocity (or energy) and also the reactivity will change from constant to slowly varying with temperature. An expansion of $\sigma^* v = \sigma^* \sqrt{E}$ into a McLaurin series in terms of $\sqrt{E}$ can help with numerical estimates. (A McLaurin series is a special case of a Taylor series, centered around zero.) For the present case, this is

$$\sigma^* v \approx s(0) + \left.\frac{\mathrm{d}s}{\mathrm{d}\sqrt{E}}\right|_{E=0} \sqrt{E} + \frac{1}{2} \left.\frac{\mathrm{d}^2 s}{\mathrm{d}E}\right|_{E=0} E + \cdots \tag{4.18}$$

and a series of integrals

$$\begin{aligned} \langle \sigma^* v \rangle &= s(0) \sqrt{\frac{4}{\pi (\mathrm{k_B} T)^3}} \\ &\times \int_0^\infty \left(\sqrt{E} + \left.\frac{\mathrm{d}s}{\mathrm{d}\sqrt{E}}\right|_{E=0} \frac{E}{s(0)} + \frac{1}{2} \left.\frac{\mathrm{d}^2 s}{\mathrm{d}E}\right|_{E=0} \frac{E^{3/2}}{s(0)} + \cdots \right) \mathrm{e}^{-E/(\mathrm{k_B}T)} \, \mathrm{d}E \\ &= s(0) \left(1 + \left.\frac{\mathrm{d}s}{\mathrm{d}\sqrt{E}}\right|_{E=0} \frac{2}{s(0)} \sqrt{\frac{\mathrm{k_B} T}{\pi}} + \frac{3}{4} \left.\frac{\mathrm{d}^2 s}{\mathrm{d}E}\right|_{E=0} \frac{\mathrm{k_B} T}{s(0)} + \cdots \right) \end{aligned} \tag{4.19}$$

is obtained to approximate the reactivity of non-resonant neutron captures. The same analytic solutions as for the integrals in Equations (4.13)–(4.15) are used to arrive at the last line of Equation (4.19). The parameters $s(0)$, $\mathrm{d}s/\mathrm{d}\sqrt{E}$, and $\mathrm{d}^2 s/\mathrm{d}E$

can be determined experimentally and are found compiled in tabular form. Note that such an expansion works best when stellar effects (see Section 4.1.3) are negligible or do not vary too strongly across the relevant energy range and therefore $\sigma^* v = \sigma v = \sigma_{\mathrm{lab}} v$.

In connection with neutron captures for the *s* process (see Section 8.2) another quantity has been used: the *Maxwellian Averaged Cross section (MACS)*. Its introduction is motivated by the fact that the reactivity is constant for s-wave neutron captures. The reactivity is then factorized into a velocity and a factor which has the dimension of an area, similar to a cross section,

$$\langle \sigma^* v \rangle_{\mathrm{s}} = \langle \sigma^* \rangle_{\mathrm{s}} v_{\mathrm{T}} = \sigma_{\mathrm{s}}^*(v_{\mathrm{T}}) v_{\mathrm{T}} = \mathrm{const.} \tag{4.20}$$

The *thermal velocity* v_{T} is the most probable velocity as given by the velocity distribution under the integral in Equation (4.8). Because of the maximum at $E = \mathrm{k_B} T$ and $v = \sqrt{2E/\mu_{\mathrm{red}}}$ (note that the velocity distribution is v times the Maxwell–Boltzmann distribution, not just the Maxwell–Boltzmann distribution), it can be expressed as function of temperature,

$$v_{\mathrm{T}} = \sqrt{\frac{2\mathrm{k_B}T}{\mu_{\mathrm{red}}}}. \tag{4.21}$$

Thus the MACS can be derived from Equations (4.8) and (4.9) by dividing by Equation (4.21):

$$\begin{aligned} \langle \sigma^* \rangle = \frac{\langle \sigma^* v \rangle}{v_{\mathrm{T}}} &= \frac{4}{\sqrt{\pi}} \frac{1}{v_{\mathrm{T}}^2} \int_0^\infty v \sigma^*(v, T) \left(\frac{v}{v_{\mathrm{T}}} \right)^2 \mathrm{e}^{-(v/v_{\mathrm{T}})^2} \, \mathrm{d}v \\ &= \frac{2}{\sqrt{\pi}} \frac{1}{(\mathrm{k_B}T)^2} \int_0^\infty E \sigma^*(E, T) \mathrm{e}^{-E/(\mathrm{k_B}T)} \, \mathrm{d}E. \end{aligned} \tag{4.22}$$

For pure s-waves and negligible stellar effects, not only is the product $\langle \sigma^* \rangle v_{\mathrm{T}}$ constant but also $\langle \sigma^* \rangle = \sigma^*(v_{\mathrm{T}})$. This means that the MACS is equal to the cross section measured at v_{T} (or at the thermal energy $\mathrm{k_B}T = \mu_{\mathrm{red}} v_{\mathrm{T}}^2/2$, respectively). For other energy dependences of the cross section $\langle \sigma^* \rangle v_{\mathrm{T}}$ is not constant (compare with Equations (4.12)–(4.15)). Still, the MACS is related to the cross section at thermal velocity by a simple, temperature-independent factor: for example, $\langle \sigma^* \rangle = 2\sigma^*(v_{\mathrm{T}})/\sqrt{\pi}$ for $\sigma^* = \mathrm{const}$ and $\langle \sigma^* \rangle = 1.5\sigma^*(v_{\mathrm{T}})$ for $\sigma^* \propto v$ (p-waves). Again, note that these relations were originally introduced in the context of laboratory measurements where stellar effects are neglected and $\sigma^* \approx \sigma_{\mathrm{lab}}$ is assumed. Correction factors have to be applied when the stellar effects are not negligible (see Section 4.1.3).

4.1.2.2 Relevant Energies for Non-resonant Reactions with Charged Particles

So far, definitions well suited for the non-resonant interaction of neutrons with atomic nuclei were presented. Contrary to reactions with neutrons, the interaction of two positively or two negatively charged particles is strongly affected by Coulomb repulsion. To better assess how this affects the energy range contributing to the reactivity integral the *astrophysical S-factor* $\tilde{S}$ has been introduced. The reaction cross section is factorized as

$$\sigma(E) = \frac{1}{E}\tilde{S}(E)\mathrm{e}^{-2\pi\eta_{\mathrm{S}}} = \frac{1}{E}\tilde{S}(E)\mathrm{e}^{-b_{\mathrm{G}}/\sqrt{E}}. \tag{4.23}$$

In nuclear astrophysics often the S-factor $\tilde{S}$ is given instead of the cross section. The basic motivation for the definition of the S-factor is that in the absence of resonances it will be a slowly varying function with energy. The factor $1/E$ accounts for the energy dependence of the cross section due to the deBroglie wavelength (see Section 3.6). At least a part of the strong energy dependence of the cross section stemming from the Coulomb barrier is removed in the S-factor because of the exponential factor containing the *Sommerfeld parameter*

$$\eta_{\mathrm{S}} = \frac{Z_{\mathrm{a}} Z_{\mathrm{A}} \mathrm{e}_{\mathrm{C}}^{2}}{\hbar v}. \tag{4.24}$$

The factor $2\pi\eta_{\mathrm{S}}$ is called the *Gamow factor* and is related to the barrier penetrability. In practical units, $2\pi\eta_{\mathrm{S}} = 11.29 Z_{\mathrm{a}} Z_{\mathrm{A}} \sqrt{\mu_{\mathrm{red}}/E}$, with the interaction energy E in keV and Z_{a}, Z_{A} being the charge numbers of the interacting nuclei. In nuclear astrophysics, the Coulomb barrier penetration often is expressed in terms of the *Gamow parameter*

$$b_{\mathrm{G}} = 2\pi\eta_{\mathrm{S}}\sqrt{E} = \frac{\pi}{\hbar} Z_{\mathrm{a}} Z_{\mathrm{A}} \mathrm{e}_{\mathrm{C}}^{2} \sqrt{2\mu_{\mathrm{red}}}, \tag{4.25}$$

which is $b_{\mathrm{G}} = 0.989 Z_{\mathrm{a}} Z_{\mathrm{A}} \sqrt{\mu_{\mathrm{red}}}$ $\mathrm{MeV}^{1/2}$ in practical units, with the reduced mass μ_{red} given in m_{u} as usual. The square of b_{G} has the dimension of an energy and b_{G}^{2} is called the *Gamow energy*.

Inserting the parameterization of the cross section given in Equation (4.23) into the definition of the reactivity $\langle\sigma^{*}v\rangle$ as provided by Equation (4.9), assuming absence of stellar effects (see, e.g., Equation (4.42) and its discussion), and using the Gamow parameter, it becomes obvious that the integral extends over a product of two exponentials,

$$\begin{aligned}\langle\sigma v\rangle &= \sqrt{\frac{8}{\pi\mu_{\mathrm{red}}}}\left(\frac{1}{kT}\right)^{3/2}\int_{0}^{\infty}\tilde{S}(E)\mathrm{e}^{-b_{\mathrm{G}}/\sqrt{E}}\mathrm{e}^{-E/(k_{\mathrm{B}}T)}\,\mathrm{d}E\\ &= \sqrt{\frac{8}{\pi\mu_{\mathrm{red}}}}\left(\frac{1}{kT}\right)^{3/2}\int_{0}^{\infty}\tilde{S}(E)\mathrm{e}^{-\left(\frac{b_{\mathrm{G}}}{\sqrt{E}}+\frac{E}{k_{\mathrm{B}}T}\right)}\,\mathrm{d}E.\end{aligned} \tag{4.26}$$

With a weak energy dependence of the astrophysical S-factor the energy region contributing to the integral is given by the product of the two exponential functions, i.e., the high-energy tail of the Maxwell–Boltzmann distribution and the low-energy part of the penetrability through the Coulomb barrier. Since the two have an opposite dependence on the energy their product forms a peak, the so-called *Gamow peak*. The width and location of the Gamow peak depends on the charges Z_a, Z_A of the interacting particles and on the plasma temperature T. The Gamow peak is located at higher energy than the peak of the Maxwell–Boltzmann distribution. Increasing charge or increasing temperature shifts the Gamow peak to even higher energies. This is illustrated in Figures 4.1 and 4.2.

Studying reactions involving light nuclei it has become customary to approximate the Gamow peak by a Gaussian,

$$e^{-\left(\frac{b_G}{\sqrt{k_B T}}+\frac{E}{k_B T}\right)} \approx e^{-\left(\frac{E-\mathcal{G}}{\Delta_G/2}\right)^2}, \tag{4.27}$$

to obtain a simple formula for the relevant energy window within which the cross sections have to be determined, the *Gamow window*. For example, from the Gaussian approximation the location $\mathcal{G}$ and width Δ_G (as 1/e width of the Gaussian) of the Gamow window is obtained as

$$\mathcal{G} = \left(\frac{b_G k_B T}{2}\right)^{2/3}, \tag{4.28}$$

$$\Delta_G = \frac{4}{\sqrt{3}}\sqrt{\mathcal{G} k_B T}. \tag{4.29}$$

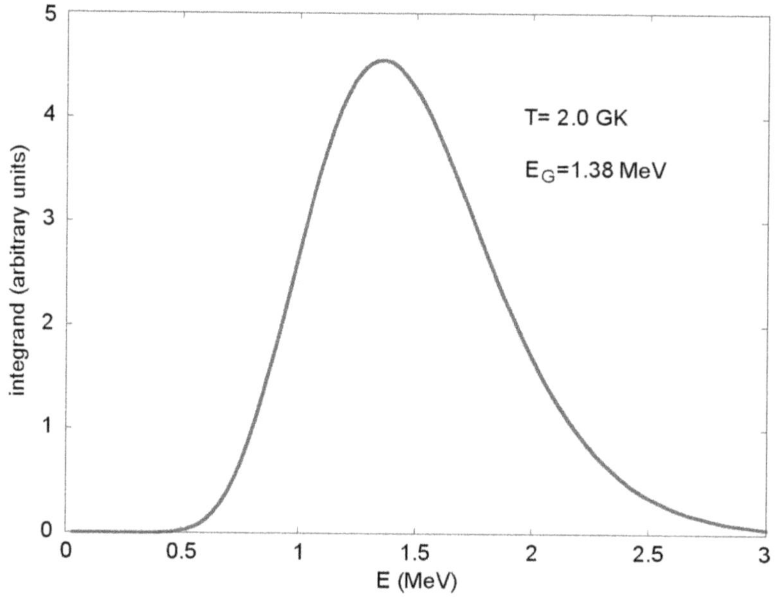

Figure 4.1. (Animation) Gamow peak for different temperatures at fixed charge $Z_a Z_A = 12$. The position $\mathcal{G}$ of the peak shifts to higher energy with increasing temperature. Also the height of the peak, which determines the magnitude of the rate, is very sensitive to the temperature. Animation available online at https://doi.org/10.1088/978-0-7503-1149-6.

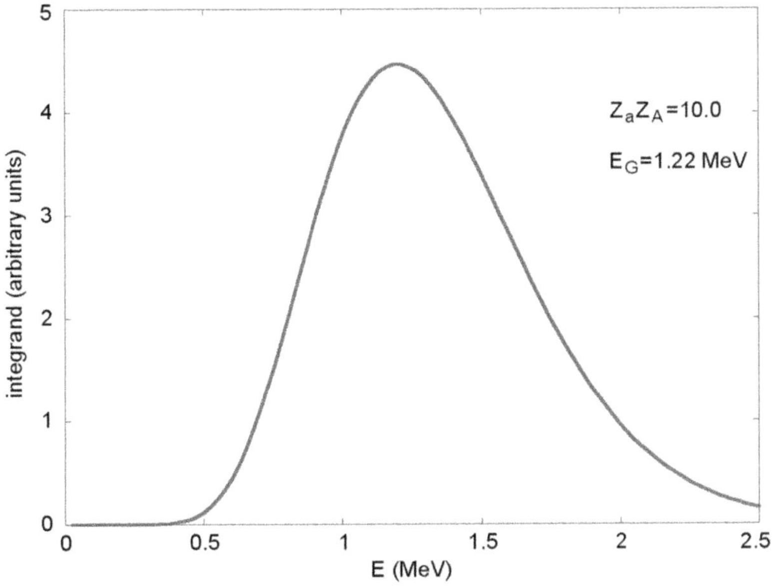

Figure 4.2. (Animation) Gamow peak for different charges at fixed temperature $T_9 = 2$. The position $\mathcal{G}$ of the peak shifts to higher energy when the product $Z_a Z_A$ increases. The height of the peak decreases with increasing $Z_a Z_A$ as the Coulomb barrier increases. Note that the fractional charges used to obtain a smooth animation actually are unphysical. Animation available online at https://doi.org/10.1088/978-0-7503-1149-6.

In practical units, this is

$$\mathcal{G} = 0.122\,04\left(\mu_{\mathrm{red}} Z_a^2 Z_A^2 T_9^2\right)^{1/3} \text{ MeV}, \tag{4.30}$$

$$\Delta_G = 0.236\,82\left(\mu_{\mathrm{red}} Z_a^2 Z_A^2 T_9^5\right)^{1/6} \text{ MeV}, \tag{4.31}$$

again inserting the reduced mass in m_u and noting that T_9 denotes the temperature in GK.

4.1.2.3 Resonant Rates

When the stellar cross section or the astrophysical S-factor displays a stronger and more complicated dependence on energy, the reactivity is obtained by numerical integration over the S-factor. Regardless of whether the reaction is with neutrons or with charged projectiles this is necessary, for example, for cross sections showing resonances and in particular resonances interfering with each other or with a direct reaction background (see Section 3.6.4). There are two limiting cases, however, which can be more easily dealt with: very narrow and very broad resonances without interference.

The Breit–Wigner formula to compute the cross section σ_{BW} for an isolated resonance is given in Equation (3.108). For a narrow resonance $\hat{\Gamma}_{tot} \ll E_{res}$ and $\hat{\Gamma}_{tot} \ll \Delta_G$. In this case the Maxwell–Boltzmann distribution can be assumed constant across the width of the resonance and taken out of the integral,

$$\langle\sigma v\rangle_{\rm narrow} \approx \sqrt{\frac{8}{\pi\mu_{\rm red}}}\left(\frac{1}{kT}\right)^{3/2} E_{\rm res}e^{E_{\rm res}/(k_B T)} \int_0^\infty \sigma_{\rm BW}\, dE. \tag{4.32}$$

Also the total and partial widths appearing in Equation (3.108) can be taken to be constant across the resonance width and the deBroglie wavelength $\lambda\!\!\!^{-}$ is also calculated at $E_{\rm res}$. This leads to

$$\begin{aligned}\langle\sigma v\rangle_{\rm narrow} &\approx \sqrt{\frac{8}{\pi\mu_{\rm red}}}\left(\frac{1}{kT}\right)^{3/2} E_{\rm res}e^{E_{\rm res}/(k_B T)} 2\pi\lambda\!\!\!^{-2}\omega_{\rm BW}\frac{\hat{\Gamma}^{\rm Aa}\hat{\Gamma}^{\rm Bb}}{\hat{\Gamma}_{\rm tot}} \\ &= \hbar^2\left(\frac{2\pi}{\mu_{\rm red}k_B T}\right)^{3/2} \omega_{\rm BW}\gamma_{\rm BW}e^{-E_{\rm res}/(k_B T)}.\end{aligned} \tag{4.33}$$

The product

$$\omega_{\rm BW}\frac{\hat{\Gamma}^{\rm Aa}\hat{\Gamma}^{\rm Bb}}{\hat{\Gamma}_{\rm tot}} = \omega_{\rm BW}\gamma_{\rm BW} \tag{4.34}$$

is called *resonance strength* and has the dimension of energy ($\omega_{\rm BW}$ is the usual combined spin factor as defined in Equation (3.109)). An expression for Equation (4.33) in practical units is

$$N_A\langle\sigma v\rangle_{\rm narrow} \approx \frac{1.54\times10^{11}}{(\mu_{\rm red}T_9)^{3/2}}\omega_{\rm BW}\gamma_{\rm BW}e^{-11.6045E_{\rm res}/T_9}\ {\rm cm^3\,mol^{-1}\,s^{-1}} \tag{4.35}$$

with $\gamma_{\rm BW}$ given in MeV. Examination of the width ratio in Equation (4.33) reveals an interesting fact. If the larger of the two widths appearing in the numerator $\hat{\Gamma}^{\rm Aa}\hat{\Gamma}^{\rm Bb}$ also dominates the total width $\hat{\Gamma}_{\rm tot}$ in the denominator, then the resonance strength is determined by the *smaller* width as the larger width cancels between numerator and denominator. This is similar to what was found in Section 3.7.1.2 for the sensitivity of cross sections in the Hauser–Feshbach model, which also depend on a ratio of transmission coefficients (or widths, respectively). Having the major contribution to the total width coming from a single width is encountered often, especially in nuclear astrophysics. For example, the neutron width will be larger than all other particle or photon widths well above the neutron threshold. On the other hand, charged particle widths are often even smaller than photon widths because at astrophysical energies they are strongly suppressed by the Coulomb barrier. The question of which width determines the energy dependence of the cross section and thus the reactivity will also play a role in the discussion of the limitations of the Gamow peak notion further down below.

A resonance is called broad when $\hat{\Gamma}_{\rm tot}/E_{\rm res} > 0.1$. In general, the energy dependence of a broad resonance has to be taken into account by numerically integrating the product of cross section and Maxwell–Boltzmann distribution. If the resonance width, however, is larger than the energy window contributing to the reaction rate integral (for charged particle reactions this is the Gamow window), $\hat{\Gamma}_{\rm tot} \gg \Delta_G$, then the S-factor will be slowly varying within the energy window. Similarly, the S-factor will only be slowly varying if the Gamow window is located far away from the resonance and only the wing of the resonance extends into it. Broad resonances have to be considered even when their nominal resonance energy is outside this energy window because the resonance wing can still extend into it. The wing of a broad resonance contributes similarly to a direct component. This can be seen by realizing that the partial widths appearing in Equation (3.108), the Breit–Wigner formula for single resonances, will only be weakly energy dependent. For a reaction with positive reaction Q-value the relative energy in the exit channel $E_{\mathrm{Bb}} = E_{\mathrm{Aa}} + Q_{\mathrm{nuc}}$ will be larger than the one in the entrance channel E_{Aa}. Therefore the relative variation $(E_{\mathrm{Bb}} + \Delta E_{\mathrm{Aa}})/E_{\mathrm{Bb}} = (E_{\mathrm{Aa}} + \Delta E_{\mathrm{Aa}} + Q_{\mathrm{nuc}})/(E_{\mathrm{Aa}} + Q_{\mathrm{nuc}})$ will be smaller than the relative variation of the relative energy in the entrance channel $(E_{\mathrm{Aa}} + \Delta E_{\mathrm{Aa}})/E_{\mathrm{Aa}}$. This may also lead to a weaker energy dependence of the width in the exit channel than in the entrance channel. Similarly, the total width will be weakly energy dependent. (For reactions with negative Q-value a similar argument can be made with the roles of entrance and exit channel exchanged; see, however, Section 4.1.2.4.) In this case, the width in the exit channel $\hat{\Gamma}^{\mathrm{Bb}}$ can be assumed to be constant and the expression for the cross section reduces to the form

$$\begin{aligned}\sigma(E_{\mathrm{Aa}}) &\approx \pi\lambdabar \hat{\Gamma}^{\mathrm{Aa}}(E_{\mathrm{Aa}})\frac{\hat{\Gamma}^{\mathrm{Bb}}}{(E_{\mathrm{Aa}} - E_{\mathrm{res}})^2 + \hat{\Gamma}^2_{\mathrm{tot}}/4} \\ &= F(E_{\mathrm{Aa}})\frac{\hat{\Gamma}^{\mathrm{Bb}}}{(E_{\mathrm{Aa}} - E_{\mathrm{res}})^2 + \hat{\Gamma}^2_{\mathrm{tot}}/4} = F(E_{\mathrm{Aa}})C,\end{aligned} \tag{4.36}$$

with an energy-dependent factor $F(E_{\mathrm{Aa}})$ and another factor C with very weak energy dependence for $|E_{\mathrm{Aa}} - E_{\mathrm{res}}| \gg 0$. The factor $F(E_{\mathrm{Aa}})$ has a similar energy dependence as a non-resonant cross section, such as a direct reaction (Sections 3.7.2 and 4.1.2.1), and therefore the resulting reactivity looks like the ones derived above using the concept of the astrophysical S-factor or making use of dominating partial waves. Particularly for s-wave neutron capture, often there is a $1/v$ contribution at thermal energies through the tails of higher lying s-wave resonances. In fact, wings of broad resonances and contributions of direct reactions are often (accidentally) confused or (consciously) mixed in literature. Despite of the similarity in properties, these two types of contributions to the reaction rate are caused by completely different reaction mechanisms (see Section 3.7). Furthermore, contrary to the direct reaction part, the properties of broad resonances (such as resonance energy and width) can also be determined outside of the astrophysical energy window, which is useful in experimental investigations. On the other hand, it should be noted that broad resonances located below the reaction threshold (implying they formally have a negative resonance energy E_{res}) can also contribute to the reaction cross sections. The wings of these *subthreshold resonances* can extend to energies above the threshold and into the Gamow window. The experimental determination of their resonance properties is not directly possible but has to be inferred via other reaction channels.

Note that the above discussion of resonant rates assumes that stellar effects, discussed in Section 4.1.3, can be neglected.

4.1.2.4 Limitations of the Gamow Peak Concept

As explained in Section 4.1.3.1, the stellar rate has to be calculated by integrating over a stellar cross section, which can be different than the laboratory cross section. Accordingly, the derived relevant energy window depends on whether the stellar cross section or laboratory cross section is used. Since the Gamow window usually is used to determine the energy range in which measurements are to be carried out, usually the laboratory cross section is applied. In principle, the stellar Gamow window can also be calculated but this requires to numerically evaluating the function in the reaction rate integral. Approximation formulae are not available in this case, unless the thermal excitation of target states is negligible (see Section 4.1.3). In addition to the complication arising from stellar effects, there are further limitations when using the notion of the Gamow peak. These are discussed below.

Although helpful in many respects, the simple Gaussian approximation of the Gamow window $\mathcal{G} \pm (\Delta_G/2)$ shown in Equations (4.28), (4.29) is valid for some but not all cases because it is oversimplified. Moreover, the Gamow peak concept itself, as given in Equation (4.26), cannot be applied in all cases. Several assumptions enter the description:

1. The energy dependence of the reaction cross section is determined by the entrance channel.
2. The S-factor is a weakly and smoothly varying function of energy.
3. The penetration through the Coulomb barrier can be approximated by the Gamow parameter given in Equation (4.25).
4. If using Equations (4.28) and (4.29): The shape of the Gamow peak can be approximated by a symmetric Gaussian.

The Gamow window concept was originally conceived in the context of charged-particle reactions occurring in solar hydrogen burning (see Chapter 7). For such reactions with light nuclei the above assumptions are valid in the absence of resonances. For reactions with heavier nuclei, however, which also have larger charge number Z and thus higher (and thicker) Coulomb barriers, assumptions 3 and 4 become increasingly invalid. For thick Coulomb barriers and interaction energies well below the barrier height, the resulting peak exhibits a strong asymmetry. Furthermore, the WKB approximation leading to Equations (4.24) and (4.25) as measures of the barrier penetration (see Section 3.5) becomes inaccurate for thick barriers. This results in an incorrect prediction of both $\mathcal{G}$ and Δ_G when using Equations (4.25)–(4.29) for heavier nuclei. Figure 4.3 shows the difference between the location of the actual peak in the integrand and the predicted $\mathcal{G}$ for a range of mass numbers.

Obviously, assumption 2 does not hold for cross sections exhibiting individual resonance features. For narrow resonances the Gamow window is split into the contributions of the individual resonances. Narrow resonances with resonance energies within the Gamow window as defined for slowly varying S-factor

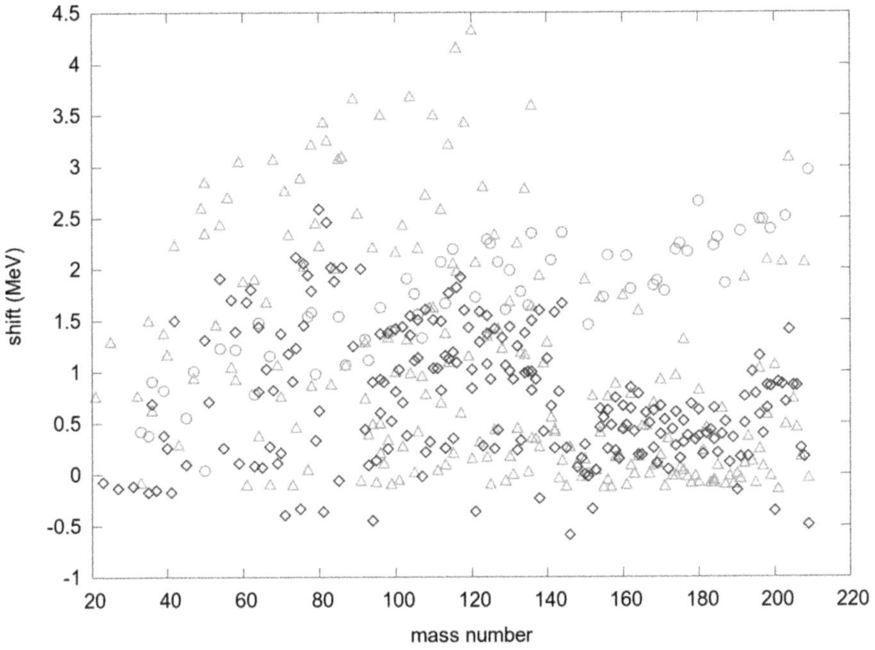

Figure 4.3. Shift of the actual position of the peak in the reaction rate integral relative to the location of the Gamow peak $\mathcal{G}$ predicted with Equation (4.28), as function of mass number. Shown are the shifts at $T_9 = 2$ for (n,p) (red circles), (n,α) (blue diamonds), and (p,α) (green triangles) reactions on stable nuclei. The predicted $\mathcal{G}$ for the neutron-induced reactions has been computed from the reverse reaction and translated to the energy scale of the forward reaction using the reaction Q-value (see text).

(Equation (4.26)), however, will contribute most. It was also shown above that with a Gamow window in the wing of broad resonances, the usual formalism still holds. It has to be noted, however, that resonances of the same spin and parity may interfere with each other. Likewise, resonances can interfere with a direct reaction component with the same partial wave contribution (see Section 3.7.1.1). If such interference effects occur, the reaction rate cannot be determined by simply adding resonant and non-resonant contributions. Rather, it has to be computed by coherently adding the contributions according to quantum mechanics to consider constructive and destructed interference. In this case, numerical integration is necessary and the predicted Gamow window provides only a rough estimate of the energies from which the dominating contributions may be originating.

Often overlooked is the required validity of assumption 1 when deriving a Gamow window as shown in Equations (4.25)–(4.29). The problem is trivially evident when trying to calculate the Gamow window for a neutron-induced reaction with charged particles in the exit channel, such as, for example, (n,p) or (n,α) reactions. There is no Coulomb barrier in the entrance channel but still the reaction cross section is strongly suppressed toward lower energies due to the Coulomb barrier in the exit channel. The obvious solution in this case is to determine the Gamow window for the reverse reaction, that is (p,n) and (α,n), respectively, and then to convert the obtained energy range via the reaction Q-value. This can be done because of the reciprocity of forward and reverse cross sections and rates (see Section 4.1.3). For reactions involving photons, however, this is not always advisable, as it is not guaranteed that the particle widths appearing in the reaction are actually the smallest widths determining the energy dependence of the cross section. For the importance of the smallest width, see Section 3.6 and the discussion of Equations (4.33) and (4.35), as well as of Equation (3.132).

The most general recipe is to determine the reaction channel in which the smallest width $\hat{\Gamma}$ is appearing and to calculate the Gamow window in this channel. In (α,γ) or (p,γ) reactions at astrophysical energies, the charged particle widths are strongly energy-dependent because of the Coulomb suppression. The height of the Coulomb barrier depends on the charge of the target nucleus. For heavier nuclei, the charged particle width usually is smaller than $\hat{\Gamma}_\gamma$ and the Gamow peak concept can be applied. For light nuclei, however, $\hat{\Gamma}_\gamma < \hat{\Gamma}_\alpha$ or $\hat{\Gamma}_\gamma < \hat{\Gamma}_\mathrm{p}$ in some cases (also depending on the reaction Q-value which determines the relative energies appearing in the entrance and exit channel). In fact, this is the regular case in standard nuclear physics where reactions at higher than astrophysical energies are studied. Going to lower energy $\hat{\Gamma}_\alpha$, $\hat{\Gamma}_\mathrm{p}$ will become smaller than $\hat{\Gamma}_\gamma$. Whether this is the case at astrophysically relevant energies remains to be determined for each individual case. The Gamow peak concept can only be applied when this condition is fulfilled. If $\hat{\Gamma}_\gamma$ remains smaller than the particle width, significant contributions to the reaction rate integral may also come from outside the energy window estimated by Equations (4.28) and (4.29), for example from narrow resonances. This dependence on partial widths in the entrance and exit channel also explains why different reactions with the same configuration in the entrance channel may have different relevant energy windows, for example, the (α,n), (α,γ), or (α,p) reactions on the same target nucleus. Summarizing the recipe for ensuring the validity of assumption 1:

- If the reaction is a direct reaction, choose a reaction channel of interacting particles (no photons) to determine the Gamow window and convert to the desired channel, if necessary.
- If the reaction is a compound reaction:
    1. Compare partial widths in the entrance and exit channel of the reaction.
    2. If the smallest width is the γ width: Gamow window has to be determined numerically.
    3. If the smallest width is a particle width, use Equation (4.26) or the approximation of Equation (4.27) in the channel of this width and convert channel energies, if necessary.

Finally, it should be remembered that the width of the Gamow window as given in Equation (4.29) is based on the 1/e-width of the peak, which actually comprises only about 60% of the total contribution to the rate integral by the peak. To obtain better accuracy in the reaction rate it is necessary to know the cross section in a larger energy range, for example, $2\Delta_G$. Due to the asymmetry of the Gamow peak especially for heavier nuclei it is advisable to refer to numerical investigations of the integrand shape. The Gamow window as given by Equations (4.28), (4.29) can serve only as a crude estimate of the astrophysically relevant energies.

4.1.3 Stellar Effects on Reaction Cross Sections and Rates

4.1.3.1 Stellar Cross Section

In an astrophysical plasma, most nuclei quickly (on the timescale of nuclear reactions and scattering) reach thermal equilibrium with all plasma components. This leads to a thermal excitation of nuclei and thus to a fraction of all nuclei being in excited states instead of being in the ground state (see Section 3.4). The population of a nuclear level is given by a *Boltzmann factor*,

$$[2J_\mu + 1]\mathrm{e}^{-E_\mu^\mathrm{x}/(\mathrm{k_B}T)} = g_\mu \mathrm{e}^{-E_\mu^\mathrm{x}/(\mathrm{k_B}T)}, \tag{4.37}$$

with spin factor $g_\mu = 2J_\mu + 1$ as defined in Equation (3.82) and excitation energy E_μ^x of level μ. This relation can be derived from a Saha equation (see also Equation (7.39)). Note that here and in all of Section 4.1.3 round brackets signify arguments of functions whereas arithmetic brackets are shown as square and curly brackets. This is to avoid confusion between function arguments and multiplication factors.

The *relative population* $\mathcal{P}_\mu$ is normalized by the sum of Boltzmann factors, which is the nuclear partition function G,

$$\mathcal{P}_\mu(T) = \frac{g_\mu \mathrm{e}^{-E_\mu^\mathrm{x}/(\mathrm{k_B}T)}}{\sum_\mu g_\mu \mathrm{e}^{-E_\mu^\mathrm{x}/(\mathrm{k_B}T)}} = \frac{g_\mu \mathrm{e}^{-E_\mu^\mathrm{x}/(\mathrm{k_B}T)}}{G(T)}. \tag{4.38}$$

Often, the partition function normalized to the ground state G_0 is used instead of the regular partition function G. It is defined as

$$G_0(T) = \frac{G(T)}{g_0} = \frac{1}{g_0}\left[\left\{\sum_\mu g_\mu \mathrm{e}^{-E_\mu^\mathrm{x}/(\mathrm{k_B}T)}\right\} + \int_{E_\mu^{\mathrm{x}}{}_\mathrm{last}}^{\infty} \sum_{J,\pi} g_J \mathrm{e}^{-\varepsilon/(\mathrm{k_B}T)} \tilde{\rho}_\mathrm{nuc}(\varepsilon, J, \pi)\, \mathrm{d}\varepsilon\right], \tag{4.39}$$

with the ground state being identified as $\mu = 0$, the first excited state as $\mu = 1$, and so on. Equation (4.39) shows additionally how the computation can be extended beyond the energy of the highest known discrete energy level μ^last by using an integration of a nuclear level density $\tilde{\rho}_\mathrm{nuc}$ (being a function of energy ε, spin J, and parity π) over a range of excitation energies ε. Although the product of the Boltzmann factor and $\tilde{\rho}_\mathrm{nuc}$ does not have a trivial energy dependence, it has been shown that for the application of Equation (4.39) at temperatures $T \leqslant 10$ GK it is sufficient to integrate only up to $E_{\rm max} \approx 25$ MeV. Temperatures above 10 GK are encountered in some explosive astrophysical events, in accretion disks, and in the formation of neutron stars and black holes. Nuclear transformations in such environments are described in reaction equilibria between several or all possible reactions, replacing full reaction networks by simplified abundance equations (see Section 4.3 for details) which contain the partition functions. At such high temperatures, a straightforward application of Equation (4.39) would overestimate the partition function because particle emission into the continuum has to be taken into account. This can be treated by approximated correction factors to $\tilde{\rho}_{\rm nuc}$ and extending the integration to $E_{\rm max} \gg 35$ MeV.

Since excited states of target nuclei are populated in a stellar plasma, the stellar reactivity $\langle\sigma^*v\rangle$ can be expressed as a sum of individual reactivities $\langle\sigma v\rangle^{\mu}_{\rm Aa}$, each for a reaction starting on a given, different excited state. Here, the index μ is not a mathematical exponent but identifies the starting level. A scheme of the energetics and the transitions between nuclear levels in the involved nuclei is shown in Figure 3.5. Here and below the shorthand

$$\sigma^{\mu}_{\rm Aa} = \sum_{\nu} \sigma^{\mu\nu}_{\rm Aa} \tag{4.40}$$

and

$$\langle\sigma v\rangle^{\mu}_{\rm Aa} = \sum_{\nu} \langle\sigma v\rangle^{\mu\nu}_{\rm Aa} \tag{4.41}$$

is used. Note that no asterisks have been used in the above reactivities as they are not reactivities at stellar conditions but involve cross sections for reactions starting on an individual level labeled μ. When combining them to a total stellar reactivity, these individual reactivities are weighted by the relative population factor $\mathcal{P}$ of each excited state μ as defined in Equation (4.38), and thus

$$\begin{aligned}\langle\sigma^*v\rangle_{\rm Aa} &= \mathcal{P}_0\langle\sigma v\rangle^{0}_{\rm Aa} + \mathcal{P}_1\langle\sigma v\rangle^{1}_{\rm Aa} + \mathcal{P}_2\langle\sigma v\rangle^{2}_{\rm Aa} + \cdots = \sum_{\mu} \mathcal{P}_\mu\langle\sigma v\rangle^{\mu}_{\rm Aa} \\ &= \left[\frac{8}{\mu^{\rm Aa}_{\rm red}\pi}\right]^{1/2} [{\rm k_B}T]^{-3/2} \sum_{\mu}\left[\mathcal{P}_\mu \int_0^\infty \sigma^{\mu}_{\rm Aa} E^{\rm A}_{\mu} {\rm e}^{-\frac{E^{\rm A}_{\mu}}{{\rm k_B}T}}\, {\rm d}E^{\rm A}_{\mu}\right].\end{aligned} \tag{4.42}$$

This means that projectiles with Maxwell–Boltzmann distributed energies are acting on each level μ separately, as illustrated by Figure 4.4. Since each excited state is exposed to the full energy range, each integral in Equation (4.42) has its own energy scale, ranging from zero to infinity but shifted relative to each other by the excitation energy of the level $E^{\rm x}_{\mu}$ (note the different ${\rm d}E^{\rm A}_{\mu}$ in each integral of the sum). This can be seen more explicitly when inserting the definition of the population factor from Equation (4.38), which leads to

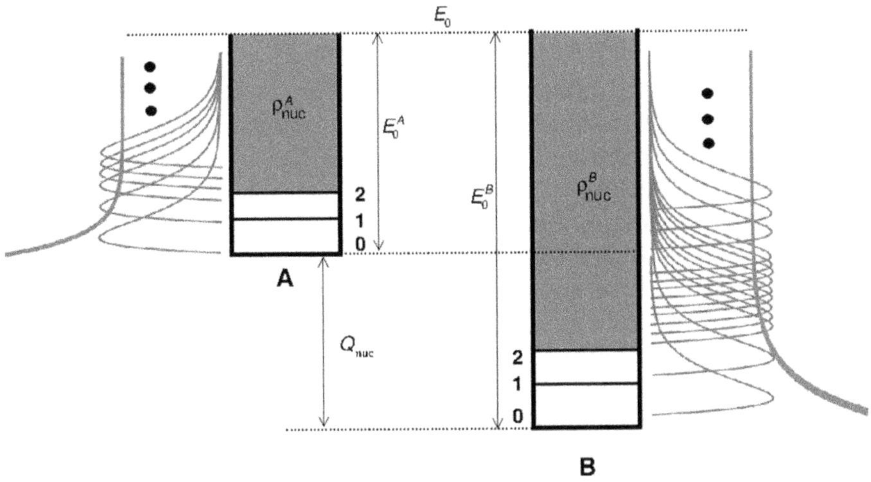

Figure 4.4. Sketch of how Maxwell–Boltzmann distributed particle energies act on excited states 0, 1, 2, ... in nuclei A and B connected by a reaction A + a ↔ b + B with positive reaction Q-value $Q_{\rm nuc}$. The notation is the same as used in Figure 3.5. The Maxwell–Boltzmann distributions for each excited states are sketched in red, the weight of each distribution is given by an exponentially decaying population factor $\mathcal{P}$, indicated by the thick brown line falling off toward higher excitation energy. In each channel the sum of the weighted integrals over all distributions (Equation (4.42)) is equal to a single integral over a sum of distributions shifted down to the ground state (Equation (4.47)) and weighted by the linear weight shown in Equation (4.55) and Figure 4.5.

$$\begin{aligned}\sum_\mu\left[\mathcal{P}_\mu\int_0^\infty \sigma^\mu_{\rm Aa}E^{\rm A}_\mu {\rm e}^{-\frac{E^{\rm A}_\mu}{{\rm k_B}T}}\,{\rm d}E^{\rm A}_\mu\right] &= \sum_\mu\int_0^\infty \mathcal{P}_\mu\sigma^\mu_{\rm Aa}E^{\rm A}_\mu {\rm e}^{-\frac{E^{\rm A}_\mu}{{\rm k_B}T}}\,{\rm d}E^{\rm A}_\mu\\ &= \sum_\mu\int_0^\infty \frac{g^{\rm A}_\mu {\rm e}^{-E^{\rm x}_\mu/{\rm k_B}T}}{g^{\rm A}_0 G^{\rm A}_0}\sigma^\mu_{\rm Aa}E^{\rm A}_\mu {\rm e}^{-\frac{E^{\rm A}_\mu}{{\rm k_B}T}}\,{\rm d}E^{\rm A}_\mu \qquad (4.43)\\ &= \frac{1}{G^{\rm A}_0}\sum_\mu\int_0^\infty \frac{g^{\rm A}_\mu}{g^{\rm A}_0}\sigma^\mu_{\rm Aa}E^{\rm A}_\mu {\rm e}^{-\frac{E^{\rm A}_\mu+E^{\rm x}_\mu}{{\rm k_B}T}}\,{\rm d}E^{\rm A}_\mu,\end{aligned}$$

where $G^{\rm A}_0$ is the normalized partition function as defined in Equation (4.39). Because of this energy shift of the Maxwell–Boltzmann distributions it is not straightforward to define the quantity $\sigma^*(E, T)$ to be used in Equation (4.9).

Before proceeding to the derivation of the stellar cross section $\sigma^*(E, T)$, it is useful to consider a theoretical quantity called *effective cross section* $\sigma^{\rm eff}$. The effective cross section is a sum over all energetically possible transitions of initial levels to final levels. It includes all the transitions shown by arrows in Figure 3.5 and therefore sums over all final levels ν *and* initial levels μ

$$\sigma^{\rm eff}_{\rm Aa}(E^{\rm A}_0) = \sum_\mu\sum_\nu \frac{g^{\rm A}_\mu}{g^{\rm A}_0}\frac{E^{\rm A}_\mu}{E^{\rm A}_0}\sigma^{\mu\nu}_{\rm Aa}\left(E^{\rm A}_\mu\right) = \sum_\mu \frac{g^{\rm A}_\mu}{g^{\rm A}_0}\frac{E^{\rm A}_\mu}{E^{\rm A}_0}\sigma^\mu_{\rm Aa}\left(E^{\rm A}_\mu\right). \qquad (4.44)$$

The relative center-of-mass energy of a transition proceeding from level μ is denoted by $E_\mu^\mathrm{A} = E_0^\mathrm{A} - E_\mu^\mathrm{x}$, with $\sigma_\mathrm{Aa}^{\mu\nu}(E_\mu^\mathrm{A}) = \sigma_\mathrm{Aa}^{\mu}(E_\mu^\mathrm{A}) = 0$ for $E_\mu^\mathrm{A} < 0$. The first term ($\mu = 0$) in the sum over μ is just the laboratory cross section $\sigma_\mathrm{lab} = \sigma^0$. Note that the effective cross section is only a function of energy, like the usually defined cross section, and does not depend on temperature, contrary to the stellar cross section.

The effective cross section is, of course, unmeasurable. Its usefulness becomes apparent later, when we combine Equation (4.38) with definition of the astrophysical reaction rate in Equation (4.9). An interesting property of the effective cross section is that it conserves reciprocity between forward and reverse effective cross section. When interchanging the labels μ and ν, and replacing label A by B, the quantity $\sigma_\mathrm{Bb}^\mathrm{eff}$ is obtained for the reverse direction commencing on levels ν, in complete analogy to the definition in Equation (4.44). The well-known reciprocity relation of nuclear reactions (derived in Section 3.6) connects the cross sections of a forward reaction ($\sigma_\mathrm{Aa}^{\mu\nu}$) and its reverse reaction ($\sigma_\mathrm{Bb}^{\nu\mu}$) between single states μ, ν in initial and final nucleus, respectively. Making use of Equation (3.95) it is straightforward to show that the two effective cross sections obey the reciprocity relations

$$\sigma_\mathrm{Bb}^\mathrm{eff} = \frac{g_0^\mathrm{A} g_\mathrm{a} \mu_\mathrm{red}^\mathrm{Aa} E_0^\mathrm{A}}{g_0^\mathrm{B} g_\mathrm{b} \mu_\mathrm{red}^\mathrm{Bb} E_0^\mathrm{B}} \sigma_\mathrm{Aa}^\mathrm{eff}, \tag{4.45}$$

$$\sigma_\mathrm{Bb}^\mathrm{eff} E_0^\mathrm{B} = \frac{g_0^\mathrm{A} g_\mathrm{a} \mu_\mathrm{red}^\mathrm{Aa}}{g_0^\mathrm{B} g_\mathrm{b} \mu_\mathrm{red}^\mathrm{Bb}} \sigma_\mathrm{Aa}^\mathrm{eff} E_0^\mathrm{Aa}, \tag{4.46}$$

although they include transitions between all energetically accessible excited states in target and final nucleus. The relative energies of the transitions proceeding on the ground states of the two target nuclei for forward and reverse reaction are denoted by E_0^A and E_0^B, respectively.

As discussed in Section 3.6, laboratory reactions commencing on nuclei in the ground state populate a number of final states ($\sigma_\mathrm{lab} = \sigma^0 = \sum_\nu \sigma^{0\nu}$) and thus the reciprocity relation does not apply. A stellar cross section includes additional reactions on excited states of the target nucleus but not all transitions contribute equally, as in the effective cross section. Nevertheless, also a reciprocity relation between forward and reverse stellar rates can be found. The important prerequisite is that these are indeed *stellar* rates, that means rates obtained by folding the *stellar* cross sections with the appropriate projectile energy distribution. Rates obtained from laboratory cross sections are not connected by a reciprocity relation unless they only include transitions between a single level in the entrance channel and another single level in the exit channel.

Returning to Equations (4.42) and (4.43), and continuing with the derivation of the stellar cross section, we have to proceed as follows. In order to obtain an expression similar to the original definition of the reactivity with a single integral, the energy scale of each term in the sum in Equation (4.43) has to be shifted to fit a common energy scale. Only in this case can the sum of integrals be replaced by a single integral. The common energy scale could be chosen freely but a natural choice is the one in which the ground state has zero excitation energy. Transforming $dE_\mu^A \to dE_0^A$ (with $E_0^A = E_\mu^A + E_\mu^x$) yields

$$\begin{aligned}
&\frac{1}{G_0^A}\sum_\mu \int_0^\infty \frac{g_\mu^A}{g_0^A}\sigma_{Aa}^\mu\left(E_\mu^A\right)E_\mu^A e^{-\frac{E_\mu^A+E_\mu^x}{k_B T}}\, dE_\mu^A \\
&= \frac{1}{G_0^A}\sum_\mu \int_{E_\mu^x}^\infty \frac{g_\mu^A}{g_0^A}\sigma_{Aa}^\mu\left(E_0^A - E_\mu^x\right)\left[E_0^A - E_\mu^x\right]e^{-\frac{E_0^A}{k_B T}}\, dE_0^A \\
&= \frac{1}{G_0^A}\sum_\mu \int_0^\infty \frac{g_\mu^A}{g_0^A}\sigma_{Aa}^\mu\left(E_\mu^A\right)\left[E_0^A - E_\mu^x\right]e^{-\frac{E_0^A}{k_B T}}\, dE_0^A.
\end{aligned} \tag{4.47}$$

The argument of the function σ is given in round brackets whereas the square brackets act as the usual arithmetic brackets. In the last line above, the lower limit of the integration was reset to zero. This is allowed because cross sections at negative energies do not give any contribution to the integral. It is mathematically equivalent to exchange summation and integration in Equation (4.47) and so we obtain

$$\begin{aligned}
\langle\sigma^* v\rangle_{Aa} &= \left[\frac{8}{\mu_{red}^{Aa}\pi}\right]^{1/2}[k_B T]^{-3/2}\frac{1}{G_0^A}\sum_\mu \int_0^\infty \frac{g_\mu^A}{g_0^A}\sigma_{Aa}^\mu\left(E_\mu^A\right)\left[E_0^A - E_\mu^x\right]e^{-\frac{E_0^A}{k_B T}}\, dE_0^A \\
&= \left[\frac{8}{\mu_{red}^{Aa}\pi}\right]^{1/2}[k_B T]^{-3/2}\frac{1}{G_0^A}\int_0^\infty \left\{\sum_\mu \frac{g_\mu^A}{g_0^A}\sigma_{Aa}^\mu\left(E_\mu^A\right)E_\mu^A\right\}e^{-\frac{E_0^A}{k_B T}}\, dE_0^A \\
&= \left[\frac{8}{\mu_{red}^{Aa}\pi}\right]^{1/2}[k_B T]^{-3/2}\frac{1}{G_0^A}\int_0^\infty \sigma_{Aa}^{eff}E_0^A e^{-\frac{E_0^A}{k_B T}}\, dE_0^A = \frac{\langle\sigma^{eff}v\rangle_{Aa}}{G_0^A}.
\end{aligned} \tag{4.48}$$

The last line was obtained by realizing that the expression in the curly brackets is identical to $\sigma_{Aa}^{eff}E_0^A$, with the effective cross section from Equation (4.44). Thus, the weighted sum over many Maxwell–Boltzmann distributions acting on the thermally populated excited states is reduced to a single distribution acting on an effective cross section and divided by the normalized partition function. In terms of relevant physics, this means that the Boltzmann factor in the population probability is offset by shifting down each Maxwell–Boltzmann distribution to the same relative energy.

Comparing Equation (4.48) to Equation (4.9) it becomes clear how the stellar cross section is related to the effective cross section:

$$\sigma^*_{\mathrm{Aa}}(E,\,T) = \frac{\sigma^{\mathrm{eff}}_{\mathrm{Aa}}(E)}{G_0^{\mathrm{A}}(T)}. \tag{4.49}$$

The temperature dependence of σ^* comes from the normalized partition function G_0 whereas the effective cross section σ^{eff} is conveniently independent of T.

4.1.3.2 Reciprocity of Stellar Rates

Equation (4.48) also allows to see easily that stellar rates obey a reciprocity relation because the effective cross sections do. The reactivity $\langle\sigma^* v\rangle_{\mathrm{Bb}}$ for the reverse reaction has to obey a similar relation as the one for the forward reaction $\langle\sigma^* v\rangle_{\mathrm{Aa}}$ given in Equation (4.48). For the reverse reactivity this is

$$\langle\sigma^* v\rangle_{\mathrm{Bb}} = \left[\frac{8}{\mu^{\mathrm{Bb}}_{\mathrm{red}}\pi}\right]^{1/2} [\mathrm{k_B}T]^{-3/2}\frac{1}{G_0^{\mathrm{B}}}\int_0^\infty \sigma^{\mathrm{eff}}_{\mathrm{Bb}} E_0^{\mathrm{B}} \mathrm{e}^{-\frac{E_0^{\mathrm{B}}}{\mathrm{k_B}T}}\,\mathrm{d}E_0^{\mathrm{B}}. \tag{4.50}$$

This expression is derived in the same manner as Equation (4.48) but by starting from thermally populated excited states in the final nucleus B. With the help of Equation (4.46) the reverse reactivity can be expressed in terms of the forward reactivity:

$$\frac{\langle\sigma^* v\rangle_{\mathrm{Bb}}}{\langle\sigma^* v\rangle_{\mathrm{Aa}}} = \frac{1+\tilde{\delta}_{\mathrm{Bb}}}{1+\tilde{\delta}_{\mathrm{Aa}}}\frac{g_0^{\mathrm{A}} g_{\mathrm{a}}}{g_0^{\mathrm{B}} g_{\mathrm{b}}}\frac{G_0^{\mathrm{A}}}{G_0^{\mathrm{B}}}\left[\frac{\mu^{\mathrm{Aa}}_{\mathrm{red}}}{\mu^{\mathrm{Bb}}_{\mathrm{red}}}\right]^{3/2} \mathrm{e}^{-\frac{Q^{\mathrm{Aa}}_{\mathrm{nuc}}}{\mathrm{k_B}T}}, \tag{4.51}$$

where $Q^{\mathrm{Aa}}_{\mathrm{nuc}} = E_0^{\mathrm{B}} - E_0^{\mathrm{A}}$ is the reaction Q-value of the forward reaction.

The relations Equations (4.45) and (4.46) are a consequence of time invariance of nuclear reactions and also apply when the projectile or ejectile is not a particle but a photon. Therefore there is also a reciprocity relation between a stellar capture rate and its reverse, the photodisintegration rate. Its derivation, however, involves an approximation and is discussed in Section 4.1.5. The resulting reciprocity relation is

$$\frac{\lambda^*_{\mathrm{C}\gamma}}{\langle\sigma^* v\rangle_{\mathrm{Aa}}} = \frac{1}{1+\tilde{\delta}_{\mathrm{Aa}}}\frac{g_0^{\mathrm{A}} g_{\mathrm{a}}}{g_0^{\mathrm{C}}}\frac{G_0^{\mathrm{A}}}{G_0^{\mathrm{C}}}\left[\frac{\mu^{\mathrm{Aa}}_{\mathrm{red}}\mathrm{k_B}T}{2\pi\hbar^2}\right]^{3/2} \mathrm{e}^{-\frac{Q^{\mathrm{Aa}}_{\mathrm{nuc}}}{\mathrm{k_B}T}}, \tag{4.52}$$

with $\lambda^*_{\mathrm{C}\gamma}$ being the photo-reactivity for $\gamma + \mathrm{C} \to \mathrm{A} + \mathrm{a}$ under stellar conditions (see Section 4.1.5).

The above reciprocity relations are very important for the application in reaction networks (see Section 4.2). Employing the expressions in Equations (4.50) and (4.52) avoids numerical inconsistencies in network calculations that may arise when forward and reverse rates are calculated separately (or even from different sources). The proper balance between the two reaction directions can only be achieved in such a treatment. Furthermore, simplified equations for reaction equilibria (see Section 4.3) can be derived that prove important in the modeling and understanding of nucleosynthesis at high temperature.

In both reciprocity relations, Equations (4.51) and (4.52), the exponential $e^{-Q_{\rm nuc}^{\rm Aa}/(k_B T)}$ appears. This implies that for positive reaction Q-value $Q_{\rm nuc} > 0$ the reactivity $\langle \sigma^* v \rangle_{\rm Bb}$ of the reverse reaction is strongly suppressed. The suppression depends also on the temperature T appearing in the exponential and therefore is smaller at higher temperature. To be able to compare the *rates* of forward and reverse reaction, it has to be remembered that the reactivities in both channels have to be multiplied by the number densities or abundances, respectively, of the interacting nuclei to obtain the rate, see Equation (4.9). Therefore the ratio of forward and reverse rate is also affected by how many reactands are available in each reaction channel. For reactions with particles in the entrance and exit channel, usually this is a minor effect compared to the exponential suppression factor. A different behavior can be found for capture reactions and their reverse reactions, the photodisintegrations. Since the photodisintegration rate depends only linearly on density (see Equation (4.97)), the ratio of forward and reverse rate becomes density-dependent. High density favors capture reactions, lower density increases the relative importance of photodisintegration. Therefore capture rates may become faster than photodisintegrations even for moderately negative capture Q-value at sufficiently high plasma density. On the other hand, neglecting the density effect or $Q_{\rm nuc}^{\rm Aa} > 0$, a rule of thumb is that photodisintegration starts dominating over capture when

$$k_B T \gtrsim Q_{\rm nuc}^{\rm Aa}/24. \tag{4.53}$$

This holds for temperatures $T \gtrsim 1$ GK due to the additional $T^{3/2}$ dependence of the photodisintegration rate in Equation (4.52) and allows for a sufficient number of photons in the high-energy tail of the Planck distribution. In practical units, Equation (4.53) is $2.0682 T_9 \gtrsim Q_{\rm nuc}^{\rm Aa}$, with T_9 in GK and $Q_{\rm nuc}^{\rm Aa}$ in MeV.

The reciprocity relations for stellar rates given in Equations (4.51) and (4.52) depend not only on the reciprocity relation for single nuclear transitions derived in Section 3.6 but also on the assumption that the excited states in all participating nuclei are occupied according to the population factor $\mathcal{P}$ as defined in Equation (4.38). This is called *detailed balance* and is a valid assumption for most astrophysical plasmas. Most nuclides reach thermal equilibrium very rapidly (faster than the reactions transforming them) through thermal collisions and interactions with photons and other plasma components. However, there are some nuclei which exhibit long-lived isomeric states (well-known examples are ^{26}Al, ^{176}Lu, and ^{180}Ta). Due to the specific spin and parity values of these *isomers* and the quantum mechanical selection rules presented in Section 1.7, they cannot be easily populated or de-populated through electromagnetic transitions. At sufficiently high temperature they may still get into equilibrium, sometimes through couplings to intermediate states, but the relevant transitions have to be carefully included. This can be achieved through an internal reaction network, not connecting different nuclei but rather including the different levels within one nucleus. Levels not being in thermal equilibrium can be included in regular reaction networks (see Section 4.2) in such a manner as if they were a different nuclide. Also in this case, however, the reaction rates for populating and depopulating each level have to be known explicitly. This also applies when ensembles of excited states are in equilibrium but the different ensembles within a nucleus are not. Then each ensemble can be treated as a separate species in a reaction network and the reactions connecting the ensembles have to be included explicitly, similarly to a QSE (see Section 4.3.2).

4.1.3.3 Importance of Thermally Excited States

A frequently made mistake is to assume that the importance of excited states in the calculation of the stellar rate is given by their Boltzmann population factors as defined in Equation (4.37). This is a misconception, as can be seen when explicitly writing down the stellar cross section by using Equation (4.44),

$$\sigma^{*}_{\mathrm{Aa}}(E, T) = \frac{1}{G_0^{\mathrm{A}}(T)} \sum_{\mu} \sum_{\nu} \frac{g_\mu^{\mathrm{A}}}{g_0^{\mathrm{A}}} \frac{E - E_\mu^{\mathrm{x}}}{E} \sigma_{\mathrm{Aa}}^{\mu\nu}\left(E - E_\mu^{\mathrm{x}}\right). \tag{4.54}$$

As before, the individual cross sections $\sigma_{\mathrm{Aa}}^{\mu\nu}$ for transitions $\mu \to \nu$ are evaluated at an energy $E - E_\mu^{\mathrm{x}}$, with $\sigma_{\mathrm{Aa}}^{\mu\nu} = 0$ for $(E - E_\mu^{\mathrm{x}}) \leqslant 0$. Instead of being weighted by the Boltzmann factor, the weight of an individual excited-state cross section in the stellar cross section is

$$\mathcal{W}_\mu(E) = \frac{E - E_\mu^{\mathrm{x}}}{E} = 1 - \frac{E_\mu^{\mathrm{x}}}{E}. \tag{4.55}$$

This is also weighting its contribution to the reaction rate integral. The weight w_μ is energy dependent but since the largest contribution of the cross section to the integral comes from around the Gamow energy $\mathcal{G}$, the effective weight can be approximated by $w_\mu(\mathcal{G})$. The exponentially decaying Boltzmann factors have been replaced by weights linearly decaying with excitation energy E_μ^{x} because of (partially) canceling the Boltzmann factor by a shifted Maxwell–Boltzmann distribution. This is illustrated in Figures 4.4 and 4.5. Similar considerations apply to photodisintegration reactions but, as derived in Section 4.1.5, a slightly different weight (Figure 4.5 and Equation (4.102)) is obtained because shifted Planck distributions do not fully cancel with the population factors.

The linear weight implies that levels at higher excitation energy potentially contribute more to the rate than would be expected from the Boltzmann factor. Neglecting the spin weights g_μ and assuming similar reaction cross sections on each excited state for reactions involving particles (not photons), levels up to

$$E_\mu^{\mathrm{x}} \approx \frac{2}{3}\mathcal{G} \tag{4.56}$$

have to be considered to include 90% of the levels with non-negligible weight. Depending on the nucleus and considered reactions, some levels with large spin values and/or large reaction cross sections at higher excitation energy may require

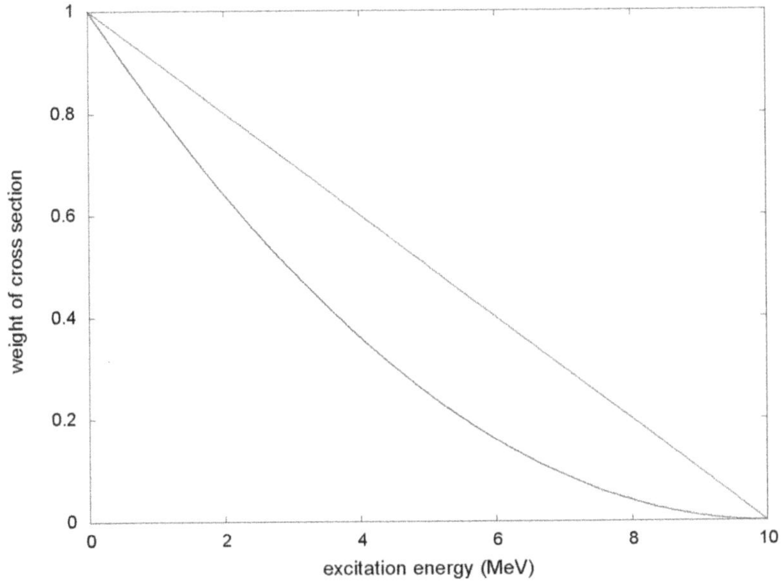

Figure 4.5. Weight of an excited state cross section as the function of excitation energy for a relative interaction energy of $E_0 \approx \mathcal{G} = 10$ MeV. Compared are the weights in a reaction channel with a particle (diagonal red line) and with a photon (curved blue line). (See Section 4.1.5 for further discussion of photodisintegration reactions.)

the inclusion of almost all levels up to very close to $E_\mu \approx \mathcal{G}$. This is especially true for s-wave neutron capture, in which the cross section increases toward lower energy and thus is larger at the shifted energy $E - E_\mu^{\mathrm{x}}$ than at E.

According to Equation (4.42), the actual *contribution* $^*X^\mu$ *of a level* μ *to the stellar rate or reactivity* can be calculated using

$$^*X^\mu = \frac{\mathcal{P}_\mu \langle \sigma v \rangle^\mu}{\langle \sigma^* v \rangle} = \frac{g_\mu \mathrm{e}^{-E_\mu^{\mathrm{x}}/(\mathrm{k_B}T)}}{G} \frac{\langle \sigma v \rangle^\mu}{\langle \sigma^* v \rangle} = \frac{g_\mu}{g_0} \frac{1}{G_0} \mathrm{e}^{-\frac{E_\mu^{\mathrm{x}}}{\mathrm{k_B}T}} \frac{\langle \sigma v \rangle^\mu}{\langle \sigma^* v \rangle}. \tag{4.57}$$

It also follows from Equation (4.42) that

$$\sum_\mu {}^*X^\mu = 1 \text{ with } {}^*X^\mu \leqslant 1. \tag{4.58}$$

The same applies for photodisintegration reactions with

$$^*X^\mu = \frac{\mathcal{P}_\mu \lambda_{\mathrm{C}\gamma}^\mu}{\lambda_{\mathrm{C}\gamma}^*} = \frac{g_\mu \mathrm{e}^{-E_\mu^{\mathrm{x}}/(\mathrm{k_B}T)}}{G} \frac{\lambda_{\mathrm{C}\gamma}^\mu}{\lambda_{\mathrm{C}\gamma}^*} = \frac{g_\mu}{g_0} \frac{1}{G_0} \mathrm{e}^{-\frac{E_\mu^{\mathrm{x}}}{\mathrm{k_B}T}} \frac{\lambda_{\mathrm{C}\gamma}^\mu}{\lambda_{\mathrm{C}\gamma}^*}. \tag{4.59}$$

For the ground-state contribution ($\mu = 0$) this reduces to

$$^*X^0(T) = \frac{1}{G_0(T)} \frac{\langle \sigma v \rangle^0}{\langle \sigma^* v \rangle(T)}, \tag{4.60}$$

$$ {}^{*}X^{0}(T) = \frac{1}{G_0(T)} \frac{\lambda^{0}_{C\gamma}}{\lambda^{*}_{C\gamma}(T)}. \tag{4.61} $$

Obviously, the combined contribution ${}^{*}X^{\text{exc}}$ of reactions on all excited states (not including the ground state) is given by

$$ {}^{*}X^{\text{exc}}(T) = 1 - {}^{*}X^{0}(T). \tag{4.62} $$

The *ground-state contribution* ${}^{*}X^{0}$ to the stellar reaction rate is of special interest when an experimental determination of a stellar reaction rate is desired. Only when the ground-state contribution is large, a measurement of the ground-state cross section σ_{lab} can constrain significantly the actual stellar rate. Otherwise the resulting, remaining uncertainty in a rate will not be the experimental uncertainty of the cross section but a combination of the experimental uncertainty and the uncertainties stemming from the prediction of the excited state cross sections using theory. Assuming all uncertainties in the excited state cross sections are coming from theory and are uncorrelated to the ground-state uncertainties, the combined total uncertainty U^{*}_{tot} in the stellar rate would be

$$ U^{*}_{\text{tot}}(T) = U_{\text{exp}} + \left[U^{\text{exc}}_{\text{th}} - U_{\text{exp}}\right]\left[1 - {}^{*}X^{0}(T)\right], \tag{4.63} $$

using the experimental uncertainty factor U_{exp} and the theory uncertainty factor (for the total uncertainty of the excited state contributions) $U^{\text{exc}}_{\text{th}}$. (*Uncertainty factors* are used to define an interval containing the "true" rate r^{*} as $r^{*}/U_{\text{tot}} \leqslant r^{*} \leqslant r^{*}U_{\text{tot}}$; this is often applied in astrophysics instead of the ± notation in experimental physics because then it becomes easier to discuss variation factors of physical properties.) Since ${}^{*}X^{0}$ depends on the plasma temperature, also the resulting uncertainty factors are temperature dependent.

Older literature often quotes a *stellar enhancement factor*, which is defined as

$$ F_{\text{SEF}} = \frac{\langle \sigma^{*} v \rangle}{\langle \sigma v \rangle^{0}}. \tag{4.64} $$

As can be easily seen when comparing to Equation (4.60), this definition neglects the normalized partition function G_0 and thus does not provide information on the actual contribution of the ground state to the stellar rate. Even for $F_{\text{SEF}} = 1$ it cannot be concluded that reactions from excited states do not contribute, as it just implies $\langle \sigma^{*} v \rangle = \langle \sigma v \rangle^{0}$ but does not show why this is the case. Varying contributions of ground state and excited states in $\langle \sigma^{*} v \rangle$ still could sum up to result in a similar value as for $\langle \sigma v \rangle^{0}$.

As derived above, the weights of the excited state contributions decrease linearly with the relative energy in the given reaction channel, see Equation (4.55). This leads to Equation (4.56) as an estimate of how many excited states to include in the reaction rate calculation. A stronger suppression of the excited-state contributions may arise for charged particles. Reactions on excited states occur at lower relative energy $E - E_\mu^\mathrm{x}$ as shown in Equation (4.54). In the presence of a Coulomb barrier in the entrance channel which affects the cross section, an additional exponential suppression term appears, significantly reducing contributions of levels at higher excitation energy (and thus lower relative interaction energy in the a + A system). Using the astrophysical S-factor defined in Equation (4.23), casting Equation (4.54) as

$$\sigma^*_\mathrm{Aa}(E, T) = \frac{1}{G_0^\mathrm{A}(T)} \sum_\mu \sum_\nu \frac{g_\mu}{g_0} \frac{1}{E} \tilde{S}^{\mu\nu}_\mathrm{Aa}\left(E - E_\mu^\mathrm{x}\right) \mathrm{e}^{-b_\mathrm{G}/\sqrt{E - E_\mu^\mathrm{x}}} \tag{4.65}$$

explicitly shows the additional, exponential suppression term. Therefore the energy-dependent weight $\mathcal{W}_\mu^\mathrm{Coul}$ of the S-factors in the presence of a Coulomb barrier is

$$\mathcal{W}_\mu^\mathrm{Coul}(E) = \frac{1}{E} \mathrm{e}^{-\frac{b_\mathrm{G}}{\sqrt{E - E_\mu^\mathrm{x}}}}, \tag{4.66}$$

to be compared to Equation (4.55). It should be noted that the same caveats apply as discussed in Section 4.1.2.4 regarding the limitation of the Gamow peak concept. Only when the behavior of the cross section is determined by the entrance channel, the above weights based on the action of the Coulomb barrier in the entrance channel are applicable. When in doubt, it is safer to compute the actual contributions from Equation (4.57).

Obviously, the excited state contributions to the stellar rate are not symmetrical with respect to the entrance and exit channel of a reaction. The excited states are different in nuclei A and B and also the relative energies are different. It is easily shown that for reactions with positive reaction Q-value the ground-state contribution is larger in the entrance channel than in the exit channel, ${}^*X^0_\mathrm{Aa} > {}^*X^0_\mathrm{Bb}$. This is because

$$E_0^\mathrm{B} = E_0^\mathrm{A} + Q_\mathrm{nuc} \tag{4.67}$$

and thus the relative energy with the strongest contributions to the reaction rate integral is higher, that is

$$\mathcal{G}' = \mathcal{G} + Q_\mathrm{nuc}. \tag{4.68}$$

Therefore the weights $\mathcal{W}_\nu$ in the system b + B decline more slowly as they are encompassing a larger energy range (see Figure 4.4). This explains the commonly known rule of thumb (the so-called *Q-value rule*) that a reaction should always be measured in the direction of positive Q-value if the aim is to constrain the largest possible fraction of the stellar rate.

It is interesting to note that the Coulomb suppression of the excited-state weights as shown in Equation (4.65) may affect the above Q-value rule. When the Coulomb barrier in the exit channel of a reaction with positive Q-value is larger than in the entrance channel (e.g., in (n,α) reactions), it could be expected that the weights $\mathcal{W}_\nu$ may become so suppressed that only a few excited states contribute in the b + B system and perhaps fewer than in the a + A system. This effect, however, is compensated by the fact that in this case also $\mathcal{G}$ has to be determined in the exit channel because of the smaller partial width at astrophysical energies. The only real exceptions are charged-particle captures for which the charged-particle widths are smaller than the γ width at astrophysical energies (see also the discussion of the applicability of the Gamow peak concept in Section 4.1.2.4) and thus the relevant energy range is indeed given by the channel also showing a Coulomb suppression of the weights. This leads to the interesting situation that for captures always the reactions in capture direction exhibit larger ${}^*X^0$, even when the Q-value is negative. This is of particular importance for (α,γ) and (γ,α) reactions on intermediate and heavy nuclei in the range of spontaneous α emitters, as for these $Q_{\mathrm{nuc}}^{(\alpha,\gamma)} < 0$, but also for proton captures close to the proton dripline. In summary, to investigate the reaction direction with the largest possible ground-state contribution to the stellar rate, the direction with positive Q-value should be chosen, except for reactions with photons in the entrance or exit channel, for which the capture reaction should be studied. The relation between capture and photodisintegration rates is further discussed in Section 4.1.5.

Thermal population of excited states also affects reactions mediated by the weak interaction, such as electron captures, β^+-, and β^--decays. The definition of their rates is given in Section 4.1.6. For example, the β-decay half-life $t_{1/2}$ of a nucleus will be changed relative to its ground-state half-life when the ground state becomes depopulated and excited states with different decay half-lives are populated. Thus, the decay "constant" $\lambda_{\mathrm{dec}} = \ln(2)/t_{1/2}$ is actually temperature-dependent,

$$\lambda_{\mathrm{dec}}^*(T) = \sum_\mu \mathcal{P}_\mu(T)\lambda_{\mathrm{dec}}^\mu, \tag{4.69}$$

where $\lambda_{\mathrm{dec}}^\mu = 1/\tau_{\mathrm{life}}^{\beta,\mu}$ and $\tau_{\mathrm{life}}^{\beta,\mu}$ is the decay lifetime of the excited state. Similar considerations also apply to other reaction types, such as electron capture and neutrino-induced reactions. In electron- or neutrino-capture reactions, populated excited states may also affect the rate by blocking or allowing transitions to otherwise available or unavailable, respectively, states. A further stellar effect is the change of available phase space for emitting or capturing electrons. This is discussed in Section 4.1.4.

4.1.3.4 Thermally Excited States and Resonance Properties

In the derivation of the stellar reactivity, a relative shift of the contributing Maxwell–Boltzmann distributions was employed, resulting in Equation (4.48) and in a stellar cross section as given in Equation (4.54). How does this affect resonant cross sections? Equation (3.108) gives the Breit–Wigner formula for a reaction cross section with n isolated, non-interfering resonances. Combining the definitions given in Equations (4.44) and (3.108) for a single resonance with spin J leads to

$$\begin{aligned}\sigma_{\mathrm{BW}}^{\mathrm{eff}} &= \sum_{\mu}\frac{g_\mu}{g_0}\frac{E_\mu}{E_0}\sigma_{\mathrm{BW}}^{\mu} = \frac{\pi g_J}{g_0 g_{\mathrm{a}} E_0}\sum_{\mu}\frac{E_\mu}{\mathcal{K}_\mu^2}\frac{\Gamma^\mu \hat{\Gamma}^{\mathrm{Bb}}}{\left[E_\mu - E_{\mathrm{res}}^{\mu}\right]^2 + [\hat{\Gamma}_{\mathrm{tot}}/2]^2}\\ &= \frac{\pi g_J}{g_0 g_{\mathrm{a}} E_0}\sum_{\mu}\frac{\hbar^2}{2\mu_{\mathrm{red}}^{\mathrm{Aa}}}\frac{\Gamma^\mu \hat{\Gamma}^{\mathrm{Bb}}}{[E_0 - E_{\mathrm{res}}^{0}]^2 + [\hat{\Gamma}_{\mathrm{tot}}/2]^2}\\ &= \frac{\pi}{\mathcal{K}_0^2}\frac{g_J}{g_0 g_{\mathrm{a}}}\frac{\sum_\mu \Gamma^\mu \hat{\Gamma}^{\mathrm{Bb}}}{[E_0 - E_{\mathrm{res}}^{0}]^2 + [\hat{\Gamma}_{\mathrm{tot}}/2]^2}\\ &= \frac{\pi}{\mathcal{K}_0^2}\frac{g_J}{g_0 g_{\mathrm{a}}}\frac{\hat{\Gamma}^{\mathrm{Aa}}\hat{\Gamma}^{\mathrm{Bb}}}{[E_0 - E_{\mathrm{res}}^{0}]^2 + [\hat{\Gamma}_{\mathrm{tot}}/2]^2}.\end{aligned} \tag{4.70}$$

This results in

$$\begin{aligned}\langle\sigma^* v\rangle_{\mathrm{BW}} &= \sum_{i=1}^{n}\langle\sigma^* v\rangle_{\mathrm{BW}}^{i}\\ &= \left[\frac{8}{\mu_{\mathrm{red}}^{\mathrm{Aa}}\pi}\right]^{1/2}[\mathrm{k_B}T]^{-3/2}\frac{1}{G_0^{\mathrm{A}}}\int_0^\infty\left[\sum_{i=1}^{n}\sigma_{\mathrm{BW},i}^{\mathrm{eff}}\right]E_0^{\mathrm{A}}\mathrm{e}^{-\frac{E_0^{\mathrm{A}}}{\mathrm{k_B}T}}\,\mathrm{d}E_0^{\mathrm{A}}\\ &= \left[\frac{8\pi}{\mu_{\mathrm{red}}^{\mathrm{Aa}}}\right]^{1/2}\frac{1}{G_0^{\mathrm{A}}[\mathrm{k_B}T]^{3/2}}\frac{1}{\mathcal{K}_0^2 g_{\mathrm{a}} g_0}\\ &\quad\times\int_0^\infty\left[\sum_{i=1}^{n} g_i\frac{\hat{\Gamma}_i^{\mathrm{Aa}}(E_0^{\mathrm{A}})\hat{\Gamma}_i^{\mathrm{Bb}}(E_0^{\mathrm{A}})}{[E_0^{\mathrm{A}} - E_{\mathrm{res}}^{i}]^2 + [\hat{\Gamma}_{\mathrm{tot}}(E_0^{\mathrm{A}})/2]^2}\right]E_0^{\mathrm{A}}\mathrm{e}^{-\frac{E_0^{\mathrm{A}}}{\mathrm{k_B}T}}\,\mathrm{d}E_0^{\mathrm{A}}\end{aligned} \tag{4.71}$$

for the stellar rate of n isolated resonances without interference. Note that $\sum_\mu \Gamma^\mu = \hat{\Gamma}^{\mathrm{Aa}}$ has been used in the last lines of Equations (4.70) and (4.71).

As mentioned in Section 3.7.1.2, simplifications can be made depending on the resonance widths. The expression for a narrow resonance is given in Equation (4.33). The *stellar* reactivity of a narrow resonance is given by

$$\langle\sigma^* v\rangle_{\mathrm{narrow}} = \left[\frac{2\pi}{\mu_{\mathrm{red}}^{\mathrm{Aa}}\mathrm{k_B}T}\right]^{3/2}\frac{\hbar^2}{G_0^{\mathrm{A}}}\frac{g_J}{g_{\mathrm{a}} g_0}\frac{\hat{\Gamma}^{\mathrm{Aa}}(E_{\mathrm{res}}^0)\hat{\Gamma}^{\mathrm{Bb}}(E_{\mathrm{res}}^0)}{\hat{\Gamma}_{\mathrm{tot}}(E_{\mathrm{res}}^0)}\mathrm{e}^{-\frac{E_{\mathrm{res}}^0}{\mathrm{k_B}T}}, \tag{4.72}$$

where the resonance energy E_{res}^0 is given relative to the ground state of the target nucleus A. How does this compare to the ground state reactivity $\langle\sigma v\rangle^0$ usually derived from a laboratory measurement? The stellar reactivity can be recast as

$$\begin{aligned}\langle\sigma^*v\rangle_{\text{narrow}} &= \langle\sigma v\rangle^0_{\text{narrow}}\frac{1}{G_0^{\text{A}}}\left[1+\sum_\mu\frac{\Gamma^\mu(E_\mu)\hat{\Gamma}^{\text{Bb}}(E_\mu)\hat{\Gamma}_{\text{tot}}(E_0)}{\Gamma^\mu(E_0)\hat{\Gamma}^{\text{Bb}}(E_0)\hat{\Gamma}_{\text{tot}}(E_\mu)}\right] \\ &\approx \langle\sigma v\rangle^0_{\text{narrow}}\frac{1}{G_0^{\text{A}}}\left[1+\sum_\mu\frac{\Gamma^\mu(E_\mu)}{\Gamma^0(E_0)}\right],\end{aligned} \tag{4.73}$$

with

$$\langle\sigma v\rangle^0_{\text{narrow}} = \left[\frac{2\pi}{\mu^{\text{Aa}}_{\text{red}}\text{k}_{\text{B}}T}\right]^{3/2}\hbar^2\frac{g_J}{g_{\text{a}}g_0}\frac{\Gamma^0(E^0_{\text{res}})\hat{\Gamma}^{\text{Bb}}(E^0_{\text{res}})}{\hat{\Gamma}_{\text{tot}}(E^0_{\text{res}})}\text{e}^{-\frac{E^0_{\text{res}}}{\text{k}_{\text{B}}T}}. \tag{4.74}$$

Equation (4.74) is identical to Equation (4.33), just with clarification that the energies refer to the ground state of target nucleus A. The second line in Equation (4.73) was obtained by neglecting the energy dependence of $\hat{\Gamma}^{\text{Bb}}$ and $\hat{\Gamma}_{\text{tot}}$. This is a valid assumption provided the reaction has a sizable, positive Q-value (see the discussion of the energy dependence of widths before Equation (4.36)). As can be seen from Equation (4.73), the contributions from excited states vanish quickly when $\hat{\Gamma}$ is strongly energy dependent and vanishes fast with decreasing energy (remember that $E_\mu = E_0 - E^{\text{x}}_\mu$). This is certainly true for reactions between charged particles and at low resonance energies. In this case, the only modification of the ground-state reactivity comes from the factor $1/G_0$ which also introduces a temperature dependence. In reactions in which the cross section is determined by neutron widths, on the other hand, resonant transitions from excited states may even dominate a resonant stellar rate. This may also be the case for charged-particle reactions at higher resonance energies.

The wings of broad resonances can also contribute significantly to the rate even when the resonance energy is outside the relevant energy window for the rate, as discussed in Section 4.1.2.3. Sometimes the values for Γ^0, $\hat{\Gamma}^{\text{Bb}}$, and $\hat{\Gamma}_{\text{tot}}$ are known experimentally at the resonance energy. Then a frequently used approach in experimental nuclear physics is to apply Equation (4.36) and to assume that $\hat{\Gamma}^{\text{Bb}}$ and $\hat{\Gamma}_{\text{tot}}$ are approximated by energy-independent values and the energy-dependence of Γ^0 is only due to a barrier penetration factor derived from an optical model. Even if the energy dependence of $\hat{\Gamma}^{\text{Bb}}$, $\hat{\Gamma}_{\text{tot}}$ is accounted for explicitly, this type of extrapolation does not include the stellar enhancement and is only valid for laboratory cross sections. The stellar rate must be calculated from a weighted sum of resonant contributions, as shown in Equation (4.71), both for the value at the resonance energy and in the extrapolation. It follows from Equations (3.108) and (4.71) that the only difference in the energy dependence, however, stems from the width in the entrance channel, where Γ^0 has to be replaced by $\hat{\Gamma}^{\text{Aa}}$. The additional transitions to excited states in the target nucleus can be measured in principle. If they are not available, $\hat{\Gamma}^{\text{Aa}}$ at the resonance energy has to be predicted from theory. Also the extrapolation is more involved because $\hat{\Gamma}^{\text{Aa}}$ will have a different energy dependence than Γ^0. The same methods can be used as in the extrapolation of Γ^0 but they have to be applied to all contributing transitions separately. In most of the cases the rate has to be determined by numerical integration of Equation (4.71) as extrapolations are not necessarily less complicated.

4.1.4 Electronic Plasma Effects

The plasma temperature in the astrophysical sites where nuclear reactions occur is so high that the nuclei are fully ionized and embedded in a cloud of free electrons. This situation affects reactions and decays and has to be considered when preparing a rate to be used in astrophysical reaction networks. Similar to the treatment of stellar cross sections affected by thermally populated excited states, as discussed in Section 4.1.3.1, these corrections also have to be modeled theoretically and experimental rates have to be corrected for them.

Electron captures and β decays are affected not only by the thermal excitation of the nuclei as shown in Section 4.1.3.3 but also by the electrons surrounding the nuclei. A nucleus in an atom can be converted by, e.g., capture of an electron from the atomic K-shell. In the plasma, electrons are captured from the free electron cloud and this may alter the half-life of a nucleus considerably. A well-known example for this is the decay of ^{7}Be. Its electron-capture lifetime at conditions in the center of the Sun ($\tau_{\rm life}^{\rm ec} = 140$ d) is almost double the one under terrestrial, non-ionized conditions ($\tau_{\rm life}^{\rm ec} = 77$ d), where K-shell electrons can be captured.

Also the β^- decay lifetimes are modified by a change in the electron emission probability. The plasma electrons reduce the phase space available for emission to the continuum and thus increase $\tau_{\rm life}^{\beta^-}$. On the other hand, bound-state decay, that means the placement of the emitted electron into a low-lying atomic shell, becomes possible even when it is forbidden in an atom because of the occupation of available electron shells. Similar considerations apply to charged-current reactions with electron neutrinos.

Nuclear reactions with charged nuclei are affected by *electron screening*. Electrons in the vicinity of the nucleus shield part of its charge and thus effectively lower its Coulomb barrier. Theoretical predictions of cross sections and rates always assume bare nuclei without any electrons and therefore have to be corrected for screening. The magnitude of the screening is strongly dependent on the temperature T, density ρ, and composition $Y = \sum_i Y_i$ (see Equation (1.15)) of the plasma. At high density, electron screening can increase the rate by several orders of magnitude and lead to *pycnonuclear* burning at much lower temperatures than otherwise needed to ignite nuclear burning. Moreover, dynamical effects due to the fast movement of electrons may also play a role, although there is discussion on whether this is important when all reactants are in thermodynamic equilibrium. It is apparent that the theoretical treatment of screening is complicated and we are far from complete microscopic descriptions. The proper inclusion of screening effects remains a challenging problem in plasma physics.

Under most conditions, the screened reactivity can be decomposed into the regular stellar reactivity and a screening factor

$$\langle \sigma^* v \rangle_{\text{screened}} = C_{\text{scr}}(T, \rho, Y)\langle \sigma^* v \rangle, \tag{4.75}$$

which describes the modified barrier penetrability through a screened Coulomb potential. This implies that the Coulomb potential $V_{\text{C}}^{\text{screened}}$ seen by the reaction partners can be described by a sum of the bare Coulomb potential V_{C} and an effective screening potential U_{scr}

$$\begin{aligned} V_{\text{C}}^{\text{screened}}(r, T, \rho, Y) &= V_{\text{C}}(r) + U_{\text{scr}}(r, T, \rho, Y) \\ &= \frac{Z_{\text{A}} Z_{\text{a}} e_{\text{C}}^2}{r} + U_{\text{scr}}(r, T, \rho, Y). \end{aligned} \tag{4.76}$$

Then the screening factor acquires the form

$$C_{\text{scr}} = \mathrm{e}^{-U_{\text{scr}}/(\mathrm{k_B} T)}. \tag{4.77}$$

The challenge is in the determination of $U_{\text{scr}}(T, \rho, Y)$ depending on the plasma conditions.

It has to be noted that Equation (4.75) applies to non-resonant reactivities. It also applies to resonant reactivities provided $\hat{\Gamma}^{\text{Aa}} \gg \hat{\Gamma}^{\text{Bb}}$, that means when the partial width in the entrance channel is larger than the partial width in the exit channel (see Equation (4.72)). A more complicated form has to be applied for $\hat{\Gamma}^{\text{Aa}} \ll \hat{\Gamma}^{\text{Bb}}$, which cannot be described with a simple factor C_{scr} as defined by Equation (4.77) (see the Further Reading section for literature providing more details).

An often used static approximation, being appropriate for early burning phases of stars, is *weak screening* in the Debye–Hückel model. Weak screening assumes that the average Coulomb energy of each nucleus is much smaller than the thermal energy, i.e., $Ze_{\text{C}} V_{\text{C}} \bar{n}^{1/3} \ll \mathrm{k_B} T$. Then a nucleus will be surrounded by a polarized sphere of charges, with a radius

$$R_{\text{D}} = \sqrt{\frac{\mathrm{k_B} T}{4\pi e_{\text{C}}^2 \rho \mathrm{N_A} \zeta}} \tag{4.78}$$

with

$$\zeta = \sum_i \left(Z_i^2 + Z_i \mathcal{D}_{\text{e}}\right)^2 Y_i, \tag{4.79}$$

where the sum runs over all charged plasma components and $\mathcal{D}_{\text{e}}$ is the electron degeneracy. A numerical expression for the *Debye radius* in practical units is

$$R_{\text{D}} = 2.812 \times 10^{-7} \sqrt{\frac{T_9}{\rho \zeta}}. \tag{4.80}$$

Inserting the plasma temperature T_9 in GK and the plasma density ρ in g cm^{-3}, Equation (4.80) yields R_{D} in cm. Weak screening implies that—and only is applicable when—the Debye radius is much larger than the average distance between neighboring nuclei.

The modified Coulomb potential around the nucleus with charge $Z_A e_C$ is given by

$$V_C^D = \frac{Z_A e_C}{r} e^{-\frac{r}{R_D}}. \tag{4.81}$$

Note the exponential factor that leads to a quicker reduction of the Coulomb potential than usual and results in complete screening ($V_C^D \to 0$) of the central charge outside the Debye radius ($r \gg R_D$). In other words, an electron cloud at R_D is shielding the potential (almost) completely. The modified potential can be expanded in a McLaurin series around $r = 0$,

$$V_C^D = \frac{Z_A e_C}{r}\left[1 - \frac{r}{R_D} + \frac{r^2}{2R_D^2} - \cdots\right]. \tag{4.82}$$

To first order, the modified Coulomb barrier for an incoming projectile with charge $Z_a e_C$ is

$$V_C^{\text{screened}} = Z_a e_C V_C^D \approx \frac{Z_a Z_A e_C^2}{r} - \frac{Z_a Z_A e_C^2}{R_D}. \tag{4.83}$$

As expected, the screened Coulomb barrier looks like the normal Coulomb barrier but shifted to lower energy by a constant. By comparison to Equation (4.76), the appropriate screening potential is identified as

$$U_{\text{scr}}(r) = U_{\text{scr}}(0) = U_{\text{scr}}^0 = -\frac{Z_a Z_A e_C^2}{R_D}. \tag{4.84}$$

The same potential U_{scr}^0 also enters the screening factor C_{scr} given in Equation (4.77). Assuming $|U_{\text{scr}}^0| \ll k_B T$ (or $T \gg 10^5 \rho^{1/3} \zeta$, with T in K and ρ in g cm^{-3}) because of weak screening, C_{scr} can also be expanded and this yields to first order

$$C_{\text{scr}}^{\text{weak}} \approx 1 + \frac{U_{\text{scr}}^0}{k_B T} \tag{4.85}$$

as screening factor in the case of weak screening. In practical units this is

$$C_{\text{scr}}^{\text{weak}} \approx 1 + 0.188 Z_a Z_A \sqrt{\frac{\rho \zeta}{T_6^3}}, \tag{4.86}$$

with the plasma temperature T_6 given in MK.

Strong screening appears in high density plasmas (at densities well beyond those encountered in stellar burning phases) and implies that the average Coulomb energy is much larger than $k_B T$. In this case it is more appropriate to use the ion-sphere model instead of the Debye–Hückel approximation. The ion-sphere model is equivalent to the Wigner–Seitz model used in condensed matter theory. Due to the Coulomb repulsion overcoming the thermal motion of the nuclei, the positively charged nuclei occupy spatial cells devoid of other positive charges. The strong screening factor can be approximated in practical units as

$$C_{\mathrm{scr}}^{\mathrm{strong}} \approx \exp\left\{\frac{0.205}{T_6}[\rho Y_e]^{1/3}[[Z_a + Z_A]^{5/3} - a^{5/3} - A^{5/3}]\right\}. \tag{4.87}$$

This assumes that the density ρ is inserted in g cm^{-3}.

More difficult to treat is *intermediate screening*, when neither the assumptions of weak screening apply, nor the ones of strong screening. Complicated treatments yield approximate solutions for astrophysical application, such as interpolation formulae between weak and strong screening. Dynamical effects in screening, that means how the electron cloud follows the motion of charged nuclei, may also play a role. Dynamical effects are considered to be negligible in weak and strong screening regimes but may become important in the intermediate regime. Their treatment is difficult, as hydrodynamic motion of all plasma components has to be followed and few attempts have been made so far to estimate the impact. Nucleosynthesis calculations usually neglect dynamical screening effects.

Physically, in the weak screening regime the nuclei act as classical ionized gas and therefore the effect of the background plasma can be calculated using the Debye–Hückel theory. In the strong screening regime, the background nuclei are frozen into a Coulomb lattice but the reacting nuclei can be considered free. At even higher densities a third regime is encountered: both the background nuclei and the reacting nuclei are bound in a Coulomb lattice. Then the definition of reaction rates has to be adapted as the motion of the nuclei cannot be described by a free gas anymore. Reactions in this regime are called pycnonuclear reactions. The pycnonuclear rate at zero temperature is given by

$$r_{\mathrm{Aa}}^{\mathrm{pyc}} = 3.9 \times 10^{46}\rho Y_e \left[\frac{A_a A_A}{A_a + A_A}\right]^2 \left[\frac{Z_a Z_A}{Z_a + Z_A}\right]^4 f_\lambda^{7/4}\, \tilde{S} e^{-2.638/\sqrt{f_\lambda}} \tag{4.88}$$

or

$$r_{\mathrm{Aa}}^{\mathrm{pyc}} = 4.76 \times 10^{46}\rho Y_e \left[\frac{A_a A_A}{A_a + A_A}\right]^2 \left[\frac{Z_a Z_A}{Z_a + Z_A}\right]^4 f_\lambda^{7/4}\, \tilde{S} e^{-2.516/\sqrt{f_\lambda}}. \tag{4.89}$$

The two versions correspond to two different approximations of the lattice potential. Again, the density should be inserted in g cm^{-3} and the astrophysical S-factor $\tilde{S}$ in MeV barn. The factor f_λ is defined as

$$f_\lambda = \frac{A_a + A_A}{2A_a A_A Z_a Z_A}\left[\frac{\rho Y_e}{Z_A}\frac{1}{1.3574 \times 10^{11}}\right]^{1/3}. \tag{4.90}$$

Introducing a temperature dependence yields a correction factor to the zero temperature rate of

$$\frac{r_{\mathrm{Aa}}^{\mathrm{pyc}}(T)}{r_{\mathrm{Aa}}^{\mathrm{pyc}}} = 1 + \frac{0.043}{\sqrt{f_\lambda}}[1 + 1.2624\mathrm{e}^{-8.7833 f_\beta^{3/2}}]^{-1/2} \times \exp\left\{-7.272\beta^{3/2}\frac{1.2231}{\sqrt{f_\lambda}}\mathrm{e}^{-8.7833 f_\beta^{3/2}}[1 - 0.631\mathrm{e}^{-8.7833 f_\beta^{3/2}}]\right\} \tag{4.91}$$

or

$$\frac{r_{\mathrm{Aa}}^{\mathrm{pyc}}(T)}{r_{\mathrm{Aa}}^{\mathrm{pyc}}} = 1 + \frac{0.043}{\sqrt{f_\lambda}}[1 + 1.2624\mathrm{e}^{-8.7833 f_\beta^{3/2}}]^{-1/2} \times \exp\left\{-7.272\beta^{3/2}\frac{1.2231}{\sqrt{f_\lambda}}\mathrm{e}^{-8.7833 f_\beta^{3/2}}[1 - 0.631\mathrm{e}^{-8.7833 f_\beta^{3/2}}]\right\}, \tag{4.92}$$

again for two approximations. The function f_β is

$$f_\beta = 0.032234 f_\lambda \left[\frac{A_{\mathrm{a}}A_{\mathrm{A}}}{A_{\mathrm{a}} + A_{\mathrm{A}}} Z_{\mathrm{a}}^2 Z_{\mathrm{A}}^2 \frac{7.6696 \times 10^{10}}{T}\right]^{2/3}, \tag{4.93}$$

with the temperature T given in K as usual. The pycnonuclear rate is extremely sensitive to the density, with a dependence of

$$r_{\mathrm{Aa}}^{\mathrm{pyc}} \propto \rho^{19/12}\mathrm{e}^{-C/\rho^6}, \tag{4.94}$$

with C being a constant.

Pycnonuclear reactions may become important only at densities above about 10^{12} g cm^{-3}. Therefore they are not relevant in hydrostatic burning of regular stars. Nuclear burning in accreted matter on the surface of neutron stars, however, partially proceeds at pycnonuclear conditions in the neutron star crust (see also Section 9.4.4).

Another completely different type of screening is observed in nuclear experiments in the laboratory. In experiments using gas or solid targets, nuclei are present in atoms, molecules, or metals, each with specific electron charge distributions around the nucleus. Although completely different from plasma screening, this type of screening has to be understood because it is especially important at the low interaction energies of astrophysical relevance. The *measured* reaction cross section $\sigma_{\mathrm{lab}}^{\mathrm{screened}}$ has to be corrected to obtain the bare cross section σ_{lab}, which can be compared to theory or used to determine the rate. Atomic screening, for example, can be treated in the adiabatic approximation, leading to

$$\frac{\sigma_{\mathrm{lab}}^{\mathrm{screened}}(E)}{\sigma_{\mathrm{lab}}(E)} = \frac{E}{E + U_{\mathrm{scr}}^{\mathrm{e}}}\mathrm{e}^{\frac{\pi\eta_{\mathrm{S}} U_{\mathrm{scr}}^{\mathrm{e}}}{E}} \tag{4.95}$$

with the Sommerfeld parameter η_{S} from Equation (4.24). The screening potential $U_{\mathrm{scr}}^{\mathrm{e}}$ in this approximation is not related to the one introduced in Equation (4.76). It is given by the difference in the electron binding energy of the target atom and the atom made from target atom plus projectile $U^{\rm e}_{\rm scr} = \mathcal{B}^{\rm Aa}_{\rm e} - \mathcal{B}^{\rm A}_{\rm e}$. In light systems, the velocity of the atomic electrons is comparable to the relative motion between the nuclei. Therefore a dynamical model is more appropriate, which would also give temperature-dependent screening potentials. However, the adiabatic approximation provides an upper limit on the expected screening effect on the cross section.

For some cases the screening potential can be determined experimentally by comparing calculated laboratory cross sections without screening (i.e., for bare nuclei) to measured cross sections in reactions which are assumed to be well understood. Nevertheless, there seem to be discrepancies between theory and laboratory determinations of $U^{\rm e}_{\rm scr}$, the latter often yielding much larger values of $U^{\rm e}_{\rm scr}$. Some of them have been explained by incorrect further nuclear data (such as stopping powers in the target) entering the data analysis for the determination of the experimental cross sections. Others remain puzzling, especially regarding cross sections of nuclei implanted in metals which are controversially discussed. Thus, the *laboratory screening* still does not seem to be well understood for now.

4.1.5 Reactions with Photons

The general definition of the reaction rate given in Equation (4.2) also applies when one of the reaction partners is a photon. Similar to the definition given in Equation (4.42) of the stellar reactivity as a weighted sum of reactivities on thermally populated excited states, the stellar *photo-reactivity* $\lambda^*_{\rm C\gamma}$ is expressed as a weighted sum of individual contributions from γ-transitions starting from an excited state,

$$\lambda^*_{\rm C\gamma} = \mathcal{P}_0\lambda^0_{\rm C\gamma} + \mathcal{P}_1\lambda^1_{\rm C\gamma} + \cdots = \sum_\mu \mathcal{P}_\mu\lambda^\mu_{\rm C\gamma}. \qquad (4.96)$$

Then the stellar photodisintegration rate is given by

$$r_{\rm C\gamma} = \bar{n}_{\rm C}\lambda^*_{\rm C\gamma}(T) = Y_{\rm C}\rho{\rm N_A}\lambda^*_{\rm C\gamma}(T) = Y_{\rm C}\rho{\rm N_A}\sum_\mu \mathcal{P}_\mu\lambda^\mu_{\rm C\gamma} \qquad (4.97)$$

for the reaction $\gamma + {\rm C} \rightarrow {\rm A} + {\rm a}$. The photodisintegration rate $r_{\rm C\gamma}$ only includes one number density (or abundance), contrary to the two-body rate derived in Section 4.1.2, which included the number densities (or abundances) of target and projectile nuclei. As shown below, the photon number density is accounted for in the integration to determine the photo-reactivity. Therefore $r_{\rm C\gamma}$ only linearly depends on density.

The quantity $\lambda^*_{\rm C\gamma}$ plays the same role as the reactivity $\langle\sigma^* v\rangle$ given in Equations (4.8)–(4.11) for two-body reactions but has different dimensions. The photo-reactivity $\lambda^*_{\rm C\gamma}$ gives the number of photodisintegration reactions per time and per target nucleus and is independent of density whereas the two-body reaction rate per time and target nucleus $\lambda^{*,{\rm A}}_{\rm trans}$, as defined in Equation (4.11), does depend on the plasma density. Both quantities depend on plasma temperature.

Again, the index μ in Equations (4.96) and (4.97) is not a mathematical exponent but identifies the excited state in the target nucleus. The importance of the individual contributions of excited states to the stellar rate is discussed in Section 4.1.3.3. As most photodisintegrations have a negative reaction Q-value, many more contributions have to be summed than for a capture rate (see Equation (4.68) and its discussion) and each individual contribution is smaller, including the ground-state contribution ${}^{*}X^{0}_{C\gamma}$ defined in Equation (4.61). The Gamow energy for photodisintegration can be calculated from Equation (4.68) as long as the particle width is smaller than the γ-width, which is fulfilled at astrophysically relevant energies (note that capture reactions proceeding on excited states have smaller relative particle energies than reactions commencing on the ground state). Otherwise there is no well-defined Gamow peak. The Coulomb suppression of excited state contributions, discussed in the paragraph after Equation (4.68), guarantees even for positive photodisintegration Q-value that ${}^{*}X^{0}_{C\gamma} \ll {}^{*}X^{0}_{Aa}$ when charged particles are involved in the reaction channel A + a. Table 4.1 lists a comparison between ${}^{*}X^{0}_{C\gamma}$ and ${}^{*}X^{0}_{Aa}$ for a few examples. For comparison, Figures 4.6 and 4.7 provide an overview of ${}^{*}X^{0}_{Aa}$ for (n,γ) and (α,γ) reactions, respectively, on stable or long-lived, naturally occurring nuclides.

The photo-reactivity for photodisintegration of nuclear level μ is found by application of Equation (4.2). For photons the relative velocity between the nucleus and the projectile is always the speed of light, $|\vec{v}_C - \vec{v}_\gamma| = c$. The thermodynamic distribution $d\tilde{n}_C$ of the nuclei is given by the Maxwell–Boltzmann distribution as shown in Equation (1.58) and also used in Equations (4.3) and (4.4). The one for the projectile distribution $d\tilde{n}_\gamma$ is given by the photon number density from Equation (1.68). The reaction cross section appearing in Equation (4.2) has to be replaced by the photodissociation cross section $\sigma^{\mu}_{C\gamma}(E_\gamma)$ of nuclear level μ, which only depends on the photon energy E_γ but not on the momentum of the nuclei C. Using this fact and the fact that the relative velocity between projectiles and nuclei is a constant, the integration over the Maxwell–Boltzmann distribution can be performed immediately, recovering the number density $\tilde{n}_C$ of the target nuclei according to Equation (1.55). This explains the appearance of $\tilde{n}_C$ in Equation (4.97). The remaining integral folds the cross section with the Planck energy distribution of the photons, yielding the photo-reactivity

Table 4.1. Comparison of Ground-state Contributions ${}^{*}X^{0}$ for Selected A + neutron $\leftrightarrow$ C + γ Reactions at $T = 2.5$ GK

A	${}^{*}X^{0}_{A}$	${}^{*}X^{0}_{C\gamma}$	A	${}^{*}X^{0}_{A}$	${}^{*}X^{0}_{C\gamma}$	A	${}^{*}X^{0}_{A}$	${}^{*}X^{0}_{C\gamma}$
^{85}Sr	0.771	0.000 59	^{185}W	0.0788	0.000 49	^{197}Pt	0.0396	0.0018
^{89}Zr	0.98	0.000 34	^{184}Re	0.0148	0.000 21	^{196}Au	0.0815	0.000 35
^{95}Zr	0.875	0.0061	^{186}Re	0.0356	0.000 24	^{195}Hg	0.0433	0.000 43
^{93}Mo	0.992	0.0043	^{185}Os	0.0318	0.000 16	^{197}Hg	0.066	0.000 84
^{141}Nd	0.737	0.0028	^{189}Pt	0.0537	0.000 069	^{203}Hg	0.551	0.0088
^{154}Gd	0.0914	0.0012	^{191}Pt	0.0541	0.000 11	^{203}Pb	0.719	0.0059

Note. Reactions are identified by the target nucleus A of the neutron capture reaction.

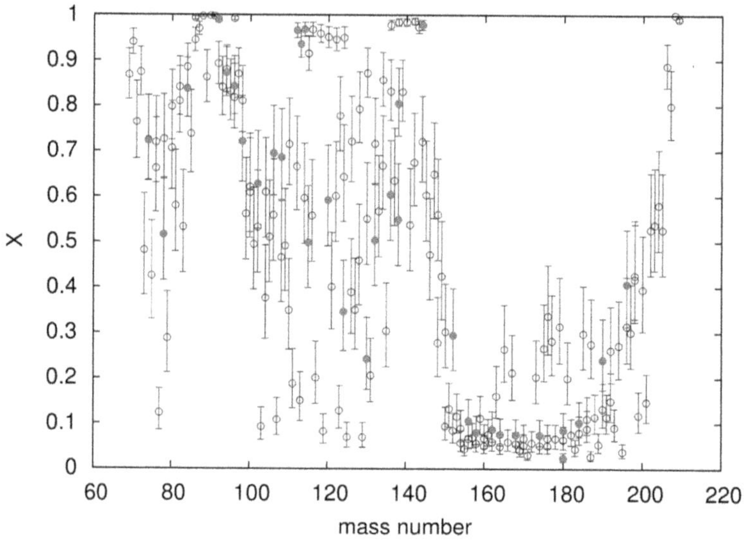

Figure 4.6. Ground-state contributions to stellar (n,γ) rates for natural nuclides at 2.5 GK. The red, filled symbols indicate p-nuclides (see Section 8.4). (Figure from Rauscher et al. 2013. © 2013 IOP Publishing Ltd. Reproduced with permission. All rights reserved.)

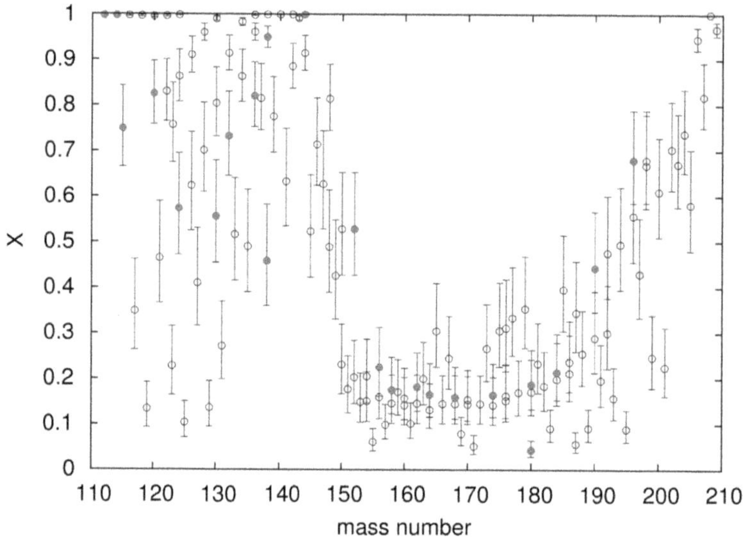

Figure 4.7. Ground-state contributions to stellar (α,γ) rates for natural nuclides at 2 GK. The red, filled symbols indicate p-nuclides (see Section 8.4). (Figure from Rauscher et al. 2013. © 2013 IOP Publishing Ltd. Reproduced with permission. All rights reserved.)

$$\lambda^{\mu}_{C\gamma}(T) = \frac{1}{\pi^2 c^2 \hbar^3} \int_0^{\infty} \sigma^{\mu}_{\gamma}(E_\gamma) \frac{E_\gamma^2}{e^{E_\gamma/(k_B T)} - 1}\, dE_\gamma. \tag{4.98}$$

Strictly mathematically, a *stellar* photodisintegration cross section as used in Equation (4.2) and defined in Equations (4.49) and (4.54) cannot be written down easily. This is because the Boltzmann factor (Equation (4.37)) contains an exponential factor $1/\exp(E/(k_B T))$ whereas the integral in Equation (4.98) contains $1/(\exp(E/(k_B T)) - 1)$. Thus, it is not possible to directly define an effective cross section and to arrive at a simple integral by shifting the Planck distributions acting on the populated excited states as it was done for particle-induced reactions in Equations (4.47) and (4.48). This mathematical transformation is only possible when approximating $\exp(E_\gamma/(k_B T)) - 1 \approx \exp(E_\gamma/(k_B T))$ in Equation (4.98). Then an effective photodisintegration cross section analogous to the one given in Equation (4.44) can be defined. It is given by

$$\sigma^{\mathrm{eff}}_{C\gamma}(E_0^C) = \sum_\mu \sum_\nu \frac{g_\mu^C}{g_0^C}\left[\frac{E_\mu^C}{E_0^C}\right]^2 \sigma^{\mu\nu}_{C\gamma}(E_\mu^C) = \sum_\mu \frac{g_\mu^C}{g_0^C}\left[\frac{E_\mu^C}{E_0^C}\right]^2 \sigma^{\mu}_{C\gamma}(E_\mu^C). \tag{4.99}$$

The energies in the system $\gamma + \mathrm{C}$ are defined similarly to the ones in Equation (4.44), that is $E_0^C = E_\mu^C + E_\mu^x$ with the excitation energies E_μ^x of nuclear levels in the nucleus C. After performing the same mathematical transformation as shown in Equation (4.48) the stellar photodisintegration rate can be expressed by a single integral containing the effective cross section,

$$r_{C\gamma} = \tilde{n}_C \lambda^{*}_{C\gamma}(T) \approx \frac{\tilde{n}_C}{\pi^2 c^2 \hbar^3} \frac{1}{G_0^C} \int_0^{\infty} \sigma^{\mathrm{eff}}_{C\gamma}(E) \frac{E^2}{e^{E/(k_B T)}}\, dE. \tag{4.100}$$

The *stellar photodisintegration cross section* then is obtained as

$$\sigma^{*}_{C\gamma}(E,T) = \frac{1}{G_0^C(T)} \sum_\mu \sum_\nu \frac{g_\mu^C}{g_0^C}\left[\frac{E - E_\mu^x}{E}\right]^2 \sigma^{\mu\nu}_{C\gamma}(E - E_\mu^x). \tag{4.101}$$

As before, $\sigma^{\mu\nu}_{C\gamma} = 0$ for $(E - E_\mu^x) \leqslant 0$. As seen already for reactions involving particles as projectile, also the cross sections $\sigma^{\mu\nu}_{C\gamma}$ for individual transitions $\mu \to \nu$ are not weighted by the Boltzmann factor of Equation (4.37), as one would assume naively, but rather by a weight

$$\mathcal{W}^{\gamma}_{\mu}(E) = \left[\frac{E - E_\mu^x}{E}\right]^2 = \left[1 - \frac{E_\mu^x}{E}\right]^2. \tag{4.102}$$

The importance of excited state contributions was discussed following Equation (4.55). Comparing Equations (4.55) and (4.102), it is found that $\mathcal{W}^{\gamma}_{\mu}$ decreases slightly faster with increasing excitation energy than $\mathcal{W}_\mu$ (see Figure 4.5). Nevertheless, contributions from excited states still play an important role because of the usually negative photodisintegration Q-value and the energy dependence of

$\sigma^{\mu\nu}_{C\gamma}$ (see Section 3.7.1.2 around Figure 3.10 for a discussion of the importance of photon energies in capture and photodisintegration reactions).

In practice, the stellar photodisintegration cross section σ^*_γ does not have to be constrained by a measurement because the photodisintegration rate can be calculated from the reactivity of the capture reaction. The introduction of the effective cross section allows to relate the stellar capture reactivity $\langle\sigma^* v\rangle_{\mathrm{Aa}}$, which assumes a form as given in Equation (4.48), to the photo-reactivity $\lambda^*_{C\gamma}$ as defined by Equation (4.100). The reciprocity relation between captures and photodisintegrations is derived similarly to Equation (4.51), making use of the effective cross sections for the particle channel given in Equation (4.44) and of the effective photodisintegration cross section from Equation (4.99). Individual transitions are related by the reciprocity relation given in Equation (3.101). The resulting relation between *stellar* capture- and photo-reactivities is given in Equation (4.52).

How large is the error stemming from the approximation involved in the derivation of Equation (4.99)? Although mathematically unsound, it turns out that setting $\exp(E/(\mathrm{k_B}T)) - 1 \approx \exp(E/(\mathrm{k_B}T))$ is a good approximation for the calculation of the rate integrals appearing in astrophysics and introduces an error of less than a few percent for astrophysically relevant temperatures and rate values. Contributions to the value of the integral turn out to be negligible at the low energies where the Planck distribution and the Maxwell–Boltzmann distribution differ considerably. This is assured with a sufficiently large and positive $Q^{\mathrm{Aa}}_{\mathrm{nuc}}$ for the capture reaction, which causes the integration over the Planck distribution to start not at zero energy but rather at a sufficiently large threshold energy. For small (of the order of $Q^{\mathrm{Aa}}_{\mathrm{nuc}} \leqslant \mathrm{k_B}T$) or negative Q-value it can be guaranteed by vanishing cross sections $\sigma^{\mu\nu}_{C\gamma}$ for individual transitions at low relative energy due to, e.g., a Coulomb barrier. The approximation may not be valid for s-wave neutron captures with very small or negative Q-value but the required correction still is only a few % as can be shown in numerical comparisons between photodisintegration rates calculated with the two versions of the denominator. Such comparisons can be found in literature. They show that captures with low Q-value, either at the dripline or in the vicinity of closed shells, exhibit the largest errors but those do not exceed 10% at low temperature. Nuclei at the driplines, however, are synthesized only at higher temperature. For $T \geqslant 1$ GK only a small number of nuclei shows errors larger than 4%. Overall, for the majority of reactions across the nuclear chart the error introduced in the reverse rate due to the approximation of the Planck distribution is much less than 1%.

4.1.6 Reactions with Leptons

The fact that leptons have considerably less mass than nucleons and atomic nuclei can be exploited when applying the definition in Equation (4.2) of the reaction rate to reactions of nuclei with electrons, positrons, neutrinos, or anti-neutrinos. In the center-of-mass system (see Section 1.6.1) the velocity of the nucleus is negligible with respect to the lepton velocity and thus the relative velocity, determining the interaction energy and cross section, is the one of the lepton, $|\vec{v}_{\mathrm{nucleus}} - \vec{v}_{\mathrm{lepton}}| \approx |\vec{v}_{\mathrm{lepton}}|$. This is a similar situation to the photodisintegrations discussed in Section 4.1.5 and implies that the integration over the $d\bar{n}_C$ can be performed separately, yielding the number density $\bar{n}_C$ of the nuclei C. The remaining integral in the lepton rate

$$r_C^{\text{lepton}} = \bar{n}_C \int_0^\infty \sigma^*_{\text{lepton}}(v_{\text{lepton}}) v_{\text{lepton}} \, d\bar{n}_{\text{lepton}} = \bar{n}_C \lambda^*_{\text{lepton}}(\rho Y_e, T) \qquad (4.103)$$

is solved by integrating the cross section over the appropriate thermodynamic distribution. This has to be chosen among the Maxwell–Boltzmann (Section 1.5.2) or the Fermi distribution (Section 1.5.4), with the Fermi distribution for a partially or fully degenerate (Section 1.5.4.1) or a relativistic (Section 1.5.4.2) or ultra-relativistic (Section 1.5.5) particle gas. The density and temperature of the plasma determine which distribution to apply. For example, electrons can be treated as an ideal gas in the early burning phases of stars but the electron gas becomes degenerate from C and O burning onwards because of the increased density. Neutrino-scattering cross sections become non-negligible only at densities above $\rho > 10^{12}$ g cm^{-3}, when they have to be described as a degenerate, relativistic Fermi gas or as an ultra-relativistic gas, depending on the temperature.

The form of the lepton rates is similar to the one of the photodisintegration rates (Equation (4.100)), with a number density of nuclei multiplied by a reactivity. The lepton reactivities, however, are a function of the plasma temperature T and the electron abundance Y_e (determining the neutron enrichment of the plasma, see Table 1.1) due to the integration over $\bar{n}_{\text{lepton}}$. Thus, the dependence of the lepton-reactivity $\lambda^*_{\text{lepton}}$ differs from the one of the photo-reactivity, which only depends on T. In both cases, the dependence on temperature stems on one hand from the T-dependence of the underlying thermodynamic distribution and, on the other hand, from the fact that the stellar reactivity includes T-dependent contributions from excited states, as shown explicitly in Equation (4.96). Especially at the high temperatures of late stellar evolution and in explosive nucleosynthesis these strongly affect also rates mediated by the weak interaction. In electron- or neutrino-capture reactions, populated excited states may also affect the rate by blocking or allowing transitions to otherwise available or unavailable, respectively, states. A further stellar effect is the change of available phase space for emitting or capturing electrons. This is discussed in Section 4.1.4.

4.1.7 Decay Rates, Half-lives, and Lifetimes

An equation similar to Equations (4.100) and (4.103) is obtained for stellar reaction rates describing β^-- and β^+-decay of a nuclear species A,

$$r_{\text{dec}} = \bar{n}_A \lambda^*_{\text{dec}}(T) = Y_A \rho N_A \lambda^*_{\text{dec}}(T). \qquad (4.104)$$

In general, reaction rates and reactivities are time-dependent. In standard, terrestrial β-decay, the quantity λ_{dec} is taken to be independent of time and temperature, and is called *decay constant*. The well-known law of radioactive decay is recovered from the differential equations describing the change in number density of nuclei of type A decaying into nuclei of type B:

$$\begin{aligned}\frac{d\bar{n}_A}{dt} &= \dot{\bar{n}}_A = -r_{dec} = -\bar{n}_A\lambda_{dec}, \\ \frac{d\bar{n}_B}{dt} &= \dot{\bar{n}}_B = +r_{dec} = +\bar{n}_B\lambda_{dec}.\end{aligned} \tag{4.105}$$

As nuclei of type A are consumed in the decay A → B at the given rate, nuclei of type B appear at the same rate. The solution of these coupled differential equations leads to

$$\begin{aligned}\bar{n}_A(t) &= \bar{n}_A(t_0)e^{-\lambda_{dec}\Delta t}, \\ \bar{n}_B(t) &= \bar{n}_A(t_0)(1 - e^{-\lambda_{dec}\Delta t}),\end{aligned} \tag{4.106}$$

with $\Delta t = t - t_0$ being the time since the decay reaction started at time t_0. Using the definition of the abundance from Equation (1.14) it becomes obvious that the abundances Y_A, Y_B of the two nuclear species follow the same law as the number densities,

$$\begin{aligned}Y_A(t) &= Y_A(t_0)e^{-\lambda_{dec}\Delta t}, \\ Y_B(t) &= Y_A(t_0)(1 - e^{-\lambda_{dec}\Delta t}).\end{aligned} \tag{4.107}$$

Note that both equations contain the initial abundance of species A on the right. Before the reaction sets in, the abundances are constant. After sufficient time has elapsed, depending on λ_{trans}, the abundance of species B has increased by the initial value of the abundance of species A, which in turn has been destroyed completely.

In radioactive decay, often the *half-life* $t_{1/2}$ is quoted instead of the decay constant, especially for measurements. It is the time elapsed until half of the initial number of nuclei $\mathcal{N}_A(t_0) = \bar{n}_A\rho$ (assuming constant density, see also Equation (1.7)) has decayed in a sample. Solving

$$\frac{\mathcal{N}_A(t_0)}{2} = \mathcal{N}_A(t_0)e^{-\lambda_{dec}t_{1/2}} \tag{4.108}$$

for $t_{1/2}$, the relation between half-life and decay constant is obtained as

$$\lambda_{dec} = \frac{\ln 2}{t_{1/2}}. \tag{4.109}$$

The value of λ_{dec} (with dimension 1/time, see also Equation (4.11)) is related to a characteristic time

$$\tau_{life} = \tau_{life}^{dec} = \frac{1}{\lambda_{dec}}, \tag{4.110}$$

the average *lifetime* of a nucleus of type A against destruction by the given reaction (a decay in the present case), converting it into a nucleus of type B. This lifetime is a good estimate for the time it takes a decay (or a reaction) to significantly affect a nuclear abundance. The lifetime is related to the half-life by

$$\tau_{\rm life}^{\rm dec} = \frac{t_{1/2}}{\ln 2}. \tag{4.111}$$

The lifetime is longer than the half-life. This can be understood easily by considering that the lifetime measures the time until the abundance has dropped to 1/e of the initial abundance, whereas the half-life gives the elapsed time until the initial abundance has been halved. An example of the change in $\mathcal{N}_{\rm A}$, $\mathcal{N}_{\rm B}$ with time and of the relation between decay constant, lifetime, and half-life is shown in Figure 4.8.

Thermal population of excited states as discussed in Section 4.1.3 also affects the β-decay half-life $t_{1/2}$ of a nucleus. It is changed relative to its ground-state half-life when the ground state becomes depopulated and excited states with different decay half-lives are populated. Thus, the decay "constant" $\lambda_{\rm dec} = \ln(2)/t_{1/2}$ becomes temperature-dependent,

$$\lambda_{\rm dec}^*(T) = \sum_\mu \mathcal{P}_\mu(T)\lambda_{\rm dec}^\mu, \tag{4.112}$$

where $\lambda_{\rm dec}^\mu = 1/\tau_{\rm life}^{{\rm dec},\mu}$ and $\tau_{\rm life}^{{\rm dec},\mu}$ is the decay lifetime of the excited state.

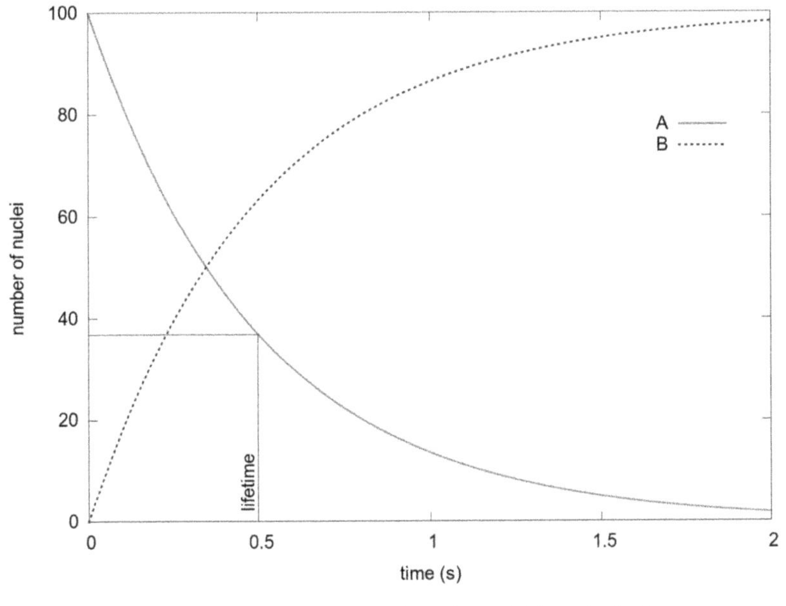

Figure 4.8. Sketch of the temporal evolution of the number of nuclei $\mathcal{N}_{\rm A}$ and $\mathcal{N}_{\rm B}$ of two nuclear species A and B, respectively, connected by a single reaction with $\lambda_{\rm trans} = 2\ {\rm s}^{-1}$. Initially (at time $t = 0$ s), $\mathcal{N}_{\rm A}(0) = 100$ and $\mathcal{N}_{\rm B}(0) = 0$. After the lifetime $\tau_{\rm life} = 1/\lambda_{\rm trans}$ has elapsed, there are 100/e nuclei of type A left. Note that the lifetime of a nucleus is longer than its half-life $t_{1/2}$. The half-life is found in this plot at the intersection of the two curves when half of the nuclei of type A have been converted to type B and thus $\mathcal{N}_{\rm A}(t_{1/2}) = \mathcal{N}_{\rm B}(t_{1/2}) = \mathcal{N}_{\rm A}(0)/2$.

The rate for spontaneous decay is an example for a reaction rate involving only a single body, i.e., a single nucleus being transformed without the action of a second particle. The decays per time only depend on the number of nuclei present and the reactivity $\lambda^*_{\rm dec}$. In fact, half-lives and lifetimes can be applied for any type of reaction, not just decays. The reactivities for reactions with photons $\lambda^*_{C\gamma}$ (Equation (4.100)) and with leptons $\lambda^*_{\rm lepton}$ (Equation (4.103)) have a similar form as the one for decays $\lambda^*_{\rm dec}$ and thus can be directly used in the above Equations (4.110) and (4.111) instead of $\lambda^*_{\rm dec}$. For other two-body reactions, the rate per target nucleus $\lambda^*_{\rm trans}$ given in Equation (4.11) has to be used instead.

When a nuclide is destroyed by more than one reaction, its resulting lifetime $\tau_{\rm life,\Sigma}$ is shorter, given by the reciprocal of the sum of the rates per nucleus,

$$\tau_{\rm life,\Sigma} = \frac{1}{\sum_i \lambda^*_{{\rm trans},i}}. \tag{4.113}$$

In this case, a *branching ratio* between the competing reactions can be defined. The branching b_i into the ith reaction channel then is given by

$$b_i = \frac{\lambda^*_{{\rm trans},i}}{\sum_j \lambda^*_{{\rm trans},j}}, \tag{4.114}$$

with $\sum_i b_i = 1$. Obviously, the branchings b_i can be temperature-dependent.

Using the rates per target nucleus for all possible processes destroying a specific nuclide in two-body reactions and reactions with photons (Equations (4.11) and (4.100), also including the abundances of projectiles in two-body reactions) as well as by decays allows the construction of so-called rate field plots. In a *rate field* all possible destruction rates are compared for each individual nuclide at given conditions (temperature and projectile number density) and the dominating destruction process is indicated, for example, by an arrow or by color. This allows to have a quick estimate of possible reaction pathways at given conditions, provided that the nuclide is actually reached by the reaction flow. The rate field could be viewed as an analogue to a dry river bed. Note, however, that this river bed is changing according to (mainly) temperature and (also) projectile number density. A further advantage of the rate field is that it allows to decide which reaction type should be preferentially studied even when the actual reaction flow is not known yet. Examples for rate field plots are given in Sections 8.4 and 9.3.3.

The simple relations above, however, only hold when the reactivities are constant. In a stellar plasma this is not the case most of the time because rates depend on plasma temperature and density which, in turn, are affected by the nuclear reactions in the plasma and the hydrodynamic evolution. Even decays showing constant half-life under terrestrial conditions depend on temperature (and thus implicitly also on time) in a stellar plasma (see Equation (4.112)). Furthermore, the abundance of a nuclear species may depend on more than one reaction and the simple connection of two species by one reaction, as used above, will not suffice. In the general case, a larger system of coupled differential equations has to be used. This is discussed in detail in Section 4.2. When the lifetime of a nucleus is determined by several competing destruction and production reactions, the definition given in Equation (4.110) becomes useless. A better measure for a typical *production or destruction timescale* of a nuclear species k is directly using the relative abundance change and is given by

$$\tau_\Delta = \left| \frac{Y_k}{\mathrm{d}Y_k/\mathrm{d}t} \right| = \left| \frac{Y_k}{\dot{Y}_k} \right|. \tag{4.115}$$

When dealing with nucleosynthesis processes involving a large number of nuclides it can be informative to define average timescales for a given type of reaction or decay,

$$\begin{aligned} \bar{\tau}_{\mathrm{life}}^{\mathrm{dec}} &= \frac{\sum_k Y_k}{\sum_k Y_k \lambda^*_{\mathrm{dec},k}} \quad \text{for decays,} \\ \bar{\tau}_{\mathrm{life}}^{\mathrm{reac}} &= \frac{\sum_k Y_k}{\sum_k Y_k \lambda^*_{\mathrm{trans},k}} \quad \text{for reactions.} \end{aligned} \tag{4.116}$$

These are averaged over all nuclides k (note that the sums run over nuclides and not reactions) with their individual rates for a specific reaction or decay. For example, $\bar{\tau}_{\mathrm{life}}^{(\mathrm{n},\gamma)}$ is for the averaged lifetime against neutron capture and $\bar{\tau}_{\mathrm{life}}^{\beta^-}$ is the averaged lifetime with respect to β^- decay.

4.1.8 Reaction Flow and Energy Generation

Knowing the abundance changes by reactions and decays it is possible to define a flow (time-integrated flux) between the various isotopes in a reaction network. Assuming a reaction connecting two nuclides A and B, the flow $\bar{\mathcal{F}}_{\mathrm{A}\to\mathrm{B}}$ between them in the time interval $\Delta t = t_2 - t_1$ is given by

$$\bar{\mathcal{F}}_{\mathrm{A}\to\mathrm{B}}(\Delta t) = \int_{t_1}^{t_2} \left. \frac{\mathrm{d}Y_\mathrm{A}}{\mathrm{d}t} \right|_{\mathrm{A}\to\mathrm{B}} \mathrm{d}t = \int_{t_1}^{t_2} \lambda_{\mathrm{trans}}^{*,\mathrm{A}\to\mathrm{B}}(t)\, Y_\mathrm{A}(t)\, \mathrm{d}t. \tag{4.117}$$

Note that $\mathrm{d}Y/\mathrm{d}t$ may be varying with time and therefore also within the time interval Δt. As explained in Section 4.1.7, $\lambda_{\mathrm{trans}}^{*,\mathrm{A}\to\mathrm{B}}$ is the reactivity appropriate to the reaction connecting the two nuclear species.

Whenever there is a non-negligible possibility that also the reverse reaction from B to A is possible, the *net reaction flow* $\bar{\mathcal{F}}_{\mathrm{net}}$ is computed by subtracting the reverse flow from the forward flow,

$$\bar{\mathcal{F}}_{\mathrm{net}}(\Delta t) = \bar{\mathcal{F}}_{\mathrm{A}\to\mathrm{B}}(\Delta t) - \bar{\mathcal{F}}_{\mathrm{B}\to\mathrm{A}}(\Delta t). \tag{4.118}$$

Furthermore, the energy released (or consumed) by a reaction is simply given by its reaction Q-value Q_{nuc} (see Equation (3.93) for the definition of the Q-value). The energy release per mass and time by a single reaction is related to its reaction rate r_r and the Q-value by

$$q_{\mathrm{nuc},r} = \frac{r_{\mathrm{r}} Q_{\mathrm{nuc}}}{\rho}. \tag{4.119}$$

In order to account for the total energy generation by all reactions in a given volume of stellar plasma, the total change in mass (at constant density) can be used to calculate the released energy using

$$\frac{\mathrm{d}E}{\mathrm{d}t} = -V\sum_i \left.\frac{\partial \tilde{n}_i}{\partial t}\right|_\rho m_i \mathrm{c}^2 = -\rho \mathrm{N_A} V \sum_i \frac{\mathrm{d}Y_i}{\mathrm{d}t} m_i \mathrm{c}^2. \tag{4.120}$$

The sum runs over all nuclide types i present in the volume. The minus sign is required because a decrease in mass indicates stronger nuclear binding and thus an energy release. To obtain the total energy generation per mass, the specific volume defined in Equation (1.8) is inserted into Equation (4.120), yielding

$$q_{\mathrm{nuc}} = \left.\frac{\mathrm{d}E}{\mathrm{d}t}\right|_V = -\mathrm{N_A}\sum_i \frac{\mathrm{d}Y_i}{\mathrm{d}t} m_i \mathrm{c}^2. \tag{4.121}$$

This includes energy released in various forms, such as electromagnetic radiation (γ-rays) and as kinetic energies of hadrons and leptons. For the energy generation term in stellar models, the neutrino energy is taken out of the total released energy, $q = q_{\mathrm{nuc}} - q_\nu$, because it does not contribute during the hydrostatic stages of stellar evolution (see Equation (2.68) and Section 2.7.1, Chapter 6). Neutrino energy deposition is important, however, at very high densities close to nuclear densities in the collapse of the core of a massive star and its subsequent explosion as a supernova (see Section 9.3).

4.2 Nuclear Reaction Networks

4.2.1 Definition

Summarizing the previous sections, the change in abundance of a nuclide C is related to the change in number density by

$$\dot{Y}_{\mathrm{C}} = \frac{1}{\rho \mathrm{N_A}} \dot{\tilde{n}}_{\mathrm{C}}. \tag{4.122}$$

Taking the example of a two-body reaction between two nuclides A and B as source of the nuclide C the abundance change is expressed as

$$\dot{Y}_{\mathrm{C}} = \frac{1}{\rho \mathrm{N_A}} r^*_{\mathrm{AB}}, \tag{4.123}$$

where r^*_{AB} is the stellar two-body reaction rate (Equations (4.9) and (4.10))

$$r^*_{\mathrm{AB}} = \frac{\tilde{n}_{\mathrm{A}}\tilde{n}_{\mathrm{B}}}{1+\tilde{\delta}_{\mathrm{AB}}}\langle\sigma^* v\rangle = \frac{1}{1+\tilde{\delta}_{\mathrm{AB}}} Y_{\mathrm{A}} Y_{\mathrm{B}} \rho^2 \mathrm{N_A^2} \langle\sigma^* v\rangle \tag{4.124}$$

with the reaction rate per particle pair $\langle\sigma^*v\rangle$ for the reaction A + B → C. Note that the exponent of the density dependence of the abundance rate-of-change $\dot{Y}$ is lower than the one for the number density rate-of-change $\dot{n}$, due to the relation between abundance and number density according to the definition in Equation (1.14). Even if the above reaction was the only one possible, also the abundances Y_A and Y_B were time-dependent and this had to be considered by a coupled set of differential equations for $\dot{Y}_A$, $\dot{Y}_B$, and $\dot{Y}_C$, including the appropriate rates for the connecting reactions. A very simple such set of differential equations is shown in Equation (4.105) (not for a two-body process but for a nuclear decay).

In the most general case, the change of abundances Y with time due to nuclear processes is traced by coupled differential equations for all reactions affecting all nuclides of interest. This is called a *reaction network*. A general reaction network can be written as

$$\dot{Y}_i = \frac{1}{\rho \mathrm{N_A}}\dot{n}_i = \frac{1}{\rho \mathrm{N_A}}\left\{\sum_j {}^1_iK_j\, {}^1_ir_j + \sum_j {}^2_iK_j\, {}^2_ir_j + \sum_j {}^3_iK_j\, {}^3_ir_j + \cdots\right\}, \tag{4.125}$$

where $1 \leqslant i \leqslant n_{\mathrm{net}}$ numbers the nucleus type, ${}^1_ir_j$ is the jth rate for destruction or creation of the ith species without a nuclear projectile involved as described in Sections 4.1.5–4.1.7 (one-body rates including photodisintegration, lepton capture, spontaneous decay), and ${}^2_ir_j$ is the rate of the jth reaction involving a nuclear projectile and creating or destroying nucleus i (two-body rates described in Section 4.1.2). Explicitly included in the sum are also three-body reactions where nucleus i is produced or destroyed together with two other (or similar) nuclei, denoted by ${}^3_ir_j$. Such reactions and reactions with even more participants (denoted by ... above) are unlikely to occur at astrophysical conditions and are usually neglected (for the special case of the triple-α reaction, see Chapter 7). The quantities ${}^1_iK_j$, ${}^2_iK_j$, and ${}^3_iK_{jk}$ are positive or negative integer numbers specifying the amount of nuclei i produced or destroyed, respectively, in the one-, two-, three-body process, respectively. As defined in Section 4.1, the rates contain the abundances of the interacting nuclei. Rates of type ${}^1_ir_j$ depend on one abundance (or number density), rates ${}^2_ir_j$ depend on the abundances of two species, and rates ${}^3_ir_j$ on three. All of these rates also depend on plasma temperature.

Equation (4.125) can also be written differently, not by numbering the reactions affecting each nuclear species i but by summing over nuclear species k, l,... and identifying the reactions by the interacting particles k, l,.... This yields

$$\dot{Y}_i = \sum_k {}^1_i\bar{K}_k\, {}_i\lambda^*_{\mathrm{trans},k} Y_k + \sum_{k,l} \frac{{}^2_i\bar{K}_{k,l}}{1+\tilde{\delta}_{kl}}\, \rho \mathrm{N_A}\, {}^i\langle\sigma^*v\rangle_{k,l} Y_k Y_l + \cdots, \tag{4.126}$$

with the integer numbers ${}^1_i\bar{K}_k$, ${}^2_i\bar{K}_{k,l}$,... again specifying the number of nuclei produced or destroyed in one-body and two-body reactions, respectively. Note that the sums run over all nuclides k, l, which also includes nuclide i. Reactivities for one-body rates are represented by $\lambda^*_{\mathrm{trans}}$ and the appropriate quantities as given in

Sections 4.1.5–4.1.7 have to be used in their place. The involved abundances can be directly seen when using the form of Equation (4.126). In the full reaction network there are as many equations as there are types of nuclides or nucleons appearing in all considered reactions. With n_{net} nuclides, this becomes a set of n_{net} coupled differential equations.

4.2.2 Simple Example

As an example for the application of Equations (4.125) and (4.126), the following coupled reactions shall be considered:

$$
\begin{aligned}
&(\mathrm{I}) && {}^{1}\mathrm{H}(\mathrm{p,e^{+}}\nu_{\mathrm{e}})^{2}\mathrm{H},\\
&(\mathrm{II}) && {}^{2}\mathrm{H}(\mathrm{p},\gamma)^{3}\mathrm{He},\\
&(\mathrm{III}) && {}^{3}\mathrm{He}(^{3}\mathrm{He,2p})^{4}\mathrm{He}.
\end{aligned}
$$

Except for leptons (which are not followed explicitly in this example), there is one differential equation for each nuclide or nucleon appearing in any of the reactions. The differential equation describing the change in abundance of a specific species has to sum over all reaction rates affecting the abundance. As there are four species in the above example, four equations are obtained:

$$
\begin{aligned}
\dot{Y}_{\mathrm{p}} &= -\frac{2}{2}\rho \mathrm{N_A}\langle\sigma^{*}v\rangle_{\mathrm{I}}\, Y_{\mathrm{p}}^{2} - \rho \mathrm{N_A}\langle\sigma^{*}v\rangle_{\mathrm{II}}\, Y_{\mathrm{p}} Y_{^{2}\mathrm{H}} + \frac{2}{2}\rho \mathrm{N_A}\langle\sigma^{*}v\rangle_{\mathrm{III}}\, Y_{^{3}\mathrm{He}}^{2}\\
&\quad \left\{+ 2\lambda_{\mathrm{e^{+}}}^{*} Y_{^{2}\mathrm{H}} + \lambda_{\mathrm{C}\gamma}^{*} Y_{^{3}\mathrm{He}} - 2\rho^{2}\mathrm{N_A^{2}}\langle\sigma^{*}v\rangle_{\mathrm{III'}}\, Y_{\mathrm{p}}^{2} Y_{^{4}\mathrm{He}}\right\},\\
\dot{Y}_{^{2}\mathrm{H}} &= \frac{1}{2}\rho \mathrm{N_A}\langle\sigma^{*}v\rangle_{\mathrm{I}}\, Y_{\mathrm{p}}^{2} - \rho \mathrm{N_A}\langle\sigma^{*}v\rangle_{\mathrm{II}}\, Y_{\mathrm{p}} Y_{^{2}\mathrm{H}}\\
&\quad \left\{+ \lambda_{\mathrm{C}\gamma}^{*} Y_{^{3}\mathrm{He}} - \lambda_{\mathrm{e^{+}}}^{*} Y_{^{2}\mathrm{H}}\right\},\\
\dot{Y}_{^{3}\mathrm{He}} &= \rho \mathrm{N_A}\langle\sigma^{*}v\rangle_{\mathrm{II}}\, Y_{\mathrm{p}} Y_{^{2}\mathrm{H}} - \frac{2}{2}\rho \mathrm{N_A}\langle\sigma^{*}v\rangle_{\mathrm{III}}\, Y_{^{3}\mathrm{He}}^{2}\\
&\quad \left\{+ 2\rho^{2}\mathrm{N_A^{2}}\langle\sigma^{*}v\rangle_{\mathrm{III'}}\, Y_{\mathrm{p}}^{2} Y_{^{4}\mathrm{He}} - \lambda_{\mathrm{C}\gamma}^{*} Y_{^{3}\mathrm{He}}\right\},\\
\dot{Y}_{^{4}\mathrm{He}} &= \frac{1}{2}\rho \mathrm{N_A}\langle\sigma^{*}v\rangle_{\mathrm{III}}\, Y_{^{3}\mathrm{He}}^{2}\\
&\quad \left\{- \rho^{2}\mathrm{N_A^{2}}\langle\sigma^{*}v\rangle_{\mathrm{III'}}\, Y_{\mathrm{p}}^{2} Y_{^{4}\mathrm{He}}\right\}.
\end{aligned}
$$

For instructive purposes, the ratios ${}^{2}_{i}\bar{K}_{k,l}/1 + \tilde{\delta}_{kl}$ were left in the above equations, even when they cancel. The curly brackets enclose contributions from the reverse rates. The reverse reaction of reaction (III) (here denoted as III′) is negligible under all nucleosynthesis conditions. Likewise the positron-induced reverse reaction of (I) is negligible in all stellar burning phases. Due to the reciprocity of stellar rates (see Equations (4.51) and (4.52)), the rates of the remaining reverse reactions also can be expressed with the help of the reactivities of the forward reactions:

$$\dot{Y}_{\mathrm{p}} = -\rho \mathrm{N_A} \langle \sigma^* v \rangle_{\mathrm{I}} Y_{\mathrm{p}}^2 - \rho \mathrm{N_A} \langle \sigma^* v \rangle_{\mathrm{II}} Y_{\mathrm{p}} Y_{^2\mathrm{H}} + \rho \mathrm{N_A} \langle \sigma^* v \rangle_{\mathrm{III}} Y_{^3\mathrm{He}}^2$$

$$\left\{ + \frac{G_0^{^2\mathrm{H}}}{G_0^{^3\mathrm{He}}} \left[\frac{\mu_{\mathrm{red}}^{\mathrm{p}^2\mathrm{H}} \mathrm{k_B} T}{2\pi\hbar^2} \right]^{3/2} \mathrm{e}^{-\frac{Q_{\mathrm{nuc}}^{\mathrm{p}+^2\mathrm{H}}}{\mathrm{k_B}T}} \langle \sigma^* v \rangle_{\mathrm{II}} Y_{^3\mathrm{He}} \right\},$$

$$\dot{Y}_{^2\mathrm{H}} = \frac{1}{2} \rho \mathrm{N_A} \langle \sigma^* v \rangle_{\mathrm{I}} Y_{\mathrm{p}}^2 - \rho \mathrm{N_A} \langle \sigma^* v \rangle_{\mathrm{II}} Y_{\mathrm{p}} Y_{^2\mathrm{H}}$$

$$\left\{ + \frac{G_0^{^2\mathrm{H}}}{G_0^{^3\mathrm{He}}} \left[\frac{\mu_{\mathrm{red}}^{\mathrm{p}^2\mathrm{H}} \mathrm{k_B} T}{2\pi\hbar^2} \right]^{3/2} \mathrm{e}^{-\frac{Q_{\mathrm{nuc}}^{\mathrm{p}+^2\mathrm{H}}}{\mathrm{k_B}T}} \langle \sigma^* v \rangle_{\mathrm{II}} Y_{^3\mathrm{He}} \right\},$$

$$\dot{Y}_{^3\mathrm{He}} = \rho \mathrm{N_A} \langle \sigma^* v \rangle_{\mathrm{II}} Y_{\mathrm{p}} Y_{^2\mathrm{H}} - \rho \mathrm{N_A} \langle \sigma^* v \rangle_{\mathrm{III}} Y_{^3\mathrm{He}}^2$$

$$\left\{ - \frac{G_0^{^2\mathrm{H}}}{G_0^{^3\mathrm{He}}} \left[\frac{\mu_{\mathrm{red}}^{\mathrm{p}^2\mathrm{H}} \mathrm{k_B} T}{2\pi\hbar^2} \right]^{3/2} \mathrm{e}^{-\frac{Q_{\mathrm{nuc}}^{\mathrm{p}+^2\mathrm{H}}}{\mathrm{k_B}T}} \langle \sigma^* v \rangle_{\mathrm{II}} Y_{^3\mathrm{He}} \right\},$$

$$\dot{Y}_{^4\mathrm{He}} = \frac{1}{2} \rho \mathrm{N_A} \langle \sigma^* v \rangle_{\mathrm{III}} Y_{^3\mathrm{He}}^2.$$

This set of reactions comprises the pp-I chain of solar hydrogen burning (see Section 7.2.1). The Q-value of the reaction p + ^{2}H → ^{3}He is $Q_{\mathrm{nuc}}^{\mathrm{p}+^2\mathrm{H}} = 5.5$ MeV. Therefore the exponential factor $\exp[-Q_{\mathrm{nuc}}^{\mathrm{p}+^2\mathrm{H}}/(\mathrm{k_B}T)]$ is tiny at solar conditions ($T \approx 15$ MK) and all reverse reactions in the curly brackets can be neglected. Whether reverse reactions can be neglected in a reaction network has to be checked for each application as it depends on the reaction Q-value of the involved reactions and on the plasma temperature. Usually, at such high temperatures encountered, for example, in explosive burning the reverse reactions have to be taken into account. On the other hand, at low s-process temperatures (see Section 8.2) only neutron captures are considered without their reverse (γ,n) reactions.

4.2.3 Solution Methods

Due to the nature of the involved reactions, vastly different timescales appear when considering a large range of different reactions in a reaction network. This results in a non-linear, stiff set of differential equations. In addition, the complete set required for solving the coupled equations comprises as many equations (n_{net}) as there are nuclei acting as reaction partners. Therefore an equation matrix of size $n_{\mathrm{net}} \times n_{\mathrm{net}}$ has to be solved. Nucleosynthesis processes include thousands of nuclides and tens of thousands reactions. Even with modern computers, this still makes it impossible to fully couple such a reaction network to a full set of hydrodynamic equations (see Section 2.9) as would be required for a complete modeling of nucleosynthesis in a given astrophysical site. The usual approach to alleviate this problem is to couple a reduced network, containing only the reactions which are most important for energy generation, to the hydrodynamics. All other reactions, which may be interesting in terms of nucleosynthesis but not for energy generation, are carried in a separate network without direct feedback to hydrodynamics (this is sometimes called operator splitting). Because it is not directly coupled to hydrodynamics, it can be evolved at different timesteps than the network coupled to the hydrodynamic equations. Such a decoupled network can also be used in a *post-processing* approach, in which initial abundances are processed by a standalone network code using a time-dependent density and temperature from a parameterization or extracted from a previously run hydrodynamic simulation.

Systems of stiff differential equations are preferentially solved by using implicit methods. An example of such an implicit method is Euler backward differentiation (but several other methods also are in use). The general problem is to solve

$$\dot{\vec{Y}}(t) = \vec{f}\,(t,\ \vec{Y}(t)), \tag{4.127}$$

for the solution abundance vector $\vec{Y}$ of length n_{net}. The system is evolved from time step t_{n+1} backwards to time t_n:

$$\vec{Y}(t_{n+1}) - \vec{Y}(t_n) = [t_{n+1} - t_n]\vec{f}\,(t_{n+1},\ \vec{Y}(t_{n+1})). \tag{4.128}$$

This leads to the system

$$\frac{\vec{Y}(t_{n+1}) - \vec{Y}(t_n)}{t_{n+1} - t_n} - \vec{f}\,(t_{n+1},\ \vec{Y}(t_{n+1})) = \vec{G}(\vec{Y}(t_{n+1})) = 0. \tag{4.129}$$

The root $\vec{G} = 0$ can be found, for example, by the Newton–Raphson method.

Other methods frequently employed to solve nuclear reaction networks are the *Bader–Deuflhard method* and *Gear's backward differentiation technique*, which are semi-implicit methods. Gear's technique is harder to implement but offers computational speed advantages over the Bader–Deuflhard method, which takes more time steps even when the solution already has converged. Both approaches use adaptive time steps and are considerably faster than explicit solution methods, though.

In all the mentioned methods at least one matrix inversion is required per time step taken. Thus, the computational speed of matrix inversion is crucial in determining the total time to solve the reaction network. The matrix inversion can be sped up by using the fact that the network matrix is symmetric and a *sparse matrix*. Although its size is $n_{\mathrm{net}} \times n_{\mathrm{net}}$, most of the matrix entries are zero due to the nature of the interacting particles and the reactions included. Most of the reactions are with neutrons, protons, and α-particles. This gives rise to tri-banded, diagonal entries plus the symmetric entries for the light projectiles. Special mathematical methods can be applied to efficiently invert such a matrix, considerably speeding up the required computational time.

4.2.4 Parameterization of Reaction Rates

Once the astrophysical reaction rates have been determined as shown in the preceding sections, a convenient and efficient way to include their values in a nuclear reaction network has to be found, for any temperature encountered in a stellar or explosive environment. A straightforward way would be to use tabulated values of reactivities and to interpolate table values between given temperatures.

This involves several pitfalls because, also depending on the actual size of the temperature grid, this can be quite taxing on computational resources regarding working memory and speed. Since the reaction network equations have to be solved many times per hydro timestep (see above in Section 4.2.3) each interpolation step adds additional computational overhead. Fine temperature grids lead to large tables which have to be carried in computer memory. A too coarse temperature grid may jump over details in the temperature evolution of a rate and may give rise to numerical artifacts depending on the interpolation method. Moreover, the treatment of reactions outside the tabulated temperature range has to be carefully defined.

More widely used in reaction network codes are parameterized rates. To put such parameterized rates directly into network codes is possible but advantageous only for a few important rates, also because of flexibility issues. It becomes difficult to update the rates or to do quick comparisons between different rate evaluations. In the past, a handful of resonant rates important for energy generation in hydrostatic burning have been implemented in such a way in stellar models. For example, the expressions for resonant rates as shown in Section 4.1.2.3 could be directly hardcoded. Also in this case, however, the expressions have to be simplified as much as possible (for example by combining several resonance terms) to save on computational time.

The majority of rates are incorporated into reaction networks as a compromise between memory usage, computation time, and accuracy by using parameterized fits to reactivities with the fit parameters read in from a file at the beginning of the calculation run. The function(s) used for fitting are chosen to minimize computational effort while still guaranteeing a correct physical behavior of the rates across a wide temperature range. Only few "standard" parameterizations have been used in the past decades. Two extensively used parameterizations are given below for illustration. They fit stellar reactivities $N_A\langle\sigma^* v\rangle$ for two-body reactions and yield the reactivity in units of cm^3 $mole^{-1}$ s^{-1}.

The WFHZ parameterization (Holmes et al. 1976; Woosley et al. 1978) uses different fitting functions depending on the type of reaction. For neutron captures,

$$N_A\langle\sigma^* v\rangle_{(n,\gamma)} = a_0\left[\frac{T_9}{0.348}\right]^{a_1} e^{a_2(T_9-0.348)+a_3(T_9-0.348)^2}, \tag{4.130}$$

with the temperature T_9 in GK and the fit parameters a_0–a_3. (Note that for $T_9 \approx 0.348$ the energy $k_B T \approx 30$ keV.) For (n,p) and (n,α) reactions, either

$$N_A\langle\sigma^* v\rangle = a_0 \exp[a_1 T_9 + a_2 T_9^2 + a_3 T_9^3] \tag{4.131}$$

is used, or the more complicated expression

$$N_A\langle\sigma^* v\rangle = a_0 \exp\left[\frac{11.605 Q_{nuc}}{T_9 + a_4} - \frac{\tau_{fit}}{(T_9 + a_5)^{1/3}}\left(1 + a_1 T_9 + a_2 T_9^2 + a_3 T_9^3\right)\right] \tag{4.132}$$

with the additional fit parameters a_4, a_5. The parameter $\tau_{\rm fit}$ reflects the Coulomb barrier penetration. It can be calculated as $\tau_{\rm fit} = 4.2487(Z_1^2 Z_2^2 \mu_{\rm red})^{1/3}$, with $\mu_{\rm red}$ being the reduced mass of the interacting particles, but is left as a free fit parameter for some reactions with small reaction Q-value. Reactivities of reactions without a neutron in both entrance and exit channel were parameterized as

$$\mathrm{N_A}\langle\sigma^*v\rangle = T_9^{-2/3}\exp\left[a_0 - \frac{\tau_{\rm fit}}{T_9^{1/3}}(1 + a_1T_9 + a_2T_9^2 + a_3T_9^3)\right]. \tag{4.133}$$

The analytic functions used in the WFHZ parameterization force the fits to exhibit physical behavior also at low temperature where the penetration through the Coulomb barrier is the dominating influence. It is less flexible regarding the inclusion of isolated resonances and therefore mostly suited for fitting reactivities of intermediate and heavy nuclei without pronounced resonance features. More recently, a hybrid approach has been suggested to improve this point, combining fits with table interpolation (Rauscher et al. 2002). In this approach, the above functions (in particular, Equation (4.133)) are used to fit the reactivities. This fit is intended to contain the bulk of the temperature dependence of the rate. Additionally, the logarithms of the ratios of the actual reactivity to the one predicted by the fitting function are carried as a table in the computer. They can be interpolated much more accurately than the reactivity itself.

The currently most widely used type of parameterization is the seven-parameter REACLIB format (Rauscher & Thielemann 2000) which fits all two-body reactivities (but also, for example, photodisintegration and decay rates) with the function

$$\mathrm{N_A}\langle\sigma^*v\rangle = \exp[a_0 + a_1T_9^{-1} + a_2T_9^{-1/3} + a_3T_9^{1/3} + a_4T_9 + a_5T_9^{5/3} + a_6\ln(T_9)]. \tag{4.134}$$

This functional dependence can also accommodate resonant behavior and it is often possible to fit the contributions of several resonances with a single expression as shown above. When several strong, isolated resonances are contributing, it is sometimes necessary to fit the reactivity as a sum of several expression of the type given in Equation (4.134). This should be avoided as much as possible, though, because it significantly increases the computing time when evaluating the expressions. The drawback of higher flexibility is that one has to pay much attention during the fitting process to get physical behavior outside the fitted temperature range. This is feasible in most cases but requires a careful and sophisticated approach to fitting the reactivity. In any case, the temperature range in which the fit is valid has to be strictly adhered to when evaluating the fitted functions.

Regardless of how the rates are implemented (tables or parameterizations), only the rate for one reaction direction is input whereas the reverse rate is computed from the reciprocity relations in Equations (4.51) and (4.52). This is necessary to avoid numerical instabilities when, for example, computing forward and reverse reactivities from fits or tables from different sources. The reaction direction to fit is always the exoergic one because using an endoergic reaction ($Q_{\rm nuc} < 0$) as basis for the reciprocity relation would enhance any fit inaccuracies due to the exponential dependence of the reciprocity relation on the reaction Q-value (see Section 4.1.3.2). A possible exception to this Q-value rule are endoergic charged-particle capture reactions obtained from experiment due to the suppression of thermally excited state contributions as explained in Section 4.1.3.3. When using the WFHZ parameterization the reverse reactivities have to be calculated directly using the reciprocity relations in Equations (4.51) and (4.52). The numerical prefactor, also containing spins and masses of the interacting particles, is carried as an additional parameter $a_{\mathrm{rev}}^{\mathrm{WFHZ}}$. When using the REACLIB format, on the other hand, also the reverse reactivities can be easily computed from the expression in Equation (4.134), only a few parameters have to be adapted. The numerical prefactor can be included in a revised parameter a_0^{rev}. The second parameter is adapted as $a_1^{\mathrm{rev}} = a_1 - 11.6045Q_{\mathrm{nuc}}$, where Q_{nuc} is for the forward reaction. All other parameters remain the same, except when computing a photodisintegration reactivity from a capture. In this latter case also the last parameter has to be adapted, $a_6^{\mathrm{rev}} = a_6 + 1.5$. The reactivity obtained with Equation (4.134) and this adapted parameter set finally has to be multiplied by the ratio of the normalized partition functions for target nucleus and final nucleus to obtain the reactivity of the reverse reactivity.

Another temperature-dependent quantity required for the calculation of the reverse reactivity is the normalized nuclear partition function (Equation (4.39)). Again, it can be carried in computer memory as a table or by applying a fitting procedure. For example,

$$G(T_9) - G^{\mathrm{low}}(T_9) = \exp\left(\frac{a_0 + a_1T_9 + a_2T_9^2 + a_3T_9^3}{T_9}\right) \tag{4.135}$$

was used by Holmes et al. (1976); Woosley et al. (1978). For most partition functions good fits were obtained with $a_3 = 0$. A separate contribution of low-lying states $G^{\mathrm{low}}(T_9)$ had to be added to accommodate a few cases that created difficulties in fitting them with the above expression. For these few cases,

$$G^{\mathrm{low}}(T_9) = \sum_{\mu \geqslant 0, E_\mu^{\mathrm{x}} \leqslant E_{\max}} \frac{g_\mu}{g_0} \mathrm{e}^{-E_\mu^{\mathrm{x}}/(k_{\mathrm{B}}T)}, \tag{4.136}$$

with $E_{\max} = 0.0862$ MeV. For all other cases, $G^{\mathrm{low}}(T_9) = 1$ and the exponential on the right-hand side of Equation (4.135) fits the combined contribution of excited states (not of the ground state) to the partition function.

A final note on this subject: When discussing accuracy of rates and when comparing different rate determinations and their impact on abundances, regardless of whether they are based on experiment or theory, attention has to be paid to the fact that not only experimental or theoretical uncertainties enter the rates but also the accuracy of a fit in the temperature range of interest has to be taken into account. This is why, on the one hand, reactivity fits have to be performed with the utmost care and, on the other hand, fits should not be used outside of the recommended temperature range.

4.3 Simplification of Reaction Networks due to Equilibria

4.3.1 Introduction

Why should reaction networks be simplified when computers are becoming faster and faster? Simplifications are not only saving computing time but are also instructive because they enable us to study nucleosynthesis properties which are independent of details in the hydrodynamic evolution of the system or even, as shown below, independent of individual reaction rates. They usually go along with a restriction to only types of reactions in the network that are actually necessary, instead of blindly evolving a large system of differential equations. Such an approach is not always feasible but considerable understanding of nucleosynthesis has been gained in the past through such means by circumventing the necessity of computationally intensive calculations. Such simplifications remain important today because it is still impossible to couple multi-dimensional hydrodynamic simulations to full reaction networks. Furthermore, restriction to the essential often provides a much better insight into the physical processes than a brute-force full network calculation. Here we are concerned with the simplifications because it has to be understood when it is necessary to know astrophysical cross sections and reaction rates, and when the knowledge of nuclear spins and masses suffices.

There are several ways to simplify reaction networks. As shown in Section 4.2.2 above, a network can be simplified by leaving out reactions which are too slow to significantly contribute to abundance changes, such as reverse reactions at low temperature. Also reactions being so fast in comparison to others that they can be assumed to be instantaneous can be taken out of the network and their net result can be implemented in a "virtual" reaction combining the result of the fast reaction with a reaction in a previous step (for example, for the decay of ^{8}Be in the last step of the pp-III chain in hydrogen burning, discussed in Section 7.2.1). This can also lead to a natural limitation of the extension of the network. An example for this is the s-process network shown in Section 8.2. A very important type of simplification is possible when all the reactions are fast. Fast reactions require the network solver routine to apply extremely tiny timesteps, leading to long computing times. As shown below, using the reciprocity relations (Equations (4.51), (4.52)) allows to simplify the full reaction network as defined in Equation (4.125) when forward and reverse rates are comparably fast.

4.3.2 Nuclear Statistical (Quasi-)Equilibrium

Equilibrium between a reaction and its reverse is established when the rates of both reaction directions are fast enough to affect the abundances significantly in the relevant time frame. Due to the exponential suppression factor in the reciprocity relation of stellar rates, usually this only occurs at high temperature when the suppression factor $\exp[-Q_{\text{nuc}}/[k_B T]] \approx 1$, leading to forward and reverse rate being close to each other. For the equilibration of capture and photodisintegration rates the criterion given in Equation (4.53) can be applied, requiring lower temperature due to the additional $T^{3/2}$ factor in the reciprocity relation. For each of the two types of reaction (particles other than photons in all channels or only photons in one channel), this implies that reactions with lower reaction Q-value will come into equilibrium earlier with rising temperature and drop out later when the temperature is falling, assuming the number densities remain more or less constant. Comparing reactions with similar Q-value between the two types, reactions with only photons in one channel will be equilibrated already at lower temperature.

Using the reciprocity relations in Equations (4.51) and (4.52), respectively, and assuming $\dot{Y}_A = \dot{Y}_B = \dot{Y}_C = 0$ it is trivial to show that the resulting equilibrium abundances Y_A, Y_B, and Y_C obey

$$\frac{Y_A Y_a}{Y_B Y_b} = \frac{g_0^A g_a G_0^A}{g_0^B g_b G_0^B} \left[\frac{\mu_{\mathrm{red}}^{\mathrm{Aa}}}{\mu_{\mathrm{red}}^{\mathrm{Bb}}} \right]^{3/2} e^{-Q_{\mathrm{nuc}}^{\mathrm{Aa}}/(k_B T)} \tag{4.137}$$

for a reaction A + a ↔ B + b and

$$\frac{Y_A Y_a}{Y_C} = \frac{g_0^A g_a G_0^A}{g_0^C G_0^C} \left[\frac{\mu_{\mathrm{red}}^{\mathrm{Aa}} k_B T}{2\pi\hbar^2} \right]^{3/2} e^{-Q_{\mathrm{nuc}}^{\mathrm{Aa}}/(k_B T)} \tag{4.138}$$

for a reaction A + a ↔ C + γ. The individual rates do not appear anymore in the relation between the abundances. Nevertheless, they still determine whether there is equilibrium allowing the application of the above equations. Note that in reality the equilibrium may be shifted with time, leading to a change in the equilibrium abundances, because T may be time-dependent. The above relations also could have been derived by using the chemical potential and the thermodynamic relation given in Equation (1.29).

Depending on the plasma density, above $T \approx$ 4–5 GK all reactions (with the exception of the ones mediated by the weak interaction) achieve equilibrium. The resulting equilibrium abundances could be derived by iterative application of Equations (4.137) and (4.138). A more convenient and elegant solution is to utilize the concept of chemical potentials introduced in Section 1.4. In a so-called chemical equilibrium the concentrations or number densities of the reaction partners in equilibrium can be deduced using their chemical potentials μ_c^{tot}, see Equations (1.29) and (1.31). For a set of nuclear reactions connecting atomic nuclei, this is called *nuclear statistical equilibrium* (NSE).

A nucleus X with proton number Z and neutron number N can be transformed by adding or removing a neutron or a proton. In equilibrium this means that

$$\begin{aligned} \mu_{c,X}^{\mathrm{tot}} + \mu_{c,p}^{\mathrm{tot}} &= \mu_{c,X'}^{\mathrm{tot}}, \\ \mu_{c,X}^{\mathrm{tot}} + \mu_{c,n}^{\mathrm{tot}} &= \mu_{c,X''}^{\mathrm{tot}}, \end{aligned} \tag{4.139}$$

with nucleus X′ having one proton more than nucleus X and nucleus X″ having one neutron more. Whether this is even possible, that means whether the resulting nuclei actually can be formed at the given conditions, is automatically accounted for by the thermodynamic definition of μ_c^{tot} (see Section 1.4). In such a manner any nucleus can be created from nucleons by sequential neutron and proton captures. Since in NSE all reactions are in equilibrium, also all such capture sequences are in equilibrium and it is sufficient to consider only the initial and final configuration, N neutrons + Z protons $\leftrightarrow$ X. Using chemical potentials, this gives

$$N\mu_{\mathrm{c,n}}^{\mathrm{tot}} + Z\mu_{\mathrm{c,p}}^{\mathrm{tot}} = \mu_{\mathrm{c,X}}^{\mathrm{tot}}. \tag{4.140}$$

Nucleons and nuclei in the stellar plasma are an ideal gas and follow the Maxwell–Boltzmann distribution. Their chemical potentials $\mu_{\mathrm{c},i}$ are given by Equation (1.57) and therefore

$$\mu_{\mathrm{c},i}^{\mathrm{tot}} = \mu_{\mathrm{c},i} + m_i\mathrm{c}^2 = \mathrm{k_B}T\ln\left\{\frac{\bar{n}_i h^3}{G_i}[2\pi m_i \mathrm{k_B}T]^{-3/2}\right\} + m_i\mathrm{c}^2. \tag{4.141}$$

Inserting this in Equation (4.140), rewriting the term in the square brackets, and replacing $\bar{n}_i$ by $Y_i\rho\mathrm{N_A}$ yields

$$\begin{aligned}
&N\left\{\mathrm{k_B}T\ln\left\{\frac{Y_\mathrm{n}\rho\mathrm{N_A}}{g_\mathrm{n}}\left[\frac{2\pi\hbar^2}{m_\mathrm{n}\mathrm{k_B}T}\right]^{3/2}\right\} + m_\mathrm{n}\mathrm{c}^2\right\}\\
&+ Z\left\{\mathrm{k_B}T\ln\left\{\frac{Y_\mathrm{p}\rho\mathrm{N_A}}{g_\mathrm{p}}\left[\frac{2\pi\hbar^2}{m_\mathrm{p}\mathrm{k_B}T}\right]^{3/2}\right\} + m_\mathrm{p}\mathrm{c}^2\right\}\\
&= \mathrm{k_B}T\ln\left\{\frac{Y_\mathrm{X}\rho\mathrm{N_A}}{G_\mathrm{X}}\left[\frac{2\pi\hbar^2}{m_\mathrm{X}\mathrm{k_B}T}\right]^{3/2}\right\} + m_\mathrm{X}\mathrm{c}^2.
\end{aligned} \tag{4.142}$$

Reordering all mass terms to one side of the equation and dividing by $\mathrm{k_B}T$ leads to

$$\begin{aligned}
&\ln\left\{\frac{Y_\mathrm{X}\rho\mathrm{N_A}}{G_\mathrm{X}}\left[\frac{2\pi\hbar^2}{m_\mathrm{X}\mathrm{k_B}T}\right]^{3/2}\right\} - N\ln\left\{\frac{Y_\mathrm{n}\rho\mathrm{N_A}}{g_\mathrm{n}}\left[\frac{2\pi\hbar^2}{m_\mathrm{n}\mathrm{k_B}T}\right]^{3/2}\right\}\\
&+ Z\ln\left\{\frac{Y_\mathrm{p}\rho\mathrm{N_A}}{g_\mathrm{p}}\left[\frac{2\pi\hbar^2}{m_\mathrm{p}\mathrm{k_B}T}\right]^{3/2}\right\}\\
&= \frac{1}{\mathrm{k_B}T}\left[Nm_\mathrm{n} + Zm_\mathrm{p} - m_\mathrm{X}\right]\mathrm{c}^2.
\end{aligned} \tag{4.143}$$

The right-hand side of this equation is related to the binding energy $\mathcal{B}_\mathrm{X}$ (see Equation (3.5)) of nucleus X,

$$\frac{1}{\mathrm{k_B}T}\left[Nm_\mathrm{n} + Zm_\mathrm{p} - m_\mathrm{X}\right]\mathrm{c}^2 = \frac{\mathcal{B}_\mathrm{X}}{\mathrm{k_B}T}. \tag{4.144}$$

Using $g_n = g_p = 2$, $m_n \approx m_p \approx m_u$, $A = N + Z$, $m_X \approx Am_u$ and solving for Y_X, the first NSE equation

$$Y_X = G_X[\rho N_A]^{A-1}\frac{A^{3/2}}{2^A}\left[\frac{2\pi\hbar^2}{m_u k_B T}\right]^{3[A-1]/2} e^{B_X/(k_B T)}\, Y_n^N\, Y_p^Z \tag{4.145}$$

is obtained.

Equation (4.145) contains three unknowns, the abundance of the nucleus X and the ones of free neutrons and protons, Y_n and Y_p (to the power of N and Z, respectively). Therefore two more equations are required to determine the abundances. As a second equation, mass conservation can be used. This is simply found by combining the normalization of the mass fractions as given in Equation (1.12) and the definition of the abundance from Equation (1.14) and reads

$$\sum_i A_i Y_i = 1. \tag{4.146}$$

The third equation is obtained by demanding charge conservation. Then the number of protons (free or inside nuclei) is set by the electron abundance Y_e. For convenience, Equation (1.16) is repeated here as the third NSE equation:

$$Y_e = \sum_i Z_i Y_i. \tag{4.147}$$

The sums in the second and third NSE equation run over all species of nuclei in the plasma, including neutrons and protons. Note that individual reaction rates are not required, only binding energies. Nevertheless, reaction rates implicitly determine which nuclei are participating in the equilibrium. The higher the rates at given density, the lower the temperature at which equilibrium is reached. With a time-dependent T evolution, this means that the rates determine whether equilibrium is reached earlier with increasing T or the freeze-out happens later with decreasing T. Also note that reactions mediated by the weak interaction are not included in the equilibrium and Y_e may be time-dependent. If such reactions are not negligible, a separate reaction network including the rates discussed in Sections 4.1.6 and 4.1.7 is necessary to track the change in electron abundance Y_e over time.

Inspection of Equation (4.145) already offers some valuable insights into the abundance patterns to expect in NSE. On one hand, the term $[\rho N_A]^{A-1}$ favors the formation of nuclei with large A at higher density. On the other hand, the term $[2\pi\hbar^2/m_u k_B T]^{3[A-1]/2}$ is inversely proportional to the temperature and therefore nuclei with low mass number will be favored at high temperature. This can be understood intuitively because photodisintegration reactions will be fast at high temperature. The exponential dependence on binding energy further modulates the abundance pattern by strongly enhancing abundances of strongly bound nuclei. While temperature and density shift the range of produced nuclei between light and heavy nuclei, the value of the electron abundance shifts the distribution between the neutron- and proton-rich side of the nuclear chart by setting the line Z/A along which nuclei are produced preferentially. That means that when varying Y_e (or alternatively the neutron excess η_n, Equation (1.18)) while leaving T and ρ constant, abundances peak for nuclei with $Z/A = Y_e$, as is nicely seen in Figure 4.9. Along a chosen Z/A line, nuclei with the highest binding energies receive the highest abundances. For conditions of $\rho \approx 10^7$–10^8 g cm^{-3} and $T \geqslant 3.5$ GK the distribution peaks at ^{56}Ni for $Y_e = 0.5$, for $0.47 \leqslant Y_e \leqslant 0.485$ the abundance peak moves toward ^{58}Ni and ^{54}Fe. Furthermore, $0.46 \leqslant Y_e \leqslant 0.47$ yields mainly ^{56}Fe, $0.45 \leqslant Y_e \leqslant 0.43$ produces ^{58}Fe, ^{54}Cr, ^{50}Ti, and ^{64}Ni, and the range below $Y_e = 0.43$–0.42 makes abundant ^{40}Ca.

When all abundances in the network obey the above relations, full NSE is achieved. In this case, no reaction rates have to be known. In another case, however, more or less extended groups of nuclei are in statistical equilibrium and the relative abundances within a group can be described by equations similar to Equation (4.145). The different groups are connected by comparatively slow reactions not being in equilibrium, which determine the abundance level of one group with respect to another group (see Figure 4.10). The rates of these slow, connecting reactions have to be known explicitly. This is called *quasi-statistical equilibrium* (QSE). It appears in various kinds of high-temperature burning, such as hydrostatic oxygen and silicon burning in massive stars and different explosive scenarios. The triple-α reaction $\alpha + \alpha \rightarrow {}^8\text{Be} + \alpha \rightarrow {}^{12}\text{C}$ (see Chapter 7) is one important example of such a slow reaction often not participating in a reaction equilibrium. It is very sensitive to the density and only will get into equilibrium for $\rho \geqslant 2 \times 10^8$ g cm^{-3}.

The abundances of nuclides i with charge number Z_i, neutron number N_i, mass number A_i, and binding energy $\mathcal{B}_i$ in a QSE group can be related to the abundance of

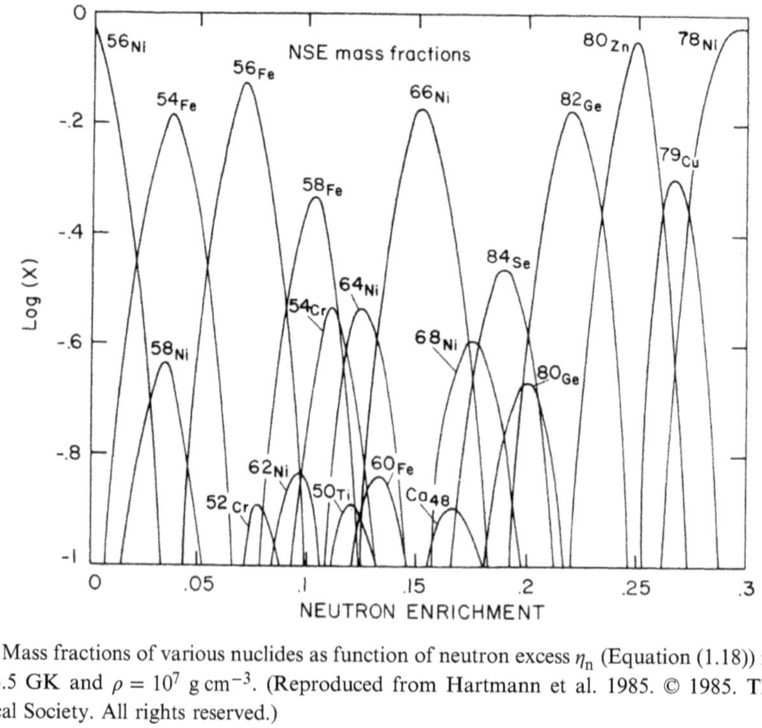

Figure 4.9. Mass fractions of various nuclides as function of neutron excess η_n (Equation (1.18)) in a full NSE with $T = 3.5$ GK and $\rho = 10^7$ g cm^{-3}. (Reproduced from Hartmann et al. 1985. © 1985. The American Astronomical Society. All rights reserved.)

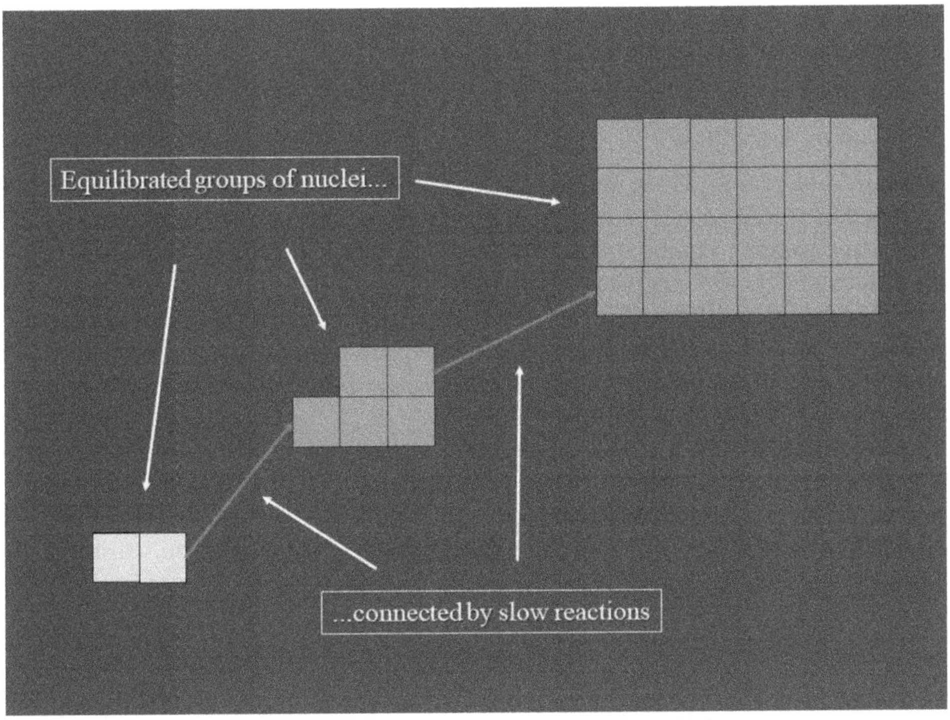

Figure 4.10. (Animation) Sketch of the concept of the quasi-statistical equilibrium (QSE); nuclei within QSE groups are equilibrated, the reactions connecting the groups are not in equilibrium. Animation available online at https://doi.org/10.1088/978-0-7503-1149-6.

a nucleus X with charge number Z_X, neutron number N_X, mass number A_X, and binding energy $\mathcal{B}_X$, which is part of the same QSE group (and therefore in equilibrium with the other nuclides in the group) but is also produced in a reaction not in equilibrium. The nucleus X acts as a gateway connecting the QSE group to other nuclides not participating in the same equilibrium. These other nuclides could be part of another QSE group or part of a nuclide region which is not in equilibrium at all. The rate for producing the nuclide X from outside the QSE group has to be known and explicitly included in the network whereas the abundances of the other nuclides within the group can be determined by (index i marks nuclei being in the same QSE group)

$$\frac{Y_i}{Y_X} = \frac{G_i}{G_X}[\rho N_A]^{A_i+A_X} 2^{A_X-A_i}\left[\frac{2\pi\hbar^2}{m_u k_B T}\right]^{3[A_i-A_X]/2} e^{(\mathcal{B}_X-\mathcal{B}_i)/(k_B T)} Y_p^{Z_i-Z_X} n^{N_i-N_X}. \quad (4.148)$$

Nucleosynthesis under extreme conditions, such as encountered in some explosive scenarios, involves exotic nuclei far from stability. According to the above, reaction rates are not needed for all of them because reaction equilibria are established at such extreme conditions. Required are nuclear masses (to determine Q-value or binding energies) as well as spectroscopic information and nuclear level densities (for the calculation of the partition functions G).

4.3.3 Capture Equilibria

A special kind of QSE is the (n,γ)–(γ,n) equilibrium or *waiting point approximation*, often used in r-process calculations (see Section 8.3). This is nothing else than a QSE within an isotopic chain, where neutron captures and (γ,n) reactions are in equilibrium under very neutron-rich conditions ($\bar{n}_\mathrm{n} \gtrsim 10^{20}$ cm^{-3}) and $T \approx$ 1–2 GK. In the r-process only neutron captures and their inverse reactions have to be considered, as well as β^- decays (with possible subsequent neutron emission). The decays are not in equilibrium and determine the timescale with which matter is processed from small Z to the heaviest nuclei.

Equilibration within a chain occurs on the timescale of the capture and photodisintegration reactions which range from about 10^{-22} s to 10^{-16} s, depending on whether direct or compound processes are involved (see also Section 3.7). The simplified network for an isotope chain includes only neutron captures and photodisintegrations,

$$\frac{\mathrm{d}Y_\mathrm{A}}{\mathrm{d}t} = \frac{1}{\rho \mathrm{N_A}}[r_{\mathrm{C}\gamma} - r_\mathrm{Aa}] = Y_\mathrm{C}\lambda^*_{\mathrm{C}\gamma} - \rho \mathrm{N_A}\langle\sigma^* v\rangle_{\mathrm{n+A}} Y_\mathrm{n} Y_\mathrm{A}, \tag{4.149}$$

because β-decays are much slower and can be neglected when only considering the abundance distribution within an isotopic chain. The nuclides A (with mass number A and neutron number N) and C (with mass number $A + 1$ and neutron number $N + 1$) are connected by the reaction A(n,γ)C (and its reverse reaction). When (n,γ)–(γ,n) equilibrium is achieved, the abundances of nuclei within an isotopic chain do not change anymore and therefore $\dot{Y}_\mathrm{A} = \dot{Y}_\mathrm{C} = 0$. Expressing $\lambda^*_{\mathrm{C}\gamma}$ using the reciprocity relation in Equation (4.52), it is found that the abundances of two neighboring isotopes are connected by

$$\frac{Y_\mathrm{C}}{Y_\mathrm{A}} = \bar{n}_\mathrm{n}\frac{G_\mathrm{C}}{2G_\mathrm{A}}\left[\frac{A+1}{A}\right]^{3/2}\left[\frac{2\pi\hbar^2}{m_\mathrm{u}k_\mathrm{B}T}\right]^{3/2} e^{S_\mathrm{n}^\mathrm{C}/(k_\mathrm{B}T)}. \tag{4.150}$$

There is an exponential dependence on the neutron separation energy $S_\mathrm{n}^\mathrm{C} = Q_\mathrm{nuc}^{\mathrm{A(n,\gamma)C}}$ of C. Also in this type of equilibrium there is no dependence on the individual capture or photodisintegration rates. The r-process flow to higher elements, however, depends on the β^--decay rates which connect the isotopic chains and are not in equilibrium. They are very slow compared to the rates in equilibrium and that is why "waiting points" are established, which are just the nuclei (usually only one or two within a chain) with the highest abundances according to Equation (4.150). The r-process cannot proceed until they decay and their decay rates have to be known as they are not in equilibrium. The waiting point(s) in two neighboring isotopic chains do not have to be contiguous and therefore the notion of an r-process "path" similar to the s-process path is not valid (see Figure 4.11). The nucleosynthesis flow through equilibrated isotope chains in the r-process is further discussed in Section 8.3.

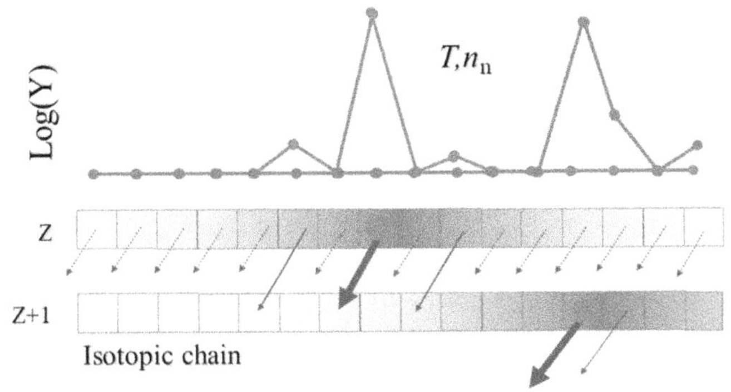

Figure 4.11. Illustration of the waiting point concept in (n,γ)–(γ,n) equilibrium. A few nuclei within an isotope chain receive the highest abundance. The decay of these nuclei (blue arrows) determines the flow into the next isotope chain where again an equilibrium with waiting points is established. The waiting points in neighboring chains are not necessarily contiguous as they depend on the neutron separation energies of the nuclides in the chain. The r-process "path" is an idealization, imagined to connect nuclides with the highest abundances in each chain.

A similar equilibrium, but between proton captures and (γ,p) reactions, is reached in the so-called rp-process (rapid proton-capture process) on the surface of mass-accreting neutron stars (see Section 9.4.4). In an rp-process, the waiting points are established close to the proton dripline. Abundances in (p,γ)–(γ,p) equilibrium are described by a formula similar to Equation (4.150) but with $\bar{n}_n$ replaced by the proton number density $\bar{n}_p$, and S_n^C replaced by the proton separation energy S_p^C.

4.3.4 Steady Flow

Considering a sequence of reactions, the net flow through the chain is determined by the slowest reaction. When the reaction sequence is initiated, the abundances of all nuclides linked by the reactions will change. After a time longer than the reaction timescales of all reactions except for the slowest one, an abundance equilibrium is reached. This can be visualized by filling a pool with a drain from a slow tap. The water level finally adjusts to a constant level when the amount of water flowing in from the tap balances the amount flowing out through the drain. This is called a *steady flow equilibrium.*

After equilibrium has been reached, and as long as steady flow is upheld, there is no need to solve a full reaction network. Rather, the steady flow abundances of the involved nuclei are constant and related to the ratios of their net destruction rates (or, equivalently, production rates as these have to be the same). For illustration, assume a chain of reactions A → B → C → D → ···, with the slowest reaction being A → B. Assuming all net reactions are in steady flow implies that they are the same and that $\dot{Y}_B = \dot{Y}_C = \dot{Y}_D$. Then

$$\frac{Y_B}{Y_C} = \frac{\langle \sigma^* v \rangle_{C \to D}}{\langle \sigma^* v \rangle_{B \to C}} \quad (4.151)$$

and

$$\frac{Y_C}{Y_D} = \frac{\langle\sigma^*v\rangle_{D\to\ldots}}{\langle\sigma^*v\rangle_{C\to D}}. \tag{4.152}$$

The slowest rate sets the abundance of B through

$$\dot{Y}_A = -\dot{Y}_B = -r_{A\to B} \tag{4.153}$$

as given by Equation (4.126). As all connected rates have to be the same in steady flow, instead of $r_{A\to B}$ any of the other rates could be used. Note that the calculation of the rate, however, requires the knowledge of abundances and therefore the choice of reaction depends on which abundances are initially known. Which rates and abundances have reached steady flow equilibrium and which have not can be estimated with the help of the lifetimes and timescales as given in Equations (4.110) and (4.115).

Steady flow considerations are helpful when investigating hydrostatic hydrogen burning of stars through the pp-chains and the CNO cycles (see Chapter 7). In the past they have also been used for sequences of neutron captures in the s-process on nuclei in between magic numbers. The fact that separate steady flows can be assigned to each mass region between closed shells has been termed *local approximation* in s-process studies (see Section 8.2). Also the flow between isotope chains through β^--decays in the r-process (Section 4.3.3 and Figure 4.11) can be modeled as a steady flow from lighter to heavier nuclei. Such an approach can successfully reproduce the observed r-process abundance pattern (see Section 8.3).

Further Reading

Arnett, D. 1996, Supernovae and Nucleosynthesis (Princeton, NJ: Princeton University Press)
Brown, L. S., & Sawyer, R. F. 1997, ApJ, 489, 968
Clayton, D. D. 1984, Principles of Stellar Evolution and Nucleosynthesis (Chicago, IL: University of Chicago Press)
Cowan, J. J., Thielemann, F.-K., & Truran, J. W. 1991, PhR, 208, 267
Debye, P., & Hückel, E. 1923, PhyZ, 24, 125
Debye, P., & Hückel, E. 1923, PhyZ, 24, 305
Fowler, W. A. 1974, QJRAS, 15, 82
Fowler, W. A., Caughlan, G. R., & Zimmerman, B. A. 1967, ARA&A, 5, 525
Fowler, W. A., Caughlan, G. R., & Zimmerman, B. A. 1975, ARA&A, 13, 69
Gasques, L. R., Afanasjev, A. V., Aguilera, E. F., et al. 2005, PhRvC, 72, 025806
Gupta, S. S., & Meyer, B. S. 2001, PhRvC, 64, 025805
Hartmann, D., Woosley, S. E., & El Eid, M. F. 1985, ApJ, 297, 837
Hix, W. R., & Thielemann, F.-K. 1996, ApJ, 460, 869
Holmes, J. A., Woosley, S. E., Fowler, W. A., & Zimmerman, B. A. 1976, ADNDT, 18, 305
Iliadis, C. 2015, Nuclear Physics of Stars (2nd ed.; New York: Wiley)
José, J. 2016, Stellar Explosions (Boca Raton, FL: CRC Press)
Lang, K. R. 1974, Astrophysical Formulae (Berlin: Springer)
Longland, R., Martin, D., & José, J. 2014, A&A, 563, A67
Rauscher, T. 2011, IJMPE, 20, 1071

Rauscher, T. 2012, ApJS, 201, 26
Rauscher, T., & Thielemann, F.-K. 2000, ADNDT, 75, 1
Rauscher, T., Dauphas, N., Dillmann, I., et al. 2013, RPPh, 76, 066201
Rauscher, T., Heger, A., Hoffman, R. D., & Woosley, S. E. 2002, ApJ, 576, 323
Rembges, F., Freiburghaus, C., Rauscher, T., et al. 1997, ApJ, 484, 412
Salpeter, E. E. 1954, AuJPh, 7, 373
Salpeter, E. E., & van Horn, H. M. 1969, ApJ, 155, 183
Sawyer, R. F. 2010, PhRvL, 104, 191103
Shaviv, N. J., & Shaviv, G. 2001, ApJ, 558, 925
Shoppa, T. D., Koonin, S. E., Langanke, K., & Seki, R. 1993, PhRvC, 48, 837
Timmes, F. X. 1999, ApJS, 124, 241
Ward, R. A., & Fowler, W. A. 1980, ApJ, 238, 266
Woosley, S. E., Fowler, W. A., Holmes, J. A., & Zimmerman, B. A. 1978, ADNDT, 22, 371

Part II

Stellar Evolution and Nucleosynthesis

Thomas Rauscher

Chapter 5

Introduction

As already laid out in the Preface, Part II of this book provides an overview of nucleosynthesis in various astrophysical sites. It also introduces the most important nucleosynthesis processes that historically were inferred from elemental and isotopic abundance distributions without initially referring to specific sites, the s-, r-, and p-process. Since simulations to model astrophysical sites as well as nuclear structure and nuclear reactions constantly improve, the understanding of how, where, and when specific elements or their isotopes were/are created also is subject to continuing development. Inherent in such an evolution of ideas and descriptions is a certain volatility, not in facts but in models and model-based deductions that are modified accordingly to the progress in knowledge. The critical re-evaluation of hypotheses and models in the light of new (experimental or observational) facts or theoretical advances is at the core of science. Therefore the picture presented in the following chapters is just a snapshot of the current state of knowledge. This is not to imply, of course, that all the presented processes and galactic histories might be replaced by a completely different set of explanations. They are, after all, soundly based on the physical principles presented in Part I of this book, to which frequent reference will be made. They are also footed in an increasing number of nuclear experiments and astronomical observations. While specific details may still be uncertain, the general picture is well established for the majority of nuclear processes in stars and stellar explosions. Nevertheless, in this golden era of multi-messenger astronomy and astrophysics it is exciting to realize that a wealth of new insights may be just a few years (or even months) away, insights that may also change our current understanding of some details concerning the origin of the elements. It is in this sense that I expect a higher probability for the necessity of frequent updates to this Part II than to Part I of this book.

Nucleosynthesis studies aim at understanding the origin and relative amounts (abundances) of nuclides (elements and their isotopes) found in the Universe and,

doi:10.1088/2514-3433/ab8737ch5

more specifically, of their relative amounts (abundances). Astronomical observations using spectral analysis can identify elements (and sometimes isotopes) in interstellar matter and on the surface of stars. Cosmochemical investigations study isotope patterns found in meteorites, which have preserved material from the early solar system. Certain meteorites also contain grains ejected in stellar winds of nearby AGB stars or from supernovae. The detailed knowledge of relative abundances is the foundation for nucleosynthesis research. Through an analysis of the details of the abundance patterns nuclear astrophysics is able to infer the processes responsible for the formation of the elements. This is not always straightforward because several processes and sites may contribute to the abundance of a specific nuclide in the course of the galactic history. The solar system abundances contain contributions from many stellar generations and many explosive events.

Foremost, the natural abundance distribution in the solar system is of special interest. Their quantitative investigation commenced in the late 19th century but a detailed abundance analysis still is a topic of current research. Figure 5.1 shows the relative abundances in the molecular cloud from which the solar system (Sun and planets) formed 4.56 Gyr ago. This is a snapshot of the composition of the local

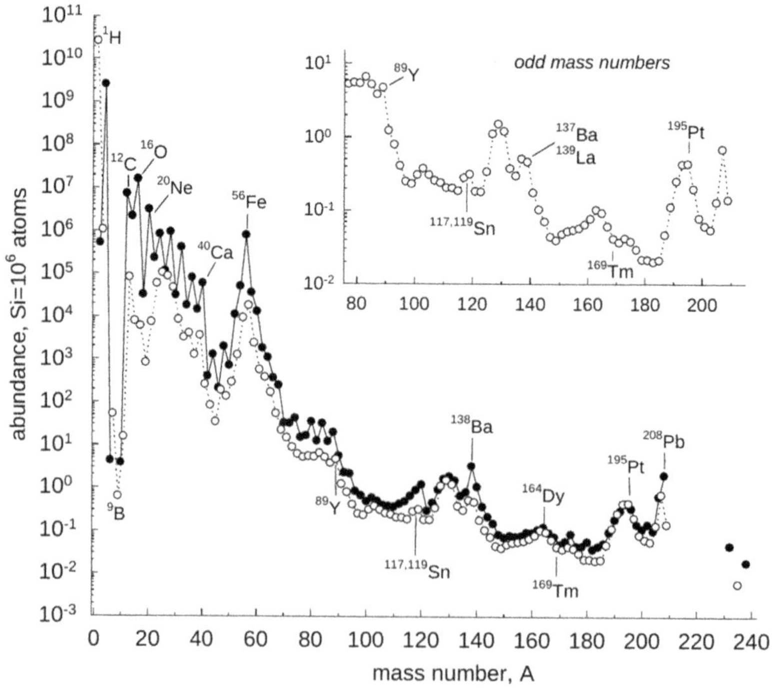

Figure 5.1. Solar system abundances of the nuclides 4.56 Gyr ago, at the formation of the solar system, as function of nuclear mass number. Nuclides with even and odd mass numbers are plotted separately with full and open circles, respectively. Note the logarithmic abundance scale. The full and dotted lines are drawn to guide the eye. (Reprinted by permission from Springer Nature: Springer from Lodders, K. 2010, Solar System Abundances of the Elements, in Principles and Perspectives in Cosmochemistry, ed. A. Goswami & E. B. Reddy (Berlin: Springer).)

interstellar medium at that time. Until today 280 naturally occurring nuclides have survived in 83 elements. The elements Tc and Pm only have short-lived isotopes (compared to the age of the solar system) and therefore are not present naturally anymore. Some isotopes of Th and U, although unstable, are sufficiently long-lived to have survived in the solar system: ^{232}Th, ^{234}U, ^{235}U, and ^{238}U.

In the following chapters, various nucleosynthesis sites and processes are described, most of which have contributed to a particular range of nuclides. Even without that background knowledge, the imprint of nuclear physics can readily be seen in the solar abundance pattern. There is a clear odd–even staggering, paralleling the odd–even variation in the nuclear binding energies (see Section 3.2). Furthermore, very strongly bound nuclides like ^{1}H and ^{4}He are very abundant whereas weakly bound nuclides, easy to destroy already at low plasma temperatures, like Li, Be, B have very low abundances. While H, He, and traces of Li, originate from nucleosynthesis in the early universe (Chapter 10), the elements up to iron are made in hydrostatic stellar burning (Chapter 7) and partially in two types of supernovae (Sections 9.3, 9.4.3). A strong drop in abundance is seen beyond the Fe-peak. The majority of the nuclides beyond iron are synthesized in neutron-capture processes because charged-particle captures cannot efficiently overcome the higher Coulomb barriers (see Sections 8.2, 8.3). The exceptions are the so-called p-nuclides (Section 8.4), which may be produced in a combination of hot and very proton-rich environments, and ^{138}La and ^{180}Ta thought to be created in a ν-process (Section 9.3.4).

An overview of the sites and nucleosynthesis processes potentially contributing to the solar system abundances is given in the following chapters. Also reactions and processes contributing to energy generation in stars and explosive events are summarized. This is not intended to be an in-depth review of the current state-of-the-art in each topic. The intention is rather to provide an introduction, allowing the reader a quick familiarization with current questions and to obtain a foundation to start working on scientific investigations in these areas.

AAS | IOP Astronomy

Essentials of Nucleosynthesis and Theoretical Nuclear Astrophysics

Thomas Rauscher

Chapter 6

Stellar Evolution

The physics foundations for understanding the properties of a stellar object are provided in Chapters 1 and 2. Here, a brief summary of stellar structure and stellar evolution is given in the light of what was discussed in the previous chapters. This is by no means supposed to be a detailed review but is intended to explain merely the environment in which nuclear reactions occur and their connection to properties of a star. For more details on the modeling and the current understanding of stellar structure and evolution, the reader is referred to specialized textbooks and reviews (see also the suggested further references at the end of the chapter).

The stellar structure equations presented in Section 2.9 not only allow to model the interior structure of a star with given mass and composition, they also permit to study its evolution with time. Among the timescales discussed in Section 2.9.2 the nuclear timescale τ_{nuc} determines how long a hydrostatic equilibrium can be upheld when converting matter into energy to compensate for the energy loss through the surface of a star. This timescale is determined by two factors, the amount of material to be converted by nuclear reactions and the luminosity of the star. Stars with larger mass have larger luminosity and thus burn their nuclear fuel faster. From the modeling presented in Chapter 2 the *mass–luminosity relation* for stars on the main sequence of the Hertzsprung–Russell diagram (HRD, i.e., stars which convert protons to α-particles as their energy source) is given by $L \propto M^{a}$, with $a = 3.2$ for stellar masses $M \leqslant \mathrm{M}_{\odot}$ and $a = 3.88$ for $M > \mathrm{M}_{\odot}$ (the change is caused by the switch in dominant H-burning mode from pp-chains to CNO cycles, see Section 7.2). The mass available for conversion to energy, however, only scales with M (only about 10%–15% of the available fuel mass is used up, however). Therefore the time spent on the main sequence is strongly decreasing with increasing mass of the star, with $M^{-2.2}$ and $M^{-2.88}$, respectively. Such a strong mass dependence can also be found for the time spent in other burning phases.

doi:10.1088/2514-3433/ab8737ch6

A *burning phase* is defined by the main nuclear "fuel" converted by nuclear reactions to sustain hydrostatic equilibrium. For example, in hydrogen burning (often abbreviated to H-burning) ^{4}He nuclei are built up from the protons (hydrogen nuclei) abundantly present in the stellar interior. This conversion provides the dominant contribution to the energy generation. Other reactions and nucleosynthesis processes can also run in parallel but may only weakly contribute to the release of energy or even absorb energy in endothermic reactions. With decreasing abundance of the main fuel in the stellar core, the nuclear burning zone moves outwards and forms a spherical shell. The burning mode is thus changing from *core burning* to *shell burning*. Since the nuclear reactions are heating another part of the star in shell burning than in core burning, the star adjusts to the change within the dynamical timescale. The core starts contracting under its own weight until it establishes conditions suited to ignite another burning phase. Since light nuclei with lower Coulomb barriers are consumed first, the next burning phase can only proceed at higher core temperature and density, suited to overcome higher Coulomb barriers. Thus, the sequence of burning phases is basically given by a sequence of higher and higher Coulomb barriers that have to be overcome (see, however, Sections 7.5 and 7.7 for exceptions). The burning phases with their main reaction sequences are discussed in detail in Chapter 7. The newly ignited core burning then releases energy in addition to the shell burning at larger radii. Again, the system has to adjust until a new hydrostatic equilibrium is achieved and core burning can proceed for a duration given by the appropriate nuclear timescale τ_{nuc}. This sequence of events (core burning → move to shell burning → core contraction → ignition of core burning in new burning phase) can be repeated several times, depending on the initial mass of the star. The initial mass (and the mass loss from the surface of the star during its evolution) determine the mass of the cores created from the "ashes" of the burning phases. The core mass and density determine whether the core is stable (i.e., forming a white dwarf) or contracting under its own weight and igniting another burning phase. A star with initial mass between 10 and about 130 $M_\odot$ can complete all possible burning phases in hydrostatic equilibrium. When reaching Si-burning, its interior exhibits a complicated structure with several concentric shells of nuclear burning.

It should also be remembered that the size of the volume within which the nuclear energy is released plays an important role in the appearance of convective structures. In Section 2.7.4 it was shown how a large temperature gradient triggers convection. Therefore, when a large amount of energy is released within a small volume, then convection will occur. This is the case in thin shells of nuclear burning above which convective cells form. Since the volume of the shell burning regions roughly scale with stellar mass (i.e., stars with lower mass have thinner shells), shell burning in low-mass stars gives rise to large convection zones. Also core burning can be radiative or convective, depending on how much energy is released in the core region. This depends on the particular nuclear reactions taking place but also on the stellar mass, which determines the luminosity required to uphold hydrostatic equilibrium. A convection zone can also appear when the opacity in a region of a star is larger than in the region beneath it. Since opacity is a function of temperature (see Section 2.7.2) this can occur in the outer layers of a star. In such a region radiation cannot effectively transport off the energy deposited by radiation from below and this starts convection. This is the case for the Sun, which has a radiative core but a large convection region further outside, reaching the surface. As a consequence of all of the above, the interior of stars on the main sequence of the HRD (see Figure 6.1) exhibit different convective regions depending on their mass. Stars with less than 0.25 $M_\odot$ are fully convective from the core to the surface. Stars with $0.25 \lesssim M \lesssim 1.2\ M_\odot$ have a radiative core and a convective envelope (outer region of the star where the opacity increases due to partially ionized hydrogen atoms) reaching the surface. Stars with more than 1.2 $M_\odot$ burn hydrogen via the strongly T-dependent CNO cycles (see Section 7.2), releasing a large amount of

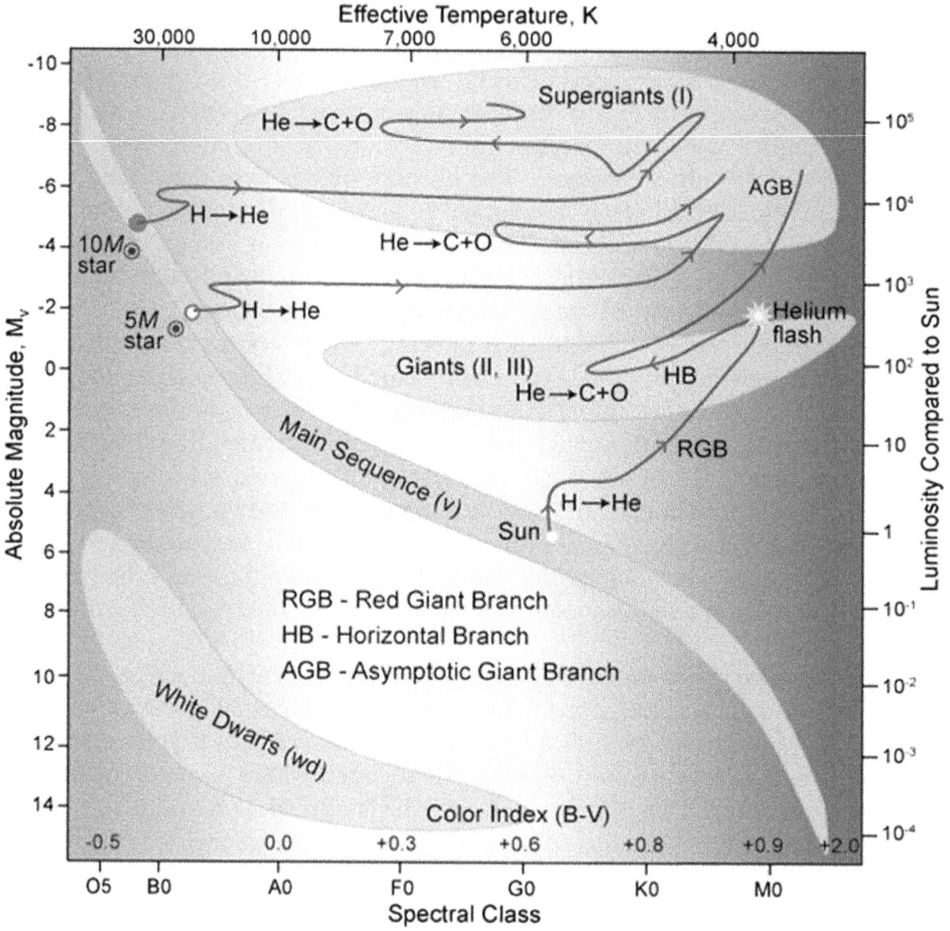

Figure 6.1. Sketch of the Hertzsprung–Russell diagram, showing stellar luminosity and absolute magnitude versus effective temperature and spectral class, respectively. Note the logarithmic scales on the luminosity and temperature axes and that temperature increases from right to left. Also shown are evolutionary tracks of stars with different initial mass. (Figure taken from Wiescher 2009; image by R. Hollow, Commonwealth Science and Industrial Research Organisation (CSIRO), Australia, adapted by Carin Cain.)

energy in their core. Therefore they have radiative cores but no convective envelopes, as the opacity change is less pronounced at larger internal radii. After completion of core H-burning, however, the situation is complicated by additional convection zones due to shell burning. It should be remembered that convection not only transports energy but also matter (see Section 2.8) and thus mixes abundances from different regions of the star. This may also affect nuclear burning and nucleosynthesis when fresh fuel (protons, α-particles, ...) is mixed down into regions where it was previously depleted.

The adaptation of the internal structure of a star of given mass proceeding through different burning stages also alters the observables measurable from the outside, such as luminosity and surface temperature (and the inferred radius from Equation (1.4)). Therefore a star will move through the HRD with time. Figure 6.1 is showing such evolutionary tracks for several stellar masses overlaid onto the HRD. Remembering what was said above concerning timescales, it is obvious that each star completes its track in a (vastly) different period of time, from several 10^9 yr for low-mass stars to a few 10^6 yr or even less for stars with masses exceeding several tens of solar masses. Table 6.1 compares the time spent in H-burning for stars with various initial mass. A star also does not trace its track through the HRD at constant pace. It will spend much more time in those parts of the track which refer to stable, hydrostatic burning phases. Switching from one burning phase to another happens via the horizontal branch in the HRD on a comparatively short timescale, by astronomical standards, of a few 10^2–10^4 yr.

The times spent in different evolutionary stages also explain the clustering of stars into various regions of the HRD. Since the HRD is a snapshot of the stellar population it is more likely to find stars in evolutionary phases with long duration. The largest number of stars is grouped in the main sequence, these are stars in H-burning, the longest stable burning phase. The position along the main sequence is initially given by the stellar mass. Other pronounced groups are the red giants and

Table 6.1. Time Spent in H-burning Depending on Initial Mass

Stellar Mass ($M_\odot$)	Time on Main Sequence (Years)
0.40	2×10^{11}
0.80	1.4×10^{10}
1.00	10^{10}
1.70	2.7×10^9
2.25	5×10^8
3.00	2.2×10^8
5.00	6×10^7
9.00	2×10^7
16.00	10^7
25.00	7×10^6
40.00	$\approx 10^6$

supergiants (see Figure 6.1). They are in He-burning, the second longest burning phase. Low mass stars do not experience any further burning stages while the advanced burning stages of massive stars only take a tiny fraction of the time spent in H- and He-burning. The number of low mass stars also vastly exceeds the one of massive stars. Together with the generally much shorter lifetimes of massive stars, this explains the "thinning out" in the HRD toward higher luminosity and the lack of pronounced groups in that region. There is another group of hot objects with low luminosity, indicating hot objects with small radii: these are the white dwarfs. Their properties and their cooling lines in the HRD are discussed in Section 2.5.

Stars (and their planets) are formed from a contracting molecular cloud. Star formation is a complicated process and not yet fully understood. Nevertheless, some important insights can be gained by applying basic physical principles. According to the virial theorem derived in Section 2.9.1 a molecular cloud contracting due to its own gravitational field will heat up. This increases the gas pressure, which counteracts the contraction. Therefore a small cloud may not be susceptible to small disturbances and will never collapse. The critical cloud mass required to initiate a collapse is called the *Jeans mass* M_{Jeans} (named after Sir James Jeans, 1877–1946). The *Jeans criterion* for the stability of a self-gravitating, non-rotating, homogeneous particle cloud with total mass M reads

$$M < M_{\text{Jeans}} = C\sqrt{\frac{1}{\rho}\left(\frac{\mathrm{k_B}T}{G\bar{m}}\right)}, \tag{6.1}$$

where $\bar{m}$ is the average particle mass, ρ the mass density, and T the temperature of the cloud. The prefactor C has different values depending on further details in the assumptions when deriving the criterion. Using the virial theorem, i.e., working with an equilibrium between energies, the constant becomes $C = \sqrt{375/(4\pi)}$. Assuming, on the other hand, equality of gas pressure and gravitational pressure in the center of the cloud, the constant turns out to be $C = \sqrt{6/\pi}$. Starting with the contraction of a cloud from a given radius and considering the sound speed in an ideal gas yields $C \approx 6.27$. All of these derivations are based on strongly simplified assumptions but also a more sophisticated treatment leads to a dependence on temperature, density, and particle mass as shown in Equation (6.1).

For illustration, based on pressure equilibrium a cloud of 10 $\mathrm{M}_\odot$ single-atomic hydrogen gas with $\rho = 10^{-17}$ kg m^{-3} (such a cloud would have a diameter of 1.65 ly and contain about 6000 atoms cm^{-3}) collapses for $T \leqslant 10$ K. The equation of state of such an ideal, mono-atomic gas is a polytrope with polytropic exponent $\tilde{\Gamma} = 5/3$ (see Section 2.3 and Table 2.1). A mixture of radiation and matter actually has a polytropic exponent closer to 4/3 (see Section 2.6). The polytropic behavior of the cloud is also affected by the ability to cool by radiation during contraction. This sensitively depends on the detailed composition of the cloud and its ionization state but generally leads to an effective polytropic exponent closer to $\tilde{\Gamma} = 1$. Using a polytrope in the derivation of Equation (6.1) yields

$$M_{\text{Jeans}} \propto T^{3/2}\rho^{-1/2} \propto \rho^{(3/2)(\tilde{\Gamma}-1)}\rho^{-1/2} \propto \rho^{(3/2)(\tilde{\Gamma}-4/3)}. \tag{6.2}$$

During the cloud collapse the density ρ increases. It is clear from Equation (6.2) that M_{Jeans} increases with increasing density for $\tilde{\Gamma} > 4/3$. For $\tilde{\Gamma} < 4/3$, however, M_{Jeans} decreases with increasing density. Therefore smaller, dense regions may decouple from the general collapse of the parent cloud and collapse faster on their own timescale. This leads to fragmentation of the original cloud. Hoyle suggested hierarchical fragmentation, i.e., further and further fragmentation until a fragment cannot collapse further either because it has become too small or because nuclear burning ignites and stabilizes the fragment. In such a manner ten thousands of stars are formed in the collapse of an originally gigantic molecular cloud. Typical giant molecular clouds are initially roughly 100 ly (9.5×10^{14} km) across and contain up to 6×10^6 $\mathrm{M}_\odot$ (1.2×10^{37} kg). Cloud collapse can also be triggered by external events, such as a collision between molecular clouds (caused by their motion within a galaxy or due to a collision of galaxies) or, more frequently, by shockwaves induced by supernova explosions.

The range of fragment masses finally becoming a star can be calculated using the *initial mass function* (IMF) $\mathcal{F}_{\mathrm{IMF}}$. It is defined as the amount of mass locked up in stars with masses in the interval $[M, M + \mathrm{d}M]$, all formed at a specific time in a given volume. Then the number of stars $\mathrm{d}N$ with masses in the range $[M, M + \mathrm{d}M]$, formed at a specific time in the given volume, is given by

$$\mathrm{d}N = \frac{\mathcal{F}_{\mathrm{IMF}}(M)}{M}\mathrm{d}M. \tag{6.3}$$

The IMF is normalized,

$$\int \mathcal{F}_{\mathrm{IMF}}(M)M\,\mathrm{d}M = 1. \tag{6.4}$$

Usually the IMF is assumed to be a generally valid function across all star-forming regions of a galaxy. The classical IMF is the *Salpeter IMF* using a power law,

$$\mathcal{F}_{\mathrm{IMF}}(M) \propto \left(\frac{M}{\mathrm{M}_\odot}\right)^{-1.35}, \tag{6.5}$$

giving the relative number of stars with given M. Although the Salpeter IMF has been derived semi-empirically using observations of main-sequence stars in the solar neighborhood, it has been found to apply across the Galaxy and seems to be a universal scaling law. Modern IMFs have mainly made adaptations to the number of very low-mass stars and of very massive stars ($\gtrsim 100$ $\mathrm{M}_\odot$). The steep drop of the IMF toward massive stars and the much shorter lifetimes of such stars explains their scarcity in the galaxies and also why they are lacking in the HRD.

Cloud fragmentation, and thus star formation, depends on many parameters, not only on the initial mass and rotation. Gas turbulence and magnetic fields affect the formation process. Also important is the composition (metallicity) of the molecular cloud because the ability to cool by radiation depends on the types of molecules in the cloud. Compression waves from external shocks further complicate fragmentation and star formation. For these reasons, it is difficult to give a theoretical upper mass limit for fragments becoming a star. The Eddington limit (see Section 2.7.2.4) poses a severe restriction on the formation of stellar objects beyond about 130 $M_{\odot}$. An object exceeding the Eddington luminosity during its contraction would not permit further matter accretion and would rather shed mass. It was already pointed out in Section 2.7.2.4, however, that the approximations made in the derivation of the Eddington limit may leave room for higher luminosities. Observations of star clusters, which have formed from a single molecular cloud, seem to point to an upper limit of about 150 $M_{\odot}$. In 2010, however, the star R136a1 in the RMC 136a cluster has been determined to have 265 $M_{\odot}$, shifting this limit by almost a factor of two (Crowther et al. 2010). It has been suggested, however, that stars larger than 150 $M_{\odot}$ in RMC 136a were created through the collision and merger of massive stars in close binary systems, instead of directly forming and therefore leaving the canonical mass limit intact (Banerjee et al. 2012). The first stars to form after the Big Bang may have been larger, up to 300 $M_{\odot}$ or more, due to their low metallicity. It is thought that these stars may not have been able to achieve hydrostatic equilibrium (see also Section 7.2 and the description of pair-instability supernovae below) and more quickly collapsed and exploded than expected for supermassive stars with higher metallicity.

The history of a contracting cloud fragment becoming a star is outlined in the following. The gravitational binding energy released in the contraction is partly radiated away and partly heats up the fragment, according to the virial theorem (Section 2.9.1). In the early collapse phase the cloud fragment is optically thin and the temperature increases only moderately. The collapse can then be approximated by free fall (Equation (2.154)). A strong increase in temperature occurs when the opacity increases due to ionization and the density increase. This increases the gas pressure which counteracts the contraction due to the gravitational force. From then on the contraction will proceed on a Kelvin–Helmholtz timescale (Equation (2.160)) and the luminosity of the object is drawn from the release of gravitational binding energy. This is called a protostar and its evolutionary track in the HRD is called the *Hayashi line*, which runs almost vertically from high luminosity to higher temperature toward the main sequence, except for very massive stars which start their life already almost on the main sequence. When temperature and density in the center of the protostar reach the conditions required to ignite H-burning, the stars establishes hydrodynamic equilibrium, implying that the contraction is halted and the luminosity and temperature remain at an almost constant level for as long as the nuclear reactions in the core can supply the energy lost through the surface. Therefore the ignition of nuclear burning does not effectively heat the star but rather prevents further contraction and thus precludes a further temperature increase! There is a brief halt in contraction even before the onset of H-burning. The interstellar medium contains a small amount of deuterium. It can be consumed by the reaction $p + d \rightarrow {}^3He$ already at a temperature much lower than the H-burning temperature (i.e., before $p + p \rightarrow d$ is possible). Therefore there is a short episode in the *pre-main sequence* (PMS) evolution of the star in which deuterium burning very briefly ($\lesssim$100 yr) prevents further contraction by establishing a quasi-hydrostatic equilibrium.

Finally, the protostar contracts further until it ignites H-burning (Section 7.2) and establishes long-term hydrostatic equilibrium, thereby becoming part of the main sequence. The theoretical concept of a star having just reached this stage while still basically having the original composition of the molecular cloud it formed from (except for the deuterium depletion) is called a *zero-age main sequence* (ZAMS) star. During H-burning, the protons in the stellar core are incorporated into ^{4}He nuclei, thereby continuously changing the composition. The composition change, in turn, also affects the energy generation and thus the luminosity, of the star so that it slowly moves "upwards" in the HRD. This explains the width of the main sequence.

The minimal mass for igniting H-burning is about 0.08 $M_\odot$. Protostars below this mass limit never ignite H-burning and are stable just because of the gas pressure. They are called *brown dwarfs*. Planets which are gas giants such as Jupiter ($\approx$0.001 $M_\odot$) or Saturn ($\approx$0.0003 $M_\odot$) do not even ignite deuterium burning, This is one of several possible criteria to distinguish planets from brown dwarfs.

The subsequent evolution and final fate of a star depend on its initial metallicity (i.e., its initial composition, see Equation (1.13)), mass, and the mass loss from its surface during its life. Table 6.2 summarizes the endpoints of stellar evolution for stars with non-zero metallicity as a function of ZAMS mass. Figure 6.4 shows more details regarding mass loss and final remnant as a function of initial mass. What happens at the end of each burning phase is determined by the mass of the core built up from the "ashes" of that burning phase. If it is stable under its own weight (stabilized by non-degenerate or degenerate gas pressure, see also Section 1.5), there will be no further burning phase and the remnant will be made from the hot core of the star plus expanding shells of hot gas blown off from the outer layers of the star. Both the core and the ejected gas layers cool off eventually, their electromagnetic emission leaving the optical wavelength spectrum. Stars with less than 0.6–0.8 $M_\odot$ (depending on metallicity, rotation, and the detailed model used) do not ignite He-burning and therefore leave a white dwarf composed mainly of He as a remnant. Stars up to 0.1 $M_\odot$ are fully convective and almost all of the star is converted into He. They barely eject any mass shells. Core He-burning ignites in stars with 0.6–0.8 $M_\odot$. Stars with less than about 2.3 $M_\odot$, however, do not establish a stable, hydrostatic He-burning phase. The pressure in their cores is dominated by a degenerate e^--gas

Table 6.2. Final Evolutionary Stages of Stars with Non-zero Metallicity as a Function of Initial Mass

Mass Range ($M_\odot$)	Final Stage	Remark
$0.08 \lesssim M \lesssim (0.6\text{–}0.8)$	He WD	Only H-burning
$(0.6\text{–}0.8) \lesssim M \lesssim 2.3$	Planetary Nebula, C/O WD	H-burning, core He-flash
$2.3 \lesssim M \lesssim 8$	Planetary Nebula, C/O WD	H-burning, He-burning, AGB phase
$8 \lesssim M \lesssim 10$	ccSN, neutron star	Final burning phases under degenerate conditions during core collapse (e^--capture SN)
$10 \lesssim M \lesssim (20\text{–}25)$?	ccSN, neutron star	All burning phases under hydrostatic conditions
$130 \gtrsim M \gtrsim (20\text{–}25)$?	ccSN, black hole	All burning phases under hydrostatic conditions

(Section 1.5.4). The pressure of a degenerate Fermi gas does not depend on temperature (Equation (1.84)). This results in a thermonuclear runaway after the ignition of He-burning, as the core cannot regulate the energy generation rate by expanding. An increase in temperature through the nuclear reactions leads to a further increase in the energy generation by nuclear reactions, which, in turn, further heats the core, and so on. Therefore the temperature rises rapidly until the degeneracy of the electrons is lifted (because their thermal energy allows them to populate higher lying energy states). When the core becomes non-degenerate the EOS changes and the pressure adjusts to the temperature, leading to a sudden expansion of the core. This is called a *core He-flash* and ends the He-burning in this type of star. A part of the core and the outer layers of unburnt hydrogen and helium are ejected, while the remaining core settles and concludes its evolution as a white dwarf mainly composed of C and O, the "ashes" of He-burning. The Sun will experience this fate in about 4.5×10^9 yr.

It may appear surprising that He-burning ignites in a degenerate core with a mass smaller than the Chandrasekhar mass (Equation (2.32)). Such a core would be expected to be stable and not contracting any further. The nuclear burning at the bottom of the H-shell surrounding the He core not only heats the core but also constantly adds mass to it. As shown in the mass–radius relation in Equation (2.31) and in Figure 2.3, an increase in mass leads to a decrease in radius, which in turn releases gravitational binding energy. The combined heating from the H-shell and the shrinking core allows to reach a temperature sufficient to ignite He-burning in cores with masses as low as $\approx 0.45\ M_\odot$.

Equation (2.55) has shown that more massive stars undergo the same burning phase at lower central density than less massive stars. This is why the He-burning core of stars with more than about $2.3\ M_\odot$ is not degenerate and they can sustain hydrostatic equilibrium. When protons are exhausted in the core, shell H-burning is established, first as a thick shell, followed by shell narrowing. A convective envelope is developing above the thin shell, leading to an expansion of the star. Because the effective (surface) temperature of the star is decreased whereas the radius has considerably increased, this phase is called the *red giant phase*. While the star is still burning protons in a shell around a He core, it moves up along the *red giant branch* (RGB; branching off from the main sequence) in the HRD. Once He-burning ignites in the core, the star is shrinking again at almost constant luminosity, covering an almost horizontal path in the HRD toward higher effective temperature. As long as there is sufficient He in the core, stable core He-burning can continue. As previously with H-burning also in He-burning the nuclear burning zone moves outwards in a shell surrounding the core composed of C and O. At this stage energy generation is occurring in two concentric shells, a shell burning protons and another one, deeper inside the star, with He-burning. The two shell burning regions again give rise to convection and an expansion of the star, moving it into the *asymptotic giant branch* (AGB) in the HRD. The AGB approaches a hypothetical extension of the RGB.

A new phenomenon appears in AGB stars, which is important for both stellar evolution and nucleosynthesis: *He-shell flashes*. He-burning in the shell of AGB stars does not proceed stably as in a quiescent hydrostatic burning phase. Instead, the burning shell is undergoing a long series of thermonuclear explosions on a timescale of decades spaced by quiet phases with a duration of the order of millennia. The mechanism responsible for these explosive flashes is described in Section 7.3.2. The flashes lead to the development of large convective zones and a rapid expansion of the star. After the flash, the star shrinks again to almost its original radius. Therefore in the HRD the star moves up and down the AGB. The pulsations induced by the He-shell flashes cause increased mass loss from the surface of the star. Expanding shells of hot gas can be observed around such AGB stars, each ejected by a pulse. The convection appearing during the pulses reach from the He-burning shell to the surface of the star, thus moving material from the He-burning region to the surface, where it is partly ejected in a later pulse (see also Section 8.2). The star continually loses mass due to the ejection from the surface in each pulse (Figure 7.4). This affects the working of the He-shell burning and eventually the star enters a post-AGB phase ejecting the remaining surface layers in a superwind. Again, the remnant is a CO white dwarf surrounded by a *planetary nebula* created in the superwind phase. Planetary nebulae are not connected to planets in any way. The term is a historic artifact from 19th century astronomy.

Stars with masses above about 8 $M_{\odot}$ continue through further burning phases after He-burning. These are listed in Table 7.1 and described in Chapter 7. As shown in Equation (2.55), stars with higher masses undergo each specific burning phase at lower central density than stars with lower mass. This is illustrated by the density–temperature diagram shown in Figure 6.2. It is similar to the diagram shown in

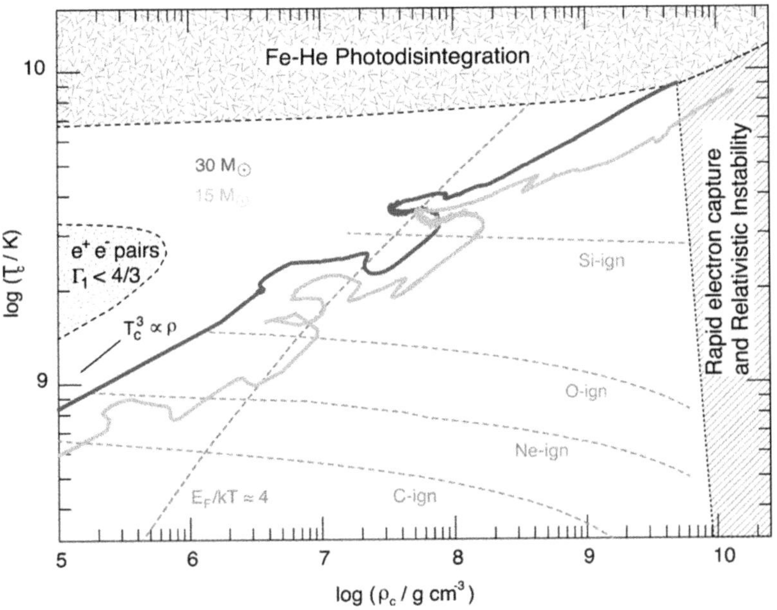

Figure 6.2. Central temperature–density tracks for a 15 $M_{\odot}$ and a 30 $M_{\odot}$ star with solar metallicity going through advanced burning phases. (Reproduced from Paxton et al. 2015. © 2015. The American Astronomical Society. All rights reserved.)

Figure 2.4 but with the added tracks of the core conditions of two stars with different masses. The track of the star with lower mass is further to the right in this diagram, this means at higher density for the same temperature. The variations in core temperatures with mass for a given burning phase are small because they are mainly set by the threshold for igniting the burning phase. The dominating nuclear reaction rates depend exponentially on temperature and therefore a small increase in temperature is already sufficient to provide the energy generation required to maintain hydrostatic equilibrium in a more massive star. It can be seen that the slope of the tracks follows well the law expressed in Equation (2.56). The small wiggles are due to switches from central to shell burning and the following adjustment of the internal structure of the star. The change in slope in the later burning phases is due to the change in EOS when approaching the region of degeneracy (see also Figure 2.4).

The zone of degeneracy is at the right edge (high density) in Figures 2.4 and 6.2. Since the evolutionary tracks have a different slope than the boundary between the non-degenerate and the degenerate region, later burning phases may proceed under degenerate conditions even when earlier burning phases were not affected by degeneracy. As discussed above, He-burning is already under degenerate conditions in stars up to 2.3 $M_\odot$. Due to the decrease in core density, the density–temperature tracks are shifted to the left in Figure 6.2 when the stellar mass (and thus the core mass) is increasing. Stars with less than 8 $M_\odot$ do not reach any further burning phases beyond He-burning and thus any later theoretical core degeneracy is unimportant. Stars between 8 and 10 $M_\odot$, however, develop a degenerate core during C-burning, mainly composed of O, Ne, and Mg. Instead of becoming a white dwarf, however, C-burning in a shell around the core heats and adds mass, leading to core contraction, as discussed above and shown in Figure 2.3, accompanied by an increase in density. The Fermi energies E_F and momenta (see Equation (1.83)) in the degenerate electron gas are increased with increasing density according to Equation (1.78). For example, $E_F = 0.75$ MeV for $\bar{n}_e = 10^7$ g cm^{-3} and $E_F = 4.7$ MeV for $\bar{n}_e = 10^9$ g cm^{-3}. Around 10^{10} g cm^{-3} the electrons reach energies sufficient to overcome the negative Q-value for electron captures (see Section 3.9) on ^{24}Mg and ^{20}Ne, which are products of C-burning. The loss of electrons reduces the pressure and also changes Y_e. This leads to a collapse of the core and core Ne-, O-, and Si-burning occurs during the collapse. Because of the short collapse timescale, these burning phases do not consume their respective fuel completely. Similarly to core collapse in more massive stars, a part of the core finally turns into a neutron star while outer layers are ejected. This is called an *electron-capture supernova*. The mass limit for progenitor stars undergoing such an evolution is also sensitive to the actual values of the electron-capture rates. The nucleosynthetic signature of these events, however, is quite different from the one of more massive stars because of the incomplete late burning phases, lack of late shell burning, and the low-density envelope, mainly consisting of only H, He, and tiny amounts of C and O.

Stars with non-zero metallicity and masses of about 10 $M_\odot$ and higher fully complete all burning phases up to and including Si-burning under non-degenerate conditions with total fuel exhaustion in the core. Figure 6.3 shows a *Kippenhahn diagram* for such a 25 $M_\odot$ star. A Kippenhahn diagram displays the full stellar evolution, including the changes in interior structure, in one figure. The horizontal axis logs time whereas the vertical axis presents a cut through the 1D model of the star, showing zones of energy generation and consumption as well as convection zones. Note that the vertical axis uses Lagrangian mass coordinates (Equation (2.4)) instead of a radius. The labels indicate how much mass is enclosed below, with the center of the star at the origin. The outer 18 $M_\odot$ of the star are not shown as this is the unburnt envelope, mainly consisting of H and He. Also shown are the abundances of main fuels in the core. The sequence of the core burning phases can be identified easily. Also easy to see is how core burning moves outwards to become shell burning and how convective regions form above thin shell burning zones.

Massive stars complete all burning phases: H-, He-, C-, Ne/O-, and Si-burning. In Si-burning, core mainly composed of Fe and Ni is formed. Once this core starts contracting, there is no further hydrostatic equilibrium phase possible because fusion of Fe/Ni nuclei cannot generate the required energy. This can be understood by the binding energy curve shown in Figure 3.1, which has a maximum in the Fe–Ni region. Therefore the core continues to contract. The density increase pushes E_{F} up

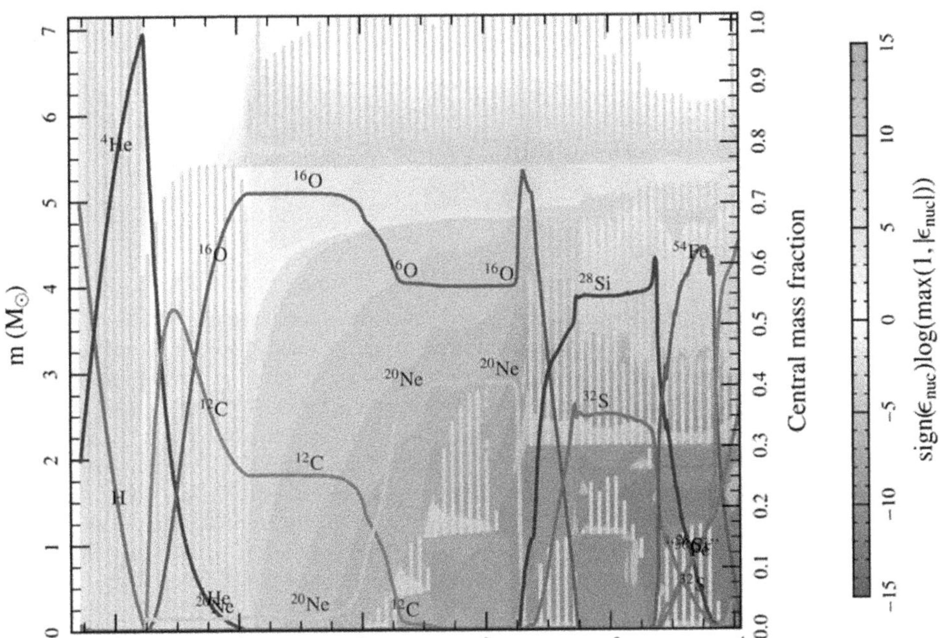

Figure 6.3. Kippenhahn diagram for a 25 $M_\odot$ star with initial solar metallicity. The color shades indicate regions of net energy generation (red) and consumption (blue). The hatched regions are the convection zones. Also shown is the temporal evolution of abundances in the core. (Reproduced from Paxton et al. 2011. © 2011. The American Astronomical Society. All rights reserved.)

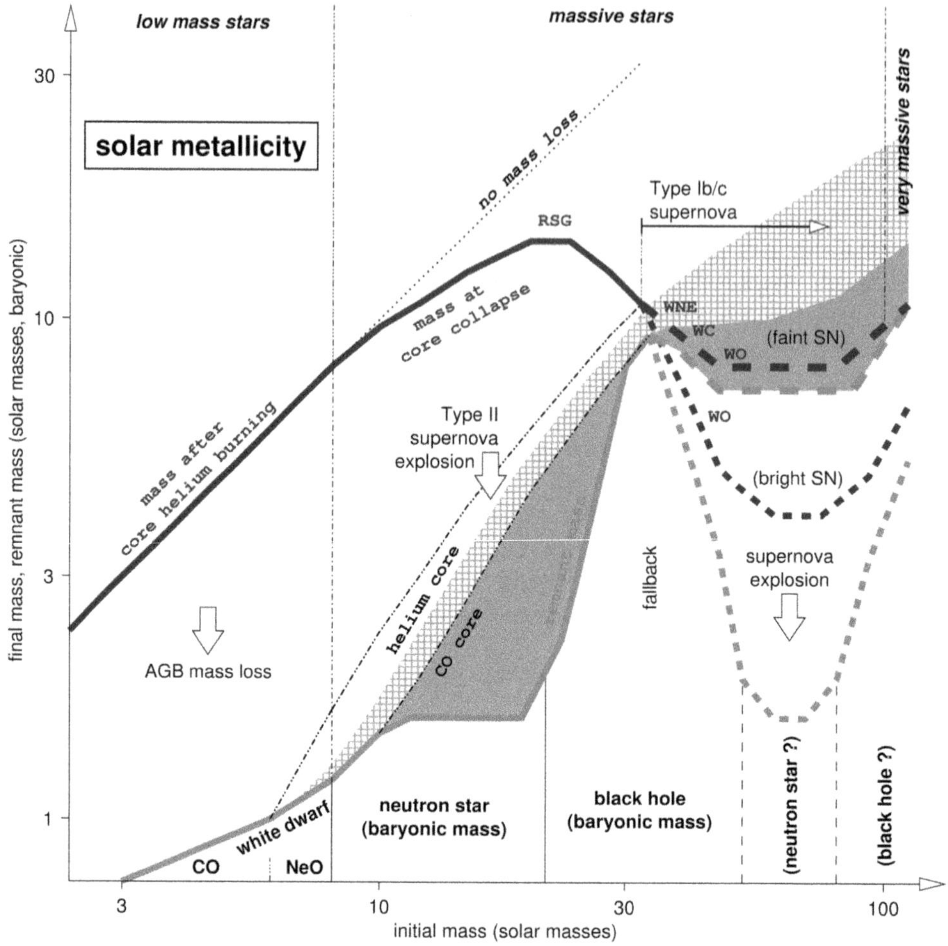

Figure 6.4. Final fate of stars with solar metallicity as function of initial mass. For the supernova classification scheme, see Section 9.2. Mass loss reduces the mass of the envelope (blue curve) until, for a mass above about 33 $M_\odot$ the He core is uncovered before the star reaches core collapse. At this point the star becomes a *Wolf–Rayet star*. Such stars have very strong stellar winds. Two estimates for the Wolf–Rayet mass-loss rate are shown: The short-dashed red and blue lines are for a high mass-loss rate. Here a "window" of initial masses may exist around 50 $M_\odot$, where neutron stars are still formed. For a low Wolf–Rayet mass-loss rate (long-dashed red and blue lines) the final mass at core collapse is higher and no neutron stars are produced. Then only black holes are formed above about 21 $M_\odot$. The labels RSG, WE, WC, and WO indicate the type of the last mass-loss phase and also the (spectral) type of the star when it explodes. (Reprinted figure with permission from Woosley et al. 2002. Copyright 2002 by the American Physical Society.)

to energies permitting electron captures. This results in a loss of central pressure and the contraction turns into a collapse of the core, which can only be halted when a transition to another equation-of-state occurs. Close to nuclear density of 2×10^{14} g cm^{-3} most of the protons have been converted to neutrons by electron captures and a *neutron star* is formed, first as a hot *proto-neutron star*, which is stabilized by the pressure of the degenerate nucleon gas. Essentially, the Pauli exclusion principle is responsible for the stiffness of nuclear matter and also keeps the neutron star stable. The formation of the proto-neutron star, which cools by neutrino emission, leads to an ejection of the outer layers of the star. This is observable as a supernova event. Thus, the remnant of such a star is a neutron star surrounded by an expanding, extremely hot shell of ejected material. A fraction of the products of stellar nucleosynthesis is thereby mixed into the interstellar medium and the generated shockwave may trigger further star formation (see above). Supernovae and some aspects of their nucleosynthesis are further discussed in Section 9.3.

It is apparent from Figure 6.1 that the evolutionary "loops" of massive stars when going through H- and He-burning are more "squished," i.e., more horizontal than those of low-mass stars. This is due to the degeneracy or partial degeneracy of the electron gas in the core of low-mass stars and the degenerate-like behavior of their thin He-burning shells, which lead to big luminosity fluctuations as discussed above. But also the *advanced burning phases* of massive stars from C-burning onwards are barely visible in the HRD. Their timescales are extremely short compared to those of H- and He-burning (see Table 7.1). Therefore it is very unlikely to observe a star in one of these phases, which results in a lack of such stars in the HRD. Moreover, their luminosity would not show an increase in the HRD even though their energy production is going up (see Table 7.1). This can be understood by the fact that the usual definition of luminosity (see Equation (2.66)) in the HRD refers to the energy flow through the stellar surface carried by photons, i.e., by electromagnetic waves. Starting with C-burning, however, energy loss through neutrinos becomes non-negligible. The neutrinos leave the star more or less unhindered, due to their very weak interaction with matter, and thus carry off part of the energy generated by the nuclear reactions. Therefore neutrino losses are also responsible for the much shorter timescales of the advanced burning stages (see also Table 7.1) as more nuclear energy has to be released to compensate for the neutrino losses and still uphold hydrostatic equilibrium. The total luminosity, photons plus neutrinos, is increasing to support the more compact star but the photon luminosity is only increasing slightly or not at all. Rather, the increase in neutrino luminosity $\mathcal{L}_\nu^*$ (see Equation (2.68)) is much stronger when advancing from one burning phase to the next. Other reasons for the shorter timescale of the advanced burning phases are the amount of available fuel and the fact that the reactions of the advanced burning stages do not liberate as much energy as those in H- and He-burning, due to the more modest binding energy differences of heavier nuclei, and therefore are exhausting their fuel more quickly.

In the advanced burning phases of massive stars, the neutrinos directly produced in nuclear reactions (decays or electron captures on nuclei) contribute only a small fraction of the total neutrino luminosity. There are several processes creating neutrinos at higher temperature or density. At high temperature, photons become energetic enough to allow the production of electron–positron pairs. Such pairs can annihilate again and a tiny fraction (10^{-19}) of such pair annihilations produces neutrinos by $e^- + e^+ \rightarrow \nu_e + \bar{\nu}_e$. Nevertheless, this is the dominant process for neutrino production in massive stars (before core collapse) with non-degenerate cores. Other processes include the photoneutrino process, $e^- + \gamma \rightarrow e^- + \nu_e + \bar{\nu}_e$ and, at high density and degenerate conditions, neutrino production by plasmon excitations. Neutrino production through electron captures only becomes important in Si-burning and mainly during the core-collapse phase (see also the discussion of stars in the range 8–10 $M_\odot$).

Figure 6.4 provides an overview of the final fates of stars as function of their initial mass. The mass loss during stellar evolution and the sizes of the He and C/O cores are also shown. As can be seen, around 25 $M_\odot$ the Fe/Ni core of the evolved star collapses into a black hole instead of a neutron star, either delayed by fallback onto the proto-neutron star or directly (for the larger masses). For further details, see Section 9.3. The mass limits and details of the evolution shown in Figure 6.4 and Table 6.2 are only estimates, as they sensitively depend on model details and can even vary from star to star, depending also on rotation, magnetic fields, etc. They are just specified here to give an idea of the range of final fates and remnants to be expected. For example, the exact mass limit for black hole formation still is under discussion.

Metallicity not only affects the stellar formation process, it also strongly affects the subsequent evolution of a star. For example, mass loss during hydrostatic burning strongly depends on the metallicity. The lower the metallicity, the lower the mass loss. In non-rotating, low-metallicity stars with masses above 100 $M_\odot$—as may have been formed in the early universe—the so-called *pair instability* plays a major role. As explained above, stars with more mass burn at higher core temperatures. At high mass, Figure 6.2 suggests that a temperature region is reached in which photons are energetic enough to produce a significant amount of e^-–e^+ pairs. Electrons and positrons each have a mass of 511 MeV/c^2 and applying $E = mc^2$, the peak of the thermal photon energy distribution (Planck distribution, see Equations (1.68) and (4.98)) at $E_\gamma = 2k_BT$ never reaches the required energy for pair-production at the stellar burning temperatures. Nevertheless, due to the extended tail of the photon energy distribution a larger and larger fraction of photons attains an energy equivalent to 1022 MeV/c^2. The particle–antiparticle creation removes photons from the radiation field and therefore the radiation pressure drops. Such very massive stars are mainly supported by radiation pressure, as can be inferred from Figure 2.4 and Equation (2.60). Below 100 $M_\odot$ the change in radiation pressure by pair production is not pronounced enough to affect the stability of the star. In the mass range 100–130 $M_\odot$ (sometimes the lower end of the mass range is given as low as 85 $M_\odot$), the pressure change leads to a temporary loss of hydrostatic equilibrium and the stellar core contracts. Due to rising temperature and density also the nuclear reactions increase drastically and together with the change in opacity this increases the pressure sufficiently for the core to bounce back and re-establish hydrostatic equilibrium. Simulations show that such a star can undergo a series of such pulses, which are stronger the more mass a star has. This *pulsational pair-instability* leads to an increased mass ejection from the surface of the star until it drops below 100 $M_\odot$, at which point the core is no longer hot enough for the pair instability to appear. In stars with masses from 130 $M_\odot$ to about 250 $M_\odot$ the pressure loss cannot be compensated anymore, leading to a core collapse triggering a nuclear runaway and finally a thermonuclear explosion as a *pair-instability supernova*. This leads to complete disruption of the star, ejecting the entire mass of the star and leaving a nebular remnant but neither a neutron star nor a black hole. Stars with more than about 250 $M_\odot$ directly collapse into a black hole after the onset of the pair instability because in the core collapse they reach temperatures sufficient for photodisintegration before runaway fusion can set in, thereby preventing any energy release by other nuclear reactions. It should be noted that rotation or higher metallicity prevents the appearance of the pair instability. Therefore it may be mainly important for the first generation of stars (population III) formed after the Big Bang.

Regarding the resulting nucleosynthesis, pair-instability supernovae eject a large amount of ^{56}Ni because the material in the stellar core is converted to this isotope during the collapse before the thermonuclear explosion. No elements heavier than Zn are ejected. They show very luminous and highly extended lightcurves due to the large ejected mass of ^{56}Ni. The nucleosynthesis products of core-collapse supernovae following a pulsational pair-instability, on the other hand, are limited to what was in the hydrogen envelope (H, He, possibly CNO) and some elements from the outer helium core (C, O, Ne, Mg). This is because most to all of the CO-core collapses to a black hole. There is no explosive nucleosynthesis and all iron-group elements made in the collapse are ingested by the black hole.

Further Reading

Banerjee, S., Kroupa, P., & Oh, S. 2012, ApJ, 746, 15

Crowther, P. A., et al. 2010, MNRAS, 408, 731

Diehl, R., Hartmann, D. H., & Prantzos, N. (ed.) 2018, Astrophysics with Radioactive Isotopes Astrophysics and Space Science Library, Vol. 453 (Berlin: Springer)

Paxton, B., et al. 2011, ApJS, 192, 3

Paxton, B., et al. 2015, ApJS, 220, 15

Ryan, S. G., & Norton, A. J. 2010, Stellar Evolution and Nucleosynthesis (Cambridge: Cambridge University Press)

Wiescher, M. 2009, Physi, 2, 69

Woosley, S. E., Heger, A., & Weaver, T. A. 2002, RvMP, 74, 1015

AAS | IOP Astronomy

Essentials of Nucleosynthesis and Theoretical Nuclear Astrophysics

Thomas Rauscher

Chapter 7

Hydrostatic Burning Phases

7.1 Introduction

As explained in Section 2.7, nuclear reactions have to replace the energy lost through the stellar surface (Equation (2.66)) to keep the star in hydrostatic equilibrium. How to calculate the energy release when many nuclear reactions contribute is laid out in Section 4.1.8 and specifically in Equation (4.121). Table 7.1 summarizes the possible burning phases with their main fuels and typical conditions. It was already mentioned in Chapter 6 that only stars with more than about 8 $M_\odot$ can complete the full sequence. Nuclear reactions not only produce energy but also transform elements. The main products of each burning phase and the most important nuclear processes are also shown in Table 7.1. They are discussed in more detail in the following sections.

7.2 Hydrogen Burning

In pure hydrogen gas H-burning can proceed starting from temperatures as low as about 10 MK. Slightly higher temperatures are required when other nuclides are admixed into the gas. It is said that H-burning converts hydrogen to helium. In fact, all nuclear species are completely ionized in the stellar region where nuclear reactions occur. In H-burning, free protons react to form ^{4}He nuclei (these are also called α-particles). The main source of energy in H-burning is the *net* reaction

$$4\mathrm{p} \rightarrow {}^4\mathrm{He} + 2\mathrm{e}^+ + 2\nu_\mathrm{e}. \qquad (7.1)$$

The two positrons are immediately converted to photons by electron–positron annihilation with free electrons in the stellar plasma, leading to a total reaction Q-value of $Q_{\mathrm{nuc}}^{\mathrm{H-burn}} = 26.731$ MeV. This net reaction, however, is not possible in the stellar cores because the probability of four protons interacting at the same time is too small. Instead, the net reaction can be realized in several steps involving

doi:10.1088/2514-3433/ab8737ch7

Table 7.1. Burning Phases of Stars: The Central Temperatures T_c, Central Densities ρ_c, Burning Timescales τ, and Luminosities $\mathcal{L}^*$ are for a 20 $M_\odot$ Star

		Products		T_c	ρ_c	τ	$\mathcal{L}^*$	
	Fuel	Main	Secondary	(GK)	(g cm^{-3})	(yr)	$L_\odot$	Processes
$\geq 8 M_\odot$ ↓; 0.8–$8 M_\odot$ ↓	H	He	^{14}N	0.04	4.53	8×10^6	6.3×10^4	pp-chains CNO-cycles
	He	C, O	^{18}O, ^{22}Ne s-process	0.2	968	1.2×10^6	10^5	3α-reaction ^{12}C$(\alpha,\gamma)^{16}$O s-process
	C	Ne, Mg	Na	0.9	1.7×10^5	10^3	1.4×10^5	^{12}C+^{12}C
	Ne	O, Mg	Al, P	1.6	3.1×10^6	0.6	1.5×10^5	^{20}Ne$(\gamma,\alpha)^{16}$O α+X…
	O	Si, S	Cl, Ar, K, Ca	2	5.6×10^6	1.25	1.5×10^5	^{16}O+^{16}O QSE
	Si	Fe	Ti, V, Cr, Mn, Co, Ni	3.3	4.3×10^7	11.5 days	1.5×10^5	^{28}Si+$\gamma \rightarrow$ n, p, α n+X, p+X,… QSE, NSE

two-particle reactions and decays. There are two distinct types of reaction sequences achieving the net H-burning reaction in stars: the pp-chains and the CNO-cycles. Which of these dominates the energy generation depends on whether CNO-nuclei (see below) are present and on the temperature at which H-burning proceeds. Since more massive stars burn at slightly higher core temperature, the CNO-cycles dominate in stars on the upper part of the main sequence whereas pp-chains are the main energy source for low-mass stars (like the Sun) in the lower part of the main sequence. The division is at about 1.3 $M_\odot$ but also depends on metallicity.

7.2.1 pp-Chains

The pp-chains receive their name from the *pp-reaction*, which is at the start of the reaction sequence: $p + p \rightarrow d + e^+ + \nu_e$. This is followed by proton capture on the deuteron (written as d or ^{2}H): $d + p \rightarrow {}^3\text{He} + \gamma$. (Remember that there are no deuterons present at the onset of H-burning because any deuterons initially present in the star-forming cloud were already burned during star formation, see Chapter 6). The pp-reaction as given has a reaction Q-value of $Q_{\text{nuc}} = 0.42$ MeV. The effective energy release, however, is 1.442 MeV because the annihilation of the positron with an electron from the plasma produces an additional 1.022 MeV in photons. After the production of ^{3}He several reaction paths become possible. They are numbered as pp-I chain, pp-II chain, and so on. Figure 7.1 gives an overview of the three pp-chains that are most important in hydrostatic H-burning and Table 7.2 lists some properties of the involved reactions. Note that for completing the pp-I chain, the first two reactions have to occur twice to produce the two ^{3}He required in the last step.

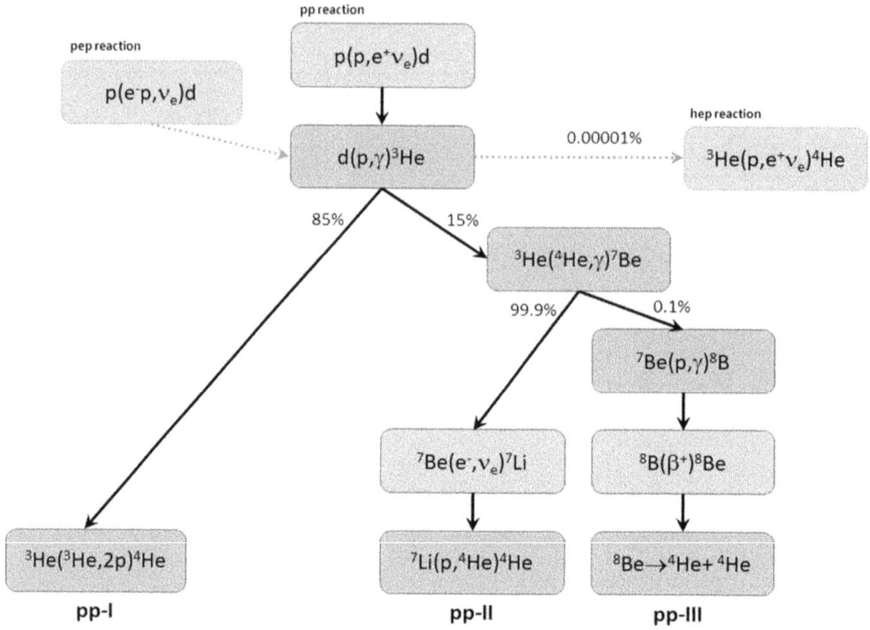

Figure 7.1. The pp-chains in stellar H-burning; shown are the pp-I, pp-II, and pp-III chain, which are the dominant source of energy in solar-type stars. Reactions releasing neutrinos (ν_e) are marked with a light green background. The branching percentages are for solar conditions.

Table 7.2. Properties of the Reactions in the pp-Chains

Reaction	Q-value (MeV)	τ_{life} (yr)	Chain
$p(p,e^+\nu_e)d$	0.420 (+1.022)	7.9×10^9	pp-I, pp-II, pp-III
$d(p,\gamma)^3He$	5.493	4.4×10^{-8}	pp-I, pp-II, pp-III
$^3He(^3He,2p)^4He$	12.859	2.4×10^5	pp-I
$^3He(^4He,\gamma)^7Be$	1.586	9.7×10^5	pp-II, pp-III
$^7Be(e^-, \nu_e)^7Li$	0.861	3.9×10^{-1}	pp-II
$^7Li(p,^4He)^4He$	17.347	1.8×10^{-5}	pp-II
$^7Be(p,\gamma)^8B$	0.135	6.6×10^1	pp-III
$^8B(\beta^+)^8Be$, $^8Be \rightarrow {}^4He + {}^4He$	18.078	3.0×10^{-8}	pp-III

Notes. Given is the reaction Q-value and the lifetime of the target nucleus against destruction by this reaction. Lifetimes are defined in Equation (4.110). The table values assume mass fractions $X_p = X_{^4He} = 0.5$, density $\rho = 100\ \mathrm{g\,cm^{-3}}$, and temperature $T = 15$ MK. Note that for clarifying the relevance for H-burning ^{4}He is used instead of α in the reaction notation.

Further note that the pp-II and pp-III chains require the presence of ^{4}He, either already initially contained in the plasma (which is the case for basically all interstellar clouds from which stars form) or produced in the pp-I chain. The reaction network equations for the pp-I chain are shown in Section 4.2.2.

It is apparent from Table 7.2 that the pp-reaction is the slowest reaction by far. Since the time to complete a reaction sequence is determined by the slowest reaction the pp-reactions determines the timescale for proton consumption. Therefore it determines the stellar luminosity and, together with the amount of consumable hydrogen, the nuclear timescale of the star (see Section 2.9.2). Although each proton in the stellar interior has a lifetime of about 10^{10} yr (a deuteron, for example, survives on average only for a second), considerable energy release is possible due to the enormous amount of protons in the volume where H-burning proceeds. In fact, the long lifetime is fortuitous for the development of life because otherwise H-burning in the Sun and in all stars would be much shorter (depending on the actual rate of the slowest reaction). The stellar luminosity would be the same as it is set by the hydrostatic condition but the available protons would be consumed at a higher rate, reducing the total lifetime of the star by about an order of magnitude (maybe to a simple multiple or less of the He-burning timescale). This may not be sufficient for the appearance of complex life.

The astrophysical S-factor for the pp-reaction as a function of center-of-mass energy E_{cm} in the p + p system is given by $\tilde{S}_{\mathrm{pp}}(E_{\mathrm{cm}}) = 3.94 \times 10^{-25} + 4.61 \times 10^{-24} E_{\mathrm{cm}} + 2.96 \times 10^{-23} E_{\mathrm{cm}}^2$ MeV barn. Extrapolated to zero energy it is $\tilde{S}_{\mathrm{pp}}(0) = (3.8 \pm 0.4) \times 10^{-25}$ MeV barn and the reactivity at 15 MK is $\langle \sigma v \rangle_{\mathrm{pp}} = 1.19 \times 10^{-43}$ cm^3 s^{-1}. This shows that the reaction cross section must be unmeasurably low. At $E_{\mathrm{cm}} = 0.5$ MeV (which is a laboratory energy of 1 MeV for the proton collision) the reaction cross section is $\sigma_{\mathrm{reac}} = 8 \times 10^{-48}$ cm^2. Perhaps surprisingly, nevertheless it can be determined theoretically quite precisely. This is because of the simple reaction mechanism and the low interaction energy between the protons. One of the two approaching protons is converted into a neutron in a virtual β^+ decay according to Equation (3.174). The neutron forms a deuteron together with the second proton. The ground state of the deuteron with spin and parity $J^{\pi_q} = 1^+$ is a state with orbital angular momentum $\ell = 0$. It is a triplet state with the spins of the neutron and the proton being parallel and simply adding up to the deuteron spin. Because of spin selection rules (see Section 1.7) the main contribution to the cross section therefore comes from a two-proton system with also $\ell = 0$. Since protons in the entrance channel are identical particles the Pauli exclusion principle requires them to have antisymmetric spins, resulting in an entrance channel spin of 0. Comparing the initial and final spins it is obvious from Section 3.9 that the transformation process of the proton must be a Gamow–Teller transition. Finally, the cross section can be factorized as

$$\sigma_{\mathrm{pp}} \propto \frac{1}{\mathcal{K}_{\mathrm{p}}^2} \frac{3}{2} f(p) g_{\mathrm{A}}^2 M_{\mathrm{nuc}}^2, \tag{7.2}$$

with the factor $3/2 = g_{\mathrm{fin}}/g_{\mathrm{ini}}/2 = (2 \times 1 + 1)/(2 \times 0 + 1)/2$ being the spin factor taking into account identical particles. The phase space volume $f(p)$ can be calculated exactly. The decay part of the reaction is characterized by the Gamow–Teller coupling constant g_{A} (see Section 3.9), which can be determined from the measurements of other decays. This leaves the nuclear matrix element, which is given by the radial overlap of the final state wave function (of the deuteron) with the initial state wave function (of the p + p system),

$$M_{\mathrm{nuc}} = \int_0^\infty \chi^*_{\mathrm{fin}}(r)\chi_{\mathrm{ini}}(r) r^2 \, \mathrm{d}r. \tag{7.3}$$

The wave functions χ_{ini} and χ_{fin} have to be determined by solving the radial Schrödinger equation as shown in Section 3.3.2. It is elaborated in Section 3.6.1 that the usual problem in obtaining the wave functions is to know the appropriate effective nuclear interaction potential (in the form of an optical potential). This usually gives rise to the largest uncertainties in the predictions. The situation is alleviated for the pp-reaction because of the Coulomb repulsion between the two protons at the low interaction energy of only 7.2 keV and the weak binding of the deuteron. The wave function χ_{fin} of the weakly bound deuteron (2.225 MeV binding energy) is peaking inside the nuclear radius of the deuteron but also extends comparatively far to larger radii. The wave function of the p + p system, on the other hand, is strongly suppressed inside the deuteron radius by the repulsive Coulomb potential. It attains non-negligible values only far outside the deuteron radius (at about three times the radius). The product under the integral in Equation (7.3) then is almost zero inside the nuclear radius and reaches a maximum around about twice the deuteron radius. This means that the major contributions to the integral come from radii where the nuclear potential is negligible and only the Coulomb potential determines the wave functions. Coulomb wave functions can be obtained with arbitrary precision (compare also Equation (3.112)) and with this the cross section in Equation (7.2) is fully determined. That the reaction effectively takes place already when the interacting nuclei are still well separated is typical for astrophysical reactions with charged reaction partners. It is especially pronounced in direct reactions with light nuclei but is also relevant in low-energy compound reactions (see Section 3.7.1).

The branching ratio out of the pp-I chain is given by the competition between two ^{3}He-consuming reactions (the ^{3}He branching), the branching between pp-II and pp-III by the competition between proton capture and electron capture on ^{7}Be (the ^{7}Be branching). Due to the same Coulomb barrier of the competing reactions in the ^{3}He branching, a strong dependence of the branching ratio on temperature is not expected. It has to be noted, however, that such a dependence nevertheless exists because the reactions show resonance features and a shift in temperature may strongly affect the reaction rate when shifting a resonance into or out of the astrophysically relevant energy window (see Section 4.1.2.2). Within the temperature range of H-burning by pp-chains, however, this is not the case. The ^{7}Be branching, on the other hand, is expected to be strongly temperature-dependent because the proton capture reaction rate exhibits a much stronger temperature dependence than the electron capture. It is important to note that the stellar electron capture rate is quite different from the terrestrial one. This is not so much due to thermal population of excited states in ^{7}Be (Equation (4.112)) but rather due to the complete ionization of the nucleus, which makes it impossible to capture an electron from an atomic shell (see Section 3.9). A free electron from the plasma has to be captured instead. This significantly alters the rate, and thus the half-life of the nucleus, with respect to the one inferred from laboratory measurements of the decay of atomic ^{7}Be. The stellar electron capture rates of light nuclei only weakly depend on temperature but strongly on the electron density. For solar conditions ($T = 15$ MK, $\rho = 150$ g cm^{-3}) the lifetime of ^{7}Be is nearly doubled compared to the terrestrial value, from $\tau_{\text{life}}^{\text{lab}} = 77$ days in the laboratory to $\tau_{\text{life}}^{\odot} \approx 140$ days in the Sun. The combined temperature dependence of the branchings results in the pp-I chain dominating in the temperature range $10 \leqslant T < 14$ MK, the pp-II chain for $14 \leqslant T \leqslant 23$ MK, and the pp-III chain for $T > 23$ MK.

Another way to produce the deuteron in the initial step of the pp-chains is the *pep-reaction*, converting a proton into a neutron by electron capture: $\text{p} + \text{e}^- + \text{p} \rightarrow \text{d} + \nu_\text{e}$. Only 0.23% of the deuterons produced at solar conditions stem from the pep-reaction with $Q_{\text{nuc}} = 1.44$ MeV. The protons have a lifetime of 1.9×10^{12} yr with respect to this improbable three-body reaction. This reaction is negligible for energy generation but its importance lies in the emission of neutrinos with 1.44 MeV which can be detected (see below).

Theoretically possible is the direct production of ^{4}He from ^{3}He in the *hep-reaction*: $\text{p} + {}^3\text{He} \rightarrow {}^4\text{He} + \text{e}^+ + \nu_\text{e}$ ($Q_{\text{nuc}} = 18.8$ MeV). The rate for this reaction is very low and so far no neutrinos from this reaction in the Sun have been detected.

Assuming the pp-chains have been active for several million years (a time comparable to the reaction timescales, Equation (4.115), of all reactions but the slowest one), the abundances of all nuclides in the chains can be determined from a steady flow as shown in Section 4.3.4, with the slowest rate obviously being the one of the pp-reaction. The three different pp-chains can be split up in three steady-flow sequences, which are all governed by the pp-rate r_{pp}. The net steady flow parameter $C_{\text{H-burn}} = -r_{\text{pp}} = \dot{Y}_\text{p} = -\dot{Y}_{^4\text{He}}$ must be a sum of steady flow parameters for each chain, $C_{\text{H-burn}} = C_{\text{pp-I}} + C_{\text{pp-II}} + C_{\text{pp-III}}$, giving the fraction of $\dot{Y}_{^4\text{He}}$ originating from each branch. The individual steady flow parameters are determined from the rates of reactions changing $Y_{^4\text{He}}$ in each branch, $C_{\text{pp-I}} = r_{^3\text{He}(^3\text{He},2\text{p})}$, $C_{\text{pp-II}} = -r_{^3\text{He}(\alpha,\gamma)} + 2r_{^7\text{Li}(\text{p},\alpha)}$, and $C_{\text{pp-III}} = 2r_{^8\text{Be}\rightarrow 2\alpha}$. The relative contribution of each branch is given by $C_{\text{pp-I}}/C_{\text{H-burn}}$, $C_{\text{pp-II}}/C_{\text{H-burn}}$, and $C_{\text{pp-III}}/C_{\text{H-burn}}$.

The energy generation is calculated using mass differences as shown in Equation (4.121) and can be expressed using the above steady flow parameter:

$$\varepsilon = \text{N}_\text{A} C_{\text{H-burn}}(4m_\text{p} - m_\alpha)\text{c}^2 = \text{N}_\text{A} C_{\text{H-burn}} Q_{\text{nuc}}^{\text{H-burn}}. \tag{7.4}$$

This includes the energy which is in neutrinos. This is actually lost because neutrinos almost freely escape the star and do not contribute to the hydrostatic pressure. Neutrino energy loss can be accounted for very simply by using steady flow in each branch and assigning the appropriate loss factor. Then the energy contributing to hydrostatic equilibrium is given by

$$\varepsilon_{\text{nuc}} = \text{N}_\text{A} C_{\text{H-burn}} Q_{\text{nuc}}^{\text{H-burn}} \left(0.98 \frac{C_{\text{pp-I}}}{C_{\text{H-burn}}} + 0.96 \frac{C_{\text{pp-II}}}{C_{\text{H-burn}}} + 0.717 \frac{C_{\text{pp-III}}}{C_{\text{H-burn}}} \right). \tag{7.5}$$

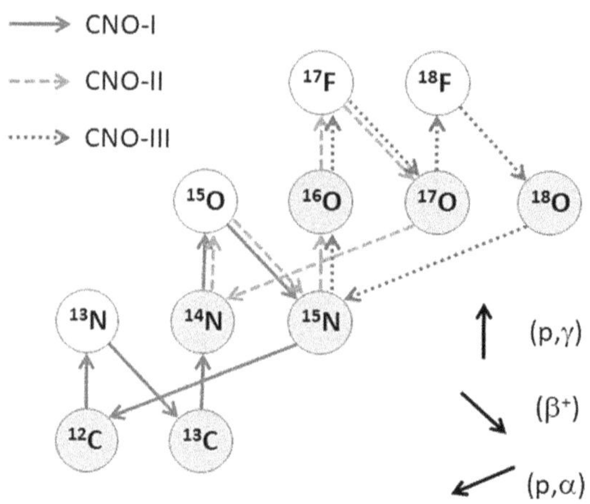

Figure 7.2. The CNO-cycles in stellar H-burning; shown are the CNO-I, CNO-II, and CNO-III cycles. Shaded circles indicate stable nuclides. The β^+ decays release a e^+ and a ν_e.

This accounts for 2% energy loss in the pp-I chain (from two neutrinos in the pp-reaction completed twice), 4% energy loss in the pp-II chain (one neutrino from the pp-reaction and one from the electron capture on ^{7}Be), and 28.3% loss in the pp-III chain (one neutrino from the pp-reaction and one from the decay of ^{8}B).

7.2.2 CNO-cycles

If any of the nuclides ^{12}C, ^{13}C, ^{15}N, or ^{15}N (all of which are stable nuclides) are already present in the stellar plasma, there is an alternative to the pp-chains to complete the net reaction of H-burning. A closed reaction loop can consume four protons to make a ^{4}He nucleus (α-particle). Starting the description of the cycle arbitrarily at ^{12}C, the reaction sequence reads $^{12}\mathrm{C}(\mathrm{p},\gamma)^{13}\mathrm{N}(\beta^+)^{13}\mathrm{C}(\mathrm{p},\gamma)^{14}\mathrm{N}(\mathrm{p},\gamma)^{15}\mathrm{O}$ $(\beta^+)^{15}\mathrm{N}(\mathrm{p},\alpha)^{12}\mathrm{C}$. This is called the *CNO-I cycle*, sometimes also referred to as *CN-cycle*. The two β^+ decays emit a total of two positrons and two electron–neutrinos, therefore the net result of this cycle is the same as shown in Equation (7.1). Note that the CNO nuclides are constantly recreated while producing ^{4}He.

Figure 7.2 shows the classical CNO-I cycle together with two further cycles, the CNO-II and CNO-III cycles. All of them include three proton captures, two β^+ decays, and a (p,α) reaction. These are in the same order in the CNO-I and CNO-II cycles: a proton capture and a β^+ decay is followed by two further proton captures and a final (p,α) reaction to close the loop. The CNO-III cycle starts with the two proton captures, followed by a β^+ decay, a further proton capture, β^+ decay, and the final (p,α) reaction. For $T \lesssim 10^8$ K the CNO-I cycle dominates. The branching into the CNO-II cycle, given by the ratio of the reaction rates for $^{15}\mathrm{N}(\mathrm{p},\gamma)^{16}\mathrm{O}$ and $^{15}\mathrm{N}(\mathrm{p},\alpha)^{12}\mathrm{C}$, is only 0.1%. Therefore the *CNO bi-cycle* (the simultaneous action of CNO-I and CNO-II) is negligible for energy generation. Nevertheless it affects the abundances of O and F isotopes on a long timescale.

The slowest reaction in the dominating CNO-I cycle is $^{14}\mathrm{N}(\mathrm{p},\gamma)^{15}\mathrm{O}$. Therefore any initial abundance in any of the other nuclides is converted into ^{14}N at which point the continuation of the cycle has to wait until the proton capture happens. Assuming that CNO-I cycling has been going on for a time longer than the lifetime of of ^{14}N against proton capture, the relative abundances in the cycle are given by the steady flow approximation (Section 4.3.4), similar to what is explained above for the pp-chains. Here, the steady flow constant (Equation (4.153)) is set by the ^{14}N(p,γ) reaction, $C_{\mathrm{CNO}} = \rho \mathrm{N_A} \langle \sigma^* v \rangle_{^{14}\mathrm{N(p},\gamma)} Y_{^{14}\mathrm{N}} Y_{\mathrm{p}}$, and the abundances of the other isotopes in the cycle are distributed according to Equations (4.151) and (4.152). The abundance change for protons and α-particles is $\dot{Y}_{\mathrm{p}} = -4C_{\mathrm{CNO}}$ and $\dot{Y}_{\alpha} = C_{\mathrm{CNO}} = \rho \mathrm{N_A} \langle \sigma^* v \rangle_{^{14}\mathrm{N(p},\gamma)} Y_{^{14}\mathrm{N}} Y_{\mathrm{p}} = \rho \mathrm{N_A} \langle \sigma^* v \rangle_{^{15}\mathrm{N(p},\alpha)} Y_{^{15}\mathrm{N}} Y_{\mathrm{p}}$, respectively.

The energy generation from the CNO-I cycle is given by

$$\varepsilon_{\mathrm{CNO}} = 0.937 \mathrm{N_A} C_{\mathrm{CNO}} Q_{\mathrm{nuc}}^{\mathrm{H-burn}}, \tag{7.6}$$

where the factor 0.937 accounts for the neutrino loss. At solar conditions the proton capture on ^{14}N is about 20 times slower than the pp-reaction and therefore the CNO-cycle does not contribute significantly to the energy generation in the Sun. This is mostly due to the higher Coulomb repulsion of the protons interacting with isotopes of C, N, and O. When the temperature is only slightly higher than the core burning temperature in the Sun, the Gamow window is shifted to energies at which the rates become much faster. This is readily seen when comparing the temperature dependence of the effective energy generation rates. Using the Gaussian approximation of the Gamow peak with position and width of the Gamow peak given by Equations (4.30) and (4.31) the reactivity of a non-resonant reaction of charged particles can be written as

$$\langle \sigma^* v \rangle = \sqrt{\frac{2}{\mu_{\mathrm{red}}}} \tilde{S}(\mathcal{G}) \frac{\Delta_{\mathrm{G}}}{(\mathrm{k_B} T)^{3/2}} \mathrm{e}^{-\tau_{\mathrm{S}}} \tag{7.7}$$

with the dimensionless reaction rate parameter

$$\tau_{\mathrm{S}} = \frac{3\mathcal{G}}{\mathrm{k_B} T} = 42.46 \left(\frac{Z_1^2 Z_2^2 \mu_{\mathrm{red}}}{T_6} \right)^{1/3}, \tag{7.8}$$

where Z_1, Z_2 are the charge numbers of projectile and target nucleus, μ_{red} their reduced mass, and T_6 the plasma temperature in 10^6 K. The height of the Gamow peak is given by $\exp(-\tau_{\mathrm{S}})$. Choosing appropriate units by inserting numerical factors yields

$$\langle \sigma^* v \rangle = 7.2 \times 10^{-19} \frac{1}{Z_1 Z_2 \mu_{\mathrm{red}}} \tilde{S}(\mathcal{G}) \tau^2 \mathrm{e}^{-\tau_{\mathrm{S}}}, \tag{7.9}$$

where the reactivity is given in cm^3 s^{-1} when the S-factor $\tilde{S}$ is inserted in keV barn. The temperature dependence is found from the fact that $\langle \sigma^* v \rangle \propto \tau_{\mathrm{S}} \exp(-\tau_{\mathrm{S}})$ and therefore

$$\langle \sigma^* v \rangle \propto T^{\tau_{\mathrm{S}} - 2/3}. \tag{7.10}$$

(Remember that the Gaussian approximation is only valid for light nuclides at low energies.) At 15 MK this yields a temperature dependence of $\langle\sigma^*v\rangle_{pp} \propto T^{3.9}$ for the pp-reaction and $\langle\sigma^*v\rangle_{^{14}N(p,\gamma)} \propto T^{20}$. At higher T the CNO temperature dependence flattens to $\propto T^{17}$ when the relevant energy window is closer to or above the peak of the Coulomb barrier. While CNO energy generation is negligible in the Sun, already in stars with about 1.3 $M_\odot$ the energy generation through pp-chains and the CNO-cycle is comparable. The CNO-cycle quickly becomes the dominating process by several orders of magnitude for higher masses.

With increasing temperature, the proton-induced reactions in the CNO-cycles speed up. Starting at $T \approx 100$ MK, the CNO-II cycle also has to be taken into account. For even higher temperature and proton densities further cycles develop (see Figure 9.15). They play a role in explosive hydrogen burning on white dwarfs and on neutron stars, discussed in Sections 9.4.2 and 9.4.4, respectively. Is there a limit to energy production in the classical CNO-cycle? When proton captures are becoming faster with increasing temperature, also $^{13}N(p,\gamma)^{14}O$ becomes faster than $^{13}N(\beta^+)^{13}C$, bypassing ^{13}C. The new sequence thus is $^{12}C(p,\gamma)^{13}N(p,\gamma)^{14}O(\beta^+)^{14}N(p,\gamma)^{15}O(\beta^+)^{15}N(p,\alpha)^{12}C$. This modified cycle is termed the *hot CNO-cycle* and also *β-limited CNO-cycle* because its energy production rate is then limited by the β-decay rates of ^{14}O and ^{15}O which have become the slowest reactions. Proton capture on ^{14}N is not a bottleneck anymore as the abundances rather accumulate in ^{14}O and ^{15}O.

7.2.3 Solar Neutrinos

Both the pp-chains and the CNO-cycles include decays emitting neutrinos. This led to the corrections for energy loss by neutrinos in Equations (7.5) and (7.6). Although the pp-reaction is slow, a large number of reactions have to occur to sustain the stellar luminosity. For the Sun, this results in a neutrino flux of 5.6×10^{10} neutrinos from the pp-reaction received on Earth per cm^2 per second. Figure 7.3 shows the detailed neutrino energy spectra for the neutrinos emitted by the pp-chains and the CNO bi-cycle in the Sun. Neutrinos emitted in a β^+ decay show a continuous energy spectrum because the energy released in the decay is distributed among the positron and the neutrino, as explained in Section 3.9. Electron capture on ^7Be leads to two final states in ^7Li, in 90% of the decays to the ground state and in 10% to the first excited state at 0.478 MeV. The neutrino carries the energy difference between initial and final state and therefore neutrinos from ^7Be decay can be observed at two discrete energies.

The ^8B neutrino flux is comparatively low but it is an important indicator of the central temperature of the Sun. The flux of neutrinos from the decay of ^8B is much more sensitive to the temperature because of two sequential reactions with charged particles leading to ^8B and the fact that the β^+ decay is practically independent of temperature. The ^8B neutrino flux $\phi_{^8B} \propto T^{25}$ whereas, for example, the flux of neutrinos from electron capture on ^7Be $\phi_{^7Be} \propto T^{11}$. The electron capture itself has a temperature dependence and there is only one charged particle reaction (after the initial pp- and d(p,γ) reactions) and their combined dependence leads to the quoted T-sensitivity. Obviously, also the number of neutrinos emitted from CNO-cycles

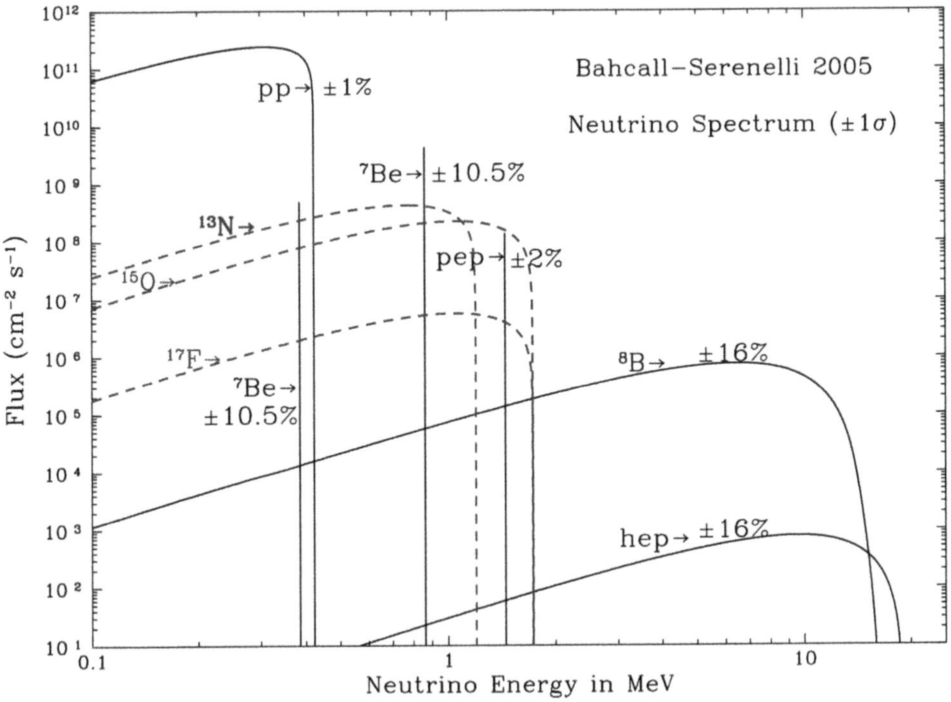

Figure 7.3. The solar neutrino flux received on Earth calculated in the standard solar model. Fluxes from neutrinos emitted from the CNO bi-cycle are shown with blue, dashed lines. Continuum fluxes are given in $cm^{-2}\ s^{-1}\ MeV^{-1}$ and discrete fluxes in $cm^{-2}\ s^{-1}$. Note the logarithmic scales on both axes. (Reproduced from Bahcall et al. 2005.)

strongly depends on temperature because of the higher Coulomb barriers ($\phi_{^{13}\mathrm{N}} \propto T^{20}$, $\phi_{^{17}\mathrm{O}} \propto T^{20}$, $\phi_{^{18}\mathrm{F}} \propto T^{23}$).

It has been a longstanding problem to explain the discrepancy between the neutrino flux predicted in the standard solar model and the measured, lower neutrino flux on Earth. Depending on the manner of neutrino detection and which neutrino energies were accessible, the observed number of neutrinos was lower by factors of 0.5 – 2/3 than expected. After more than three decades of experimental and theoretical efforts, the puzzle was solved in the early years of the 21st century: neutrinos change their flavor on the way from their source to the terrestrial neutrino detectors. Although all the neutrinos released in the pp-chains and CNO-cycles are electron–neutrinos (ν_e), only about a third of those neutrinos have retained their flavor (i.e., being ν_e) upon arrival at Earth. The remaining neutrinos have transformed into ν_μ and ν_τ. To give a detailed account of the history of the *solar neutrino problem* and of neutrino physics is beyond the scope of this textbook. Only a brief overview of the underlying physics is provided here.

In the standard model of particle physics it is assumed that neutrinos are massless particles, that is, their rest mass is zero. The flavor, identifying the type of neutrino (ν_e, ν_μ, ν_τ), can be taken as a quantum number identifying the flavor eigenstate in an abstract Hilbert space. The three unit vectors pointing to the respective flavor eigenstates form an orthonormal vector basis and therefore the type of neutrino can be specified by giving a vector in this three-dimensional flavor space, with the values of the three flavor components. For a neutrino in a flavor eigenstate two of the components will be zero. Extending the standard model of particle physics by assuming that neutrinos have a rest mass, a similar construction can be made for mass eigenstates ν_1, ν_2, ν_3 (with masses m_1, m_2, m_3, respectively) and their orthonormal basis. (A further assumption is that there are three mass eigenstates.) Again, for a neutrino being in a mass eigenstate two of the vector components in the mass space are zero. The important point is that flavor eigenstates and mass eigenstates do not have to coincide, their eigenbasis can be "rotated" with respect to each other. A neutrino with a specific flavor can be in a superposition of mass eigenstates and a neutrino of given mass can be in a superposition of flavor eigenstates. This "rotation" can be generally specified by a rotation matrix **U** connecting the two vector spaces,

$$\begin{pmatrix} \nu_e \\ \nu_\mu \\ \nu_\tau \end{pmatrix} = \mathbf{U} \begin{pmatrix} \nu_1 \\ \nu_2 \\ \nu_3 \end{pmatrix}. \tag{7.11}$$

The unitary 3 × 3 matrix **U** is called the *Pontecorvo–Maki–Nakagawa–Sakata matrix* (PMNS matrix) and its components $U_{\alpha_f,i}$ correspond to the amplitude of mass eigenstate i in flavor α_f. There are nine components but five of them can be incorporated into phases of the lepton fields, leaving four free parameters. In the usual parameterization these are three mixing angles ($\theta_{12}, \theta_{23}, \theta_{13}$) and a phase δ_{CP}. The matrix is factorized as

$$\begin{aligned} \mathbf{U} &= \begin{pmatrix} 1 & 0 & 0 \\ 0 & c_{23} & s_{23} \\ 0 & -s_{23} & c_{23} \end{pmatrix} \begin{pmatrix} c_{13} & 0 & s_{13}e^{-i\delta_{CP}} \\ 0 & 1 & 0 \\ -s_{13}e^{i\delta_{CP}} & 0 & c_{13} \end{pmatrix} \begin{pmatrix} c_{12} & s_{12} & 0 \\ -s_{12} & c_{12} & 0 \\ 0 & 0 & 1 \end{pmatrix} \mathbf{P} \\ &= \begin{pmatrix} c_{12}c_{13} & s_{12}c_{13} & s_{13}e^{-i\delta_{CP}} \\ -s_{12}c_{13} - c_{12}s_{23}s_{13}e^{i\delta_{CP}} & c_{12}s_{23} - s_{12}s_{23}s_{13}e^{i\delta_{CP}} & s_{23}c_{13} \\ s_{12}s_{23} - c_{12}c_{23}s_{13}e^{i\delta_{CP}} & -c_{12}s_{23} - s_{12}c_{23}s_{13}e^{i\delta_{CP}} & c_{23}c_{13} \end{pmatrix} \mathbf{P}. \end{aligned} \tag{7.12}$$

The above equation uses the abbreviations $s_{ij} = \sin\theta_{ij}$ and $c_{ij} = \cos\theta_{ij}$. The diagonal matrix **P** contains phases appearing when neutrinos are Majorana neutrinos, meaning they are their own antiparticles. If they are Dirac particles (with a distinction between neutrino and antineutrino), the matrix **P** simply is the identity matrix. The Majorana phases do not enter the neutrino oscillations, however, and therefore **P** is usually omitted in the discussion. They do enter in the analysis of double-β decay experiments striving to determine whether neutrinos are Majorana or Dirac particles. The values of the mixing angles and the remaining phase cannot be measured directly but are determined by fits to measurements studying the disappearance of ν_e, $\bar{\nu}_e$ and neutral current interactions (inelastic collisions of neutrinos with nuclei), including data from reactor neutrinos, atmospheric neutrinos, and solar neutrinos. As of 2019 they are (Esteban et al. 2019)

$$\begin{aligned}
\theta_{12} &= 33.82^{\circ}{}^{+0.78}_{-0.76^{\circ}},\\
\theta_{23} &= 48.6^{\circ}{}^{+1.0^{\circ}}_{-1.4^{\circ}},\\
\theta_{13} &= 8.60^{\circ}{}^{+0.13^{\circ}}_{-0.13^{\circ}},\\
\delta_{CP} &= 221^{\circ}{}^{+39^{\circ}}_{-28^{\circ}},\\
\Delta m^2_{21} &= 7.39^{+0.21}_{-0.20} \times 10^{-5}\,\mathrm{eV}^2,\\
\Delta m^2_{31} &= 2.528^{+0.029}_{-0.031} \times 10^{-3}\,\mathrm{eV}^2.
\end{aligned} \tag{7.13}$$

These values assume normal mass ordering $m_1 < m_2 < m_3$ (an inverted mass hierarchy $m_3 < m_1 < m_2$ is also considered alternatively when fitting mixing angles to data, as experiments do not provide information on the masses but only on squared mass differences $\Delta m^2_{ij} = m_i^2 - m_j^2$). The quoted uncertainties are 1σ errors. It should be noted that the value of δ_{CP} is still consistent with no CP-violation (which requires δ_{CP} to be 0° or 180°) within 2σ uncertainty. The most recent data, however, seem to show a slight preference for a small CP-violation in the neutrino sector (Esteban et al. 2019). The above formalism is similar to the one introduced for the connection between flavor eigenstates and mass eigenstates of quarks. For quarks, the corresponding rotation matrix is called the *Cabbibo–Kobayashi–Maskawa matrix* (CKM matrix). It turns out that the mixing angles for neutrinos are much larger than for quarks (which are $\theta_{12} = 3.04° \pm 0.05°$, $\theta_{23} = 2.38° \pm 0.06°$, $\theta_{13} = 0.201° \pm 0.011°$), implying a stronger relative rotation of the flavor eigenbasis with respect to the mass eigenbasis for neutrinos.

The flavor of a neutrino that is emitted or absorbed in a reaction or decay involving the weak interaction is determined by the coupling to the respective lepton (or its antiparticle). This is called a charged-current reaction. For example, the neutrinos involved in any nuclear reaction transforming a neutron or a proton are electron–neutrinos (or their antiparticles), see Section 3.9. With the flavor selected, the neutrino is in a superposition of mass eigenstates due to the relative rotation of the mass eigenbasis. Denoting the flavor eigenstate by $|\alpha_f\rangle$ and the mass eigenstate by $|\nu_i\rangle$ in Dirac notation (bra–ket notation), the superposition would be

$$|\alpha_f\rangle = \sum_{i=1}^{3} U^*_{\alpha_f,i}\,|\nu_i\rangle, \tag{7.14}$$

where $U^*_{\alpha_f,i}$ is the complex conjugate of the respective component of the PMNS matrix. (For antineutrinos, $U_{\alpha_f,i}$ is used, without the complex conjugation.) When the neutrino moves in space, the relative phases of the three mass eigenstates vary at different rates and therefore the mass mixture is changing as a function of position.

A different mixture of mass eigenstates corresponds to a different mixture of flavor states because

$$|\nu_i\rangle = \sum_{\alpha_\mathrm{f}=\mathrm{e},\mu,\tau} U_{\alpha_\mathrm{f},i}|\alpha_\mathrm{f}\rangle, \tag{7.15}$$

where $U^*_{\alpha_\mathrm{f},i}$ has to be used for antineutrinos instead of $U_{\alpha_\mathrm{f},i}$. This means that, for example, a ν_e will become a mixture of ν_e, ν_μ, ν_τ after some distance. Since the quantum mechanical phase is periodic, after traveling for some further distance, the original pure ν_e will reappear. This is why this behavior is called *neutrino (flavor) oscillation*. More specifically, the one-dimensional propagation of mass eigenstates over a distance $L = x(t) - x(t = t_0)$ during a time t is described by using plane wave solutions of the form

$$|\nu_i(t)\rangle = \mathrm{e}^{-\mathrm{i}\frac{E_i t}{\hbar}}\mathrm{e}^{p_i L}|\nu_i(t = t_0)\rangle, \tag{7.16}$$

with $E_i = \sqrt{p_i^2 c^2 + m_i^2 c^4}$ being the energy of the ith mass eigenstate and $_{pi}$ its momentum. Applying Equation (1.93) (ultra-relativistic particles) and using the customary natural units ($\hbar = 1$, c = 1) in the following, Equation (7.16) becomes

$$|\nu_i(L)\rangle = \mathrm{e}^{-\mathrm{i}\frac{m_i^2 L}{2E_\nu}}|\nu_i(t = t_0)\rangle, \tag{7.17}$$

where $E_\nu \approx p$ (ultra-relativistic) is now the total energy of the neutrino. The probability for observing a neutrino originally of flavor α_f as having flavor β_f at a distance L is

$$\begin{aligned} P_{\alpha_\mathrm{f}\to\beta_\mathrm{f}}(L) = |\langle\alpha_\mathrm{f}(t)|\beta_\mathrm{f}\rangle|^2 &= \left|\sum_i\sum_j U^*_{\alpha_\mathrm{f},i}U_{\beta_\mathrm{f},j}\mathrm{e}^{-\mathrm{i}\frac{m_i^2}{2E_\nu}L}\langle\nu_i|\nu_j\rangle\right|^2 \\ &= \left|\sum_i U^*_{\alpha_\mathrm{f},i}U_{\beta_\mathrm{f},i}\mathrm{e}^{-\mathrm{i}\frac{m_i^2}{2E_\nu}L}\right|^2 = \sum_i\sum_j U^*_{\alpha_\mathrm{f},i}U^*_{\beta_\mathrm{f},i}U_{\alpha_\mathrm{f},j}U_{\beta_\mathrm{f},j}\mathrm{e}^{-\mathrm{i}\frac{\Delta m_{ij}^2}{2E_\nu}L} \\ &= \sum_i U^2_{\alpha_\mathrm{f},i}U^2_{\beta_\mathrm{f},i} + 2\sum_{i>j} U^*_{\alpha_\mathrm{f},i}U^*_{\beta_\mathrm{f},i}U_{\alpha_\mathrm{f},j}U_{\beta_\mathrm{f},j}\cos\left(\frac{\Delta m_{ij}^2}{2E_\nu}L\right). \end{aligned} \tag{7.18}$$

The second line of the above equation was obtained by realizing that $\langle\nu_i|\nu_j\rangle = \tilde{\delta}_{ij}$ and then by writing the square over the sum as a double sum. The last line separates summations over diagonal and off-diagonal elements of the PMNS matrix.

Equation (7.18) is generally valid but it is cumbersome to explicitly specify its dependence on mixing angles. In many cases, however, it is sufficient to restrict the discussion to a two-flavor case. For example, solar neutrinos oscillate between ν_e and a mixture ν_x of ν_μ and ν_τ. Because θ_{13} is small and two of the mass eigenstates are close in mass compared to the third one, one of the two (depending on the assumed mass hierarchy) can be neglected. In this case the unitary rotation matrix $\mathbf{U}'$ contains only one mixing angle θ', for example, mixing ν_e and ν_μ,

$$\begin{pmatrix} \nu_e \\ \nu_\mu \end{pmatrix} = \begin{pmatrix} \cos\theta' & \sin\theta' \\ -\sin\theta' & \cos\theta' \end{pmatrix} \begin{pmatrix} \nu_1 \\ \nu_2 \end{pmatrix}. \tag{7.19}$$

Using the similarly simplified version of Equation (7.18) the transformation probability evaluates to

$$\begin{aligned} P_{\nu_e\to\nu_\mu}(L) &= 2\sin^2\theta'\cos^2\theta' - 2\sin^2\theta'\cos^2\theta'\cos\left(\frac{\Delta m^2}{2E_\nu}L\right) \\ &= \sin^2(2\theta')\sin^2\left(\frac{\Delta m^2}{4E_\nu}L\right). \end{aligned} \tag{7.20}$$

In practical units the argument of the second trigonometric function can be approximated as $1.27\Delta m^2 L/E_\nu$, where Δm^2 has to be inserted in eV^2, L in km, and E_ν as GeV. The oscillation length $L_0 = 4\pi E_\nu/\Delta m^2$ is the distance at which the system returns to its original state. The above also explicitly shows the prerequisite of non-vanishing neutrino masses in at least one generation because otherwise $\Delta m^2 = 0$ and there would be no oscillations. The probability shown in Equation (7.20) actually is for the appearance of ν_μ from an initial source of ν_e (or any other two-flavor case). Relevant for the measured solar neutrino flux on Earth is the survival probability of ν_e, simply given by

$$P_{\nu_e\to\nu_e}(L) = 1 - P_{\nu_e\to\nu_\mu}(L) = 1 - \sin^2(2\theta')\sin^2\left(\frac{\Delta m^2}{4E_\nu}L\right). \tag{7.21}$$

The above equations describe neutrino oscillations for propagation in vacuum. When neutrinos pass through matter, there is a certain, albeit tiny, probability that they interact with the matter. The probabilities are different depending on the flavor. In other words, the neutrinos have a different refraction coefficient or opacity depending on flavor. All three neutrino flavors interact with quarks in the same manner. All of them are scattered on electrons through exchange of a Z^0 boson (*neutral-current reactions*). Electron–neutrinos can additionally interact by exchange of $W^\pm$ bosons (*charged-current reactions*). Therefore there is an additional interaction potential V_{matter} for electron–neutrinos, leading to a difference in propagation from the other two neutrino flavors. Staying in the two-flavor model, the equation of motion can be written as differential equation involving the two mass eigenstates:

$$\mathrm{i}\frac{\mathrm{d}}{\mathrm{d}t}\begin{pmatrix} \nu_1 \\ \nu_2 \end{pmatrix} = \begin{pmatrix} \frac{m_1^2}{2p} & 0 \\ 0 & \frac{m_2^2}{2p} \end{pmatrix} \begin{pmatrix} \nu_1 \\ \nu_2 \end{pmatrix}. \tag{7.22}$$

The equation of motion for the flavor eigenstates is then obtained by applying the transformation

$$
\begin{aligned}
\mathrm{i}\frac{\mathrm{d}}{\mathrm{d}t}\begin{pmatrix}\nu_e\\ \nu_\mu\end{pmatrix} &= \mathbf{U}^{\mathrm{T}}\begin{pmatrix}\frac{m_1^2}{2p} & 0\\ 0 & \frac{m_2^2}{2p}\end{pmatrix}\mathbf{U}\begin{pmatrix}\nu_e\\ \nu_\mu\end{pmatrix}\\
&= \frac{1}{2p}\begin{pmatrix} m_1^2\cos^2\theta' + m_2^2\sin^2\theta' & \frac{\Delta m^2}{2}\sin(2\theta')\\ \frac{\Delta m^2}{2}\sin(2\theta') & m_1^2\sin^2\theta' + m_2^2\cos^2\theta'\end{pmatrix}\begin{pmatrix}\nu_e\\ \nu_\mu\end{pmatrix}.
\end{aligned}
\tag{7.23}
$$

Since only ν_e are affected by the additional interaction possibility, only the top left component of the matrix on the second line of Equation (7.23) has to be modified to account for matter interaction. According to Equation (7.16) the flavor propagates as

$$
|\nu_e\rangle = \sum_i \mathrm{e}^{-\mathrm{i}\frac{E_i t}{\hbar} - n_{\mathrm{ref}} p_i L} U_{\nu_e,i}|\nu_i\rangle, \tag{7.24}
$$

where the phase is already modified by an effective refraction index n_{ref} as is also customary in classical wave scattering. From scattering theory (see also Section 1.6) the effective refraction index is obtained from the superposition of the amplitudes of many individual scattering events and is given by

$$
n_{\mathrm{ref}} = 1 + \frac{V_{\mathrm{matter}}}{p} = 1 + \frac{2\pi\mathcal{N}}{p^2}\mathcal{F}_{\mathcal{K}}(0), \tag{7.25}
$$

with $\mathcal{N}$ being the number of scattering centers. That is half the number of electrons because only electrons with left-handed chirality contribute to the scattering of left-handed neutrinos and therefore $\mathcal{N} = \bar{n}_e/2$. The forward angle scattering amplitude $\mathcal{F}_{\mathcal{K}}(0)$ has to be derived using the high-energy approximation for neutrino–electron scattering and is given by

$$
\mathcal{F}_{\mathcal{K}}(0) = \frac{\sqrt{2}}{\pi} p g_{\mathrm{V}}. \tag{7.26}
$$

Using $E_i t/\hbar - n_{\mathrm{ref}} p_i L = E_i t/\hbar - p_i L - (n_{\mathrm{ref}} - 1) p_i L$ it is easy to realize that the impact of matter simply is to add $-\sqrt{2}\bar{n}_e g_{\mathrm{V}}$ to the top left component of the matrix on the second line of Equation (7.23). It is important to note that in consequence all related quantities, such as mixing angles and appearance or disappearance probabilities, become dependent on the electron number density $\bar{n}_e$. It is convenient to introduce new oscillation parameters $\Delta m_{\rm MSW}$ and $\theta_{\rm MSW}$ in matter, which can be determined from the equation of motion using

$$\begin{aligned} m_{1,\rm MSW}^2 \cos^2\theta_{\rm MSW} + m_{2,\rm MSW}^2 \sin^2\theta_{\rm MSW} &= m_1^2 \cos^2\theta' + m_2^2 \sin^2\theta' \\ &\quad - \sqrt{2}\bar{n}_{\rm e} g_{\rm V}, \\ \Delta m_{\rm MSW}^2 \sin(2\theta_{\rm MSW}) &= \Delta m^2 \sin(2\theta'), \\ m_{1,\rm MSW}^2 \sin^2\theta_{\rm MSW} + m_{2,\rm MSW}^2 \cos^2\theta_{\rm MSW} &= m_1^2 \sin^2\theta' + m_2^2 \cos^2\theta'. \end{aligned} \tag{7.27}$$

This yields

$$\Delta m_{\rm MSW}^2 = \Delta m^2 \sqrt{\left(\cos(2\theta') - \frac{2\sqrt{2}p\bar{n}_{\rm e}g_{\rm V}}{\Delta m^2}\right)^2 + \sin^2(2\theta')}, \tag{7.28}$$

$$\sin^2(2\theta_{\rm MSW}) = \frac{\sin^2(2\theta')}{\left(\cos(2\theta') - \frac{2\sqrt{2}p\bar{n}_{\rm e}g_{\rm V}}{\Delta m^2}\right)^2 + \sin^2(2\theta')}. \tag{7.29}$$

The modified value $\Delta m_{\rm MSW}^2$ for Δm^2 can also be interpreted as electron–neutrinos acquiring a different effective mass when traveling through matter. Similar to Equation (7.20), the transformation probability becomes

$$P_{\nu_{\rm e}\to\nu_\mu}^{\rm MSW}(L, \mathcal{N}) = \sin^2(2\theta_{\rm MSW}) \sin^2\left(\frac{\Delta m_{\rm MSW}^2}{4E_\nu} L\right). \tag{7.30}$$

As above, there is an oscillation length $L_0^{\rm MSW} = 4\pi E_\nu/\Delta m_{\rm MSW}^2$. Since the matter interaction leads to an additional phase difference between the neutrino states, also a refraction length $L_{\rm ref} = \pi\sqrt{2}/(g_{\rm V}\bar{n}_{\rm e})$ can be defined, specifying the distance over which the additional matter phase equals 2π. Inspection of the denominator in Equation (7.29) shows that $\theta_{\rm MSW}$ becomes maximal when

$$\cos(2\theta') = \frac{2\sqrt{2}p\bar{n}_{\rm e}g_{\rm V}}{\Delta m^2}. \tag{7.31}$$

This is equivalent to having $L_0 = L_{\rm ref}\cos(2\theta')$ and is a resonance where the eigenfrequency of the neutrino mixing $1/L_0$ coincides with the eigenfrequency of the medium $1/L_{\rm ref}$. Assuming a fixed neutrino energy E_ν, a resonance density

$$\bar{n}_{\rm e}^{\rm res} = \frac{\Delta m^2 \cos(2\theta')}{2\sqrt{2}E_\nu g_{\rm V}} \tag{7.32}$$

can be found. The full width at half maximum of the resonance is given by

$$\Delta_{\rm FWHM} = 2\bar{n}_{\rm e}^{\rm res}\tan(2\theta'). \tag{7.33}$$

In a medium with varying density the layer with the electron density range $\bar{n}_{\rm e}^{\rm res} \pm \Delta_{\rm FWHM}/2$ is called the resonance layer. It is remarkable that the resonance effect is very strong while L_0 and $L_{\rm ref}$ can still be large and no major oscillation effect would be expected. This is called the *Mikheyev–Smirnov–Wolfenstein effect* (MSW effect).

The resonant amplification of the neutrino oscillation is important for solar neutrinos. In the core of the Sun, electron–neutrinos are released at high $\bar{n}_{\rm e}$, close to the resonance condition, and therefore $\theta_{\rm MSW} \approx \pi/2$. This implies that the $\nu_{\rm e}$ is forced into having an almost pure mass eigenstate ν_2 (using the two-neutrino approximation as before). From the core to the surface, $\bar{n}_{\rm e}$ declines. This decline is slow enough to apply the adiabatic theorem of quantum systems, which states that there are no transitions between eigenstates if they change only slowly. The neutrino stays in the ν_2 mass eigenstate while passing through the Sun. Upon leaving the Sun the matter effect vanishes and $\theta_{\rm MSW} = \theta'$. Up to that point the mass eigenstate has not changed and therefore $|\nu_2\rangle = \sin\theta'|\nu_{\rm e}\rangle + \cos\theta'|\nu_\mu\rangle$. This means that the flavor eigenstate of the neutrino is only partially $\nu_{\rm e}$. The fraction transformed into ν_μ at this point is given by

$$|\langle\nu_\mu|\nu_2\rangle|^2 = \cos^2\theta', \tag{7.34}$$

which would be $\cos^2\theta' \approx 1$ for small θ'. On the way to the neutrino detector on Earth, there could be additional vacuum oscillations (the size of the Sun is too small in comparison with the vacuum oscillation length to have additional vacuum oscillation within the Sun outside of the resonance layer). They are small, however, when θ' is small. Therefore the MSW effect is the dominant cause for the missing $\nu_{\rm e}$ from the solar neutrino flux on Earth and thus the solution to the solar neutrino problem.

The MSW effect is also relevant for neutrinos passing dense matter in supernovae (Section 9.3) and accretion disks (Section 9.4.5).

7.3 Helium Burning

7.3.1 Reactions

When the hydrogen in the core of a star has been exhausted and hydrogen burning has moved into a shell, the mass of the He-core produced in hydrogen core- and shell-burning determines the further fate of the star. Either the core is stabilized against further contraction by the pressure of the electron gas. Then the core will remain as a He-white dwarf and no new burning phase can ignite. For stars with initial masses of about 0.5 $M_\odot$ and above, the contraction of the He-core and the additional heating of hydrogen burning in a shell around the core achieve temperatures ($(1-2)\times10^8$ K, depending on density) and densities (10^5–10^8 kg m^{-3}) sufficient to fuse ^{4}He nuclei and start the helium burning phase. As described in Chapter 6, core helium burning proceeds under degenerate conditions in stars with $M \lesssim 2.3\ M_\odot$.

Not only had the Coulomb barriers prevented the formation of heavier nuclides at lower temperature but also the lack of stable or long-lived nuclides with mass numbers $A = 5$ and $A = 8$, which is a consequence of nuclear structure. It turns out that the only viable path to go beyond ^{4}He is to make ^{12}C from ^{4}He. The net process $3\alpha \rightarrow {}^{12}$C is called the *triple-α reaction*. It has a net Q-value of $Q_{\text{nuc}} = 7.275$ MeV. As in hydrogen burning, a three-body reaction is improbable and so this net reaction is realized by two sequential two-body reactions. The first step is $^4\text{He} + {}^4\text{He} \leftrightarrows {}^8\text{Be}$ (an endothermic reaction with $Q_{\text{nuc}} = -92$ keV). As already mentioned in the discussion of H-burning, ^{8}Be dissociates spontaneously into two α-particles with a lifetime $\tau_{\text{life}}^{^8\text{Be}\rightarrow\alpha+\alpha} = 2.6 \times 10^{-16}$ s. Nevertheless, in a hot helium gas with a large number of ^{4}He nuclei, there will be a few ^{8}Be nuclei present at any given moment in time. The equilibrium abundance can be derived as shown in Section 4.3 by requiring $\dot{Y}_{^8\text{Be}} = 0$. This implies that creation and destruction rates are equal,

$$\frac{1}{2}\rho \text{N}_\text{A}\langle\sigma^*v\rangle_{\alpha+\alpha} Y_\alpha^2 = Y_{^8\text{Be}}\lambda_{\text{dec}}^{^8\text{Be}\rightarrow\alpha+\alpha}. \tag{7.35}$$

As explained in Section 4.1.7, the rate is related to the lifetime $\lambda_{\text{dec}}^{^8\text{Be}\rightarrow\alpha+\alpha} = 1/\tau_{\text{life}}^{^8\text{Be}\rightarrow\alpha+\alpha}$ which, in turn, is related to the width $\Gamma_{\text{g.s.}}^{^8\text{Be}} = \hbar/\tau_{\text{life}}^{^8\text{Be}\rightarrow\alpha+\alpha}$ of the decaying ground state according to Equation (3.49). Therefore the equilibrium abundance of ^{8}Be is given by

$$Y_{^8\text{Be}} = \frac{\hbar}{2\Gamma_{\text{g.s.}}^{^8\text{Be}}}\rho \text{N}_\text{A}\langle\sigma^*v\rangle_{\alpha+\alpha} Y_\alpha^2. \tag{7.36}$$

(The additional contribution coming from $^8\text{Be} + \alpha \leftrightarrows {}^{12}\text{C}^*$ is negligible.) At typical He-burning conditions (100 MK and 10^5 g cm^{-3}) the abundance ratio of ^{8}Be to ^{4}He is $Y_{^8\text{Be}}/Y_{^4\text{He}} \approx 5.2 \times 10^{-10}$. Although the lifetime of ^{8}Be seems extremely short, it is long compared to the time it takes an α-particle to cross the ^{8}Be nucleus, which is on the order of 10^{-19} s. This is also the typical reaction time for adding a further α to ^{8}Be in the reaction $^8\text{Be}(\alpha,\gamma)^{12}\text{C}^*$. This is the second step of the triple-α reaction, producing the carbon isotope in an excited state (indicated by the asterisk), though, not yet releasing the full reaction Q-value before it decays to the ground state. The de-excitation of ^{12}C* proceeds by emission of photon pairs or electron–positron pairs. This takes $\tau_{^{12}\text{C}^*} \approx 1.8 \times 10^{-16}$ s but is negligible compared to the other timescales in the reaction sequence when assuming equilibrium processes. The production rate of ^{12}C from ^{8}Be is

$$\begin{aligned}\dot{Y}_{^{12}\text{C}} &= \rho \text{N}_\text{A}\langle\sigma^*v\rangle_{^8\text{Be}(\alpha,\gamma)} Y_\alpha Y_{^8\text{Be}} \\ &= \rho^2 \text{N}_\text{A}^2 \frac{\hbar}{2\Gamma_{\text{g.s.}}^{^8\text{Be}}}\langle\sigma^*v\rangle_{\alpha+\alpha}\langle\sigma^*v\rangle_{^8\text{Be}(\alpha,\gamma)} Y_\alpha^3.\end{aligned} \tag{7.37}$$

The ^{8}Be abundance given in Equation (7.36) assumes that $^4\text{He} + {}^4\text{He} \leftrightarrows {}^8\text{Be}$ can be treated like a resonant reaction in which the ground state of ^{8}Be appears at a resonance energy of 92 keV with the width $\Gamma_{\text{g.s.}}^{^8\text{Be}}$. At temperatures $T < 28$ MK, however, the Gamow window for $^4\text{He} + {}^4\text{He} \leftrightarrows {}^8\text{Be}$ is not covering the resonance state and the reaction proceeds in a non-resonant manner. A more general expression would be (Nomoto et al. 1985)

$$\dot{Y}_{^{12}\mathrm{C}} = \rho^2 \mathrm{N}_\mathrm{A}^2 \frac{\hbar}{2} \int_0^\infty \frac{1}{\Gamma_{\mathrm{g.s.}}^{^8\mathrm{Be}}(E)} \sigma_{\alpha+\alpha}^*(E) E \mathrm{e}^{-E/(\mathrm{k_B}T)} \langle \sigma^* v \rangle_{^8\mathrm{Be}(\alpha,\gamma)}(E)\, \mathrm{d}E, \tag{7.38}$$

where the integration over energy in the $\alpha + \alpha$ system is written explicitly and also the energy dependences of $\Gamma_{\mathrm{g.s.}}^{^8\mathrm{Be}}$, $\sigma_{\alpha+\alpha}^*$, and $\langle\sigma^* v\rangle_{^8\mathrm{Be}(\alpha,\gamma)}$ are specified explicitly. The general equation reverts to Equation (7.37) when ^{4}He + ^{4}He ⇆ ^{8}Be is resonant and the approximation of Equation (4.33) can be applied to the integration of $\sigma_{\alpha+\alpha}^*$. It has been shown that the triple-α reactivity can be well approximated by summing over a resonant and a non-resonant contribution, $\langle\sigma^* v\rangle_{3\alpha} = \langle\sigma^* v\rangle_{3\alpha}^{\mathrm{res}} + \langle\sigma^* v\rangle_{3\alpha}^{\mathrm{nonres}}$, with the non-resonant part dominating at $T < 28$ MK and the resonant part for $T > 28$ MK (Nomoto et al. 1985).

Similar considerations apply to the reaction ^{8}Be$(\alpha,\gamma)^{12}$C* ($Q_{\mathrm{nuc}} = 7.367$ MeV), feeding the $J^{\pi_q} = 0^+$ state at $E^x = 7.654$ MeV excitation energy in ^{12}C: it can be treated as a resonant reaction for $T > 74$ MK but is non-resonant for lower temperature. The low-temperature regime is important for He-burning on white dwarfs and neutron stars but hydrostatic He-burning in stars proceeds at $T > 100$ MK, allowing to treat both the ^{4}He + ^{4}He ⇆ ^{8}Be and the ^{8}Be$(\alpha,\gamma)^{12}$C* reaction as resonant. Moreover, the two reaction steps can be treated as an equilibrium process between free α-particles, the unstable ground state of ^{8}Be, and the excited state in ^{12}C.

Assuming the validity of Equation (7.36), inserting the narrow resonance approximation (Equation (4.33)) for $\langle\sigma^* v\rangle_{\alpha+\alpha}$ leads to the *Saha equation* for the ^{8}Be abundance,

$$Y_{^8\mathrm{Be}} = \rho \mathrm{N_A} Y_\alpha^2 \hbar^3 \left(\frac{2\pi}{\mu_{\mathrm{red}}^{\alpha\alpha} \mathrm{k_B} T} \right)^{3/2} \mathrm{e}^{-E_{\mathrm{res}}^{\alpha+\alpha}/(\mathrm{k_B}T)}. \tag{7.39}$$

In the derivation of the above equation it has to be noted that $\omega_{\mathrm{BW}} = 1$ and $\gamma_{\mathrm{BW}} = \Gamma_{\mathrm{g.s.}}^{^8\mathrm{Be}}$. The value of γ_{BW} comes from the fact that there is no other way for the ground state to decay.

Also assuming ^{8}Be$(\alpha,\gamma)^{12}$C* to proceed as a resonant reaction, the excited state in ^{12}C appears as a resonance at $E_{\mathrm{res}}^{^8\mathrm{Be}+\alpha} = 287$ keV ($E_{\mathrm{res}} = E^x - Q_{\mathrm{nuc}} = E^x - S_\alpha^{^{12}\mathrm{C}}$). Again, the expression for the reactivity of a resonant reaction as given in Equations (4.33) and (4.35) is used. Here, the total width $\hat{\Gamma}_{\mathrm{tot}} = \hat{\Gamma}^\alpha + \hat{\Gamma}^{\mathrm{rad}} = \hat{\Gamma}^\alpha + \hat{\Gamma}^\gamma + \hat{\Gamma}^{\mathrm{pair}}$ is the sum of the α-width and the radiation width. The latter is comprised of the width for the emission of photon pairs (a single γ transition $0^+ \to 0^+$ to the ground state of ^{12}C is forbidden by spin selection rules, see Section 1.7) and the width for electron–positron pair emission (the positron in this pair annihilates with a plasma electron). Despite of the Coulomb barrier for the α-particle, $\hat{\Gamma}^\alpha \gg \hat{\Gamma}^{\mathrm{rad}}$ and therefore the radiation width dominates the resonance strength $\omega_{\mathrm{BW}}\gamma_{\mathrm{BW}}$ (see the discussion following Equation (4.35)). It evaluates to $\omega_{\mathrm{BW}}\gamma_{\mathrm{BW}} = (3.7 \pm 0.5) \times 10^{-3}$ eV.

Combining the narrow resonance expression for $^{8}Be(\alpha,\gamma)^{12}C^{*}$ with Equation (7.39) for $Y_{^8Be}$ gives the rate of the (resonant) triple-α reaction:

$$r_{3\alpha} = \frac{1}{2}(Y_\alpha \rho N_A)^3 \left(\frac{2\sqrt{3}\pi\hbar^2}{m_\alpha k_B T}\right)^3 \frac{\omega_{BW}\gamma_{BW}}{\hbar} e^{-(m_{^{12}C^*}-3m_\alpha)c^2/(k_B T)}. \tag{7.40}$$

Note that the mass $m_{^{12}C^*}$ of the excited $^{12}C^*$ enters in the exponential. The mass difference in the exponential is derived from the sum of the resonance energies $E_{res}^{\alpha+\alpha}$ and $E_{res}^{^8Be+\alpha}$, which can be expressed as an effective Q-value. The excitation energy of 7.654 MeV of this excited state is released in the de-excitation to the ground state of ^{12}C.

The triple-α rate has two remarkable features which set it apart from other rates appearing in hydrostatic and explosive nucleosynthesis. First, the energy production due to the triple-α rate at hydrostatic (low) temperature is extremely sensitive to temperature. It is found to be $\varepsilon_{3\alpha} \propto T^{41}$ at 100 MK, due to the resonant nature of the two sequential reactions. This is a much stronger temperature dependence than any other rate exhibits. At the much higher temperatures encountered in explosive burning, on the other hand, the rate becomes almost independent of temperature. This is addressed in Chapter 9 and shown in Figure 9.1. Second, the triple-α rate is also strongly dependent on density, $r_{3\alpha} \propto \rho^3$. Therefore it is extremely sensitive to density variations in addition to its temperature dependence, more than other (two-body) reactions which only exhibit a ρ^2 dependence (see Equation (4.10)). Both features are of importance in He-shell burning of AGB stars (Section 7.3.2). Furthermore, the density dependence is essential for burning under explosive conditions (Section 8.3 and Chapter 9), for which it determines whether α-particles can be efficiently processed into heavier nuclei.

Once ^{12}C has been formed, further reactions occur. The most important one is the capture of another ^{4}He nucleus on ^{12}C, producing oxygen in the reaction $^{12}C(\alpha,\gamma)^{16}O$ (Q_{nuc} = 7.162 MeV). This reaction also significantly contributes to the energy production in He-burning. (A further α capture on ^{16}O does not occur because $^{16}O(\alpha,\gamma)^{20}Ne$ does not have any resonances at the relevant energies and therefore its cross section is very small.) The reaction rate for $^{12}C(\alpha,\gamma)^{16}O$ is not only important for determining the energy generation, even more importantly its value sets the C/O ratio surviving after He-burning, which in turn affects the structure and evolution of the star in subsequent burning phases. With more carbon surviving, the carbon burning phase will be able to release more energy and also convection will be affected. This impacts the nucleosynthesis of many further nuclides beyond C and O. The current uncertainties in the extrapolation of the cross section to lower energy leave enough room to obtain quite different nucleosynthesis patterns in massive stars. Therefore this reaction sometimes has been alluded to being the most important reaction in nuclear astrophysics. The cross section of $^{12}C(\alpha,\gamma)^{16}O$ is not known experimentally in the Gamow window around about 300 keV because it is much too low for direct measurements up to now ($\sigma \approx 10^{-17}$ barns). The reaction proceeds through the high-energy tails of two subthreshold resonances (corresponding to excited states in ^{16}O at 7.117 MeV ($J^{\pi_q} = 1^-$) and 6.917 MeV ($J^{\pi_q} = 2^+$)) and there is also possible interference with the low-energy tail of a broad 1^- resonance at 9.585 MeV excitation energy. This makes predictions or extrapolations very difficult. The suggested values for the astrophysical S-factor at 300 keV are in the range 100–200 keV barn, with a probable preference of 150–170 keV barn. An acceptable uncertainty for constraining stellar evolution and nucleosynthesis beyond He-burning, however, would be $\lesssim$10%.

The cores of stars with non-zero metallicity also contain a small fraction of ^{14}N left over from the CNO-cycles (at solar metallicity this would be about 1.4%–2% by mass). It can be destroyed very efficiently by ^{14}N(α,γ)^{18}F(β^+)^{18}O even before the onset of proper He-burning. The ^{18}O is converted to ^{22}Ne toward the end of He-burning, when the temperature begins to increase, by ^{14}O(α,γ)^{22}Ne. The production of ^{18}O is sensitive to both nuclear physics (the still uncertain α-capture rate) and stellar physics. Concerning the latter, the treatment of convection is crucial, using the Ledoux criterion (see Equations (2.110) and (2.111), respectively) yields more ^{18}O.

Availability of the nuclide ^{22}Ne during He-burning is very important because it provides a source of neutrons via the reaction ^{22}Ne(α,n)^{25}Mg for the s-process (see Section 8.2). Free neutrons are unstable and can only participate in neutron captures when they are constantly produced by another reaction. The reaction ^{22}Ne(α,n)^{25}Mg is endothermic ($Q_{\mathrm{nuc}} = -0.482$ MeV), however, and therefore only efficient at higher temperature. Therefore it is only active toward the end of core He-burning in massive stars but not in the cores of AGB stars. In AGB stars it can provide a part of the neutron exposure during He-shell flashes, see below. Modeling the weak s-process component in massive stars is complicated by the fact that the ^{22}Ne(α,n)^{25}Mg channel competes with the ^{22}Ne(α,γ)^{26}Mg channel and their relative strength is poorly determined experimentally. Especially important is to determine the importance of a resonance at 633 keV in the ^{22}Ne + α system. The uncertainty in neutron production in massive stars is so large that it is sometimes quoted as the second largest (nuclear) problem after ^{12}C(α,γ)^{16}O. The s-process and its nucleosynthesis is discussed in more detail in Section 8.2. Here it is just mentioned in passing that not only heavier elements beyond Fe are made by the s-process but that neutron captures also produce lighter nuclides, such as ^{36}S, ^{37}Cl, ^{40}Ar, ^{40}K, and ^{45}Sc. Important for the efficiency of the s-processing is not only the neutron source but also the amount of *neutron poisons*. These are nuclides absorbing a considerable fraction of the neutrons in (n,γ) or (n,p) reactions due to either their large reaction cross section (such as ^{14}N, ^{17}O) or their large abundance in the plasma during He-burning (such as ^{12}C and ^{16}O). It is curious to note that also ^{22}Ne, ^{25}Mg, and ^{26}Mg efficiently absorb neutrons. Neutron poisons have a higher impact in neutron-poor environments, such as at low metallicity or at low temperature when the release of neutrons is inefficient. Therefore they are more significant in the interpulse phase of AGB stars (see below) than in the AGB pulse phase and in massive stars.

7.3.2 He-shell Flashes

With the exhaustion of α-particles in the center of the star, the burning zone moves outwards and becomes a burning shell. At this stage, there are two shells burning, a

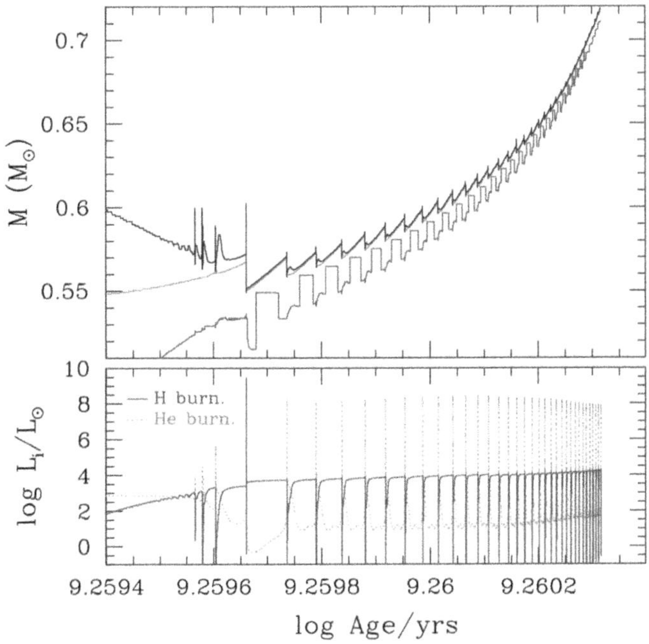

Figure 7.4. Simulation of the thermal pulsing of an AGB star with 2 $M_\odot$ and solar metallicity. Upper panel, top to bottom: Temporal evolution of the mass coordinates of the inner border of the convective envelope, the mass location of maximum energy production in the H-burning shell, and the maximum energy production within the He-core. During each interpulse period, the flat segment of the lowest line corresponds to the location of the non-convective burning of the $^{13}C(\alpha,n)^{16}O$ reaction in the ^{13}C pocket. Lower panel: Temporal evolution of the H-burning and He-burning contributions to luminosity. (Reprinted figure with permission from Käppeler et al. 2011. Copyright 2011 by the American Physical Society.)

H-burning shell further out and the He-burning shell. In stars with masses between 2.3 and about 10 $M_\odot$ (AGB and super-AGB stars) an interesting phenomenon ensues, caused by the strong density dependence of the triple-α rate. The energy generation in the volume of the He-burning shell is not sufficient to balance the mass layers on top of it and therefore hydrostatic self-regulation is impeded. Since not enough pressure can be built up, the shell is squeezed more and more, leading to a gradual increase in density. This speeds up the triple-α reaction. Due to the ρ^3 dependence and the high temperature sensitivity of the triple-α reaction the point of equilibrium between internal pressure and forces on the shell by surrounding layers is overshot and an explosive expansion of the He-shell sets in. This has a similar effect as the thermonuclear runaway in a degenerate gas encountered in the core He-flash in stars with $M < 2.3\,M_\odot$ and is called a *He-shell flash*. In the expansion, the density drops rapidly, quenching the triple-α reaction. The forceful expansion also pushes the H-burning shell outwards and extinguishes H-burning. Once the burning has ceased, the gravitational pull overcomes the outward motion and the shells move back, re-igniting H- and He-burning. This sets the stage for another He-shell flash.

AGB stars undergo a large number of such thermal pulses, where the thermonuclear flash phase lasts only a few hundred years whereas the time between pulses is a hundred to a thousand times longer (Figure 7.4). Oscillations and vibrations are induced into the stellar plasma by these pulses, leading to increased mass loss from the surface of the star. AGB stars have strong stellar winds which considerably decrease their total mass during their evolution. He-shell flashes only occur in low-mass stars because the thickness of the burning shells roughly scales with the stellar mass. In stars with more than 9–10 $M_\odot$, the He-burning shell is thick and develops sufficient pressure to maintain hydrostatic equilibrium.

The He-shell flashes have another important impact: they temporarily create large convection zones, mixing the plasma constituents across large distances and between shells within the star, also mixing fresh protons from the hydrogen layer down into the helium layer. This is important for the production of the s-process nuclei, for which AGB stars provide the main component. The detailed workings of the s-process in AGB stars, however, are somewhat complicated and even until today fully self-consistent modeling is not possible. The general mechanism is understood and outlined in the following. The s-process itself is discussed in Section 8.2.

At the elevated temperature in the He-burning region during the thermal pulse the reaction ^{22}Ne(α,n)^{25}Mg acts as a neutron source. As mentioned previously, this reaction is an efficient neutron source only at higher temperature. Thermal pulses are short by astronomical standards and the neutron irradiation during such a burst would not suffice to produce the whole range of s-process nuclides from Fe to Bi, even if many such bursts occur.

A much more efficient neutron source would be ^{13}C(α,n)^{16}O. The problem is that α-particles are abundant in the He-shell but there is not enough ^{13}C. The solution is a complicated mixing process caused by the thermal pulses and the extended convection regions formed above the He-shell during a He-shell flash. The ^{12}C created in helium burning at the bottom of the He-shell is distributed across the whole shell by the large convective region. After the He-shell flash ceases, the H-shell, which had been pushed outwards and extinguished (see above), settles back and H-shell burning restarts. In the brief period before the hydrogen ignition, protons diffuse down from the hydrogen envelope into the upper part of the helium layer. They react with ^{12}C to form ^{13}C via ^{12}C(p,γ)^{13}N(β^+)^{13}C in a region just below the H-shell burning, the so-called ^{13}C pocket. In this pocket, additionally heated by the H-shell burning, ^{13}C(α,n)^{16}O continues during the interpulse period, that is for much longer than just during a pulse. The next He-shell flash extinguishes both the H-shell burning and wipes out the ^{13}C pocket but both are re-established after the pulse. This situation is sketched in Figure 7.5. Most of the main s-process component stems from the ^{13}C pocket, with a minor contribution from the ^{22}Ne (α,n)^{25}Mg directly in the shell-flash zone.

Typical temperatures at which reactions take place are in the range of 100–700 MK, with the higher temperatures attained in the shell-flash region and lower temperature in the ^{13}C pocket. Historically, most neutron capture reactions for the s-process were experimentally investigated at an energy around 30 keV (this also motivated the introduction of the 30 keV MACS as given in Equation (4.22)), which had been derived in parameterized models as the most important interaction energy

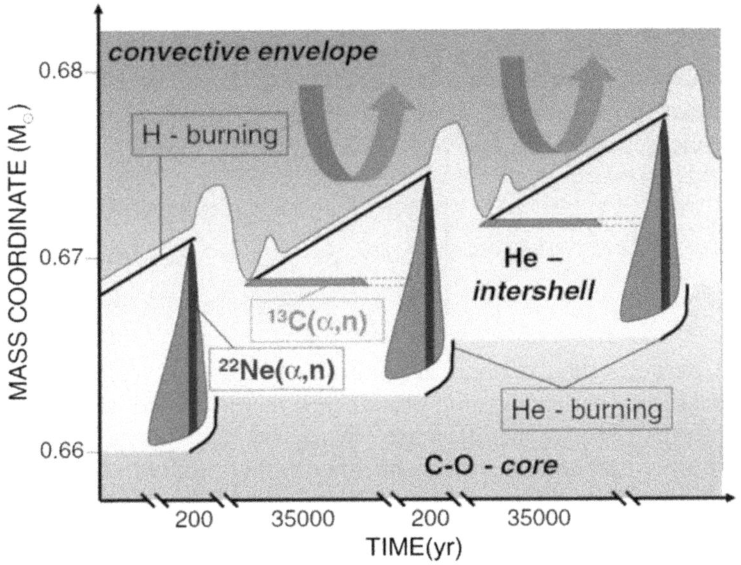

Figure 7.5. Sketch of the nuclear burning in thermal pulses of AGB stars. (Figure from Reifarth et al. 2014.

before detailed simulations revealed the interplay between shell flashes and the ^{13}C pocket.

It is interesting to note that the combined action of the extended convective zone in the thermal pulse and the convective envelope above the H-burning shell transport the products of the neutron captures to the surface of the star. When the unstable element Tc was found in the spectra of AGB stars in the mid-1950s, it was immediately obvious that this element must have been produced inside the star by an s-process because even the half-life of its most long-lived isotope (^{98}Tc, $t_{1/2} = 4 \times 10^8$ yr) is too short to have survived since the formation of the star.

7.4 Carbon Burning

As pointed out in Chapter 6, stars below about 8 $M_\odot$ do not ignite C-burning. Although each of the advanced burning stages (C-burning and beyond) commence with a well-defined process, they enable many reactions with many different nuclides once ignited. It becomes impossible to follow the energy generation and nucleosynthesis by simple parameterizations and extended nuclear reaction networks (see Section 4.2) have to be used. For the same reason it is not practicable to list all reactions occurring for each advanced burning stage. Only the most important ones are presented here.

Another point to remember is the increasing importance of neutrino losses to the energy budget of the hydrostatic star. Once a temperature of about 500 MK is exceeded, neutrino losses from pair production dominate. This was already mentioned in Chapter 6 and can be implied from the fact that the (electromagnetic) luminosities given in Table 7.1 do not further increase in the advanced burning stages.

After He-burning the most abundant nuclides in the stellar core are ^{12}C and ^{16}O. At temperatures of (6–8) $\times 10^8$ K and a density $\rho \approx 10^5$ g cm^{-3} ^{12}C nuclei start fusing. This releases nucleons and α-particles in the reactions

$$^{12}\mathrm{C} + {}^{12}\mathrm{C} \rightarrow {}^{24}\mathrm{Mg}^* \rightarrow \begin{cases} {}^{20}\mathrm{Ne} + \alpha & Q_{\mathrm{nuc}} = 4.62\ \mathrm{MeV} \\ {}^{23}\mathrm{Na} + \mathrm{p} & Q_{\mathrm{nuc}} = 2.24\ \mathrm{MeV} \\ {}^{23}\mathrm{Mg} + \mathrm{n} & Q_{\mathrm{nuc}} = -2.63\ \mathrm{MeV} \end{cases} . \tag{7.41}$$

The released particles then undergo a plethora of reactions with the nuclides either having survived from earlier burning phases or directly created in C-burning. Mainly contributing to the energy generation are ^{23}Na(p,α)^{20}Ne, ^{23}Na(p,γ), and ^{12}C(α,γ)^{16}O. Further reaction sequences with non-negligible fluxes are ^{20}Ne(α,γ)^{24}Mg, ^{23}Na(α,p)^{26}Mg(p,γ)^{27}Al, ^{21}Ne(α,n)^{24}Mg, ^{22}Ne(p,γ)^{23}Na, ^{25}Mg(p,γ)^{26}Al(β^+)^{26}Mg.

Neutron emission from the ^{12}C + ^{12}C fusion happens only rarely due to the negative Q-value. Nevertheless, it is important in changing the neutron excess η_{n} (definition given in Equation (1.18)) by permitting the sequence ^{20}Ne(n,γ)^{21}Ne (p,γ)^{22}Na(β^+)^{22}Ne(α,n)^{25}Mg(n,γ)^{26}Mg and also by the decay ^{23}Mg(β^+)^{23}Na. This allows even stars born with zero metallicity to develop $Y_{\mathrm{e}} < 0.5$ during C-burning (and later burning phases). For example, 15 M$_\odot$ and 25 M$_\odot$ stars with zero metallicity obtain $\eta_{\mathrm{n}} = 1.24 \times 10^{-3}$ and $\eta_{\mathrm{n}} = 6.8 \times 10^{-4}$, respectively. (For remarks on the general evolution of Y_{e} see Section 7.7.)

Aside from the ^{16}O still surviving from He-burning, the main products of C-burning are $^{20-22}$Ne, ^{23}Na, $^{24-26}$Mg, and ^{26,27}Al. Small amounts of ^{29,30}Si and ^{31}P are also made. The production of neutron-rich nuclides depends on the actual neutron excess.

In C-burning at low temperature but high density, a mild s-process can ensue because the sequence ^{12}C(p,γ)^{13}N(β^+)^{13}C(α,n)^{16}O(α,γ)^{20}Ne becomes possible and neutrons are released from the ^{13}C(α,n) source. The Q-value of ^{12}C(p,γ)^{13}N, however, is only $Q_{\mathrm{nuc}} = 1.93$ MeV. The connection of capture and photodisintegration rates is given in Equation (4.52). According to the rule of thumb in Equation (4.53) photodisintegration of ^{13}N by ^{13}N(γ,p)^{12}C will be faster than the β^+ decay of ^{13}N at $T \gtrsim 800$ MK and consequently ^{13}C cannot be reached. Only with the s-processing $\eta_{\mathrm{n}} = 5 \times 10^{-3}$ is possible, otherwise at the end of C-burning $\eta_{\mathrm{n}} = 2.24 \times 10^{-3}$ and $\eta_{\mathrm{n}} = 1.96 \times 10^{-3}$ for 15 M$_\odot$ and 25 M$_\odot$ stars, respectively.

The reaction cross section of ^{12}C + ^{12}C is not well constrained at subCoulomb energies. It exhibits a pronounced oscillatory behavior caused by strong, quasi-molecular resonances with spacings of a few 100 keV. This makes a prediction difficult as simple barrier penetration models cannot account for the resonance structures.

7.5 Neon Burning

Up to this burning stage, the sequence of phases is determined by increasing Coulomb barriers. The most abundant elements after C-burning are O, Ne, and Mg. The next burning phase, however, is not O-burning as expected from the Coulomb barrier criterion because the Q-value of $^{16}O(\alpha,\gamma)^{20}Ne$ is only $Q_{nuc} = 4.73$ MeV. According to Equation (4.53) α-particles are released from ^{20}Ne before $^{16}O + {}^{16}O$ can set in. The α-particles react with the nuclides present in the plasma and this leads to a rearrangement of abundances. After $^{20}Ne(\gamma,\alpha)^{16}O$ the most important reaction sequence for energy generation is $^{20}Ne(\alpha,\gamma)^{24}Mg(\alpha,\gamma)^{28}Si$. The energy yield is only about one quarter that of C-burning and therefore Ne-burning is brief. Its importance is in altering the plasma composition, not so much in establishing a long-term stable hydrostatic phase.

Further reactions of interest to nucleosynthesis are $^{25}Mg(\alpha,n)^{28}Si$, $^{26}Mg(\alpha,n)^{29}Si$, $^{26}Mg(p,n)^{26}Al$, $^{26}Mg(\alpha,\gamma)^{30}Si$, $^{27}Al(\alpha,p)^{30}Si$, and $^{30}Si(p,\gamma)^{31}P$. The net effect of Ne-burning is that the final composition is enhanced in ^{16}O, and in all the isotopes of Mg, Al, Si, and P. Additionally, ^{36}S, ^{40}K, ^{46}Ca, ^{58}Fe, $^{61,62,64}Ni$ are made, as well as traces of the unstable nuclides ^{22}Na and ^{26}Al.

As already explained in Chapter 6, only stars with more than about 10 $M_\odot$ pass through Ne-burning hydrostatically. In stars with masses in the range of about 8–10 $M_\odot$ the Fermi-energies (Equation (1.78)) of the electrons in the degenerate ONeMg cores after C-burning reach values that allow the most energetic electrons to initiate endothermic electron captures, initially mainly on ^{20}Ne ($Q_{nuc}^{EC} = -7.024$ MeV) and ^{24}Mg ($Q_{nuc}^{EC} = -5.516$ MeV). This causes a loss of degeneracy pressure and thus hydrostatic equilibrium cannot be established by Ne-burning and the following burning phases. They are passed during a core collapse and this leads to incomplete burning due to the brevity in each phase, changing the nucleosynthetic signature.

7.6 Oxygen Burning

Very soon after the onset of Ne-burning, temperature ($T \approx 2$ GK) and density ($\rho \approx 10^7$ g cm^{-3}) are sufficient to permit $^{16}O + {}^{16}O$ fusion. Similarly as in C-burning, the fusion is followed by the release of light particles,

$$^{16}O + {}^{16}O \rightarrow {}^{32}S^* \rightarrow \begin{cases} ^{28}Si + \alpha & Q_{nuc} = 9.59 \text{ MeV } (34\%) \\ ^{31}P + p & Q_{nuc} = 7.68 \text{ MeV } (56\%) \\ ^{31}S + n & Q_{nuc} = 1.45 \text{ MeV } (5\%) \\ ^{30}P + d & Q_{nuc} = -2.41 \text{ MeV } (5\%) \end{cases}. \quad (7.42)$$

The endothermic deuteron channel is only open at high temperature. When this is the case, the relative contributions of the channels are the ones given in percent. The released deuteron is immediately photodisintegrated into a neutron and proton due to the elevated temperature during O-burning. The ^{31}S created by neutron emission decays to ^{31}P by $^{31}S(\beta^+)^{31}P$, which in turn is destroyed by $^{31}P(p,\alpha)^{28}Si$.

Analogously to the ^{12}C fusion reaction, the reaction cross section of ^{16}O + ^{16}O is not well constrained at the astrophysically relevant energy of $E_{\mathrm{c.m.}} \approx 4$ MeV. Again, oscillatory behavior due to quasi-molecular resonances is expected, not reproduced by simple barrier penetration models.

The light particles released from the excited ^{32}S nuclei react with the nuclides present in the plasma, leading to a large number of possible reactions that can only be tracked in an extended reaction network. The main products of O-burning are ^{28}Si, $^{32-34}$S, ^{35,37}Cl (with ^{37}Cl produced as ^{37}Ar), ^{36,38}Ar, ^{39,41}K (^{41}K produced as ^{41}Ca), and ^{40,42}Ca. The main part ($\approx$90%) of the final mass fractions is ^{28}Si and ^{32}S.

In high-density O-burning ($\rho > 2 \times 10^7$ g cm^{-3}) the electron captures ^{33}S(e$^-$, ν_e) ^{33}P, ^{35}Cl(e$^-$, ν_e)^{35}S, and ^{37}Ar(e$^-$, ν_e)^{37}Cl increase the neutron excess to $\eta_n = (1.2\text{–}3.5) \times 10^{-2}$. In massive stars this occurs in core O-burning because shell burning generally proceeds at slightly higher temperature but lower density.

Due to the high temperatures reached in O-burning, many forward and reverse reaction rates become comparable and are fast in comparison to the burning timescale. This leads to the creation of regions ("clusters") in the nuclear chart where the abundances of nuclides are in equilibrium. This is an example of the appearance of quasi-statistical nuclear equilibrium (QSE, see Section 4.3.2). The abundances in each cluster can be computed using Equation (4.148). The clusters are connected by slower reactions that are not in equilibrium and have to be considered explicitly. This allows to simplify, and thus computationally speed up, the treatment of the reaction network as explained in Sections 4.2 and 4.3. For example, at some point ^{28}Si, ^{29}Si, and ^{30}P are connected by neutron and proton captures, respectively, equilibrated with their reverse reactions and so they form a cluster. Likewise, ^{34,35}S and ^{35,36}Cl are a cluster and their abundances are in equilibrium within that cluster but not with the ^{28}Si-cluster. With increasing temperature, more and more nuclides join the clusters and smaller clusters merge into larger ones as their connecting reactions come into equilibrium.

7.7 Silicon Burning

Like Ne-burning, Si-burning is not initiated by fusion but rather by photodisintegration of ^{28}Si. Neutrons, protons, and α-particles are "boiled off" from ^{28}Si and the resulting nuclei. The ejected particles can be recaptured quickly and establish a large QSE system. At the start of Si-burning there are two large QSE-clusters, comprising on the one hand nuclei in the mass number range $24 \lesssim A \lesssim 46$ and on the other hand Fe-group nuclides. These two clusters merge and extend already in the early Si-burning phase during an initial temperature increase and form a large cluster containing all nuclides heavier than ^{24}Mg.

Since the neutron excess η_n (or alternatively Y_e, see Equation (1.18)) crucially determines the range of nuclides produced in Si-burning and also determines the conditions for the Fe-core collapse and subsequent explosion, it is helpful to recapitulate its evolution from the initial composition of the star until the onset of Si-burning. The mass of stars born with zero metallicity is composed of about 23% ^{4}He ($\eta_n = 0$) and about 77% ^{1}H ($N - Z = -1$) and therefore initially they have a negative total neutron excess. During H-burning, $\eta_n \to 0$ ($Y_e \to 0.5$) in the burning zone because of the conversion of ^{1}H into ^{4}He. Later generation stars have an admixture of heavier nuclides ("metals," see Equation (1.13)) and therefore their also initially negative neutron excess is assuming a slightly different value from zero after H-burning. It can be estimated by taking, for example, solar metallicity $\mathcal{Z}_{\odot} \approx 0.019\,6$ and realizing that the mass fraction of ^{12}C, ^{14}N, and ^{16}O (all of them have $N = Z$) already makes up about $X_{\mathrm{CNO}} \approx 0.014$. Most other "metals" are so-called α-nuclei (which can be constructed by a series of α-particle captures starting from ^{16}O), also having $N = Z$. Only Fe-group nuclides have a positive neutron excess. Representative for the Fe-group is ^{56}Fe with $X_{^{56}\mathrm{Fe}\odot} \approx 0.001\,17$ and $N - Z = 4$. This leads to having $\eta_{\mathrm{n}} \approx 1.6 \times 10^{-4}$ ($Y_{\mathrm{e}} = 0.499\,92$) after H-burning in solar metallicity stars. The main He-burning reactions do not change this value significantly unless s-processing is possible. As explained in Section 7.3 this depends on the availability of ^{14}N. For solar composition $Y_{^{14}\mathrm{N}} \approx 10^{-3}$, leading to $\eta_{\mathrm{n}} \approx 2 \times 10^{-3}$ ($Y_{\mathrm{e}} \approx 0.499$) because $N - Z = 2$ for ^{14}N. (He-shell flashes in AGB stars further increase the neutron content of the plasma but this does not happen in massive stars.) This neutron excess is not changed much in C-burning unless in low-temperature burning, if the ^{13}C(α,n) neutron source is activated. In this case η_{n} up to 5×10^{-3} is possible. Hydrostatic Ne-burning (without electron captures in degenerate cores) does not change these numbers. As presented above, electron captures appear in high-density O-burning and thereby may increase the neutron excess up to $\eta_{\mathrm{n}} \approx 3.5 \times 10^{-2}$ ($Y_{\mathrm{e}} \approx 0.482\,5$). On the one hand this still is not far from $Y_{\mathrm{e}} = 0.5$, thereby preventing the production of very neutron-rich isotopes (a finding which will be important for the discussion in Section 8.3). On the other hand, such a value already results in a very non-solar abundance pattern and it has to be concluded that (the parts of) the material in the core of massive stars reaching this neutron excess are not ejected in a supernova explosion. Probably only products of O-shell burning are mixed into the interstellar medium.

Turning back to Si-burning, the nuclides preferentially produced in Si-burning are determined based on their binding energies and on the Y_{e} during Si-burning, as the QSE- and NSE-abundances depend on these parameters (see Equations (4.145) and (4.148)). For $Y_{\mathrm{e}} = 0.5$ this is a distribution of abundances around the peak abundance of ^{56}Ni, the Fe-group. The width of the distribution is given by the temperature. Additionally, some lighter nuclides survive which are not in equilibrium. In the stellar core, electron captures have already occurred during O-burning and the initial fuel at the ignition of Si-burning is a mixture of ^{28}Si and ^{29,30}Si in comparable amounts. Then the dominant product is not ^{56}Ni but rather ^{54}Fe or even ^{56}Fe. Details depend sensitively on the actual conditions. For $\eta_{\mathrm{n}} \lesssim 0.006$, and only for this, near solar abundances of ^{48,49}Ti, ^{51}V, ^{50,52,53}Cr, ^{55}Mn, and ^{54,56,57}Fe are produced. Also some lighter species such as ^{28}Si, ^{32}S, ^{36}Ar, and ^{40}Ca can be made but may be destroyed during the following explosion of the star (see below). The actual abundances and abundance distributions very sensitively depend not only on the exact thermodynamical conditions but also the core-collapse mechanism may impact the abundances of the innermost layers ejected in the supernova explosion of a massive star. Both Si-burning and the explosion are notoriously difficult to model and different approaches have been used, giving rise to different answers regarding the nucleosynthetic yield from these inner layers.

Another crucial parameter is the density achieved during the burning process because this determines when (and if) the triple-α reaction joins the equilibrium due to the strong density-dependence of this rate (see Equation (7.40)). As mentioned in Section 4.3.2, the triple-α rate requires densities of $\rho \geqslant 2 \times 10^8$ g cm^{-3} to get into equilibrium. This may be achieved at the lower end of the mass range of massive

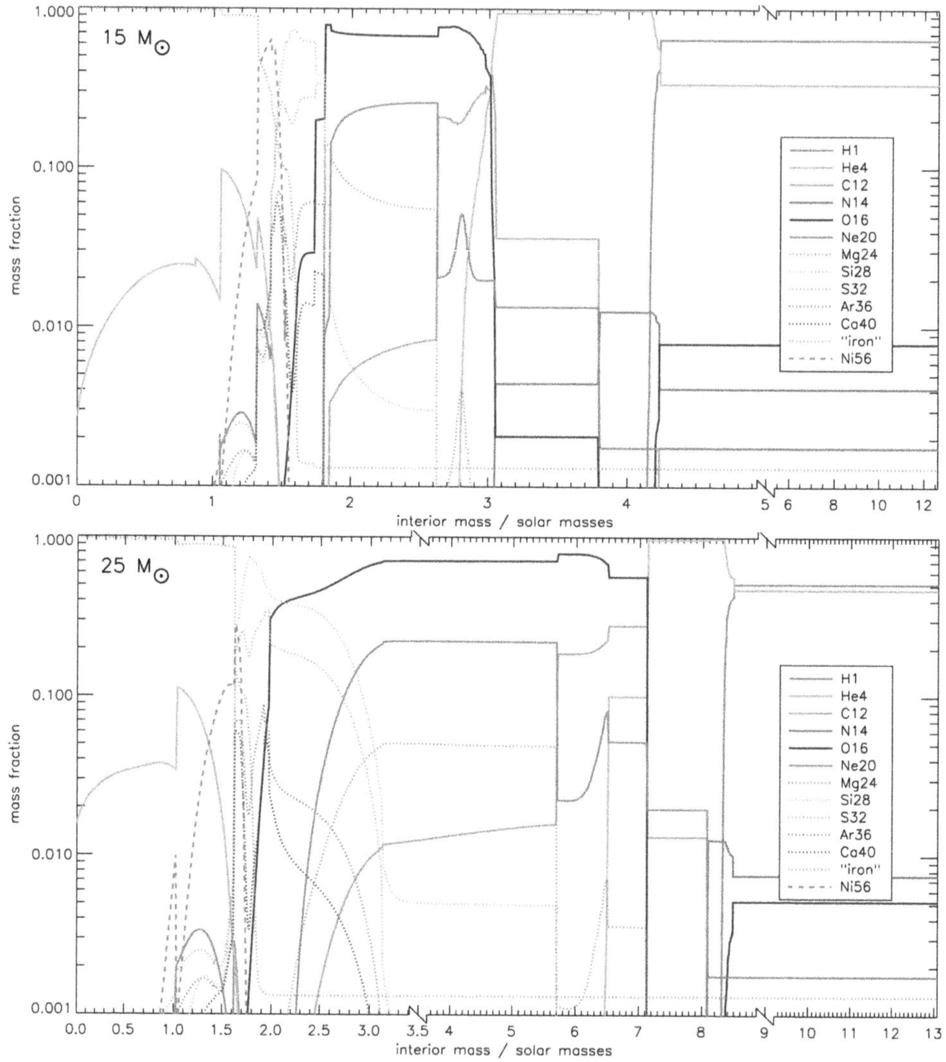

Figure 7.6. Inner structures of a 15 M$_\odot$ (top) and a 25 M$_\odot$ (bottom) star in the final Si-burning stage; note the logarithmic scale of the mass fractions. (Reprinted figure with permission from Woosley et al. 2002. Copyright 2002 by the American Physical Society.)

stars, but not for more massive stars, and only in central burning, not in shell burning. Toward the end of Si-burning and at the onset of core collapse (see Section 9.3) the temperature becomes sufficiently high to bring into equilibrium also the reactions linking Mg with Ne, C with O, and (if permitted by density) also C to He. Then a full NSE is established and the abundances will not depend on individual reaction rates anymore. They can be computed by applying Equation (4.145). For $T \lesssim 10$ GK, NSE favors nuclei with the highest binding energies at a given Y_e (or η_n). Therefore in full NSE also the lighter nuclides are converted into a distribution of Fe-group nuclides around ^{56}Ni ($\eta_n = 0$) or ^{56}Fe ($\eta_n \approx 0.07$). If more of such matter is ejected, the abundances of the Fe-group will be enhanced relative to lighter nuclides from Si-burning.

During Si-burning the mass of the Fe/Ni-core formed from the products of Si-burning is constantly increasing. The core contracts until the pressure loss from electron captures turns the contraction into a collapse. There is no energy generation possible by the fusion of Fe-group nuclei and therefore no new burning phase is ignited. This leads to a complete collapse of the core and eventually the formation of a neutron star, also causing a shockwave ploughing through the outer layers of the star and ejecting them in a supernova explosion. An overview of this process and of the explosive shell-burning caused by the shockwave is given in Section 9.3. In fact, the collapse of the core and the appearance of the shockwave occur on a timescale much shorter than the Kelvin–Helmholtz timescale (see Section 2.9.2) of the outer layers. Therefore they barely react before being hit by the supernova shock from the inside of the star.

Cuts through evolved stars in Si-burning shortly before the central core collapse are shown in Figure 7.6. The typical "onion shell" structure can be seen, where the "ashes" of each burning phase are stacked on top of each other. At the lower end of each "ash layer" is a shell burning region, further adding burnt material to the layer below. Note that the Lagrangian mass scale (see Section 2.1) is plotted differently for the outer mass layers. The outermost layers (nominally extending up to 15 $M_\odot$ and 25 $M_\odot$, respectively, but affected by mass loss) are not shown.

Further Reading

Bahcall, J. N. 1989, Neutrino Astrophysics (Cambridge: Cambridge University Press)
Bahcall, J. N., & Peña-Garay, C. 2004, NJPh, 6, 63
Bahcall, J. N., Serenelli, A. M., & Basu, S. 2005, ApJL, 621, L85
Diehl, R., Hartmann, D. H., & Prantzos, N. (ed.) 2018, Astrophysics with Radioactive Isotopes, Astrophysics and Space Science Library, Vol. 453 (Berlin: Springer)
Esteban, I., et al. 2019, JHEP, 1, 106 see also http://www.nu-fit.org for latest results
Käppeler, F., Gallino, R., Bisterzo, S., & Aoki, W. 2011, RvMP, 83, 157
Nomoto, K., Thielemann, F.-K., & Miyaji, S. 1985, A&A, 149, 239
Paxton, B., et al. 2011, ApJS, 192, 3
Paxton, B., et al. 2015, ApJS, 220, 15
Reifarth, R., Lederer, C., & Käppeler, F. 2014, JPhG, 41, 053101
Serenelli, A. 2016, EPJA, 52, 78
Woosley, S. E., Heger, A., & Weaver, T. A. 2002, RvMP, 74, 1015

AAS | IOP Astronomy

Essentials of Nucleosynthesis and Theoretical Nuclear Astrophysics

Thomas Rauscher

Chapter 8

Origin of the Elements Beyond Fe

8.1 Introduction

As explained in Chapter 7, the regular hydrostatic burning phases only lead to the production of nuclei up to the Fe-region unless neutrons are released through the reactions ^{22}Ne(α,n)^{25}Mg or ^{13}C(α,n)^{16}O. Only neutrons can efficiently build up elements with higher charge number Z because they are not affected by a Coulomb barrier. Beyond the Fe-peak, the observed elemental abundances rapidly drop. Using this fact and the characteristic double-peak structures in nuclear abundances (see below) it has been realized in 1957 (Burbidge et al. 1957; Cameron 1957a, 1957b, 2013) that two distinct neutron capture processes must have been at work for creating the majority of the nuclides above Fe, with each contributing about half. The two processes are sketched in Figure 8.1. One of the processes involves neutron captures that are slower than β^- decays and therefore its production follows the line of stability. It was termed *s-process* for "slow neutron-capture process." The other process was called *r-process*, which is short for rapid neutron-capture process. The neutron-capture rates in the r-process are much faster than any β^- decays close to stability and therefore highly neutron-rich and very short-lived isotopes are produced.

The problem is to identify the sites of these processes. Free neutrons are unstable and therefore a process involving reactions with neutrons must either occur in a neutron-rich environment, where neutron captures are fast enough to proceed before the neutrons decay, or in an environment with a neutron source providing a supply of "fresh" neutrons. An environment for slow neutron captures had been identified already in the mid-1950s by having identified unstable Tc in the spectra of AGB stars. The site of the r-process is more elusive. For a long time, it was thought that the innermost, still ejected layers of a massive star ejected in a supernova explosion are sufficiently neutron-rich to allow for an r-process. This

doi:10.1088/2514-3433/ab8737ch8

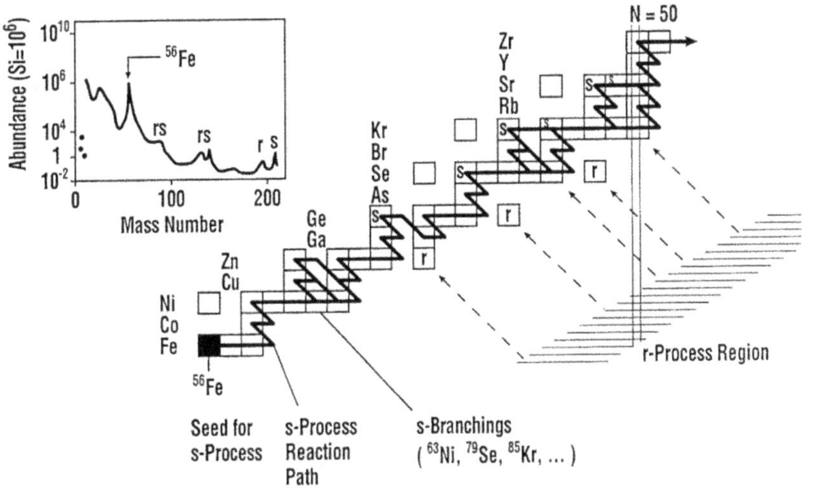

Figure 8.1. Sketch of the neutron-capture process paths responsible for the formation of the nuclei between iron and the actinides. The observed abundance distribution in the inset shows a strong decline beyond iron and characteristic twin peaks. Nuclides only reached by the s-process (s-only) or r-process (r-only) are marked by "s" and "r," respectively. There are also proton-rich isotopes that cannot be reached by either process, the so-called p-nuclides. They have to be made in a different way. See text for further details. (Reprinted figure with permission from Käppeler et al. 2011. Copyright 2011 by the American Physical Society.)

has been disputed recently because state-of-the-art simulations do not find the right combination of Y_e, temperature, and density to produce a full r-process. Very recently, heavy nuclides from an r-process have been identified in the ejecta from a merger between two neutron stars. It is still not clear, however, whether this can explain the abundances of r-process elements found in low-mass stars formed early in the Galaxy. It is also conceivable that several sites contributed to the total budget of r-process elements identified in the solar system.

Although the site of the r-process may still be unknown, the nuclear processes producing the s- and r-nuclides are well understood in principle. They are briefly outlined in the following. An in-depth review of the astrophysical simulations implementing these processes and especially of those attempting to clarify the origin of the r-process nuclides, however, is beyond the scope of this textbook. In the following presentation it is helpful to remember that both the s- and r-process can contribute to the abundance of an isotope of an element although pure s- and r-nuclides exist (*s-only* and *r-only* isotopes). Only a few elements are dominated by either process, Ba is a classical s-process element whereas Eu is a classical r-process element. The r-process also contributes 97% of the abundance of the only stable isotope of gold (^{197}Au). s-only and r-only isotopes are crucial for pinning down details in the two processes because only they allow to disentangle the contributions of the two processes and thus also the detailed conditions at which they proceed. Another clue is provided by the double-peak structure found in the abundance pattern beyond Fe (see the inset in Figure 8.1). Three double peaks can be found (the first one less pronounced), with a sharper peak located at stable nuclides with closed neutron shells and a wider peak accompanying each sharp peak at slightly lower mass number. This is an indication of two neutron capture processes at work and the locations of the peaks provide constraints on the amount of neutrons (or the neutron number density) required to build them. The nuclear shell closures with the magic neutron numbers as explained by the nuclear shell model (Section 3.3) not only affect the binding energy of a nucleus but also its reaction cross section for neutron capture. The fact that a sharp peak at nuclides with magic neutron numbers is found, shows that these nuclides were built up by a series of neutron captures along the line of stability, as shown in Figure 8.1. Every instance that the nucleosynthesis process reaches a nucleus with a magic neutron number, the abundance peaks because the flow to heavier nuclides is impeded by the smaller capture cross section. The companion peak, which is shifted to lower mass number, indicates that the nucleosynthesis process hit the magic neutron number at a lower mass number. Since the mass number is the sum of neutron and proton number, this is only possible far from stability, as in the "r-process region" sketched in Figure 8.1. It can only be reached at high neutron density with fast neutron captures. The nuclides produced in that region decay to stability when the neutron density drops at the end of the process and form the companion peak to the s-process peak. The mass number at which the r-process peak appears provides indications on neutron number density and temperature during the r-process (see below and Section 4.3.3). The fact that the r-process peak is wider than the s-process peak shows that the reaction path in the r-process is not as well defined as for the s-process, either because of a range of temperatures and densities encountered or because of a larger range of nuclides produced at each specific condition. This is further addressed in Section 8.3.

An example of the individual contributions of s- and r-process to the total solar abundances is shown in Figure 8.2. This is from a parameterized model for the s- and r-processes. The contributions to the double peaks is clearly seen, as well as the varying fractional contributions to nuclides between the peaks. Note that the contributions are shown as a function of mass number A, adding up abundances of isotopes from different elements.

There are two basic timescales (see also Section 4.1.7) in the scenario of heavy-element nucleosynthesis by neutron captures: (1) the beta-decay lifetimes, and (2) the time intervals between successive captures that are inversely proportional to the neutron capture reaction rates and the neutron flux. If the rate of neutron capture is slow compared to the relevant β^- decays, the synthesis path will follow closely the line of stable nuclides in the nuclear chart. This is the s-process with typical neutron number densities of $\approx 10^8$ neutrons/cm^3. It terminates at Bi because any further elements are unstable and decay back. On the other hand, if the rate of neutron capture is faster than the β^- decays, very neutron-rich and highly unstable nuclei will be formed. After the neutron flux has ceased, those nuclei will be transformed to stable nuclei by a sequence of β^- decays. The production of very neutron-rich nuclides and their subsequent decay is called the r-process, requiring neutron number densities of 10^{18}–10^{22} neutrons/cm^3. These values were inferred from inspection of the s- and r-process peaks, using an "inverse approach." Such an

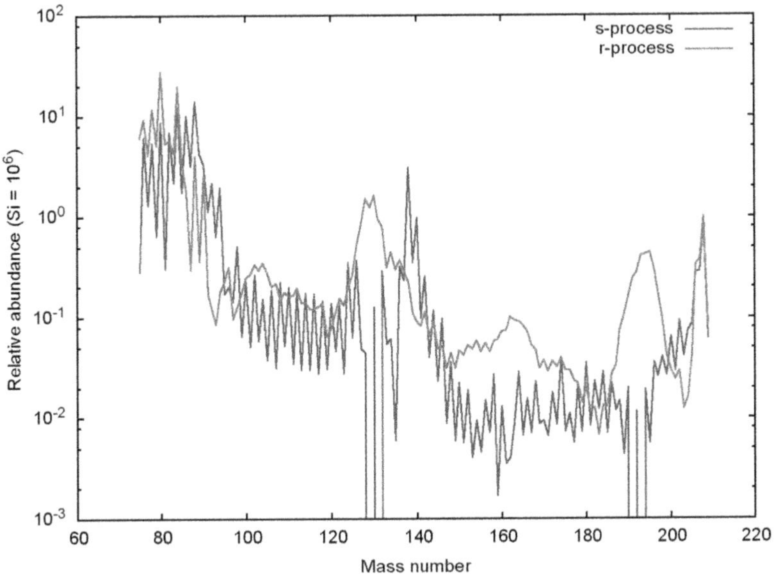

Figure 8.2. Contributions of the s- and r-process to the solar abundances.

approach is site-independent, meaning that it does not rely on the detailed modeling of a specific nucleosynthesis site (such as a core-collapse supernova) which may be difficult. Rather, it applies an inverse reasoning by analyzing abundance patterns and drawing conclusions on the required conditions, such as density, temperature, and timescales, perhaps by using simple parameterizations. Having pinned down the required conditions, the next step is to identify the astrophysical site in which they are realized. Of course, this again requires more detailed simulations but the inverse approach is helpful in selecting or ruling out certain scenarios and also to identify which knowledge of further input (such as properties of the involved nuclides) is necessary. The inverse approach also provides a better understanding of the underlying physical processes, which often are hard to extract from a "direct," computer-intensive, brute force multi-physics simulation. Historically, such inverse approaches have been used not only in the seminal papers by Burbidge et al. (1957); Cameron (1957a, 1957b, 2013) but also in a broad range of follow-up studies in the following decades, striving to clarify further details of various nucleosynthesis processes.

8.2 The s-Process

8.2.1 The Classical Model

When the unstable element Tc was found in the spectra of AGB stars in the mid-1950s, it was immediately obvious that this element must have been produced inside the star by an s-process and brought to the surface by a large convective zone because even the half-life of its most long-lived isotope (^{98}Tc, $\tau_{1/2} = 4 \times 10^8$ yr) is too short to have survived since the formation of the star. How the reactions

^{22}Ne(α,n)^{25}Mg and ^{13}C(α,n)^{16}O act as neutron sources in the thermal pulses, caused by He-shell flashes, of AGB stars has been described in Section 7.3.2. Even before the details of the mechanism were known, the required conditions to reproduce the abundances of s-only nuclides had been derived by simple parameterized models.

Simplifications of reaction networks are discussed in Section 4.3. In a fully consistent s-process treatment, the neutron sources and neutron poisons, which are in the light nuclei, have to be considered. They determine the actual neutron flux under given environmental conditions. In a parameterized, site-independent model of the s-process in the region of nuclides beyond Fe, however, the neutron number density is treated as a parameter. In this case, the network can be simplified to contain only neutron captures and β^- decays. Contrary to the equations for equilibrated chains in the r- and rp-processes, as shown in Equation (4.149), reverse reactions are negligible at the low temperatures of the s-process as even the temperatures achieved in He-shell flashes are low by nuclear physics standards. Therefore (γ,n) reactions do not have to be included, similarly to what is shown for the reaction network example in Section 4.2.2. Then the abundance change in nuclide B (with neutron number N and proton number Z) is given by

$$\begin{aligned}\dot{Y}_{\mathrm{B}} &= \rho \mathrm{N_A} \langle \sigma^* v \rangle_{\mathrm{A(n,\gamma)B}} Y_{\mathrm{A}} Y_{\mathrm{n}} - \rho \mathrm{N_A} \langle \sigma^* v \rangle_{\mathrm{B(n,\gamma)C}} Y_{\mathrm{B}} Y_{\mathrm{n}} - \lambda^*_{\beta,\mathrm{B}} Y_{\mathrm{B}} \\ &= \bar{n}_{\mathrm{n}} \langle \sigma^* v \rangle_{\mathrm{A(n,\gamma)B}} Y_{\mathrm{A}} - \bar{n}_{\mathrm{n}} \langle \sigma^* v \rangle_{\mathrm{B(n,\gamma)C}} Y_{\mathrm{B}} - \lambda^*_{\beta,\mathrm{B}} Y_{\mathrm{B}},\end{aligned} \tag{8.1}$$

with the abundance Y_{A} of nucleus A (with neutron number $N - 1$ and proton number Z) and the β^--decay rate $\lambda^*_{\beta,\mathrm{B}}$ of nucleus B. Since the β^- decay is faster than any neutron capture in most cases, instantaneous decay can be assumed. This means that any unstable nucleus immediately decays and turns into an isotope of the next element. On the other hand, the β^--decay term can be dropped in Equation (8.1) for stable nuclides. Instantaneous decay is a good approximation for most unstable nuclides encountered in the s-process. There is a number of long-lived nuclides, however, which have lifetimes comparable to their lifetimes with respect to neutron capture. For such nuclides, the (temperature-dependent) β^--decay term has to be left in the network to account for the branching of the s-process path (see Section 8.2.2).

If the neutron exposure has lasted a span of time sufficient to produce all nuclides in a chain, then steady flow can be assumed, as defined in Section 4.3.4. This implies that $\dot{Y}_{\mathrm{B}} = 0$ and Equation (8.1) simplifies to

$$\langle \sigma^* v \rangle_{\mathrm{A(n,\gamma)B}} Y_{\mathrm{A}} = \langle \sigma^* v \rangle_{\mathrm{B(n,\gamma)C}} Y_{\mathrm{B}} = \mathrm{const.} \tag{8.2}$$

Assuming s-wave neutron capture, the relation in Equation (4.20) can be applied, replacing $\langle \sigma^* v \rangle$ by $\langle \sigma^* \rangle v_{\mathrm{T}}$, with the MACS $\langle \sigma^* \rangle$ as defined in Equation (4.22). Since the s-process abundances may stem from the superposition of neutron captures at different neutron densities, it is advantageous to consider the neutron flux ϕ_{n} instead of a constant $\bar{n}_{\mathrm{n}}$. The time-integrated neutron flux τ_{nex} is defined as

$$\tau_{\mathrm{nex}}(t) = \int_0^t \phi_{\mathrm{n}}(t')\,\mathrm{d}t' = \int_0^t \bar{n}_{\mathrm{n}}(t') v_T\,\mathrm{d}t' = v_T \int_0^t \bar{n}_{\mathrm{n}}(t')\,\mathrm{d}t'. \tag{8.3}$$

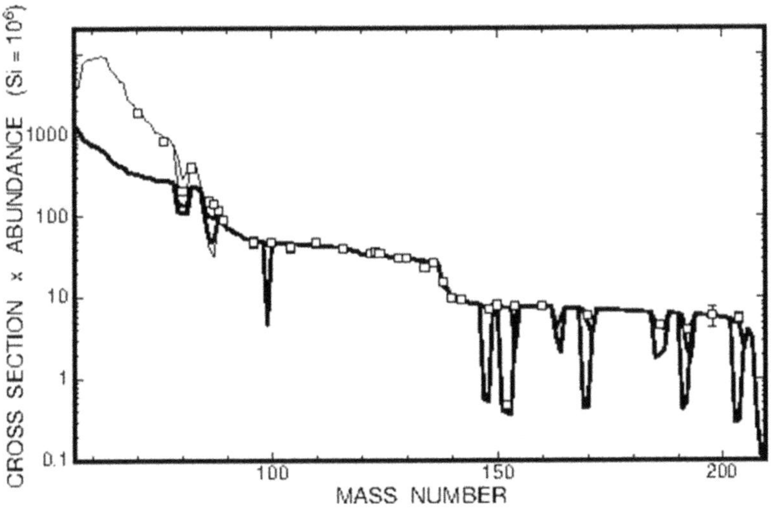

Figure 8.3. Product of neutron capture MACS and s-process abundances for s-only nuclides (open squares). Also shown are theoretical results from the classical model, applying a single component of neutron exposure (thick line) and a sum of two components (thin line). Selected s-process branchings are also shown. (Reprinted figure with permission from Käppeler et al. 2011. Copyright 2011 by the American Physical Society.)

Transforming dY/dt to $dY/d\tau_{\text{nex}}$ allows to rewrite Equation (8.2) as the *local equilibrium approximation*,

$$\langle\sigma^*\rangle_{\mathrm{A(n,\gamma)B}}\, Y_{\mathrm{A}} = \langle\sigma^*\rangle_{\mathrm{B(n,\gamma)C}}\, Y_{\mathrm{B}} = \text{const.} \tag{8.4}$$

It implies that a nucleus with a large capture cross section should have a small abundance and a nucleus with a small cross section a large abundance, as it is expected in a steady flow. It also implies that a constant value should be obtained for all products of cross sections and abundances. Figure 8.3 shows the actual situation for s-only nuclides. Instead of having a single constant value for all s-only nuclides, three regions can be distinguished: two plateaus between the magic neutron numbers (Section 3.3.1) 50, 82, and 126, and a steep increase in the product value for the few s-only nuclides below $N = 50$. At the magic neutron numbers the cross sections are small and this delays the entrance of the following nuclides into the steady flow, leading to the stepped structures seen in the figure. In between the magic neutron numbers, steady flow seems to be a good assumption. Below $A = 84$ ($N = 50$), there must be another contribution which clearly is not in steady flow.

The *classical model of the s-process* assumes that an initial seed of ^{56}Fe is irradiated with a continuous distribution of multiple neutron exposures $\rho_{\text{nex}}(\tau_{\text{nex}})$. The neutron exposure is the Laplace transform of the $\langle\sigma^*\rangle Y$ distribution,

$$\langle\sigma^*\rangle Y = \int_0^\infty \rho_{\text{nex}}(\tau_{\text{nex}})\langle\sigma^*\rangle Y \, d\tau_{\text{nex}}, \tag{8.5}$$

and ρ_{nex} has been found to have the form of an exponential distribution,

$$\rho_{\mathrm{nex}}(\tau_{\mathrm{nex}}) = \frac{f_{56}\,Y^{\odot}_{^{56}\mathrm{Fe}}}{\tau^0_{\mathrm{nex}}} \mathrm{e}^{-\tau_{\mathrm{nex}}/\tau^0_{\mathrm{nex}}}. \tag{8.6}$$

The free parameter τ^0_{nex} is the mean neutron exposure and f_{56} is the fraction of the iron seed nuclei (with solar abundance $Y^{\odot}_{^{56}\mathrm{Fe}}$) which have been irradiated by neutrons. Using such a distribution and assuming instantaneous decay of unstable nuclides allows to solve Equation (8.1) analytically. This yields

$$\langle\sigma^*\rangle_{\mathrm{B(n,\gamma)C}}\, Y_{\mathrm{B}} = \langle\sigma^*\rangle_{\mathrm{A(n,\gamma)B}}\, Y_{\mathrm{A}} \left[1 + \frac{1}{\tau^0_{\mathrm{nex}}\langle\sigma^*\rangle_{\mathrm{B(n,\gamma)C}}}\right]^{-1} \tag{8.7}$$

for the connection between the $\langle\sigma^*\rangle Y$ products for two neighboring isotopes. It can be easily seen that the local approximation in Equation (8.4) is recovered when $\tau^0_{\mathrm{nex}}\langle\sigma^*\rangle_{\mathrm{B(n,\gamma)C}}$ is large (for example, because $\langle\sigma^*\rangle$ is large between the magic numbers). For an arbitrary nuclide i in the s-process path the $\langle\sigma^*\rangle Y$ product is given by

$$\langle\sigma^*\rangle_i\, Y_i = \frac{f_{56}\,Y^{\odot}_{^{56}\mathrm{Fe}}}{\tau^0_{\mathrm{nex}}} \prod_{j=1}^{i} \left[1 + \frac{1}{\tau^0_{\mathrm{nex}}\langle\sigma^*\rangle_j}\right]^{-1}. \tag{8.8}$$

The product runs over all nuclides in the path starting from ^{56}Fe and leading up to the selected nuclide i. When the $\langle\sigma^*\rangle$ are known, the only free parameters are f_{56} and τ^0_{nex}. The result of a single fit of the two parameters to the $\langle\sigma^*\rangle Y$ product of s-only nuclides is shown by the thick line in Figure 8.3. Very good reproduction across a wide range of mass numbers (including the stepped plateaus) is achieved with 0.04% of the solar ^{56}Fe abundance as seed and on average about 15 neutrons captured by each seed nucleus. This is called the *main component of the s-process.*

The steep increase below the magic neutron number 50 requires the addition of a second component. Using the parameter $f_{56} \approx 2.5$ and allowing for only about 1.1 neutrons captured by each seed on average leads to a steeply declining $\langle\sigma^*\rangle Y$ product with increasing A, shown as a thin line Figure 8.3. This was termed the *weak component of the s-process.* Its parameters are not well constrained in the classical model because there are only six s-only nuclides below $A = 90$. In contrast to the main s-process component, steady flow equilibrium is not achieved in the weak component because the neutron exposure is too low. This has the important consequence that a particular MACS not only determines the abundance of the respective nuclide (as shown in Equation (8.4)) but also affects the abundances of all heavier nuclides following it in the s-process path as well. Thus, the whole range of nuclides produced in the weak component is very sensitive to the actual values of the neutron capture cross sections.

Despite of the initial success of the classical model, with the advent of very precise measurements of $\langle\sigma^*\rangle$ since the late 1990s discrepancies between measured abundances and the ones predicted in the classical model became apparent. They showed in abundances for s-only nuclides in the main path as well as in s-process branchings and proved that the assumption of a superposition of neutron exposures at constant temperature is too simplistic. Sophisticated models of AGB stars became available around the same time, allowing to obtain details of the neutron exposures and timescales for the s-process main component (see Section 7.3.2). The weak s-process component has been identified to originate in hydrostatic He-burning (and partially in convective C-shell burning) of massive stars (see Sections 7.3 and 7.4). Typical neutron energies were found to be around $k_B T \approx 8$ keV and 25–30 keV for interpulse and He-shell flash burning, respectively, in AGB stars, and $k_B T \approx 90$ keV in massive stars.

The s-process is an example of a secondary process in nucleosynthesis. It is stronger in stars with higher metallicity because it relies on the existence of seed nuclei. A nucleosynthesis process building on pre-existing abundances is called a *secondary process* and it synthesizes *secondary nuclides*. A *primary nuclide* is one directly produced from H and He, such as ^{12}C and ^{16}O in He-burning (Section 7.3). Primary processes modify abundances already in the early Galaxy whereas secondary processes only act at later times, based on the previously built abundances (see also Chapter 11). A precise reproduction of s-process abundances is also important to be able to determine the size of the r-process contribution required to explain the solar system abundances of nuclides which have received contributions from both processes. This helps to constrain the r-process, which is less well modeled than the s-process (see Section 8.3). It has to be cautioned, however, that the solar system abundances are the result of many events over the history of the Galaxy and therefore straightforwardly disentangling the s- and r-process contributions based on solar abundances may remain problematic (see Chapters 5 and 11).

8.2.2 Branchings in the s-Process Path

There are 15 places in the s-process path along stability where the decay lifetimes $\tau_{\text{life}}^{\beta}$ of certain nuclides are so long that they become comparable to their lifetimes $\tau_{\text{life}}^{(n,\gamma)}$ against neutron capture. These give rise to branchings in the s-process flow. Such branching features can be seen in Figure 8.3. The full list of branchings with their connected nuclides is shown in Table 8.1.

An example for the detailed reaction flow in such a branching is shown in Figure 8.4 for the ^{192}Ir branching. It can be seen that the s-only nuclide ^{192}Pt can only be reached by going through the β^- decays of ^{191}Os and ^{192}Ir. If neutron captures on these nuclides are faster than the decays, ^{192}Pt is bypassed. How much ^{192}Pt is produced in relation to the other Pt isotopes depends sensitively on the ratios $\tau_{\text{life}}^{\beta}/\tau_{\text{life}}^{(n,\gamma)}$ for ^{191}Os and ^{192}Ir. The neutron capture rates depend on the density ρ (via $\bar{n}_n$) whereas the decays do not and thus such a branching can be used to determine the (average) neutron density during the s-process. Many decays are also temperature dependent and therefore a combination of several branchings allows to determine both temperature and density. For the shown example, however, the temperature dependence is weak. In the branching around ^{192}Ir the main branch point (and the main uncertainty) is in ^{192}Ir. This is because due to the comparatively short decay lifetime of ^{191}Os a considerable fraction of it will decay before capturing a neutron. The lifetime of ^{192}Ir, on the other hand, is long enough for neutron

Table 8.1. The Isotopic Chains of the s-Process Branchings

#	Name	Involved Nuclides
(1)	^{63}Ni	$^{62,\underline{\mathbf{63}},64}$Ni, $^{63,\underline{\mathit{64}},65}$Cu, $^{64,\underline{65},66}$Zn
(2)	^{79}Se	$^{78,\underline{79},80}$Se, $^{79,\underline{\mathit{80}},81}$Br, $^{\overline{80},\underline{81},\overline{82}}$Kr
(3)	^{85}Kr	$^{84,\underline{\mathbf{85}},\underline{\mathit{85m}},86}$Kr, $^{85,\underline{86},87}$Rb, $^{\overline{86},\overline{87},88}$Sr
(4)	^{95}Zr	$^{94,\underline{\mathbf{95}},96}$Zr, $^{95,\overline{96},97}$Mo
(5)	^{99}Tc	$^{98,\underline{\mathbf{99}},64}$Tc, $^{99,\overline{100}}$Ru
(6)	^{134}Cs	$^{132,\underline{133},134}$Xe, $^{133,\underline{\mathbf{134}},\underline{135}}$Cs, $^{\overline{134},135,\overline{136}}$Ba
(7)	^{148}Pm	$^{146,\underline{147},148}$Nd, $^{\underline{147},\underline{\mathbf{148}},\underline{\mathit{148m}}}$Pm, $^{147,\overline{148},149,\overline{150}}$Sm
(8)	^{151}Sm	$^{\overline{150},\underline{\mathbf{151}},152}$Sm, $^{151,\underline{\mathit{152}},152,\underline{\mathbf{154}}}$Eu, $^{\overline{152},\underline{153},\overline{154},155}$Gd
(9)	^{163}Ho	^{163,164}Dy, $^{\underline{\mathbf{163}},\underline{\mathit{164}},165}$Ho, $^{\overline{164}}$Er
(10)	^{170}Tm	$^{168,\underline{169},170}$Er, $^{169,\underline{\mathbf{170}},\underline{171}}$Tm, $^{\overline{170},171,172}$Yb
(11)	^{176}Lu	$^{174,\underline{175},176}$Yb, $^{175,\underline{\mathbf{176m}},\underline{176}}$Lu, $^{\overline{176},177}$Hf
(12)	^{179}Hf$-$^{179}Ta	$^{179,180,\underline{\mathbf{181}},\underline{\mathbf{182}}}$Hf, $^{\underline{\mathbf{179}},\underline{180},\underline{\mathit{180m}},\underline{182}}$Ta
(13)	^{185}W	$^{184,\underline{\mathbf{185}},186}$W, $^{185,\underline{186},187}$Re, $^{\overline{186},\overline{187},188}$Os
(14)	^{192}Ir	$^{190,\underline{191},192}$Os, $^{191,\underline{\mathbf{192}},192}$Ir, $^{\overline{192},\underline{193},194}$Pt
(15)	^{204}Tl	$^{203,\underline{\mathbf{204}},205}$Tl, $^{\overline{204},\underline{205},206}$Pb

Notes. Overlined mass numbers identify stable s-only nuclides, underlined mass numbers indicate unstable nuclides. Among the unstable nuclides, the main branch point is shown in bold face. Italics indicate a competition between several decay modes (among EC, β^+, and/or β^-).

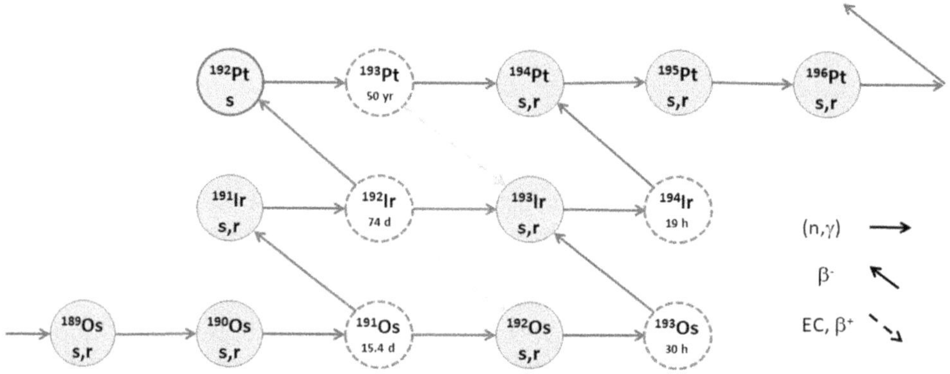

Figure 8.4. The s-process flow in the vicinity of the ^{192}Ir branching. At ^{191}Os and ^{192}Ir neutron capture and β^- decay are competing, with the main branching at ^{192}Ir. For unstable nuclides the half-lives are given. For stable nuclides it is indicated whether they receive a contribution from the r-process or are s-only.

captures to be significant. Finally, although ^{193}Pt can be considered as being stable during s-processing (i.e., as long as a neutron flux is available), any produced ^{193}Pt will decay to ^{193}Ir when the neutron irradiation has ceased.

In order to use branchings to determine the detailed conditions ($\bar{n}_n$, T) of s-process nucleosynthesis, a highly accurate and precise knowledge of neutron capture cross sections and decay half-lives is crucial. Furthermore, also the temperature dependence of the rates (also for decays) has to be known. This includes the modifications of the cross sections and lifetimes by stellar plasma effects, as discussed in Sections 4.1.3 and 4.1.7. It is to be noted that although the s-process temperature range is comparatively low, thermal population of excited states may still play a role due to the high intrinsic level-density (and low level-spacing) of intermediate mass and heavy nuclides. This is especially pronounced in the rare-earth region with their deformed nuclei. It implies that the ground-state contribution to the stellar rate (see Section 4.1.3) may be small and therefore the uncertainty in such a rate may still be dominated by theory uncertainties even when the ground-state cross section has been measured with high precision.

8.3 The r-Process

8.3.1 The Classical r-Process

In order to explain the solar abundance pattern beyond Fe, another neutron capture process had to be postulated, supplementing the s-process nucleosynthesis. The r-process accounts for additional peaks in the abundance distribution, shifted to lower mass number with respect to the s-process peaks, see Figures 8.1 and 8.2. It also explains the existence of neutron-rich isotopes of an element which are separated from the s-process path by short-lived isotopes. These are called r-only nuclides, in analogy to s-only nuclides, because they receive a contribution only from the r-process whereas both s- and r-process contributed to the majority of other nuclides. As the s-only nuclides are important for the s-process, so are the r-only nuclides important to pin down the conditions of the r-process and the properties of their extremely neutron-rich progenitors.

To reproduce the location of the r-process peaks, neutron number densities larger than 10^{20} cm^{-3} and temperatures around 1 GK are required. At such conditions, (n,γ)–(γ,n) equilibrium is established within each isotope chain. The details of this equilibrium and the resulting abundance distribution within an isotope chain are extensively discussed in Section 4.3.3. Only one or two isotopes in the chain receive non-negligible abundances. The flow into the next isotope chain with higher charge number Z is delayed until these isotopes, the waiting points, decay. Equation (4.150) gives the abundance for each isotope Y_x in an isotope chain. It does not describe the flow into the next isotope chain by β^- decays. Since the β^- decays are decoupled from the captures and photodisintegrations in an (n,γ)–(γ,n) equilibrium, another set of differential equations can be used to describe that flow. The change of total abundance in a chain is the difference between what is fed into the chain by β^- decays from the chain with $Z - 1$ and what is flowing out of the chain by β^- decays of the populated isotopes,

$$\frac{\mathrm{d}_Z Y}{\mathrm{d}t} = \left[{}_{Z-1}Y \sum_x {}_{Z-1}\tilde{\mathcal{P}}_{xZ-1}\lambda^*_{\mathrm{dec},x}\right] - \left[{}_Z Y \sum_x {}_Z\tilde{\mathcal{P}}_{xZ}\lambda^*_{\mathrm{dec},x}\right], \tag{8.9}$$

with ${}_Z\lambda^*_{\mathrm{dec},x}$ being the (stellar) decay rate of isotope x with proton number Z. The abundance ${}_Z Y_x$ of isotope x is expressed as a fraction of the composition (i.e., total abundance) ${}_Z Y = \sum_x {}_Z Y_x$ (see also Equation (1.15)),

$${}_Z Y_x = {}_Z\tilde{\mathcal{P}}_{xZ}Y. \tag{8.10}$$

The population ${}_Z\tilde{\mathcal{P}}_x = {}_Z\tilde{\mathcal{P}}_x(\tilde{n}_\mathrm{n}, T, {}_ZS_\mathrm{n}^x)$ for each isotope with neutron separation energy ${}_ZS_\mathrm{n}^x$ in the chain can be found by comparison to Equation (4.150).

Equation (8.9) is a reduced network for the r-process with as many equations as there are elements (defining the Z range) included. It can be further simplified by assuming a steady flow equilibrium as defined in Section 4.3.4. When the β^- decays have had time to establish a steady flow, then ${}_Z\dot{Y} = 0$ for each Z and the flow into and out of an isotope chain is the same,

$$ {}_{Z-1}Y\, {}_{Z-1}\lambda_\mathrm{dec}^{*,\mathrm{eff}} = {}_ZY\, {}_Z\lambda_\mathrm{dec}^{*,\mathrm{eff}} = \cdots = \mathrm{const}. \tag{8.11} $$

The above equation was written using an effective decay rate for a Z-chain,

$$ {}_Z\lambda_\mathrm{dec}^{*,\mathrm{eff}} = \sum_x {}_Z\tilde{\mathcal{P}}_x\, {}_Z\lambda_{\mathrm{dec},x}^{*}. \tag{8.12} $$

The steady flow constant can be determined from knowing ${}_ZY$ for one chain, for example, for a seed region and the abundances in the other chains follow immediately. The seed may be different depending on the actual site of the r-process. It could, for example, originate from an NSE (Section 4.3.2). It is to be noted that neither Equation (8.9) nor Equation (8.11) require the knowledge of neutron capture rates or their reverse reactions. This is a consequence of the (n,γ)–(γ,n) equilibrium. The neutron separation energies (entering via the population coefficient) and the β^--decay lifetimes have to be known, however. Assuming these as known, the remaining free parameters $\tilde{n}_\mathrm{n}$ and T can be adjusted to obtain the observed abundance distribution.

The timescale τ_Δ^r of this r-process is given by the β^- decay half-lives of the waiting points, as their sum determines the time it takes to produce the heaviest r-process nuclides from the seed nuclides,

$$ \tau_\Delta^\mathrm{r} = \frac{1}{\sum_Z {}_Z\lambda_\mathrm{dec}^{*,\mathrm{eff}}}. \tag{8.13} $$

This is not to be confused with the duration of an r-process, which is the time the required conditions are upheld. This is given by external, hydrodynamic conditions and can be much longer than τ_Δ^r. At very neutron-rich conditions, producing very short-lived nuclides, also τ_Δ^r becomes short. Then nucleosynthesis runs up to the region of fissionable nuclei and this can lead to *fission cycling*. Depending on the exact conditions and the relation of the process duration and τ_Δ^r several or many cycles can occur, in which fragments of the fissioned nuclides capture neutrons again and again to synthesize further fissionable nuclides (see Section 8.3.2).

It was found that at least three steady-flow components with $\tilde{n}_\mathrm{n} = 10^{20}, 10^{22}, 10^{24}$ g cm^{-3} at $T = 1.35$ GK are required to reproduce the three r-process abundance peaks at $A = 80, 130, 195$. In between the peaks steady flow is well established. In perfect steady-flow conditions the abundances would show similar horizontal abundance plateaus as found in the s-process (Section 8.2). In the s-process there is steady flow in the neutron captures whereas in the r-process it is in the β^- decays (*β-flow equilibrium*). As it can further be seen in Figure 8.2, there is stronger odd–even staggering in the r-process abundances than for the ones from the s-process, which is a reflection of the sensitivity to neutron separation energies within an isotope chain. Neutron separation energies are larger for nuclides with even neutron numbers, which consequently receive higher abundances. Because of this, there is also a direct correlation between abundances and $_Z\lambda_{\text{dec}}^{*,\text{eff}}$. Again similar to the s-process, the plateaus can become "tilted" for incomplete steady flow and with the superposition of more components. This has to be tracked by a dynamic network calculation (see Section 8.3.2).

Although the abundance maxima in each chain depend on $\bar{n}_\text{n}$ and T, using Equation (4.150) it can be seen that the neutron separation energy S_n has to be the same for the maxima in all chains for given $\bar{n}_\text{n}$ and T. Setting the left-hand side of Equation (4.150) to unity and rearranging the equation yields a relation for the neutron separation energy of the nuclei with the highest abundance in each isotope chain,

$$S_\text{n}(Y_\text{max}) = \frac{T_9}{5.04}\left(34.075 - \log \bar{n}_\text{n} + \frac{2 \log T_9}{2}\right). \tag{8.14}$$

The numerical constants used here imply that the neutron separation energy S_n is obtained in MeV when $\bar{n}_\text{n}$ is input in cm^{-3} and T_9 is the temperature in GK, as usual. For the three components reproducing the r-process peaks it is found that $S_\text{n} = 3.81, 3.28, 2.75$ MeV (Freiburghaus et al. 1999). The exact value does not only depend on the location of the r-process peaks for stable nuclides (which are built when the neutron-rich progenitor nuclides decay back to stability) but also on the mass model used for the unknown masses far off the line of stability. Different mass models predict a different evolution of S_n when moving away from stability. Figure 8.5 illustrates this for two models. The features in the contour lines directly affect the location of the abundance maximum in an isotope chain and therefore also the r-process abundance distribution after this has decayed back to stability. For example, the saddle point structures seen in the two examples before the $N = 50$

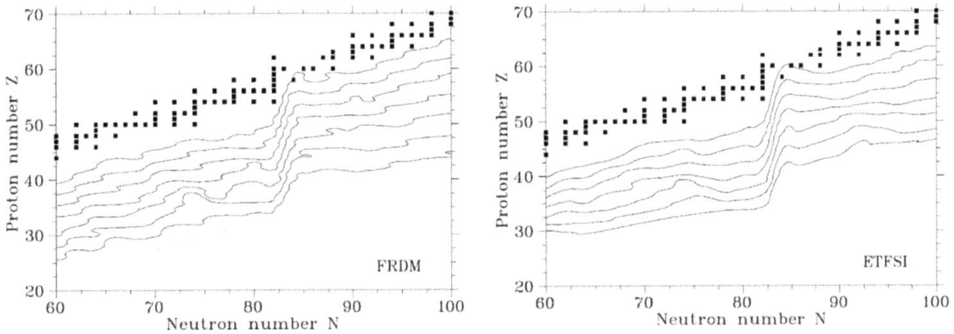

Figure 8.5. Constant neutron separation energy contours for two mass models, FRDM (left) and ETFSI-1 (right). Shown are the contours for $S_\text{n} = 7, 6, \ldots, 1$ MeV, moving away from stability, for nuclides with even neutron number. Stable nuclides are indicated by black squares. (Reproduced from Freiburghaus et al. 1999. © 1999. The American Astronomical Society. All rights reserved.)

shell closure lead to a considerable underproduction of r-process nuclides before the peaks (abundance "troughs"). This is seen in Figure 8.6 for the ETFSI-1 mass model. In comparison, if the nuclear shell structure is not as pronounced far off stability as it is close to stability (*shell quenching*) this problem would be alleviated. Figure 8.6 also shows an example for a mass model with artificial shell quenching, ETFSI-Q, based on the ETFSI-1 mass model. So far, there is no experimental evidence for a general quenching of nuclear shells for nuclides far away from stability. Therefore it is more likely that the mass models have problems to realistically describe the transition from deformed to spherical nuclides around the magic numbers, leading to spurious saddle point features. This is a nice example of how investigations in nuclear astrophysics inform and motivate improvements in nuclear physics.

8.3.2 Dynamical r-Process Calculations

The high neutron number density and temperature required to reproduce the r-process abundance peaks indicates an explosive environment, as such conditions are not achieved in hydrostatic burning. Since nuclides with short half-lives far from stability are produced, the short processing timescale ($\tau_{\Delta}^{\mathrm{r}} \lesssim 1$ s) is compatible with an explosion. Regardless of the specific site, $\bar{n}_{\mathrm{n}}$ and T will be strongly time-dependent, with an initially hot, dense phase finally cooling down. In the classical approach, instantaneous freeze-out is assumed, that means that the neutron supply is sharply cut off, the temperature drops very rapidly, and the short-lived nuclides produced in the (n,γ)–(γ,n) equilibrium directly decay to their stable descendants. Obviously, this is a crude simplification and the full nucleosynthesis details are only revealed when performing a dynamical calculation with a full reaction network, containing about 3000 nuclides and their connecting reactions. On the other hand, the fact that the solar r-abundances are already well reproduced by the classical model suggests that

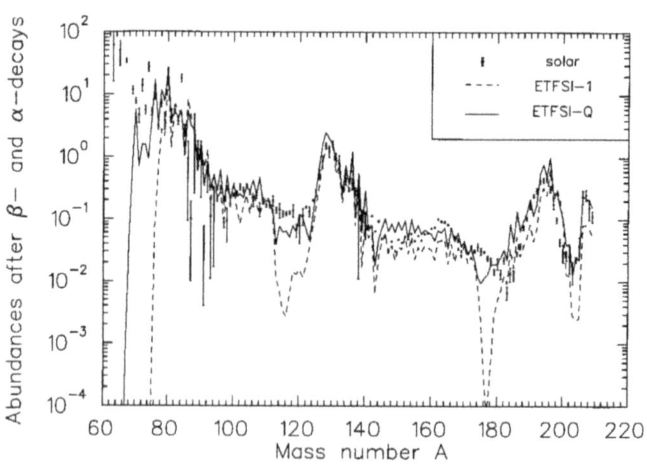

Figure 8.6. Fits to solar r-abundances with 10 different S_{n} contours for two mass models, one with artificial shell quenching (ETFSI-Q). (Reproduced from Freiburghaus et al. 1999. © 1999. The American Astronomical Society. All rights reserved.)

indeed most of the r-process nucleosynthesis takes place in equilibrium conditions and that indeed the freeze-out is fast. This has been confirmed in dynamical r-process simulations using extended reaction networks.

The first assumption to relax is the one of β-flow equilibrium. Performing time-dependent calculations, but still assuming (n,γ)–(γ,n) equilibrium (Equation (8.9)), and using the three components specified above with the (realistic) durations of 1.2, 1.6, and 2.15 s again leads to an establishment of steady flow in between the magic neutron numbers. This is because the $_Z\lambda_{\rm dec}^{*,\rm eff}$ are much smaller than the process duration, except for the longest half-lives in the progenitors of the nuclides forming the peaks because they are produced closest to stability. Considerations of more components does not change this general picture and only slightly improves the reproduction of solar abundances (Figure 8.6).

When performing full simulations with extended reaction networks, the time dependence of $\tilde{n}_{\rm n}$ and T has to be modeled. In principle, this dependence is set by the specific site and explosion type. This also sets the seed composition which, in turn, is determined by $Y_{\rm e}$. Site-independent studies leave $Y_{\rm e}$ and initial values for ρ and T as free parameters. A realistic assumption for all explosive environments is the *adiabatic expansion of hot matter*. Therein, the two parameters $\tilde{n}$ and T can be treated as related to each other by using the entropy (Equation (1.28)). The entropy at high T is the radiation entropy S_γ (Equation (1.76)) but at very high T and low ρ the pressure contribution of electron-positron pairs also has to be considered (see Chapter 6 and Figure 6.2). Therefore it is advantageous to use the expression given in Equation (1.96) for a combined entropy S accounting for ultra-relativistic bosons as well as ultra-relativistic fermions in the hot plasma. In an adiabatic (S = const) expansion, the temperature T and volume V are related by

$$T(t) = T_0 \left[\frac{V_0}{V(t)} \right]^{\tilde{\Gamma}-1}, \tag{8.15}$$

with T_0 and V_0 being the initial values of temperature and volume, respectively. In spherical symmetry,

$$V(t) = \frac{4\pi R^3(t)}{3} = \frac{4\pi}{3}[R_0 + v_{\rm ex}t]^3, \tag{8.16}$$

with the initial radius R_0 and the expansion speed $v_{\rm ex} = {\rm d}R/{\rm d}t$. Using $\tilde{\Gamma} = 4/3$ (Section 1.5.3), the temperature evolution is then given by

$$T(t) = \frac{T_0 R_0}{R_0 + v_{\rm ex}t} \tag{8.17}$$

and the density is obtained from Equation (1.96) when using the entropy S as a parameter. A choice of R_0 and $v_{\rm ex}$ is equivalent to the definition of an expansion timescale

$$\tau_{\rm ex} = \frac{R_0[{\rm e} - 1]}{v_{\rm ex}} \tag{8.18}$$

after which T has dropped to 1/e of its initial value, that means $T(\tau_{ex}) = T_0/e$. For example, $v_{ex} = 4500$ km s^{-1} corresponds to $\tau_{ex} = 50$ ms.

A comparison of the abundances obtained with the classical approach, assuming (n,γ)–(γ,n) equilibrium, and a fully dynamical approach studying nucleosynthesis in adiabatically expanding matter with a complete reaction network is shown in Figure 8.7. This is a snapshot after 1 s processing time, shortly before the freeze-out of the r-process, and the conditions were those appropriate for a low-entropy component from a neutron star merger (Section 9.4.5), corresponding to $S_n \simeq 2$ MeV, with processing running into the region of fissionable nuclides. The two calculations show similar abundance maxima in each isotope chain, showing that equilibrium is achieved. There are maximum deviations of about a factor of two for two groups of nuclides around $N = 100$ and $N = 140$ but the majority of abundances shows much better agreement. The averaged lifetimes (according to the definition in Equation (4.116)) of the nuclides against neutron captures, (γ,n) reactions, and β^- decays are compared in Figure 8.8 for the same conditions as used for Figure 8.7. It is obvious that (n,γ) and (γ,n) reactions are comparably fast and both are much faster than the decays up to a little longer than 1 s. In this phase (n,γ) and (γ,n) are also faster by many orders of magnitude than the overall process duration. These are the prerequisites for an (n,γ)–(γ,n) equilibrium as discussed in Section 4.3.3. The second phase is the freeze-out phase, after about 1 s processing time, when the neutron reactions become slower than the β^- decays. In this late phase, there is a brief opportunity for neutron captures to modify the abundances produced in equilibrium. How fast the freeze-out is depends on the detailed

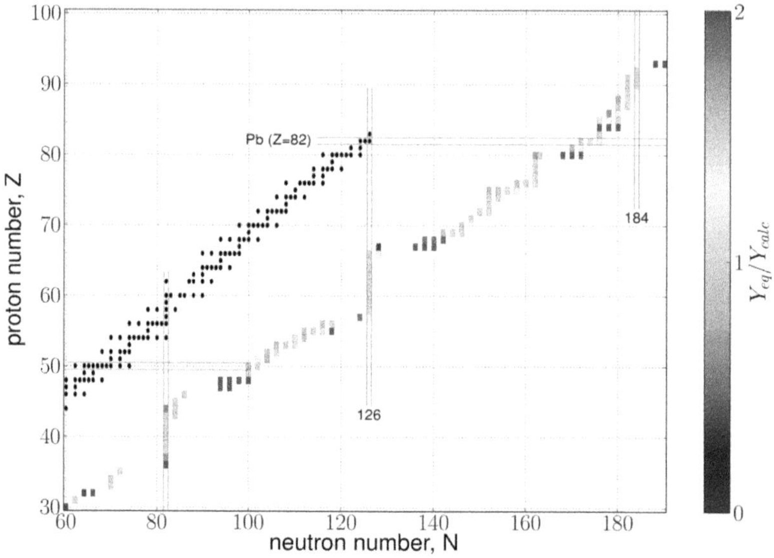

Figure 8.7. Color-coded comparison of the abundances far from stability obtained during an r-process calculated with a full network (Y_{calc}) to abundances Y_{eq} obtained from assuming (n,γ)–(γ,n) equilibrium. Stable nuclides are indicated by black dots. (Reproduced from Eichler et al. 2015. © 2015. The American Astronomical Society. All rights reserved.)

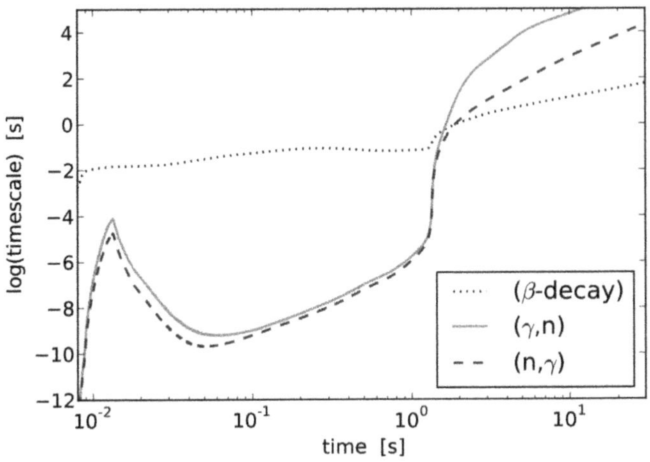

Figure 8.8. Averaged lifetimes $\bar{\tau}_{\rm life}$ during r-process nucleosynthesis. (Reproduced from Eichler et al. 2015.

hydrodynamic conditions, that is, on the entropy S and expansion timescale $\tau_{\rm ex}$. Photodisintegrations always freeze-out faster than the captures.

Another way to quantify the efficiency of the r-process is the relation between free neutrons and the amount of seed nuclei to capture neutrons. The ratio of free neutrons to seed abundances $r_{\rm seed} = Y_{\rm n}/Y_{\rm seed}$ is also set by the entropy S and expansion timescale $\tau_{\rm ex}$. Matter under explosive conditions sufficiently hot to allow for an r-process will go through a NSE phase with all reactions, including charged-particle reactions, in equilibrium (see Section 4.3.2). During the adiabatic expansion of the hot bubble, the abundances adjust to the varying temperature and density. Reactions involving charged particles (protons, α-particles) are the first ones to cease when the temperature is dropping during the expansion. Then only reactions involving neutrons and decays can act. If there is a sufficient amount of neutrons left over after charged-particle freeze-out, an r-process can be initiated on the abundance distribution present at that time, the seed. Such a freeze-out produces nuclides in the mass range $50 \lesssim A \lesssim 100$ (according to the NSE discussed in Section 4.3.2), depending on S and $Y_{\rm e}$. Especially the density plays an important role. Starting from very hot conditions with high abundances of neutrons, protons, and α-particles present (either initially or from dissociation of heavier nuclides, see Section 7.7), it depends on the density whether α-particles can be fully converted into heavier nuclides or not. This is because the bottleneck for processing the α-particles is the triple-α reaction, with its rate strongly depending on density (see also Section 7.3). According to Equations (1.76) and (1.96), high S implies low ρ (the temperature is set by the charged-particle freeze-out temperature, which depends on the Coulomb barriers and is around 3–4 GK). At low density, the triple-α rate is slow, even at high T. An alternative to the triple-α reaction would be $\alpha + \alpha + {\rm n} \rightarrow {}^9{\rm Be}$ but it is similarly depending on density. Therefore full NSE is not achieved. Rather, a QSE (see Section 4.3.2) with one QSE group with light particles (neutrons, protons, and foremost α-particles) and a second QSE group with heavier nuclides

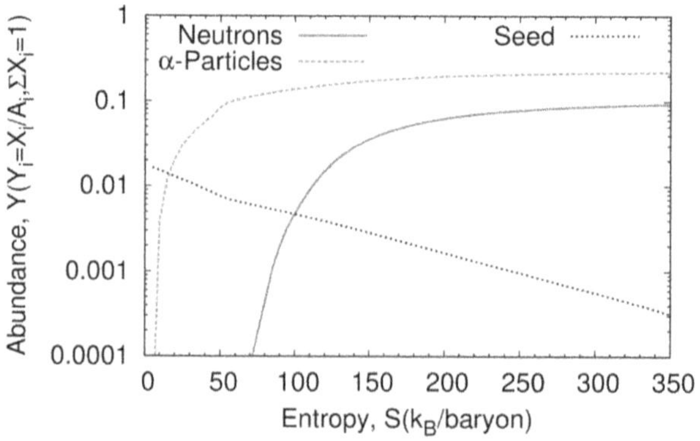

Figure 8.9. Remaining abundances after charged-particle freeze-out from NSE or QSE as function of entropy per baryon S^b for an adiabatic expansion with v_{ex} = 7500 km s^{-1} and initial $Y_e = 0.45$. (Reproduced from Farouqi et al. 2010. © 2010. The American Astronomical Society. All rights reserved.)

(the seed nuclides) is formed, with their relative abundances given by the speed of the connecting reaction(s), which basically is the triple-α reaction. Figure 8.9 shows this dependence of the abundances of neutrons, α-particles, and seed nuclides on the entropy. Therefore, at high entropy, a large fraction of the total composition still is in α-particles and the relative amount of seed nuclides is low. This has been coined *α-rich freeze-out* and is typical for conditions encountered in core-collapse supernova (see Section 9.3). A shorter τ_{ex} (i.e., a higher v_{ex}) has a similar effect as a lower density because then less time is available to assemble heavier nuclides.

The neutron-to-seed ratio r_{seed} determines how many neutrons per seed nuclide are available for capture. Thus, it determines whether an r-process can occur at all and how far up in mass number it can go. To reach the r-process peak at $A = 130$, $r_{seed} \approx 30$ is needed. To reach the actinide region is only possible with $r_{seed} \approx 150$. For $Y_e = 0.45$, this translates into entropies per baryon ($S^b = S/(N_A k_B)$) $S^b = 200$ k$_B$/baryon and 350 k$_B$/baryon, respectively. The value of r_{seed} at a given entropy also depends on Y_e, which determines how neutron-rich the NSE matter is (see Section 4.3.2). This leads to the interesting result that for very low Y_e much lower entropies and a much more restricted entropy range is required to reproduce the solar abundances across a large range of mass numbers. Figure 8.10 shows how r_{seed} depends on S and Y_e. Entropies below 0.15 k$_B$/baryon lead to a normal freeze-out from a full NSE, not an α-rich freeze-out. Then r_{seed} becomes independent of S and only depends on Y_e. This is why the lines for r_{seed} flatten toward low entropy in the right panel of Figure 8.10.

Although the implications of r_{seed} are generally valid, the above discussion focused on a "hot" r-process, in which the temperature during the r-processing is sufficiently high to sustain the (n,γ)–(γ,n) equilibrium during most of the neutron processing phase. An exploration of the parameter space with dynamical r-process simulations finds that the r-process can also operate under "cold" conditions in very neutron-rich matter. This can occur in environments where, after the initial seed building by NSE or QSE, the temperature drops very quickly so that photodisintegrations are too

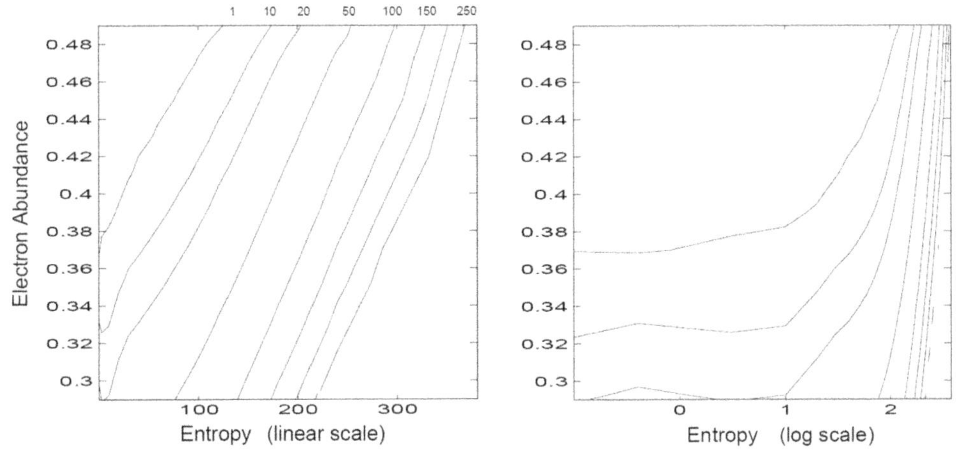

Figure 8.10. Lines of constant $r_{\rm seed}$ as function of initial S and Y_e for an adiabatic expansion timescale of 0.05 s. The right panel uses a logarithmic scale on the entropy axis. The entropy is given as entropy per baryon ($S^{\rm b}$). (Reproduced from Freiburghaus et al. 1999. © 1999. The American Astronomical Society. All rights reserved.)

slow to play a role. In the presence of high neutron number densities $\bar{n}_{\rm n}$, neutron captures then are still fast and produce very neutron-rich nuclides. The production within an isotope chain is then a competition between neutron captures and β^- decays. Toward the neutron dripline (see Section 3.2), nuclides become very short-lived and decays become faster than the neutron captures. For extremely neutron-rich conditions, nuclides at the neutron dripline could be produced but this would result in abundance patterns different than solar. Again, a simplified reaction network can be used to describe a cold r-process. It is not sufficient anymore to consider only separated isotope chains. Instead of using Equation (4.149), neutron captures and β^- decays have to be included explicitly,

$$\begin{aligned}\dot{Y}_{Z,N} = \rho {\rm N_A} \langle \sigma^* v \rangle^{\rm cap}_{Z,N-1} Y_{\rm n} Y_{Z,N-1} + \lambda^{*,\gamma}_{Z,N+1} Y_{Z,N+1} + \lambda^{*,{\rm dec}}_{Z-1,N+1} Y_{Z-1,N+1} \\ - \rho {\rm N_A} \langle \sigma^* v \rangle^{\rm cap}_{Z,N} Y_{\rm n} Y_{Z,N} - \lambda^{*,\gamma}_{Z,N} Y_{Z,N} - \lambda^{*,{\rm dec}}_{Z,N} Y_{Z,N},\end{aligned} \tag{8.19}$$

with the neutron capture reactivity $\langle \sigma^* v \rangle^{\rm cap}_{Z,N}$, the photodisintegration rate per nucleus $\lambda^{*,\gamma}_{Z,N}$, and the β^--decay rate per nucleus $\lambda^{*,{\rm dec}}_{Z,N}$ for a nucleus with proton number Z and neutron number N. Neutron capture and neutron emission by photodisintegration are related by the reciprocity relation shown in Equation (4.52). The change of the neutron abundance is given by

$$\dot{Y}_{\rm n} = \left(\sum_Z \sum_N \lambda^{*,\gamma}_{Z,N} Y_{Z,N} \right) - \left(\sum_Z \sum_N \rho {\rm N_A} \langle \sigma^* v \rangle^{\rm cap}_{Z,N} Y_{\rm n} Y_{Z,N} \right), \tag{8.20}$$

with the sums running over all nuclides in the network, not only within an isotope chain. Under cold r-process conditions, Equation (8.19) actually splits into two independent equations, one for the total abundance in an isotope chain ${}_{Z}Y$ as before and one for the total abundance in an isobar chain ${}_{A}Y$,

$$ {}_{Z}\dot{Y} = {}_{Z-1}\lambda_{\rm dec}^{*,\rm eff}\ {}_{Z-1}Y - {}_{Z}\lambda_{\rm dec}^{*,\rm eff}\ {}_{Z}Y, \tag{8.21}$$

$$ {}_{A}\dot{Y} = \bar{n}_{\rm n}\langle\sigma^{*}v\rangle_{A-1}^{\rm cap,eff}\ {}_{A-1}Y - \bar{n}_{\rm n}\langle\sigma^{*}v\rangle_{A}^{\rm cap,eff}\ {}_{A}Y. \tag{8.22}$$

Here, the effective neutron capture reactivity for a chain of isotones is defined analogously to the effective decay rate,

$$\langle\sigma^{*}v\rangle_{A}^{\rm cap,eff} = \frac{1}{{}_{A}Y}\sum_{Z}\langle\sigma^{*}v\rangle_{Z,N}^{\rm cap}\, Y_{Z,N}. \tag{8.23}$$

If the process timescale is sufficiently long compared to β^{-}-decay and neutron-capture lifetimes, ${}_{Z}\dot{Y} = {}_{A}\dot{Y} = 0$ and a steady-flow equilibrium is established also in a cold r-process. In this case, it is a combined capture and β-flow equilibrium, satisfying for each Z and A, respectively,

$$ {}_{Z-1}\lambda_{\rm dec}^{*,\rm eff}\ {}_{Z-1}Y = {}_{Z}\lambda_{\rm dec}^{*,\rm eff}\ {}_{Z}Y, \tag{8.24}$$

$$\langle\sigma^{*}v\rangle_{A-1}^{\rm cap,eff}\ {}_{A-1}Y = \langle\sigma^{*}v\rangle_{A}^{\rm cap,eff}\ {}_{A}Y. \tag{8.25}$$

It is important to note, however, that there are no waiting points as in (n,γ)–(γ,n) equilibrium and the population of the isotopes within an isotope chain is not given by Equation (4.150) anymore. Rather, there is a competition between neutron captures and β^{-} decays and thus the decays of isotopes become important which have decay lifetimes $\tau_{\rm life}^{\rm dec} \lesssim \tau_{\rm life}^{(\rm n,\gamma)}$. The flow into the next isotope chain occurs at the first (few) isotope(s) fulfilling that condition. Therefore the additional condition

$$\bar{n}_{\rm n}\langle\sigma^{*}v\rangle_{A}^{\rm cap,eff}\ {}_{A}Y \approx {}_{Z}\lambda_{\rm dec}^{*,\rm eff}\ {}_{Z}Y \tag{8.26}$$

holds for the combined capture and decay steady-flow. Nevertheless, the timescale $\tau_{\Delta}^{\rm r}$ to get to heavier nuclides again is determined by the sum of effective decay rates ${}_{Z}\lambda_{\rm dec}^{*,\rm eff}$ as given in Equation (8.13). Because of Equation (8.26), the timescale alternatively can be calculated by summing over the effective neutron capture reactivities defined in Equation (8.23). According to Equations (8.24) and (8.25), the peaks in the abundance distribution obtained in a cold r-process correspond to the nuclides with the longest β^{-}-decay lifetimes and the longest neutron-capture lifetimes in an isotope chain and an isobar chain, respectively.

Provided a large enough $r_{\rm seed}$, how far can the r-process go? As mentioned above, with a sufficient neutron supply, the r-process can synthesize nuclides also in the actinide region. Whether there is a definitive limiting endpoint still is an open question. When the r-process runs into the region of fissionable nuclides, a complicated competition between β^{-} decays, α decays, neutron captures, and several fission modes (spontaneous fission of thermally excited nuclides, neutron-induced fission, β-delayed fission) ensues. It has been shown that neutron-induced fission is the dominating fission channel. An important ingredient to calculate the rate for neutron-induced fission is the fission barrier (see Section 3.8). At r-process conditions, the fission process occurs at energies just above the fission barrier, which makes barrier tunneling unimportant and simplifies the predictions. To date, nevertheless, there are large discrepancies between the fission barriers obtained with various nuclear models. This makes it impossible to draw any definitive conclusions on whether there is an endpoint and where it would be located. The nuclides ^{232}Th, ^{234}U, ^{235}U, and ^{238}U are found naturally in the solar system and therefore the r-process should at least make progenitor nuclides of these.

The knowledge of fission properties is further important in the context of fission cycling. If r_{seed} is sufficiently large to reach fissionable nuclides the fission fragments can again capture neutrons and, provided the process timescale is long enough, again produce fissionable nuclides. Assuming that each nucleus fissions into two fragments, the mass fraction of heavy r-process nuclides is doubled with each fission cycle, $X_{\mathrm{r}} = 2^n X_{\mathrm{seed}}$ with n being the number of completed cycles. This can lead to a strong enhancement of r-process abundances when the process timescale is long. The details of how fission cycling impacts the abundance pattern, in particular close to the freeze-out phase, depends not only on the fission rates but also on the fission fragment distribution (see Section 3.8), including heavy fragments as well as neutrons. This is currently not well known.

Although the main r-processing occurs far from stability, the abundance pattern can be modified during the r-process freeze-out phase, when $\tilde{n}_{\mathrm{n}}$ and T are dropping. As shown in Figure 8.8, neutron captures can briefly become important when the (n,γ)–(γ,n) equilibrium is left. Even if the initial supply of free neutrons has been exhausted during the r-process, neutrons released by fission and by β-delayed neutron emission are important in smoothing the strong odd–even staggering in the original equilibrium abundances. In the presence of neutron emission from decay- and fission-processes, the reaction network cannot be reduced anymore to only locally include rates within an isotope (or isobar) chain. Rather, the neutrons emitted from any other nucleus have to be taken into account. Therefore, Equation (8.19) has to be extended to include the neutron-emitting processes,

$$\begin{aligned}
\dot{Y}_{Z,N} = {} & \rho \mathrm{N_A} \langle \sigma^* v \rangle^{\mathrm{cap}}_{Z,N-1} Y_{\mathrm{n}} Y_{Z,N-1} + \lambda^{*,\gamma}_{Z,N+1} Y_{Z,N+1} \\
& - Y_{Z,N} \left[\rho \mathrm{N_A} \langle \sigma^* v \rangle^{\mathrm{cap}}_{Z,N} Y_{\mathrm{n}} + \lambda^{*,\gamma}_{Z,N} \right] \\
& - Y_{Z,N} \left[\sum_{k_{\mathrm{n}}=0}^{k_{\mathrm{n,max}}} \left({}^{k_{\mathrm{n}}}\lambda^{*,\mathrm{dec}}_{Z,N} + \rho \mathrm{N_A} {}^{k_{\mathrm{n}}}\langle \sigma^* v \rangle^{\mathrm{nf}}_{Z,N} Y_{\mathrm{n}} + {}^{k_{\mathrm{n}}}\lambda^{*,\beta\mathrm{f}}_{Z,N} \right) \right] \\
& + \sum_{k_{\mathrm{n}}=0}^{k_{\mathrm{n,max}}} {}^{k_{\mathrm{n}}}\lambda^{*,\mathrm{dec}}_{Z-1,N+k_{\mathrm{n}}} Y_{Z-1,N+k_{\mathrm{n}}} \\
& + \sum_{Z'} \sum_{N'} \tilde{\Theta}(1 - \tilde{\delta}_{Z'Z} - \tilde{\delta}_{N'N}) Y_{Z',N'} \\
& \times \sum_{k_{\mathrm{n}}=0}^{k_{\mathrm{n,max}}} \left(\rho \mathrm{N_A} {}^{k_{\mathrm{n}}}\langle \sigma^* v \rangle^{\mathrm{nf}\to Z,N}_{Z',N'} Y_{\mathrm{n}} + {}^{k_{\mathrm{n}}}\lambda^{*,\beta\mathrm{f}\to Z,N}_{Z',N'} \right),
\end{aligned} \tag{8.27}$$

with the Heaviside function $\tilde{\Theta}$ as defined in Equation (1.81). Neutrons emitted in any nuclear process not only act locally around the Z,N region in the nuclear chart where they have been released but rather act globally and can be captured by any other nucleus. The evolution for the neutron abundance then becomes

$$\begin{aligned}\dot{Y}_{\mathrm{n}} &= \left[\sum_Z\sum_N \lambda_{Z,N}^{*,\gamma} Y_{Z,N}\right] \\ &- \left[\sum_Z\sum_N \left(\langle\sigma^* v\rangle_{Z,N}^{\mathrm{cap}} + \sum_{k_{\mathrm{n}}=0}^{k_{\mathrm{n,max}}} {}^{k_{\mathrm{n}}}\langle\sigma^* v\rangle_{Z,N}^{\mathrm{nf}}\right)\rho N_{\mathrm{A}} Y_{\mathrm{n}} Y_{Z,N}\right] \\ &+ \sum_Z\sum_N \left[\sum_{k_{\mathrm{n}}=1}^{k_{\mathrm{n,max}}} k_{\mathrm{n}}\left({}^{k_{\mathrm{n}}}\lambda_{Z,N}^{*,\mathrm{dec}} + \rho N_{\mathrm{A}}\, {}^{k_{\mathrm{n}}}\langle\sigma^* v\rangle_{Z,N}^{\mathrm{nf}} Y_{\mathrm{n}} + {}^{k_{\mathrm{n}}}\lambda_{Z,N}^{*,\beta\mathrm{f}}\right)\right] \times Y_{Z,N}.\end{aligned} \quad (8.28)$$

In the above equations k_{n} denotes the neutron multiplicity, meaning the number of neutrons released after a decay (β-delayed neutron emission, ${}^{k_{\mathrm{n}}}\lambda_{Z,N}^{*,\mathrm{dec}}$) or in a fission process (fission neutrons). The maximum number of released neutrons considered in the reaction network is given by $k_{\mathrm{n,max}}$. It could be different for decays and different fission types but here the same cut-off is assumed for simplicity. Two types of fission are considered in Equations (8.27) and (8.28) (see also Section 3.8): neutron-induced fission ${}^{k_{\mathrm{n}}}\langle\sigma^* v\rangle_{Z,N}^{\mathrm{nf}}$ (a captured neutron triggers fission) and β-delayed fission ${}^{k_{\mathrm{n}}}\lambda_{Z,N}^{*,\beta\mathrm{f}}$ (the daughter nucleus from the β^--decay of a nucleus with (Z,N) fissions). Not only neutron absorption or release in fission has to be accounted for but also the production of a nucleus as a fission fragment. This is denoted by ${}^{k_{\mathrm{n}}}\langle\sigma^* v\rangle_{Z',N'}^{\mathrm{nf}\to Z,N}$ and ${}^{k_{\mathrm{n}}}\lambda_{Z',N'}^{*,\beta\mathrm{f}\to Z,N}$ in Equation (8.27), for a fissioning nucleus with (Z',N') yielding a fission fragment with (Z,N). The probability for such a fragment is given by the fission fragment distribution which has to be predicted from fission theory (see also Section 3.8). Fission processes are usually assumed to act only for nuclides with $Z > 83$.

Final processing during the freeze-out is of special importance for the *rare-earth peak*. This is a wide, not very pronounced peak in the mass range around $A \approx 162$ caused by the r-process, as also clearly seen in Figure 8.2. This is a region of deformed rare-earth nuclides. The rare-earth peak is, to current knowledge, not a reflection of nuclear properties in the r-process path far from stability but is formed during the freeze-out. For low r_{seed} it stems from a competition between neutron captures (also including neutrons emitted from fission processes or from excited nuclides after β^--decay) and β^- decays. For high r_{seed}, fission fragments with mass numbers of $A \approx 160$ are produced and can directly feed into the rare-earth region. The rare-earth peak might be used to probe the conditions during r-process freeze-out but this is complicated by the fact that predictions of capture and decay rates are difficult for deformed nuclei and also the fission fragment distribution is not known well. The fission neutrons released in a strong r-process with high r_{seed} may also strongly affect the other, classical r-process peaks. A too high neutron release shifts the peaks to larger A. The position of the peaks therefore both indicate the location of the r-process "path" and provide constraints on the amount of neutrons released in the freeze-out.

8.3.3 Identifying the r-Process Site

The above overview of the r-process mechanism made it clear that properties of nuclides and their nuclear reactions far from stability have to be known. Although experimental investigations are underway, they are difficult and require new facilities for the production and study of highly unstable nuclides. An experimental approach to study *reactions* with short-lived nuclei seems even farther away. Even if this would be possible, however, the r-process environment is hot enough so that thermal excitation of nuclei in the plasma cannot be neglected (see Sections 4.1.3.3 and 4.1.5). Therefore, currently there is no data on the majority of the nuclides involved in the r-process. This also affects theory because nuclear models are not constrained well. Intermediate and heavy nuclei constitute many-body systems which are currently treated mostly in phenomenological approaches, especially when concerning reaction theory. The predictive power of such approaches when approaching the dripline is not clear. On the other hand, the number of isotopes between stability and the dripline is limited and the previous successes in reproducing the solar abundance features show that the predictions cannot be too far off. Accurate nuclear physics predictions together with hydrodynamical simulations, however, are helping to pin down the actual production site of the r-process. Moreover, adding to the difficulty in experiment and theory is the fact that nuclei in different mass ranges have to be explored, from intermediate mass nuclides in the Fe–Ni group to the heaviest nuclei which can also fission. Neutron capture reactions far from stability are not important in the (n,γ)–(γ,n) equilibrium but neutron captures may affect abundances in the freeze-out of a hot r-process. On the other hand, they are important (together with β^- decays) in a cold r-process. Also reactions on neutron-rich isotopes of light elements (involving reactions with neutrons and α-particles) are of interest because they determine the building of the seeds. Furthermore, a cold r-process may commence not from the Fe-region but from lighter elements.

The discussion above focused on conveying an understanding of the basic r-process mechanism. Parameter studies in the classical and dynamical approaches serve to better understand the nucleosynthesis details and the range of possibilities for producing abundance distributions which are similar to the ones observed in the solar system. By themselves, however, they cannot identify the actual astrophysical site of the r-process. Knowing the detailed conditions to reproduce r-abundances, educated guesses can be made to identify the site. Full hydrodynamic simulation of astrophysical sites then have to show whether these conditions can actually be realized. Such full simulations can also study the effect of varying expansion velocity, for example, because faster layers ejected in an explosion run into slower layers.

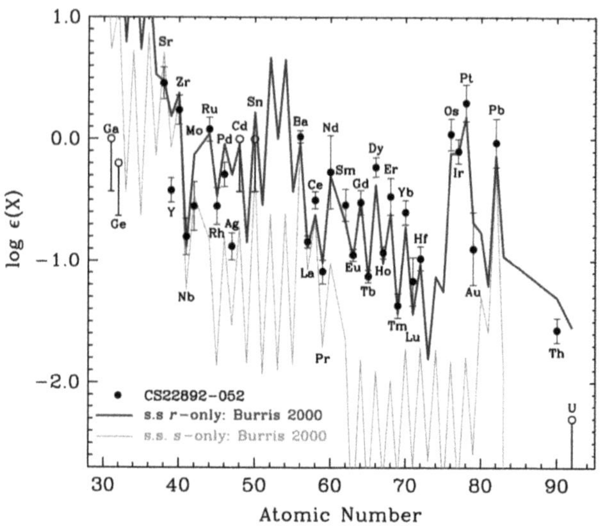

Figure 8.11. Abundances of intermediate and heavy elements in a star with low metallicity. For comparison, two lines show scaled solar abundances for the s-process and for the r-process. The s-process curve is scaled to reproduce the Ba abundance in the metal-poor star, the r-process curve is scaled to the Eu abundance. See text for further details and Equation (1.22) for the abundance notation. (Reproduced from Sneden et al. 2003. © 2003. The American Astronomical Society. All rights reserved.)

Further clues to the site and the working of the r-process come from observational determination of (elemental) abundances from stellar spectra and from cosmochemical investigations of isotopic abundances in meteoritic grains. Of special interest are abundances in old stars with low metallicity. They formed in the early Galaxy and provide a view of the composition of the interstellar medium at the time of their formation. While the solar abundances already contain contributions of many stellar generations and different production sites, stars with low metallicity contain only few contributions. Therefore they permit to see the result of single (or few) nucleosynthesis events. Accurately determining the weak line spectra of lanthanides and actinides of metal-poor stars is difficult but in the last decades, and continuing today, dedicated observational searches have focused on such old stars. An interesting picture has emerged: also old stars show similar r-process abundance patterns as the Sun, only the total amount of r-process nuclei is reduced. An example for this is shown in Figure 8.11 for one of the first metal-poor stars, CS22892-052 (a red giant star with 0.8 $M_\odot$ located in the Galactic halo), analyzed in detail for r-process elements. It is obvious that the abundances of CS22892-052 are not compatible with an s-process pattern. They fit an r-process pattern very well, however, at least from Ba to heavier elements. The Th abundance is lower than the prediction (from a calculation reproducing the solar abundance pattern of the three r-process peaks). This is not surprising because the longest-lived isotope of Th is ^{232}Th with a half-life of 1.405×10^{10} yr. This half-life is comparable to astronomical timescales and therefore a fraction of the ^{232}Th in the star has already decayed since it was produced in an r-process event and incorporated in the molecular cloud from which the star formed. Therefore the difference in measured and predicted abundance can be used to get an upper limit on the age of the star.

Since the analysis of CS22892-052, many similar stars have been found, all showing the usual r-process pattern above Ba. This very good accordance between the solar pattern and the one found in metal-poor stars is an indicator of the robustness of the process, producing the same pattern every time the r-process occurs. Below Ba, however, there is more variation, both in total abundance level compared to the total abundance level in Ba and beyond, and in the pattern itself which is not always a perfect reproduction of the solar r-pattern for these elements. This has given rise to the notion that (at least) two sites are responsible for the production of r-process nuclides. A similar idea has originated from the analysis of isotope ratios in solar system meteoritic material. One site is providing a strong, robust r-process producing nuclides with $A \gtrsim 130-140$. The other site would be producing the lighter r-process nuclides and would also not be as robust. This means that there may be more variation among the relative abundances of these lighter r-process nuclides. A cold r-process environment could be a candidate for the weak r-process, as the individual abundances produced are more sensitive to the detailed conditions than equilibrium abundances. The strong r-process either occurs more frequently or ejects more mass of r-process nuclides per event. It should also appear early in the history of the Galaxy to account for the abundance levels seen in old stars. The second r-process component might be rarer or ejecting less mass per event. If it is rarer, it would have appeared later in the Galaxy.

Whether there is only one or several sites of the r-process still is a hot topic in current research. Likewise, the search for the site of the r-process still is ongoing. The long time favorite candidates have been the innermost, still ejected layers of a massive star in its core-collapse supernova (ccSN) explosion (see Section 9.3). These layers cool down from very high temperatures while moving in a neutrino wind emitted from the nascent neutron star. The neutrino interactions render the matter neutron-rich, so that an r-process can ensue once the matter has cooled sufficiently from nuclear statistical equilibrium to allow neutron captures. Recent multi-D models of core-collapse explosions, however, have been unable to provide the entropy range required to explain the full solar r-process pattern (Section 9.3.3). Additionally, it would be problematic to avoid overproduction at mass numbers below 80 when producing all three r-process peaks in a single core-collapse event. This is because all components make nuclides below 80 and their superposition overproduces that region. A promising alternative site are neutron star mergers (Section 9.4.5) which allow nucleosynthesis in (n,γ)–(γ,n) equilibrium with much lower Y_e than in core-collapse ejecta and thus require a smaller entropy range to produce solar abundances. Since such mergers are rare events which occur late in the Galaxy, probably they cannot explain the r-process patterns found in very old stars, so-called ultra-metal-poor stars, which formed early in the Galaxy. Further alternatives for possible sites are neutron-rich material in magnetically driven jets during ccSN, which circumvent the problems in reproducing solar abundances in neutrino-driven outflows.

Not only the nuclear physics of very neutron-rich, unstable nuclides currently is not well explored but also the astrophysical uncertainties, i.e., the knowledge of mechanisms and conditions, are considerable, especially in neutrino-wind models. These are strongly coupled to the lack of our knowledge regarding the ccSN explosion mechanism. Further insights can only be brought by 3D hydrodynamics calculations with realistic neutrino transport to constrain the actual conditions, such as neutrino-wind properties (determining the evolution of the Y_e of the ejected material) as well as the behavior of the ejecta regarding fallback and wind-termination shock, which can alter the time evolution of the nucleosynthesis conditions considerably. In models of neutron star mergers, the incomplete knowledge of the nuclear equation-of-state at high temperature is a major source of uncertainty, as it determines not only the properties of the neutron stars but also their behavior in a merger event. For a further discussion of neutron star mergers and the uncertainties connected to their r-process output, see Section 9.4.5.

8.4 The p-Nuclides

8.4.1 Properties

As explained above, nucleosynthesis of elements heavier than iron is dominated by neutron capture processes. Nevertheless, there are several proton-rich nuclides which cannot be reached by neutron-capture processes, either because they are reached by the s-process or a shielded from direct contributions from the r-process by stable nuclides directly in the decay path, as illustrated in Figures 8.1 and 8.12.

In the classical works of Burbidge et al. (1957); Cameron (1957a, 1957b, 2013) 35 proton-rich nuclides between Sr and Hg were identified which were thought not to be synthesized by the neutron-induced capture reactions of the s- and r-processes. They were called "p-isotopes" by Burbidge et al. (1957) and "excluded isotopes" by Cameron (1957a, 1957b, 2013). In this book they are called p-isotopes when referring to isotopes of a specific element and *p-nuclides* in general. Their abundances are typically factors of 0.1–0.001 smaller than s- and r-abundances. Table 8.2

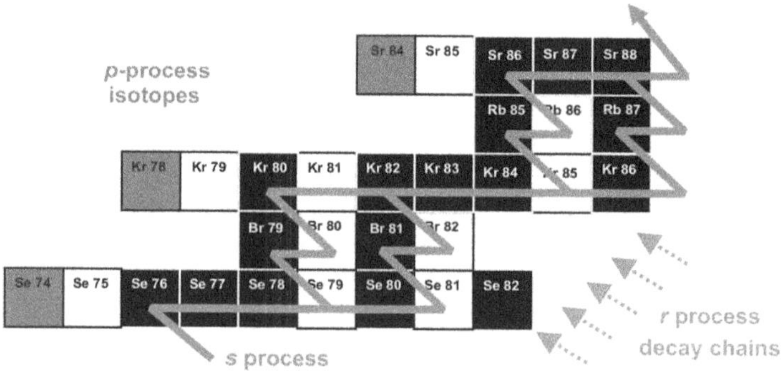

Figure 8.12. Region of the nuclear chart indicating the s-process path and decay paths of nuclides produced in the r-process. Stable isotopes are in black and red. Marked in red are the p-nuclides ^{74}Se, ^{78}Kr, and ^{84}Sr. (Figure from Rauscher et al. 2013. © 2013 IOP Publishing Ltd. Reproduced with permission. All rights reserved.)

Table 8.2. Classical p-Nuclides, Their Fraction (in %) in the Isotopic Composition of Elements (Berglund & Wiesen 2011), and Solar p-Abundances (Relative to Si = 10^6) from Lodders (2003)

Nuclide	p-Fraction in Element (%)	Solar Abundance	Comment
^{74}Se	0.89 (4)	5.80×10^{-1}	
^{78}Kr	0.355 (3)	2.00×10^{-1}	
^{84}Sr	0.56 (1)	1.31×10^{-1}	
^{92}Mo	14.53 (30)	3.86×10^{-1}	
^{94}Mo	9.15 (9)	2.41×10^{-1}	
^{96}Ru	5.54 (14)	1.05×10^{-1}	
^{98}Ru	1.87 (3)	3.55×10^{-2}	
^{102}Pd	1.02 (1)	1.46×10^{-2}	
^{106}Cd	1.25 (6)	1.98×10^{-2}	
^{108}Cd	0.89 (3)	1.41×10^{-2}	
^{113}In	4.29 (5)	7.80×10^{-3}	r-Process contribution
^{112}Sn	0.97 (1)	3.63×10^{-2}	
^{114}Sn	0.66 (1)	2.46×10^{-2}	
^{115}Sn	0.34 (1)	1.27×10^{-2}	r-Process contribution
^{120}Te	0.09 (1)	4.60×10^{-3}	
^{124}Xe	0.0952 (3)	6.94×10^{-3}	
^{126}Xe	0.0890 (2)	6.02×10^{-3}	
^{130}Ba	0.106 (1)	4.60×10^{-3}	
^{132}Ba	0.101 (1)	4.40×10^{-3}	
^{138}La	0.08881 (71)	3.97×10^{-4}	ν-Process
^{136}Ce	0.185 (2)	2.17×10^{-3}	
^{138}Ce	0.251 (2)	2.93×10^{-3}	
^{144}Sm	3.07 (7)	7.81×10^{-3}	
^{152}Gd	0.20 (1)	6.70×10^{-4}	
^{156}Dy	0.056 (3)	2.16×10^{-4}	
^{158}Dy	0.095 (3)	3.71×10^{-4}	
^{162}Er	0.139 (5)	3.50×10^{-4}	
^{164}Er	1.601 (3)	4.11×10^{-3}	
^{168}Yb	0.123 (3)	3.23×10^{-4}	
^{174}Hf	0.16 (1)	2.75×10^{-4}	
^{180m}Ta	0.01201 (32)	2.58×10^{-6}	ν-Process?
^{180}W	0.12 (1)	1.53×10^{-4}	
^{184}Os	0.02 (1)	1.33×10^{-4}	
^{190}Pt	0.012 (2)	1.85×10^{-4}	
^{196}Hg	0.15 (1)	6.30×10^{-4}	

lists these nuclides, their solar abundances, and their fractional content in the total abundance of the corresponding element. Almost all of the p-nuclides are even–even nuclei, with the exception of ^{113}In, ^{115}Sn, ^{138}La, and ^{180m}Ta. There is an indication that different nucleosynthesis processes have made these four nuclides and the remaining 31 nuclides. Improved models of the s- and r-process have suggested large contributions of the r-process to the abundances of ^{113}In and ^{115}Sn, contrary to what has been assumed in the original literature. Furthermore, ^{138}La has been found to be produced by the ν-process during a core-collapse supernova explosion (see Section 9.3.4). Also ^{180}Ta (and ^{180m}Ta) may at least have received a strong contribution from the ν-process (see Section 9.3.4).

At least one other nucleosynthesis process is needed to produce the bulk (31 nuclides plus perhaps a fraction of ^{180}Ta) of the p-nuclides. It is often generically called a *p-process* but this may be confusing because in the older literature a p-process implied proton captures (somewhat similar to the s-process). Moreover, there may not be a single process synthesizing all p-nuclides but several processes are most likely contributing and should be specified with their proper names. How can proton-rich stable isotopes be reached by a nucleosynthesis process? In general, there are three possibilities: first, by proton captures on a stable isotope of the next lower element (with $Z-1$); second, by ejecting neutrons from isotopes of the same element; third, by emission of protons and/or α-particles from isotopes of heavier elements. Proton captures directly feeding into a p-nuclide are hindered by the high Coulomb barriers and may only contribute to the lightest p-nuclides up to Ru, if at all. Higher temperatures do not help because they also increase photodisintegration. In order to synthesize heavier elements by proton captures, the proton number density has to be very high because this speeds up capture rates without affecting photodisintegration (see Equation (4.52)). In this case, very proton-rich, unstable nuclides far from stability are created, somewhat similar to an r-process but on the proton-rich side of the nuclear chart. This results in an rp- or νp-process, depending on whether neutrinos are present. They can only possibly contribute to the lower half of the p-nuclides, if at all, but to what extent they contribute to these p-nuclides is not yet fully clear. Probably their overall contribution during the Galactic history is too small to explain solar p-abundances. The rp-process is discussed in Section 9.4.4 and the νp-process in Section 9.3.3.

8.4.2 The γ-Process

A viable approach to synthesize the bulk of p-nuclides is by particle emission from thermally excited nuclei, which corresponds to the second and third possibility for making proton-rich isotopes as given above. At sufficiently high plasma temperature and not too high density, neutrons, protons, and α-particles are ejected from a nucleus. This is a photodisintegration reaction which is the reverse of a capture reaction (see Section 4.1.3.2). Such a (partial) photodisintegration process in a hot stellar plasma is the so-called *γ-process*. It was first found in the outer layers of models of massive stars during their explosion in a core-collapse supernova (see Section 9.3) but also found in models of white dwarfs exploding as type Ia supernovae (see Section 9.4.3).

Prerequisite for a γ-process is the existence of heavy nuclides as seeds. These are nuclides previously made in the s- and/or r-process and already present in the plasma when it is heated. The temperature required to release particles from a nucleus is obviously determined by the particle separation energies (see Table 3.2). As nuclei closer to the Fe-region are more strongly bound (see Section 3.2), higher temperatures (around 3 GK) are required to induce particle emission in the region of the lighter p-nuclides whereas lower temperature suffices for the region of the heaviest p-nuclides and beyond (around 2 GK). In this temperature range between 2 and 3 GK, a heavy seed distribution is first altered by (γ,n) reactions (because the neutron separation energy is lower for neutron-rich isotopes) and thus proton-rich nuclei are created. A comparison of (γ,n), (γ,p), and (γ,α) rates (per target nucleus, see Equation (4.100)) along the Mo isotope chain is shown in Figure 8.13. After several neutron emissions, (γ,p) and (γ,α) reactions start to compete with neutron emission because proton- and/or α-separation energies become lower than S_n and therefore the emission of protons and/or α-particles becomes more favorable than neutron emission. This deflects the reaction flow toward lower charge number Z, as sketched in Figure 8.14. The complicated reaction flow can only be traced in a full reaction network calculation, including about a thousand nuclides and their connecting

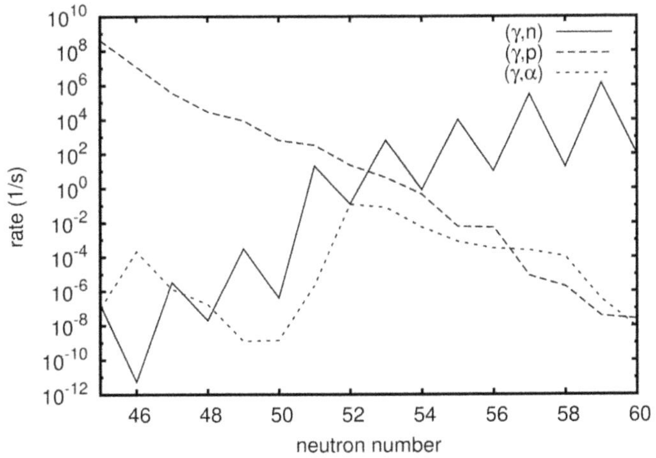

Figure 8.13. Comparison of photodisintegration rates for Mo isotopes at 2.5 GK. (Figure from Rauscher et al. 2013. © 2013 IOP Publishing Ltd. Reproduced with permission. All rights reserved.)

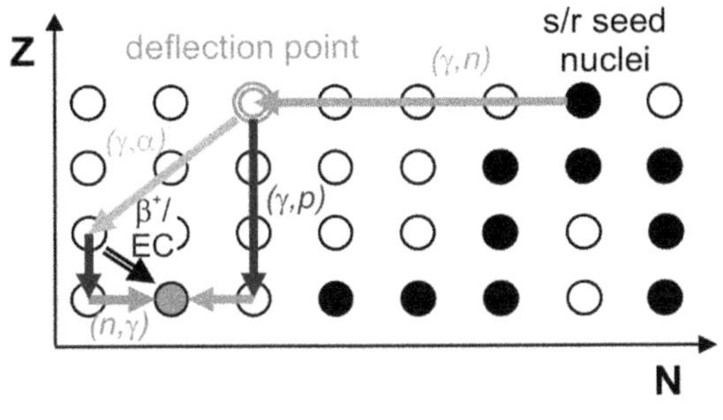

Figure 8.14. Illustrating sketch of deflection points in the γ-process. (Figure from Rauscher et al. 2013. © 2013 IOP Publishing Ltd. Reproduced with permission. All rights reserved.)

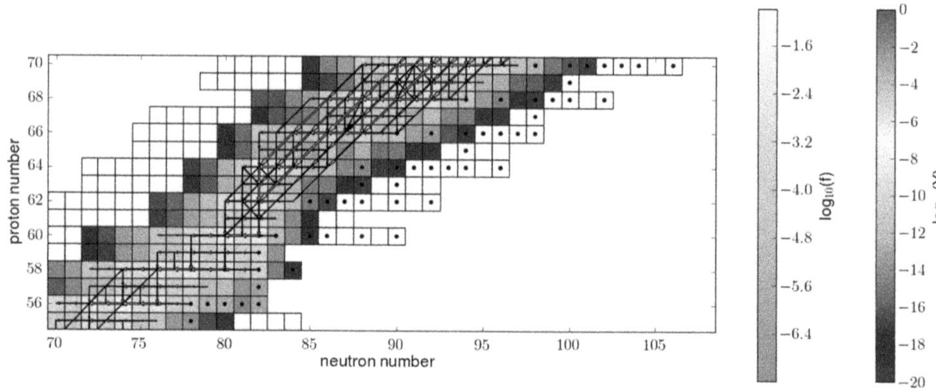

Figure 8.15. Snapshot of the reaction flow in the γ-process for $T_9 = 2.8$ GK and $\rho = 4.5 \times 10$ g cm^{-3} in a region of the nuclear chart. The dots indicate stable nuclides. The color of the boxes show the abundances and the size and color shade of the arrows give the net flow (Equation (4.118)).

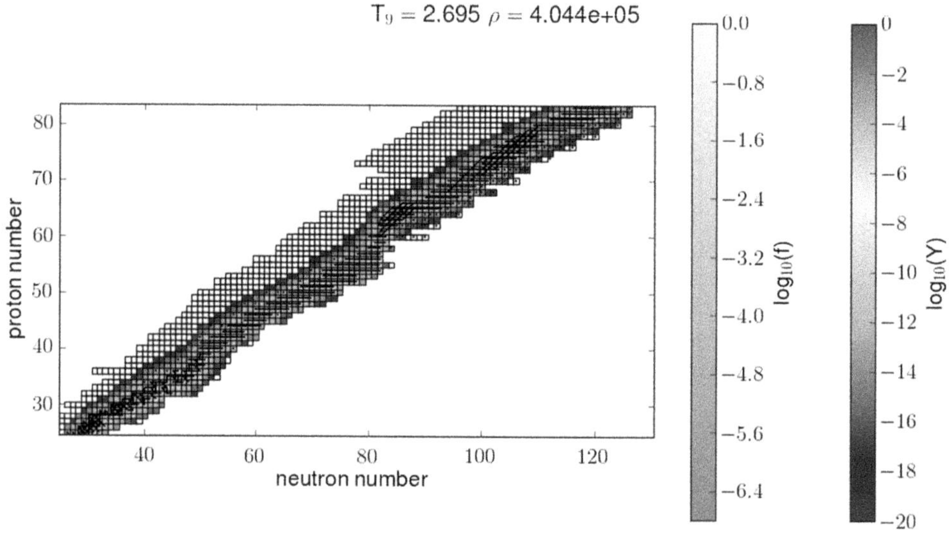

Figure 8.16. (Animation) View of the γ-process flow in a single layer of a massive star experiencing a temperature increase during the passage of the supernova shockfront. See text for details. Animation available online at https://doi.org/10.1088/978-0-7503-1149-6.

reactions. Examples for the γ-process flow are shown in Figure 8.15 and in the animated Figures 8.16 and 8.17, demonstrating the flow in a mass shell of a massive star when the supernova shockfront is passing through (see Section 9.3 for further details). In the animations it can be clearly seen how the reaction flow starts with (γ,n) reactions on stable nuclides (marked by dots) and then moves toward the proton-rich isotopes before being deflected to lower Z when (γ,p) or (γ,α) are faster than neutron emission. It should be noted that for the peak temperature in the shown layer all heavier nuclides are destroyed and only the very lightest p-nuclides

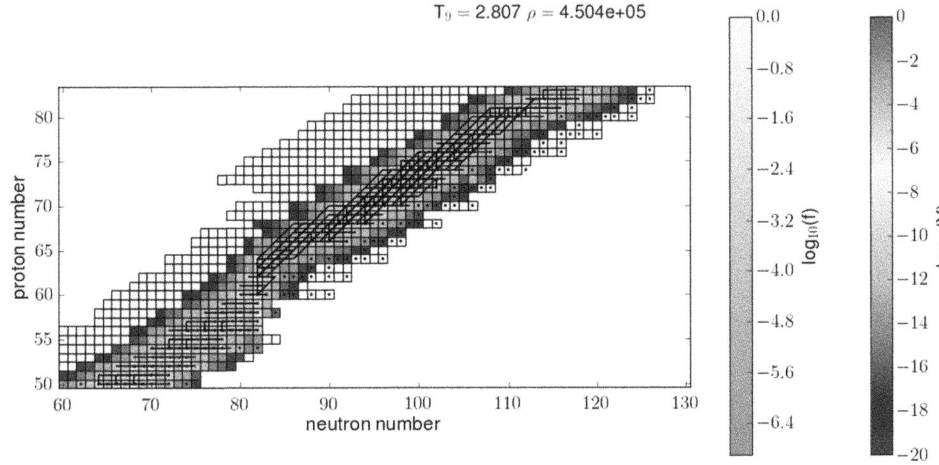

Figure 8.17. (Animation) Same as Figure 8.16 but zoomed into the region $50 \leqslant Z \leqslant 83$. Animation available online at https://doi.org/10.1088/978-0-7503-1149-6.

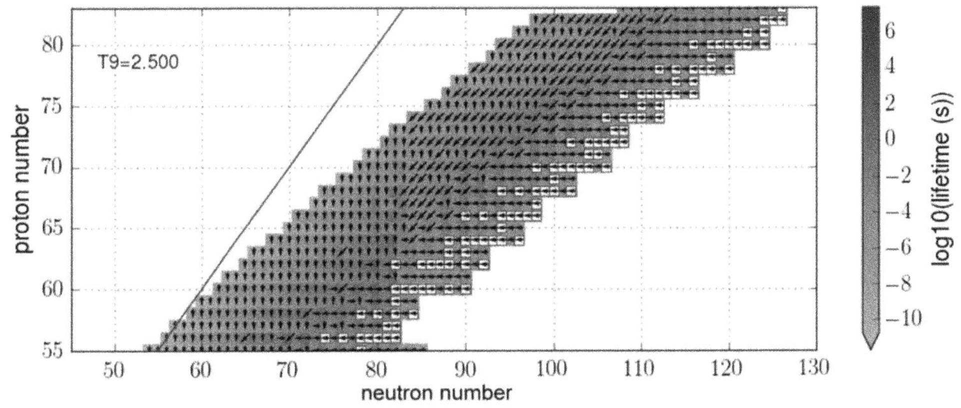

Figure 8.18. (Animation) Rate field for varying temperatures for $55 \leqslant Z \leqslant 83$. The arrows indicate the main destruction reaction for each individual nuclides. The color shade indicates the lifetime of the nucleus from the sum of all destruction processes. For naturally stable nuclides, this lifetime is not shown. They are indicated by unfilled boxes. Animation available online at https://doi.org/10.1088/978-0-7503-1149-6.

survive. This is just for illustrating the basic reaction flow. Also very noticeable in the figures is the fact that the reaction flow spreads out much more than in the s- or r-process. Instead of a narrow "path," a reaction "boulevard" is encountered. This is even more pronounced in sites with a larger range of densities, such as in thermonuclear supernovae, where the flow spreads out even wider (see Section 9.4.3).

Another way to investigate the competition between different reactions on a single nucleus is to look at rate field plots, as defined in Section 4.1.7. These types of plots do not show actual reaction flows because they neglect the abundances of the target nuclei. Instead, for making the Figures 8.18 and 8.19, the rates per target nucleus (Equations (4.11) and (4.100), also including the abundances of projectiles in

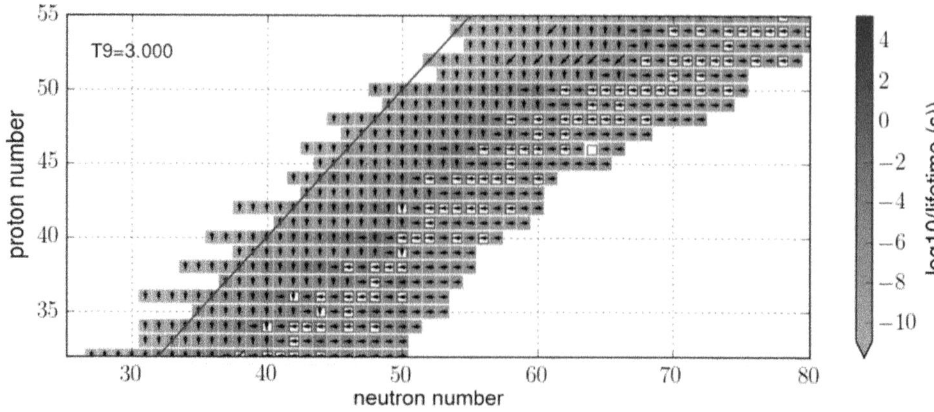

Figure 8.19. (Animation) Rate field for varying temperatures for $32 \leqslant Z \leqslant 55$. The arrows indicate the main destruction reaction for each individual nuclides. The color shade indicates the lifetime of the nucleus from the sum of all destruction processes. For naturally stable nuclides, this lifetime is not shown. They are indicated by unfilled boxes. Animation available online at https://doi.org/10.1088/978-0-7503-1149-6.

two-body reactions) of all possible destruction processes were compared for each individual nuclide and the dominating destruction process has been indicated by the "direction" of the resulting product nuclide. This allows to have a quick estimate of possible reaction pathways at given conditions, provided that the nuclide is actually reached by the reaction flow. The net lifetime of a nuclide indicated in Figures 8.18 and 8.19 was obtained by applying Equation (4.113). In the shown rate fields it is clearly seen that (γ,n) reactions dominate at and close to stability. A few Z units further toward proton-rich isotopes charged-particle emission takes over. A distinctive change is observed above $N = 82$. Above this neutron shell closure, (γ,α) is dominating among the charged-particle emissions, whereas for $N \leqslant 82$ proton emission is dominating, with few exceptions. This is due to the nuclear structure change above the closed neutron shell at $N = 82$ which reduces α-particle separation energies (see Sections 3.2 and 3.3).

The γ-process ends when the environment becomes too cool to permit photodisintegration and β-decay chains finally populate the next stable isobar encountered for a given mass number. The actual process conditions characterizing the outcome are the peak temperature reached, the duration of the high temperature condition, and the amount of the seeds available for photodisintegration. As explained above, different peak temperatures T_{peak} are required to actually affect nuclides in different mass ranges, with lower temperatures sufficient for the heavier nuclides. This constrains the acceptable temperature maxima to 2–3 GK. Process duration τ_Δ^{p} and peak temperature T_{peak} have to be fine-tuned in order to be able to reproduce the observed p-abundances. If the high temperature is held too long, all seed nuclides as well as the p-nuclides in a certain mass range would be completely destroyed. Therefore the acceptable τ_Δ^{p} is given by $\tau_\Delta^{\text{p}} < \tau_\Delta^{\text{nuc}}$, where τ_Δ^{nuc} is the destruction timescale of the nuclide of interest, as defined in Equation (4.115). (The destruction timescale can be approximated by the lifetime of a nuclide at T_{peak} according to

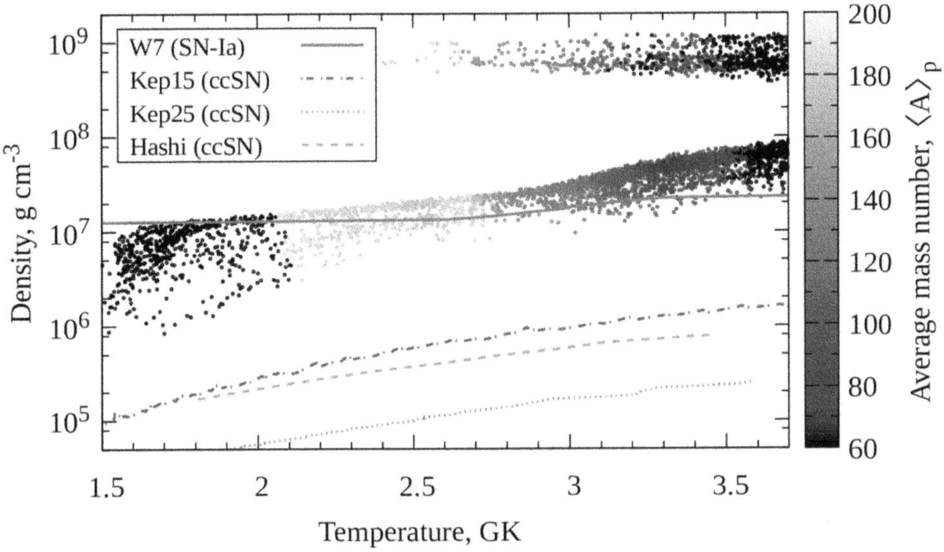

Figure 8.20. Production of p-nuclides in the γ-process in ccSN and SNIa as function of peak density and peak temperature. The two groups of dots are conditions experienced by tracer particles in the p-producing regions of a 2D thermonuclear (SNIa) supernova simulation, the W7 line is from a 1D model of a SNIa. The other lines refer to conditions in 1D ccSN models, the Kep15 and Hashi models are for a 15 $M_\odot$ progenitor star whereas Kep25 is based on a 25 $M_\odot$ progenitor. The dot color encodes the average mass number of the produced p-nuclides. As this is tightly coupled to the peak temperature achieved but independent of density, also the same average mass numbers are found in the same temperature regions for the W7, Kep15, Kep25, and Hashi models. The black dots at either end of the plotted temperature range indicate that no p-nuclides are produced (at low T) or all p-nuclides are destroyed (at high T). See text and Nishimura et al. (2018) for details. (Reproduced from Nishimura et al. 2018, by permission of Oxford University Press on behalf of the Royal Astronomical Society.)

Equation (4.113).) This limits the process duration to be on the order of seconds or less, typical for explosive environments in which also the temperature required for photodisintegration can be reached. Because of the different temperature sensitivity of different mass ranges, not all p-nuclides can be made in the same small volume of exploding plasma. Rather, an extended plasma region, experiencing a range of peak temperatures, has to be involved. Interestingly, the temperature is the main quantity to scrutinize when searching for appropriate plasma conditions, density is less important. This is a reflection of the fact that photodisintegration rates depend on the photon density, which is set by the temperature (see Equation (1.68)) and independent of the matter density ρ. Figure 8.20 illustrates this point in a comparison of p-nucleus production as function of temperature and density in several models of core-collapse supernovae (ccSN) and thermonuclear supernovae (SN Ia). The temperature range for p-nuclide production with given mass number A is well defined and independent of density. It is the same for ccSN which make the p-nuclides at much lower density as those in SN Ia.

The long-time favored site for the production of p-nuclides were ccSN because model calculations found a γ-process proceeding in the outer layers of the exploding star. As an alternative site, hot matter originating from SN Ia was suggested to overcome deficiencies in the predicted abundances of certain p-nuclides. Details of the p-nuclide production in ccSN and SN Ia, and the differences, are discussed in Sections 9.3.4 and 9.4.3, respectively. A longstanding problem is posed by the comparatively large abundances, relative to the other p-nuclides, of ^{92,94}Mo and ^{96,98}Ru. They are not produced in massive stars as there is not enough s- and r-process material present to act as seed to produce these nuclides by photodisintegration in sufficient quantities. Also the region of p-nuclides with $A < 124$ seems underproduced in explosions of massive stars but it has to be kept in mind that there are strong contributions from other processes to the In and Sn isotopes. Thermonuclear supernovae may offer a way out because in a certain type of these supernovae, additional s-processing can occur before the explosion, strongly increasing the amount of seed nuclei. Whether this is actually feasible is currently under investigation. Moreover, extremely proton-rich environments giving rise to a νp-process may also increase the abundances of the lightest p-nuclides. A definitive answer hinges on the detailed modeling of the ccSN explosion mechanism and the neutrino transport in the deepest layers of exploding stars (or on the detailed modeling of other potential sites for the νp-process). The νp-process is further discussed in Section 9.3.3. Another problematic mass region for the production of p-nuclides is $150 \leqslant A \leqslant 165$. The underproduction in this region may be cured by enhanced seeds or by modified reaction rates. The νp-process cannot contribute to this mass range. This means that either SN Ia may also produce these nuclides without rate modifications or that the nuclear physics input has to be improved. Recently, there has been a further concern raised by galactic chemical evolution studies (see Chapter 11) regarding the ability of ccSN to produce the overall solar p-abundances. It appears as if the yield of ccSN accumulated over the history of the Galaxy is not large enough to account for the observed abundance levels of any of the p-nuclides. If this is consolidated, another source of p-nuclides is definitely required, in addition to or instead of ccSN. Problematic for the identification of p-nuclide producing sites is the lack of an element with an abundance dominated by its p-isotope(s). With few exceptions, stellar spectra only provide information on elemental abundances. Even if it was possible to separate isotopic lines in principle, the abundances of p-isotopes within an element are too small with respect to other isotopes (Table 8.2), precluding a measurement. The only way to obtain information on isotopic p-compositions from outside the solar system is in the study of presolar grains, which have formed in the stellar winds of AGB stars or in the ejecta of ccSN, were subsequently embedded in the pre-solar cloud, and are found in certain types of meteorites (Lugaro 2005).

Regarding the potential requirement of improved nuclear input, it is worthwhile to remember that although the main mechanism in the γ-process is γ-induced partial nuclear disintegration, the reactions proceed quite differently than in the laboratory. The thermal excitation of nuclides is discussed in Section 4.1.3.3 and its relevance for γ-induced reactions is shown in Table 4.1 and in Figures 4.6 and 4.7. The temperatures in the γ-process are so high that the ground-state contribution to the stellar rate (as defined in Equation (4.57)) is very small. This implies that directly measuring the laboratory (γ,x) cross section of a nucleus in the ground state constrains only a small fraction of the stellar rate. A large number of transitions would have to be measured individually, which as of now may be unfeasible, especially for unstable nuclei. As discussed in Section 4.1.3.3, the ground-state contributions to the reverse reactions, i.e., the capture reactions, are much larger and therefore include fewer relevant transitions between nuclear states. Nevertheless, also for capture reactions at γ-process temperatures there are considerable contributions from reactions on thermally excited states. As explained in Section 4.2.4, in reaction networks the rate for only one reaction direction is taken from theory or experiment, anyway, and the reverse rate is calculated from the reciprocity relation given in Equation (4.52). According to the Q-value rule (Sections 4.1.3.3 and 4.2.4), the adopted rate is the one of the exoergic reaction with the possible exception for charged particle captures. The dominance of reactions on excited states implies that all the rates are based on nuclear reaction theory so far, even for stable nuclides.

8.5 The i-Process

Observations have indicated an enhancement of light s-process elements, such as Rb, Sr, Y, in "very late thermal pulses" (VLTP) in post-AGB stars or young white dwarfs, occurring when the H-burning shell already has ceased burning. At the same time, abundances of Ba and La, as elements of the second s-process peak (see Section 8.2) are not increased. Furthermore, increased abundances of Sr, Y, Zr have been found in very old, metal-poor stars (Travaglio et al. 2004). Carbon-enhanced metal-poor stars with s-process plus Eu content (CEMP-r/s stars) also show a peculiar abundance pattern not explicable by a superposition of the regular s- and r-processes. The observations could perhaps be explained by assuming an additional process with a neutron density between the ones of the s- and r-processes and exposure times much longer than explosive timescales of only a few seconds as in the r-process (see also Chapter 9).

It was suggested by Cowan and Rose (1977) that $\bar{n}_{\rm n} \approx 10^{15}$ g cm^{-3} can be achieved when mixing hydrogen into a shell convectively burning helium. If the unstable ($t_{1/2} = 9.96$ m) ^{13}N produced by ^{12}C(p,γ)^{13}N is quickly transported down into the thermal pulse region by convection while it decays to ^{13}C, it can avoid destruction by a further proton capture. Neutrons are then efficiently released by the reaction ^{13}C(α,n)^{16}O, similar to the s-process during the interpulse period (see Section 7.3.2), but within a thermal pulse reaching (2–3) $\times$ 10^8 K. The neutrons are irradiating the already present nuclides in the He-burning layer, leading to the so-called *i-process.* It proceeds through unstable, neutron-rich nuclei but remains close to the line of stability.

Several sites with unstable He-burning have been suggested beyond post-AGB VLTPs (which do not contribute significantly to the Galactic budget of light s-process elements): the core He-flash (Chapter 6), He-shell flashes (Section 7.3.2) in low-metallicity AGB stars, and super-AGB stars. Recently, He-shell flashes on rapidly accreting white dwarfs have been investigated more closely as a promising site, maybe also able to account for the galactic chemical evolution of low-mass s-process elements (Denissenkov et al. 2017). Such accreting white dwarfs give rise to novae (see Section 9.4.2) but the mass loss during each strong pulse may prevent the triggering of a thermonuclear supernova (see Section 9.4.3) despite of the rapid accretion. The i-process in such accreting white dwarfs may significantly contribute to neutron-rich isotopes of elements from Fe up to Mo (Côté et al. 2018).

Further Reading

Argast, D., Samland, M., Thielemann, F.-K., & Qian, Y.-Z. 2004, A&A, 416, 997
Arnould, M., & Goriely, S. 2003, PhR, 384, 1
Arnould, M., Goriely, S., & Takahashi, K. 2007, PhR, 450, 97
Berglund, M., & Wiesen, M. E. 2011, PApCh, 83, 397
Burbidge, E. M., Burbidge, G. R., Fowler, W. A., & Hoyle, F. 1957, RvMP, 29, 547
Cameron, A. G. W. 1957a, PASP, 69, 201
Cameron, A. G. W. 1957b, Chalk River Laboratories Report, CRL-41 (AECL-454)
Cameron, A. G. W. 2013, Stellar Evolution, Nuclear Astrophysics, and Nucleogenesis, ed. D. M. Kahl (Mineola: Dover)
Côté, B., Denissenkov, P., Herwig, F., et al. 2018, ApJ, 854, 105
Cowan, J. J., & Rose, W. K. 1977, ApJ, 212, 149
Cowan, J. J., et al. 2020, RvMP, in press; arXiv:1901.01410
Denissenkov, P. A., Herwig, F., Battino, U., et al. 2017, ApJL, 834, L10
Diehl, R., Hartmann, D. H., & Prantzos, N. (ed) 2018, Astrophysics with Radioactive Isotopes, Astrophysics and Space Science Library, Vol. 453 (Berlin: Springer)
Eichler, M., et al. 2015, ApJ, 808, 30
Farouqi, K., Kratz, K.-L., Pfeiffer, B., et al. 2010, ApJ, 712, 1359
Freiburghaus, C., et al. 1999, ApJ, 516, 381
Käppeler, F., Gallino, R., Bisterzo, S., & Aoki, W. 2011, RvMP, 83, 157
Lodders, K. 2003, ApJ, 591, 1220
Lugaro, M. 2005, Stardust From Meteorites: An Introduction to Presolar Grains (Singapore: World Scientific)
Nishimura, N., Rauscher, T., Hirschi, R., et al. 2018, MNRAS, 474, 3133
Qian, Y.-Z., & Wasserburg, G. J. 2000, PhR, 333, 77
Rauscher, T. 2012, ApJL, 755, L10
Rauscher, T., Applegate, J. H., Cowan, J. J., Thielemann, F.-K., & Wiescher, M. 1994, ApJ, 429, 499
Rauscher, T., Mohr, P., Dillmann, I., & Plag, R. 2011, ApJ, 738, 143
Rauscher, T., Dauphas, N., Dillmann, I., et al. 2013, RPPh, 76, 066201
Sneden, C., et al. 2003, ApJ, 591, 936
Travaglio, C., Rauscher, T., Heger, A., Pignatari, M., & West, C. 2018, ApJ, 854, 18
Travaglio, C., Gallino, R., Arnone, E., et al. 2004, ApJ, 601, 864
Wehmeyer, B., Fröhlich, C., Côté, B., Pignatari, M., & Thielemann, F.-K. 2019, MNRAS, 487, 1745

AAS | IOP Astronomy

Essentials of Nucleosynthesis and Theoretical Nuclear Astrophysics

Thomas Rauscher

Chapter 9

Explosive Nucleosynthesis

9.1 General Considerations

The timescale of nuclear burning in the hydrostatic burning phases discussed in Chapter 7 was determined by the amount of available fuel and the requirement to uphold hydrostatic equilibrium as described by the basic equations of stellar structure (Section 2.2), despite of continuous energy loss through the stellar surface. The structure equations set the amount of energy required to keep the star stable which, in turn, constrains the temperature for each burning phase and this also determines how quickly the available fuel is exhausted. A hydrostatic burning phase continues until almost all of the nuclei involved in energy generating reactions have been depleted (any surviving ones are consumed quickly at the onset of the next burning phase). Contrary to this behavior, nuclear burning and hydrodynamical evolution are only loosely coupled in explosive burning. The change in temperature and density is set by external conditions and not by any requirement for hydrostatic equilibrium. There may be some feedback on the hydrodynamics by the nuclear energy generation, for example by delaying the cooling process of an initially hot matter volume, but the main behavior is not determined by the availability of nuclear fuel. In consequence, temperature and density are not regulated by pressure equilibria and also the nucleosynthesis timescale is only constrained at the upper end, which is the complete depletion of the nuclei acting as fuel. Most often, however, the short hydrodynamic timescale of explosive processes lead to an incomplete consumption and thus to very different nucleosynthesis patterns than expected from hydrostatic burning. Another reason for different abundance patterns is that higher temperatures than in hydrostatic burning can be achieved, speeding up two-body and photodisintegration reactions and making them faster than decays (close to stability). This makes it possible to synthesize a wider range of unstable nuclides than in hydrostatic processes.

doi:10.1088/2514-3433/ab8737ch9

As the process duration is not given by the exhaustion of nuclear fuel, the impact of an explosive process on abundances can be estimated by comparing the process timescale $\tau_{\Delta}^{\text{process}}$ to the destruction timescale $\tau_{\Delta}^{\text{nuc}}$, as defined in Equation (4.115), of a nuclide. Only for $\tau_{\Delta}^{\text{process}} \gtrsim \tau_{\Delta}^{\text{nuc}}$ appreciable depletion of the abundance of that nuclide can occur. The timescale $\tau_{\Delta}^{\text{nuc}}$ is determined by the rates of the most important destruction reaction(s) and its dependence on temperature and density is similar to the one of the reaction(s). Inspection of Equation (4.125), together with Equations (4.9) and (4.100), reveals that $\dot{Y} = \mathrm{d}Y/\mathrm{d}t$ is exponentially dependent on T but only linearly on ρ, except for the abundances depending on the triple-α rate. Therefore the temperature is the quantity mainly determining the consumption timescale. Conversely, the minimum temperature to be reached for making an essential modification of an abundance can be specified when assuming a process duration. For example, for explosive shell burning in core-collapse supernovae (see Section 9.3) with a timescale of the order of seconds, the following minimum temperatures (assuming typical densities as in the shells of massive stars) have to be reached for an appreciable change in the abundance of the main nuclear fuel: 1.8 GK (explosive C-burning), 2.5 GK (explosive Ne-burning), 3.0 GK (explosive O-burning), 4 GK (explosive Si-burning). The He-burning timescale is special due to the peculiarities of the triple-α rate. Above 1 GK the rate is only weakly dependent on temperature, see Figure 9.1. As a consequence, also the He-destruction timescale remains basically constant for $T > 1$ GK. However, it remains being very sensitive to the matter density ρ. Considerable depletion of the ^{4}He abundance on a process timescale on the order of seconds is only possible for $\rho > 10^{5}$ g cm^{-3}. He-shells of

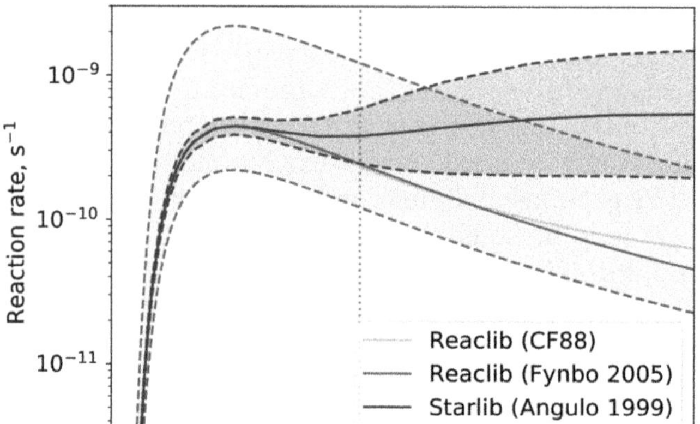

Figure 9.1. Triple-α reaction rate as function of plasma temperature; shown are the rates from three different evaluations: Angulo et al. 1999; Caughlan & Fowler 1988; Fynbo et al. 2005. The latter two rates neglect a disputed resonance included in Angulo et al. The uncertainty range for the Angulo et al. rate is recommended by Sallaska et al. 2013; whereas for the Fynbo et al. rate an uncertainty range of $r^{*}_{3\alpha}\left\{{}^{\times 2}_{/2}\right.$ is adopted. (Reproduced from Nishimura et al. 2019, by permission of Oxford University Press on behalf of the Royal Astronomical Society.)

massive stars have lower densities but α-particles are present at such densities in other environments discussed below.

The above timescale estimates under consideration of temperature and density can be applied to any environment under consideration but the details of explosive nucleosynthesis under varying hydrodynamical conditions have to be followed by coupling extended nuclear reaction networks (also including nuclides further away from stability) to the hydrodynamic equations. This is very demanding with respect to the computational effort, which is currently prohibitive for using such fully coupled, extended networks in detailed multi-D simulations. If there is no energy feedback from the nuclear reactions to the hydrodynamic evolution, nucleosynthesis can be followed in a decoupled network, only taking as input the temperature and density as function of time. The decoupled network can either be run "parallel" to the solution of the hydrodynamic equations during the simulation or by *post-processing* initial abundances with a temperature–density profile recorded from the simulation. Post-processing has been extensively used in one-dimensional, spherical symmetric models without mixing between different mass zones. More complicated matter flows are found in 2D and 3D simulations of explosive sites which render it impossible to divide the matter flow into a few larger, independent zones. Instead, *tracer particles* (short: tracers) are used. The exploding region of an object is subdivided into a large number of mass elements (several ten thousands, usually with the same mass) and the history of the averaged temperature and density in such an element is recorded as it moves through the simulation. In the post-processing, reaction network calculations are run for all the tracers having experienced temperatures required for nuclear processes and their individual results are combined (weighted by, e.g., their individual mass) to derive the total abundance change.

Site-independent studies exploring a parameter range of conditions are only possible for conditions allowing to make assumptions largely independent of the details of the hydrodynamic evolution and when well-understood abundance data sufficient for detailed comparison are available. Several examples for such an approach are given in Chapter 8. More often it is preferable to tie the investigation of explosive nucleosynthesis to a specific site or environment which automatically provides constraints on initial abundances, process timescales, temperatures, and densities to be expected. Detailed hydrodynamical models also provide the time-dependence of the relevant physical quantities. Therefore, in the following sections astrophysical sites (or regions within such sites) are presented and their nucleosynthesis is discussed. Only an overview can be provided here, without going too much into details of simulations or of competing models and interpretations. The summary given is supposed to provide an entrance point for the reader to grasp the general picture and allowing to progress into further specialized literature for the topics of greatest interest.

9.2 Classification of High-energy Phenomena

Before delving into the details of nucleosynthesis in various sites, a short introduction to astronomical nomenclature is in order. The naming of astronomical phenomena may be confusing and seem inconsistent at times. This is due to the historical context in which phenomena were discovered. Initially, a certain phenomenon is detected with the instruments available at the time and the clarification of the underlying physical mechanism often follows much later. A good example for this is the term "nova," which has been incorporated into a number of names describing a variety of phenomena caused by completely different stellar mechanisms. "Nova" was originally incepted to describe a "new" star appearing within known constellations, quickly reaching a maximal brightness, and fading again over a longer period of days, weeks or years, as seen by naked eye or with simple telescopes. In 1572 Tycho and Sophie Brahe observed such an event in the constellation Cassiopeia and coined the term "stella nova," meaning "new star." In modern terminology it was supernova SN1572 but until the 1930s there was no distinction made between what we now call (classical) novae and supernovae. Only with improvements in technology and the ability to determine absolute magnitudes and spectra, the classification became more refined. Large differences in released energy between the events was found. Walter Baade and Fritz Zwicky introduced the term "supernova" in the early 1930s for events which were much more energetic than classical novae. Rudolph Minkowski and Fritz Zwicky developed a more detailed classification scheme in the 1940s, further distinguishing several types of supernovae based on their spectra. This scheme provided the basis of the classification in use today. More recently, events even more energetic than regular supernovae were detected. These were called hypernovae in extension of the standard terminology.

Novae, in addition to being less energetic, also differ from supernovae in their lightcurve evolution. Novae reach their maximum brightness within days and then slowly fade over weeks to months. Supernovae (as well as kilonovae and hypernovae) reach their maximum within hours and the exponential decay of the lightcurve proceeds faster, depending on the amount of ejected ^{56}Ni which prevents the ejecta from faster cooling with its radioactive decay heat. Hypernovae eject large amounts of ^{56}Ni (up to 5 $M_\odot$) and their spectra show extremely broadened lines, indicating ejecta at 99% the speed of light.

All the above phenomena emit light in the visible range, even if only briefly. With the advent of X-ray and γ-ray detectors built into satellites, additional short but highly energetic bursts of electromagnetic emission were found, many without obvious optical counterpart. Again, these may originate from a variety of sources and were termed, irrespective of their origin, X-ray bursts and γ-ray bursts, respectively. These high-energy phenomena can be subdivided in several categories depending on the burst duration (ranging from a hundredth of a second to thousands of seconds), the variability of the emission during the burst, and the shape of the emission curve with time. One type of X-ray burst is thought to be caused by explosive burning on the surface of a neutron star. The underlying nuclear process is the rp-process discussed in Section 9.4.4. Such bursts are not to be confused with the long-term continuous X-ray emission from hot ejecta in supernova remnants or the γ-line emissions from the decay of unstable nuclides produced in the explosive events.

A summary of explosive phenomena is given in Table 9.1. The energy quoted in Table 9.1 comprises the sum of kinetic energy and electromagnetic emission. It does not include any neutrino emission, if present. The energy is given in units of "bethe" (B). This unit was introduced by Steven Weinberg in 2006 to honor the recently deceased Hans Bethe. It is equal to the previously commonly used practical unit "foe" which Gerald E. Brown and Hans Bethe had introduced as an abbreviation for "ten-to-the-fifty-one ergs" in their work on the supernova mechanism. Therefore $1\ \mathrm{B} = 1\ \mathrm{foe} = 10^{51}\ \mathrm{erg} = 10^{44}\ \mathrm{J}$. The rest-mass energy of the Sun is $M_{\odot}c^2 = 1787\ \mathrm{B}$.

The detailed supernova classification scheme is provided in Figure 9.2. It is based on observational details in their spectral lines. Type II-L and II-P are distinguished by the temporal evolution of their lightcurves, with type II-P showing a plateau-phase when the luminosity stays almost constant over a period of about 100 days. It is remarkable that all supernova types can be explained with the explosion of a massive star after the collapse of its Fe-core (ccSN, see Section 9.3), with the exception of type Ia. Type Ia supernovae are modeled as the complete disruption of a white dwarf (an object without a hydrogen envelope), see Section 9.4.3. This is also called a *thermonuclear supernova*. Incidentally, a thermonuclear supernova releases a similar amount of energy than a core-collapse supernova when the energy released in neutrinos is not counted.

Novae and hypernovae are also sorted into further subcategories. Based on the variability of their lightcurves, novae are classified as NA (fast novae with a rapid

Table 9.1. Overview of Explosive Phenomena

Name	$\mathcal{L}^*$ Increase	Energy (B)	Mechanism
Nova	10^2-10^5	10^{-5}	Accreting white dwarf
Kilonova		10^{-2}–10^{-1}	NS + NS merger, NS + BH merger
Supernova	10^{11}–10^{12}	≈ 1.2	ccSN, white dwarf disruption
Hypernova	$>10^{13}$	10–100	Pair-instability SN? Collapsar? Magnetar?
X-ray burst			Various
γ-ray burst			Hypernova, supernova, kilonova

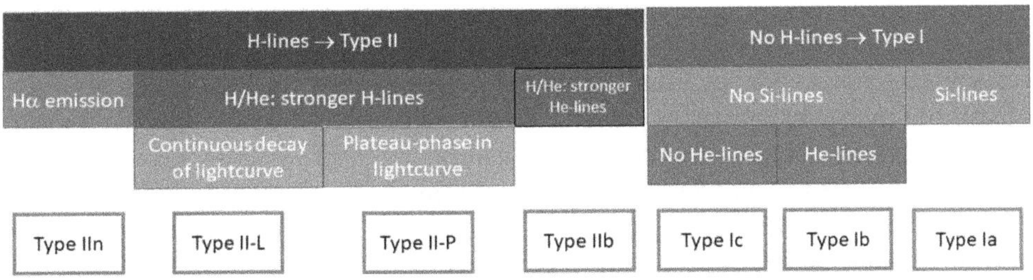

Figure 9.2. Supernova classification scheme based on observational features (line spectra, shape of the lightcurve). Only two different physical mechanisms are behind all types: type Ia supernovae are caused by the thermonuclear explosion of a white dwarf whereas all other types originate from the core collapse and subsequent explosion of a massive star.

rise in brightness and a subsequent decrease by 3 magnitudes within 100 days), NB (slow novae losing 3 magnitudes over 150 days or more), NC (very slow novae, staying at their maximum for 10 yr or more and then slowly fading), and NR (recurrent novae with more than one observed eruption, see Section 9.4.2). Hypernovae are subdivided according to their spectral properties, similar to supernovae. The term "hypernova" was first used to describe pair-instability supernovae but has subsequently evolved into encompassing a larger range of potential sources. Collapsars and magnetars are two possible mechanisms suggested to produce hypernovae. A collapsar is a fast rotating massive star which collapses into a black hole. Due to the fast rotation, a part of the stellar matter forms an accretion disk and jets. In the jets, matter is ejected at relativistic energies. In a magnetar, the jets are funneled from an accretion disk around a neutron star with a very strong magnetic field. Both phenomena also create γ-ray bursts. Further models for making hypernovae have been suggested.

The kilonova (sometimes also called mergernova or macronova) is the youngest member in the "nova family." Predicted since the early 2000s, it is caused by the merging of two neutron stars or a merger of a neutron star with a black hole. While the bulk of the neutron star matter is incorporated into a newly formed object (presumably a black hole), a smaller fraction can escape the gravitational well. This hot, dense, and neutron-rich plasma is one site of the r-process (see Sections 8.3, 9.4.5). A kilonova is also accompanied by a short γ-ray burst.

9.3 Core-collapse Supernovae

9.3.1 The Core Collapse and Formation of a Neutron Star

As explained in Chapter 6, stars with more than about 8 $M_\odot$ form a core mainly consisting of Fe and Ni from Si-burning (in short called Fe-core). Stars above 10 $M_\odot$ made this core in hydrostatic Si-burning whereas stars in the mass range $8 \lesssim M_\odot \lesssim 10$ (see Table 6.2 and the description of electron-capture supernovae in Chapter 6) form it during an already occurring collapse. Also the hydrostatically made Fe-core becomes unstable. The details of the collapse, formation of a neutron star, core bounce and subsequent supernova explosion are very intricate and especially the explosion aspect is not yet fully understood. It requires detailed hydrodynamical modeling in 3D with all the relevant physics included, including neutrino transport. As this is beyond the focus of this textbook, only a crude overview is given to convey a basic understanding of how the conditions giving rise to explosive nucleosynthesis in massive stars come about.

In the advanced stages of hydrostatic burning, the limiting masses for the cores of massive stars have to be described using Equation (2.43). Typical Fe-cores produced in Si-burning have a central Y_e of 0.42, which rises to $Y_e = 0.48$ at the outer edge of the core. Shortly before collapse, the electron entropy per baryon ranges from 0.4 in the center to 1.0 at the edge of the core of a 15 $M_\odot$ star and from 0.5 to 1.8 $M_\odot$ for a 25 $M_\odot$. This leads to limiting Chandrasekhar core masses of 1.34 $M_\odot$ and 1.79 $M_\odot$, respectively. As the core approaches and exceeds this generalized Chandrasekhar mass, there is no new source of nuclear energy which could re-establish hydrostatic equilibrium. As the core contracts further (on a Kelvin–Helmholtz timescale, see Equation (2.160), dominated by the neutrino luminosity), the density increases and it heats up. This gives rise to changes in the core based on three fundamental mechanisms: photodisintegration, electron capture, and NSE abundances shifting to heavier and heavier nuclides. As density increases, also E_F increases (see Equation (1.79)) and electrons are captured on Fe-group nuclei. This reduces Y_e and therefore also M_{Ch}, accelerating the collapse. Another way to look at this is that electrons are lost from the Fermi-gas exerting counter-pressure to the contraction and thus the pressure is reduced. Photodisintegration, predominant in more massive stars, has a similar impact on the pressure as photons are lost and radiation pressure is decreasing. It is important to note that at this point nuclides are not fully disintegrated (as originally envisaged before the 1980s) but mainly α-particles up to $X_\alpha \approx 0.1$ are forced out of Fe-group nuclides. This is because, due to the high T, the abundances are driven toward NSE abundances (Equation (4.145)) according to the current Y_e. The density, however, is not high enough to allow a fast triple-α rate and therefore free α-particles appear (see also Equation (4.148)). The combination of these effects rapidly turn the Kelvin–Helmholtz contraction into a free-fall collapse (Equation (2.154)).

Due to the shrinking Chandrasekhar mass, the inner core (with M^*_{Ch}, see Equation (2.43)) decouples from the outer part of the Fe-core. The inner core behaves like a polytrope with $\tilde{n} = 3$, for which any restoring forces vanish (see Section 2.4). Using the Lane–Emden coordinate ξ as defined in Section 2.4, it can be shown that for the contracting polytrope $\dot{r} \propto r$. This is called a *homologous collapse*. The radius at which the velocity $\dot{r}$ approaches the local sound speed $\sqrt{dP/d\rho}$ is taken to be the edge of the inner core. Matter in the inner core (collapsing at subsonic velocity) cannot "communicate" with matter in the outer core. The inner core collapses to nuclear density (the $\rho_0 \simeq 2.4 \times 10^{14}$ g cm^{-3} given in Equation (3.4)). According to the NSE equation, the increase in density in the inner core allows α-particles to be re-captured and pushes the composition to heavier and heavier nuclei, from initially $\langle A \rangle \approx 50$ to $\langle A \rangle \gtrsim 1000$. When the matter density of the nucleon "soup" outside the nuclei is comparable to the nuclear density ρ_0, the heavy nuclei merge together to start forming a *proto-neutron star*, i.e., the hot progenitor to a neutron star. When this occurs, the EOS is changed into a nuclear EOS for which the repulsion term of the strong nuclear force (see Section 3.1) dominates at extremely tiny distances between nucleons and prevents the proto-neutron star from further collapsing. The phase transition is very sudden and the inner core overshoots to several times ρ_0 and bounces back. The rebounding inner core contains about 0.7–0.8 $M_\odot$ and runs into the still collapsing outer core at about 7×10^5 km s^{-1}, creating an outward moving shock front. Up to about 1985, it was thought that this shock races through all the outer layers of the star and ejects them as a supernova. This is called the *prompt explosion mechanism*. Soon after, it was realized that the rebound energy in the shock is not sufficient to cause an explosion. Starting from the edge of the homologous core the shock front ploughs through the outer core, locally increasing its temperature. Shock energy is continuously lost by the photodisintegration of nuclei in the outer core. The energy loss can be estimated by realizing that most of the nuclei in the outer core still belong to the Fe-group with a binding energy of about 8.7 MeV/nucleon (see Section 3.2), which translates into 8×10^8 erg g^{-1} or 10^{52} erg per $M_\odot$. Typical kinetic energies in the shock after the rebound are (3–8) $\times 10^{51}$ erg (depending on the stiffness of the nuclear EOS), implying that the prompt shock spends all its energy already when moving though the outer core. It may break out of the outer core only for very small core masses (and favorable nuclear EOS) but it is not clear whether these appear in stars massive enough to initiate core collapse. The shock wave loses further energy through neutrinos which can freely escape when the shock moves to lower-density regions with $\rho < 10^{12}$ g cm^{-3}. Neutrinos of all flavors diffuse out ahead of the shock. Behind the shock, scattering between electrons and neutrinos of all flavors lowers the average neutrino energies and makes it easier for neutrinos to escape, carrying off energy, because their interaction cross sections scale with E_ν^2. This aggravates the problem of having a successful prompt explosion. After about 10 ms after core bounce the outward motion of the shock has stalled and the shock has turned into a standing accretion shock. Matter from the outer core keeps falling in at a rate of 1–10 $M_\odot$ s^{-1} and accretes on the proto-neutron star in the center.

During the hydrostatic burning phases of a star, neutrinos escape freely and lose (almost) no energy due to their very weak interaction with matter. At densities $\rho \gtrsim 10^{11}$ g cm^{-3}, however, the mean free path of neutrinos becomes so short that they make many interactions before being able to escape. This is called *neutrino trapping*. The critical density beyond which neutrinos are trapped depends on the interaction cross-sections for scattering of neutrinos on electrons, nucleons, and nuclei. When heavy nuclei are present, neutrino–nucleus scattering dominates for ν_e because its cross section scales as $\sigma \propto Z^2$. In the trapping region, electron captures and neutrino captures are in equilibrium (see Equation (3.175)) and the total lepton abundance stays constant, $Y_{lep} = Y_e + Y_\nu =$ const. The electron abundance stops decreasing and therefore also M_{Ch}^* becomes constant. Typical values found in simulations are $Y_{lep} = 0.36-0.38$, resulting in $M_{Ch}^* \approx 0.7$–$0.8\ M_\odot$. In analogy to the photosphere of a star, a *neutrino sphere* can be defined which is the surface of the last scattering events before the neutrino can escape. For ν_e, which can interact with nuclei and nucleons through charged-current and neutral-current interactions (mediated by the weak nuclear force), the sphere is located at a radius in the range of 80–100 km. The sphere for $\bar{\nu}_e$ is about 20 km further inwards because of lower interaction cross-sections. The innermost neutrino sphere is the one of ν_μ, $\bar{\nu}_\mu$, ν_τ, $\bar{\nu}_\tau$ because they do not take part in charged-current reactions with nucleons and nuclei but only in neutral-current processes, such as elastic scattering on electrons and nucleons (see also Section 7.2.3). Therefore they decouple already at higher density than ν_e and $\bar{\nu}_e$, further inside the collapsing core. The energy dependence of the neutrinos (the neutrino spectra) coming from the neutrino spheres, although not exactly thermal, can be parameterized as a Fermi–Dirac distribution (see also Section 1.5.1.2)

$$f(E_\nu) \propto \frac{E_\nu^2}{(k_B T_\nu)^3}[1 + e^{E_\nu/(k_B T_\nu)-a'}]^{-1}, \qquad (9.1)$$

where a' is an effective degeneracy parameter, related to a chemical potential. The neutrino temperature T_ν is different for the different neutrino types because of the different location of the neutrino spheres. Since T_ν is increasing toward the center of the collapse, neutrinos emitted from deeper neutrino spheres are coming from a hotter environment. The exact values depend on details of the interaction cross sections, neutrino transport, and hydrodynamics of the collapse. The neutrino temperature was estimated higher in the older literature ($k_B T_{\nu_e} = 5$ MeV, $k_B T_{\bar{\nu}_e} = 6$ MeV, $k_B T_{\nu_x} = 8$ MeV, where ν_x comprises all neutrino flavors except for $\nu_e, \bar{\nu}_e$) but more widely used in later investigations were lower values ($k_B T_{\nu_e} = 4$ MeV, $k_B T_{\bar{\nu}_e} = 5$ MeV, $k_B T_{\nu_x} = 6$ MeV). The latest values with improved neutrino interaction cross sections and detailed neutrino transport are even lower ($k_B T_{\nu_e} = 2.8$ MeV, $k_B T_{\bar{\nu}_e} = 4$ MeV, $k_B T_{\nu_x} = 4$ MeV). It should be noted that $T_\nu \approx 3.15\langle E_\nu \rangle$ is not necessarily equal to the plasma temperature T even in the trapping region. Simulations have found significant divergence from full thermal equilibrium.

The proto-neutron star contracts from an initial radius of about 50 km to a radius of 10–12 km, initially while still accreting matter and later without further accretion on a Kelvin–Helmholtz timescale of 3–10 s. The released gravitational binding energy of about 3×10^{53} erg is radiated away in neutrinos of all flavors, created by neutrino pair-production. It may be surprising to note that this is a luminosity comparable to what is found in the rest of the Universe combined at that moment, with roughly 10^{10} galaxies with on average $\mathcal{L}^*_{\text{galaxy}} \approx 10^{43}$ erg s^{-1}! The time-dependence of the total neutrino luminosity of the neutrino spheres (summed over all flavors) has been found to decrease as

$$\mathcal{L}^*_\nu = \mathcal{L}^*_{\nu,0} e^{-t/\tau_\nu}, \tag{9.2}$$

with $\mathcal{L}^*_{\nu,0} = 3 \times 10^{53}$ erg and $\tau_\nu = 3$ s (note that also the $\langle E_\nu \rangle$ are time-dependent). It was realized early on that the energy lost from the shock could be replaced if only a small fraction of the energy in neutrinos could be transferred to matter. The current idea of how to achieve a successful explosion is that about 1% of the energy released in neutrinos is transferred to matter in the neutrino trapping region. This would be sufficient to explain the observed explosion energies. The transferred energy heats the material behind the shock wave (mainly by ν_e, $\bar{\nu}_e$ whereas the contribution of other flavors is estimated to $\lesssim$20%), causing further expansion and thus reviving the stalled shock front. If enough energy is deposited, then the shock front can be pushed out of the outer core and then pass through and eject the outer layers of the star. Shock revival happens only after another 0.5–1 s due to the time neutrinos take to penetrate the trapping region and reach the neutrino sphere (random walk, see Section 1.6.2). This is the so-called *delayed explosion mechanism.* A full 3D simulation of the collapse and subsequent shock front is among the most demanding computational problems of our time. The heating efficiency by neutrinos is not only affected by their interaction cross sections but also by hydrodynamics (resolution, convection treatment, asymmetry effects) and the way the full Boltzmann (fluid) transport equations for neutrinos are solved. It has become obvious that 3D modeling is essential but in this case approximations for neutrino transport have to be used to make the simulation computationally feasible. While the simulations did not produce successful explosions in early attempts (unless unphysical and too optimistic approximations were implemented), recent years have seen progress toward achieving the required energy deposition. Further complications arise when desiring to include rotation, magnetic fields, or neutrino oscillations. Obviously, massive stars explode but a detailed understanding of how this explosion comes about still is an active and "hot" research field.

Once the shock wave overcomes the infalling matter stream and escapes the outer core, it can easily make its way to the stellar surface, imparting kinetic energy onto the outer layers. As the dynamical timescale of these outer layers (Equation (2.159)) is much longer than the time it takes for core collapse, bounce, and shock revival the hydrostatic shell burning continues and does not notice what is going on in the core. Only when the shock front arrives, density and temperature are suddenly raised and this leads to a brief episode of explosive burning while the layer is ejected from the star. Details of the explosive shell burning are given below. The ejected layers become the *supernova remnant*, which is an expanding cloud of hot plasma, eventually cooling down and forming a gas of atoms and molecules. The cloud initially emits X-rays and γ-rays and becomes visible in the optical after a few hours to days. The visible light output is still increasing for a few hours after which the luminosity decays exponentially. This is called the *supernova lightcurve*. The change in luminosity over the first several months is driven by the amount of kinetic energy in the ejecta. At later times, however, the expanding cloud cools much slower than expected from regular expansion and radiation losses. This is due to continued heating by the decay heat of also ejected radioactive nuclei. Specifically, a large mass fraction is ejected as ^{56}Ni ($t_{1/2} = 6.07$ d) which determines the lightcurve's evolution at intermediate times. After several weeks, the slope of the luminosity curve changes when the decay of ^{56}Co (the product nucleus of the ^{56}Ni decay with $t_{1/2} = 77.24$ d) takes over the task of slowing down the overall cooling. While the supernova classification (Figure 9.2) is constructed around the lightcurve in the visible range of wavelengths, it is interesting to note that lightcurves may exhibit a very different behavior at other wavelengths. For example, the X-ray emission of certain supernovae is much more uniform over time and may also extend to very long times. In another example, at ultraviolet wavelengths there is an early extremely luminous peak lasting only a few hours which does not have a visible counterpart.

The compact object in the center of the remnant cannot only be a neutron star but also a black hole. This depends on the initial mass of the collapsing Fe-core but also on the later fallback of material after the explosion. Not all the matter above the stalled and revived shock is ejected, a fraction of it is falling back. The amount of fallback also depends on the details of the shock revival and the conditions in the core and therefore is quite uncertain. The amount of ejected ^{56}Ni can provide clues, nevertheless. It is coming from the deepest, still ejected layers and the ejected amount can be inferred from the shape of the later lightcurve. Further radioactive nuclides also originating from deep layers with appropriate half-lives for later detection are, for example, ^{44}Ti, ^{57}Co, ^{60}Fe, and ^{26}Al. Their γ-emission in the decays can be observed by γ-ray telescopes in satellites. They provide a further diagnostic of the explosion mechanism and a cross check for the models.

The proto-neutron star formed in the core-collapse continues to emit neutrinos also after the stellar layers have been ejected. About 50 s after the core bounce it settles to its final radius of about 10–12 km (depending on the actual equation-of-state) but is still hot, with a core temperature of $kT \approx 0.5$ MeV. Neutrino cooling continues and after 50–100 yr the compact object has become isothermal by heat exchange between its core and the surface via conduction in the electron gas. X-rays are emitted from the surface but the cooling is dominated by neutrinos by several orders of magnitude until the neutron star reaches an age of about 3×10^5 yr. The most important neutrino cooling mechanism in compact objects (also white dwarfs) with high density, in which other heat transport mechanisms are inefficient, see Section 2.7.3, is the *Urca process*. The basic idea behind the Urca process is a reaction cycle, alternatingly capturing an electron and making a β^- decay,

$$e^- + A \rightarrow B + \nu_e,$$

followed by

$$B \rightarrow A + e^- + \bar{\nu}_e.$$

The net reaction is

$$e^- + A \rightarrow A + e^- + \nu_e + \bar{\nu}_e. \tag{9.3}$$

There is no change in composition because the nucleus A is recreated from the decay of nucleus B but the neutrino–antineutrino pair carries off energy. Since the reaction sequence in Equation (9.3) is endothermic, the energy put in neutrinos comes from the thermal environment. The energy loss rate in this process is difficult to determine. It does not only depend on temperature and density but also on the specific properties of the involved nuclides. In neutron stars, nucleons act as the catalysts. The *direct Urca process* with nucleons is

$$\begin{aligned} n &\rightarrow p + e^- + \bar{\nu}_e \\ e^- + p &\rightarrow n + \nu_e. \end{aligned} \tag{9.4}$$

The necessity of simultaneously conserving energy and momentum requires the proton-to-neutron ratio to exceed 1/8 which is far above the ratio found in neutron star matter with densities around nuclear density ρ_0. Therefore an alternative was suggested, the *modified Urca process*, in which an additional nucleon participates to allow momentum conservation:

$$\begin{aligned} n + n &\rightarrow p + n + e^- + \bar{\nu}_e \\ e^- + p + n &\rightarrow n + n + \nu_e. \end{aligned} \tag{9.5}$$

The additional nucleon in the above reaction sequences is assumed to be a neutron but it could also be a proton. The modified Urca rate is a factor 10^{-4}–10^{-5} slower than the direct Urca rate and the cooling process is correspondingly slower. In this case neutron stars can remain observable by thermal photon emission from the surface for several million years. The proton-to-neutron ratio in a neutron star scales with density, however, and therefore it is essential to know the EOS of the neutron star well to be able to tell whether the direct process is ruled out or not. It is interesting to note that the thermal radiation from a neutron star is not a blackbody radiation. The emission from the surface is considerably modified by the neutron star atmosphere, a thin layer of intermediate and heavy nuclides. This poses a severe problem to infer effective surface temperatures and radii of neutron stars.

Further properties of neutron stars are discussed in Section 9.4.5.

9.3.2 Artificial Explosions

Due to the prevailing uncertainties in the details of the explosion mechanism and the difficulties of simulating the revival and outward propagation of the shock in 3D, investigations of explosive nucleosynthesis in massive stars have a long history in using artificial explosions of (mostly 1D) models. In such models, stellar evolution is followed in a stellar model as described in Chapter 2 until the onset of the collapse of the Fe-core. At this point the simulation is halted and the stellar structure obtained so far is stored. An assumed explosion energy is then deposited into deep layers and the outward propagation of the resulting shock wave is followed in a hydrodynamic code. Crucial for obtaining realistic stellar yields is the obtained mass cut. The *mass cut* is the (idealized) mass shell (or radius, if using Euler coordinates instead of Lagrange coordinates, Section 2.1) beyond which matter is ejected, whereas matter below the mass cut falls back and becomes part of the proto-neutron star. Nucleosynthesis can be either followed simultaneously by coupling extended reaction networks to the hydrodynamic evolution or by post-processing the composition in each layer with the obtained shock wave profiles (see Sections 4.2, 4.2.3, and 9.1). There are several ways to deposit the required energy. Over the past decades, variants of approaches in these broad categories have been used:

- *Piston:* Kinetic energy is deposited into deep layers by modeling the motion of a piston placed at some radius (or mass shell) and then first moving inwards, followed by an outward movement to simulate the core collapse and bounce. Both inward and outward motion are modeled as ballistic, free-fall motion (Section 2.9.2) with a modified local gravitational acceleration to control the amount of deposited energy. The initial position of the piston is set to the edge of the Fe-core identified by a Y_e discontinuity and the innermost position is set to 500 km. The obtained mass cut is determined by the initial position of the piston (which is not treated as a free parameter) and imposed kinetic energy because some of the material in the outer layers will fall back and settle on the piston. The energy is chosen to reproduce either the observed explosion energies or the observed amount of ejected ^{56}Ni.
- *Kinetic energy bomb:* A layer at a given radius or mass coordinate (usually at 1 $M_\odot$, that is inside the Fe-core) is given an outward velocity, chosen appropriately to eject all the matter outside the core. This will also eject some core material and can also be fine-tuned to reproduce observations.
- *Thermal bomb:* An inner boundary is chosen at a fixed radius. Thermal energy (heat) is artificially injected either by increasing the luminosity at the

inner boundary or by depositing energy at a constant rate during a specified time into a region above the inner boundary, bounded by specified mass coordinates. This leads to an explosive expansion of the material above the inner boundary. A mass cut and the amount of deposited energy have to be specified.

- *Modified neutrino-energy deposition:* In an attempt to reproduce the neutrino heating found in multi-D simulations to obtain better predictions of explosion energy, mass cut, and nucleosynthesis (in particular in the deepest layers which are barely ejected and in which Y_e is modified during the explosion due to the large neutrino fluxes), recent approaches focus on implementing an effective description of neutrino-energy deposition. One such approach is the "neutrino light-bulb" in which the proto-neutron star is cut out and replaced by a fixed boundary emitting neutrinos using an analytical description for the luminosities of the various neutrino species. This triggers explosions when suitable luminosities are chosen. When using detailed Boltzmann neutrino-transport schemes, artificially increased neutrino opacities have been used to achieve higher energy deposition leading to an explosion. A common draw-back to these approaches is that the ad hoc modification of the properties of ν_e and $\bar{\nu}_e$ may lead to an inconsistent impact on Y_e, and thus on the composition, of the innermost ejecta. To avoid this, more sophisticated neutrino treatments have come up recently. Among them is the PUSH method, in which the energy in the neutrino species other than ν_e, $\bar{\nu}_e$ is made available for heating.

Applied to 1D models, all these approaches result in a comparable behavior of the shock wave in the layers outside of the Fe-core, see Equation (9.9). Differences are found especially in the innermost, barely ejected layers which are under the influence of the strongest neutrino flux from the proto-neutron star, called the *neutrino wind*. Approaches injecting kinetic energy are not suited to describe these layers whereas thermal bomb models may use appropriate entropies but still have to manually specify the mass cut. Approaches modifying the neutrino-energy deposition may be better suited for describing the innermost layers but only if they are able to use luminosities and opacities for ν_e, $\bar{\nu}_e$, which are consistent with those obtained from multi-D models and realistic nuclear physics. Some details of nucleosynthesis in the neutrino wind are presented in Section 9.3.3. Nevertheless, all results for the innermost layers coming from 1D models have to be taken with a grain of salt. Multi-D simulations show convection and mixing processes within the core and also further out behind the shock which render the notion of a well-defined mass cut untenable. This may even affect the production of p-nuclides because contributions to the p-nuclides may come from layers far below the Ne/O-layer (for a discussion of the γ-process see below and in Section 8.4).

9.3.3 Nucleosynthesis in the Neutrino Wind and the νp-Process

After shock revival and onset of the actual explosion, the proto-neutron star still accretes matter from the inner Fe-core and fallback material. After several hundreds of milliseconds up to about a second, however, accretion ceases and the proto-neutron star contracts and cools on a Kelvin–Helmholtz timescale by further emission of neutrinos of all flavors. These neutrinos continue to deposit energy, mainly by the reactions given in Equation (3.175), in the cooler layers just above the neutrino sphere(s), leading to a continuous, dilute outflow (initial mass loss rate is a few 10^{-2} $M_\odot$ s^{-1}) from the surface of the newly formed neutron star (behind the regular shock wave and therefore behind the ejecta expelled by the revived shock). This outflow is called *neutrino-driven wind* (in short: neutrino wind). Its properties sensitively depend on the neutron star equation-of-state (determining radius and mass of the nascent neutron star) and on the luminosity and spectra of the emitted neutrinos. In particular, the Y_e in the neutrino wind depends on the differences in luminosities and spectra of ν_e and $\bar{\nu}_e$. The electron abundance is given by

$$Y_e = \left[1 + \frac{L_{\bar{\nu}_e}\left(\langle E_{\bar{\nu}_e}\rangle - 2\Delta_{np} + 1.2\Delta_{np}^2/\langle E_{\bar{\nu}_e}\rangle\right)}{L_{\nu_e}\left(\langle E_{\nu_e}\rangle + 2\Delta_{np} + 1.2\Delta_{np}^2/\langle E_{\nu_e}\rangle\right)}\right]^{-1}, \tag{9.6}$$

when equilibrium between ν_e- and $\bar{\nu}_e$-captures can be assumed. Nucleosynthesis in the neutrino wind, in turn, is very sensitive to the prevailing Y_e. There are still considerable uncertainties affecting the conditions in the neutrino wind, with respect to neutrino transport but also the nuclear physics of the neutrino interactions and the neutron star itself. Moreover, neutrino spectra could also be affected by non-standard effects such as flavor oscillations (see Section 7.2.3).

For a long time, the neutrino-driven wind has been considered as the site of the r-process (see Section 8.3) because it was thought that Y_e is low and the temperature high enough to allow fast neutron captures. As explained in Section 9.3.1, however, the density attained even in the ejected part of the Fe-core is not high enough to permit a full NSE. Instead, these layers experience an *α-rich freeze-out* when expanding outwards. Basically, this is a QSE (Section 4.3.2) with one QSE group being around ^{4}He and another in Fe-group nuclides. Figure 9.5 shows that the conditions obtained in ejected material do not allow a full NSE. According to the discussion in Section 8.3.2, for each given Y_e a specific entropy range is required to allow a full r-process. Current multi-D simulations suggest that the required combination of entropy and Y_e cannot be found in the wind ejecta. For the time being, ccSN are therefore ruled out as a site for the r-process. According to the current view, the only feasible way to have an r-process in exploding massive stars is in magnetically driven jet outflows from explosions of rotating stars with high magnetic fields (starting at around 10^{12} G). The matter ejected in the jet comes from deep regions close to the surface of the neutron star and is expected to have $Y_e = 0.1$–0.15. Fast jet ejection along a polar axis avoids further interactions with neutrinos in the wind, which could increase Y_e, thus providing the environment suited for commencing with an r-process in the freeze-out phase. Details are sensitive to modeling the effects of rotation and magnetic field. It is currently being studied how efficient the r-process could be in such sites and how much they can contribute to the total r-process budget in the Galaxy (see the reviews given in the suggestions for "Further Reading" in this chapter and in Chapter 8).

A number of multi-D simulations did not only fail to provide Y_e low enough for the r-process in the neutrino-driven wind but even predicted proton-rich neutrino winds at early times ($\lesssim$1 s). This is because as long as the luminosities and average energies of ν_e and $\bar{\nu}_e$ are comparable, the neutron-proton mass difference favors a proton-rich composition (see also Equation (9.6)). As noted above, this prediction depends on details of the used neutrino transport and nuclear physics input. Therefore currently it is not yet clear how proton-rich the matter in the neutrino-wind can actually become. The Y_e of neutrino-wind material may also be time dependent because the opacity of the neutron star surface changes with respect to ν_e and $\bar{\nu}_e$ and becomes more opaque for $\bar{\nu}_e$ at later times, leading to a decrease in Y_e.

If there is a proton-rich wind component, a so-called *νp-process* can ensue. The wind ejecta quickly cool from the initially very high temperature (10 GK, with only free neutrons and protons present) in an α-rich freeze-out, creating mainly ^{56}Ni and α-particles. Because of $Y_e > 0.5$, also a large number of free protons is left. In the temperature window $1.5 \lesssim T \lesssim 3$ GK, rapid proton captures ensue on ^{56}Ni and subsequently produced nuclides. Production of heavier nuclides would be stopped at ^{64}Ge, which has an electron-capture lifetime longer than a minute. This is too long in comparison with the expansion timescale (of the order of seconds) to allow for production of an appreciable number of nuclides beyond ^{64}Ge before nuclear reactions freeze out. At νp-process conditions, however, a small number of free neutrons is continuously created by $\bar{\nu}_e$ captures on the free protons. This supply of free neutrons allows for (n,p) reactions bypassing any slow electron captures and β^+ decays, not just of ^{64}Ge, but also of other potential bottlenecks at higher mass number. A typical reaction flow is shown in Figure 9.3, showing how the "waiting points" with slow proton-capture rates and long half-lives (for example, ^{64}Ge, ^{68}Se, ^{72}Kr, ^{76}Sr) can be overcome by fast (n,p) reactions. Some outflows from the proton-capture chains can also occur before reaching a waiting point because (n,p) reactions are favored on the proton-rich side of the nuclear chart, as long as free neutrons are available.

The main nucleosynthesis flow in the νp-process is characterized by rapid proton captures in a (p,γ)–(γ,p) equilibrium with (n,p) reactions connecting the contiguous isotonic chains. The abundances within an isotonic chain in (p,γ)–(γ,p) equilibrium can be calculated by a similar relation as shown in Equation (4.150) for the (n,γ)–(γ,n) equilibrium, see Section 4.3.3. It is also discussed in Section 9.4.4. Although such an equilibrium is also achieved in the rp-process on the surface of accreting neutron stars (see Section 9.4.4), the νp-process proceeds at lower density than the rp-process. The resulting nucleosynthesis path follows the $N = Z$ line only up to the Mo region, reaching further and further into neutron-richer isotopes between Mo and Sn, moving gradually away from the $N = Z$ line. The path is pushed strongly toward stability at the Sn isotopes and above, providing a strong barrier for the efficient production of any elements beyond Sn. Decay and (n,p) reaction timescales are longer for nuclides closer to stability and the higher Coulomb barriers suppress proton captures.

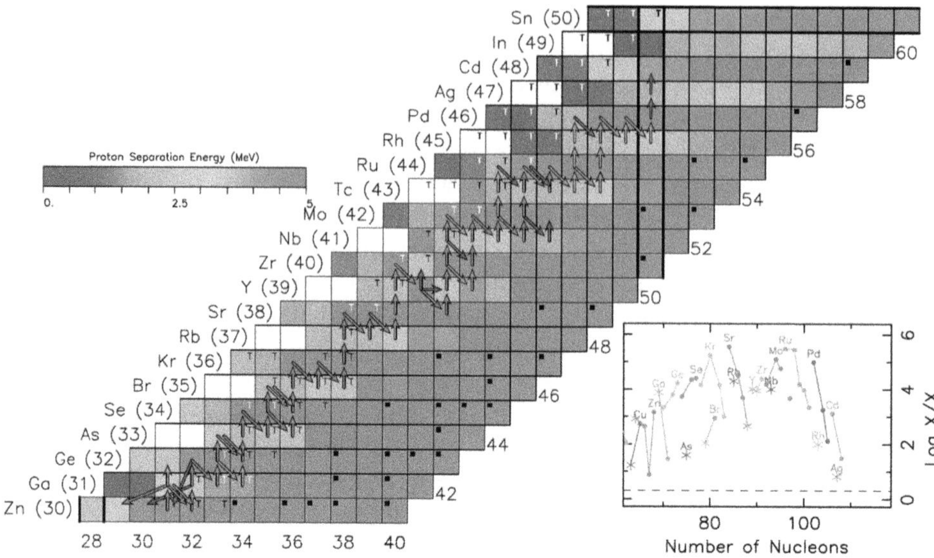

Figure 9.3. Nucleosynthesis path in the νp-process. Net nuclear flows are plotted in three strengths: red (strong), green (intermediate), and blue (weak). Stable species are represented by a filled black square in the upper left corner. Each nucleus is color-coded by the value of its S_p. Proton unbound nuclei are colored white. Nuclei with $S_p > 5$ MeV are colored gray. Nuclides with theoretical binding energy are labeled by "T." Production factors of stable nuclides (including the decays of their radioactive progenitors after the end of the νp-process) are given in the inset. (Reproduced from Pruet et al. 2006. © 2006. The American Astronomical Society. All rights reserved.)

The location of the effective νp-process path is determined by the nuclear properties giving rise to the (p,γ)–(γ,p) equilibrium and the very fast (n,p) reactions, and remains remarkably unaffected by variations of the astrophysical parameters within realistic limits such as entropy, Y_e, and expansion timescale, as long as the conditions permit the appearance of a νp-process. Whenever a νp-process occurs, the nucleosynthesis path beyond ^{56}Ni initially follows the $N = Z$ line and gradually veers off toward stability. Systematic variations of reaction rates show only small effects, if any, regarding the path location. This is a consequence of the (p,γ)–(γ,p) equilibrium in which the path is determined by nuclear mass differences (see Section 4.3.3). All these variations, however, determine how far up the path follows the $N = Z$ line before diverging, or whether it is terminated already at low Z. Consequently, it is clear that the achieved abundances within the path are also determined by these conditions. Similar to the neutron-to-seed abundance ratio $r_{\rm seed}$ introduced for r-process nucleosynthesis (see Section 8.3.2), the number ratio Δ_n of free neutrons—created by the reaction p($\bar{\nu}_e$,e$^+$)n—to seed nuclei is a good indicator for the strength of the νp-process. It is given by

$$\Delta_n = \frac{Y_p}{Y_{\rm seed}} \bar{n}_{\bar{\nu}_e} = \frac{Y_p}{Y_{\rm seed}} \int_{T_9 \leqslant 3} \lambda_{\bar{\nu}_e} \, dt, \tag{9.7}$$

where $\lambda_{\bar{\nu}_e}$ is the rate for $p + \bar{\nu}_e \rightarrow n + e^+$ and Y_{seed} is the seed abundance, i.e., the abundance of nuclei with $Z > 2$, taken at the onset of the νp-process at $T_9 = 3$. The seed abundance is in large part determined by the abundance of ^{56}Ni. The $\bar{\nu}_e$-capture rate can be estimated to

$$\lambda_{\bar{\nu}_e} \approx C \frac{L_{\bar{\nu}_e} k T_{\bar{\nu}_e}}{r^2}. \tag{9.8}$$

The rate in captures per second is obtained with a numerical prefactor $C = 0.015$ when specifying the electron antineutrino luminosity $L_{\bar{\nu}_e}$ in units of 10^{52} erg s^{-1}, the effective temperature $kT_{\bar{\nu}_e}$ in MeV, and the distance r to the $\bar{\nu}_e$-sphere in 10^8 cm. As in the r-process, a combination of Y_e and entropy S also determines Δ_n (if the triple-α rate is known, see below). Larger Δ_n implies more neutrons and therefore faster (n,p) rates and thus faster processing from ^{56}Ni to heavier species. The production of heavier nuclides, however, is very sensitive to two rates of particular importance, the triple-α rate and the rate of ^{56}Ni(p,n)^{56}Co. As explained before, the triple-α rate connects the QSE group of light nuclides to those of Fe-group nuclides and therefore determines the amount of seed nuclides available at the onset of the νp-process. A slower triple-α rate leaves more protons and thus reduces the ^{56}Ni seed. Figure 9.4 shows the resulting mass fractions for two choices of the triple-α rate, corresponding to the rates shown in Figure 9.1. The trajectory numbers in both plots refer to similar conditions regarding Y_e and S. Due to the difference in the triple-α rate, however, the same Δ_n is obtained for different combinations of Y_e and S. For example, $\Delta_n \approx 19$ (with a production peak around $A \approx 85$–90) is found in trajectory #13 ($Y_e = 0.675$, $S^b = 82.9\ k_B$ baryon^{-1}) for the Fynbo et al. (2005) rate but in trajectory #18 ($Y_e = 0.7$, $S^b = 123\ k_B$ baryon^{-1}) when using the Angulo et al. (1999) rate. Two rates are competing with the triple-α rate, ^{7}Be(α,γ)^{11}C and ^{10}B(α,p)^{13}C. Uncertainties in these rates may have a similar impact as those of the triple-α rate. The reaction ^{56}Ni(p,n)^{56}Co, on the other hand, is the major bottleneck for the further processing starting from ^{56}Ni and therefore determines the overall efficiency of the νp-process across the range of all possibly created nuclides.

Sensitivity studies exploring the parameter space spanned by Y_e and S have identified further bottlenecks in addition to the above reactions which may be

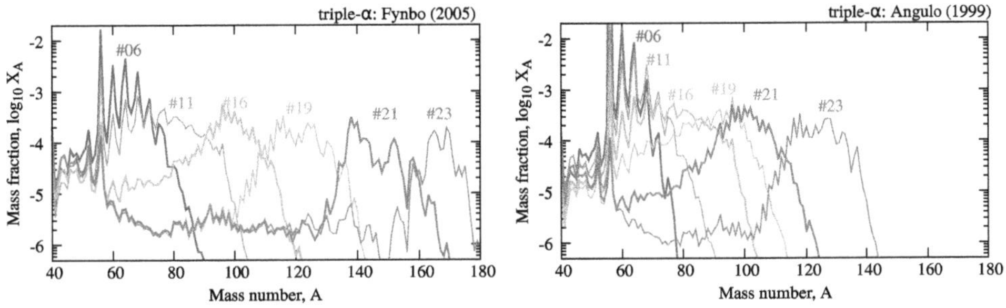

Figure 9.4. Nucleosynthesis in the νp-process for two choices of the triple-α rate, corresponding to the triple-α rates shown in Figure 9.1. (Reproduced from Nishimura et al. 2019, by permission of Oxford University Press on behalf of the Royal Astronomical Society.)

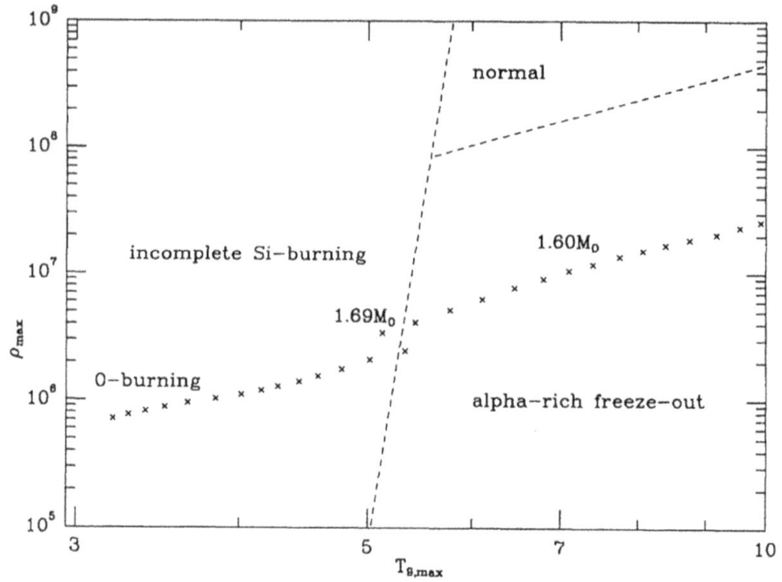

Figure 9.5. Explosive Si-burning at various combinations of peak temperature and peak density. The crosses mark conditions encountered in trajectories from ejected layers of a ccSN. (Reproduced from Thielemann et al. 1990. © 1990. The American Astronomical Society. All rights reserved.)

encountered in the path to heavier nuclides. It also has been shown that reaction cycles may form via (n,α) and (p,α) reactions, which impede the processing toward larger A.

The importance of the νp-process for nucleosynthesis lies in the fact that it could produce unstable nuclides which subsequently decay to stable, proton-rich nuclides, as shown in the inset of Figure 9.3. Thus, it could provide an explanation for high abundances of Sr, Y, Zr in metal-poor stars. It could also provide a contribution to the light p-nuclides (see Section 8.4) and thus offer a way out of the underproduction problem of light p-nuclides in the γ-process in massive stars (see Section 9.3.4). Due to the current uncertainties in modeling the neutrino luminosities, and thus the conditions in the neutrino-driven wind outflow, it remains an open question, however, whether the νp-process actually occurs in massive stars and up to which mass number A it can efficiently produce nuclides. Nevertheless, the νp-process may not be confined to massive stars but could appear also in other sites providing appropriate conditions, such as outflows from accretion disks around black holes and neutron stars.

9.3.4 Explosive Shell Burning

The situation in a massive star with an initial mass exceeding about 10 M$_\odot$ with its onion-shell structure of layers originating from different burning phases stacked on top of each other is shown in Figure 7.6. As the dynamical timescale of these outer layers (Equation (2.159)) is much longer than the time it takes for core collapse, bounce, and shock revival, the hydrostatic shell burning continued and did not notice what was going on in the core. Only when the shock front arrives, density and temperature are suddenly raised and this leads to a brief episode of explosive burning while the layer is ejected from the star. As explained in Section 9.1, the crucial quantity is the burning timescale in comparison to the time it takes the shock to pass through. The shock passes within 1 – 2 s, estimated from the hydrodynamic timescale (Equation (2.160)) $\tau_{KH} \approx 446/\sqrt{\rho} \approx 1$ s (ρ in g cm^{-3}). The burning timescale, in turn, is mainly determined by the temperature reached in the shock, except for the triple-α reaction, as already discussed. Except for the deepest layers close to the origin of the final shock wave and in the neutrino-driven wind, the temperature as function of radius can be approximated by

$$T_{\rm peak} = C\left(\frac{E_{\rm exp}}{r^3}\right)^{1/4}. \tag{9.9}$$

This equation yields $T_{\rm peak}$ in K, when the numerical coefficient is $C = 1.33 \times 10^{10}$, and the explosion energy (as kinetic energy) $E_{\rm exp}$ is given in multiples of 10^{51} erg and the radius r in multiples of 10^8 cm. For a typical explosion energy of 1.2×10^{51} erg, $T_{\rm peak} > 5$ GK is attained for $r \lesssim 3700$ km. At this temperature NSE is achieved and therefore the layers ejected from this deep region will consist of Fe-group nuclides (the dynamical timescale is too short to change Y_e from the value it had after the hydrostatic burning phase) and α-particles. As mentioned before and further illustrated by Figure 9.5, densities in the outer layers of a massive star are too low to allow a full NSE even at high temperature and therefore an α-rich freeze-out will always be found, with α-particles surviving. For temperatures below 4–5 GK the dynamic timescale set by the passing shock is too short for complete processing into an α-rich QSE, leading to incomplete Si-burning. On the other hand, at about 13,000 km from the stellar center $T_{\rm peak}$ has dropped to 2 GK. Beyond that radius, no further burning of fuels heavier than He takes place (see Section 9.1).

Figure 9.6 and 9.7 show the change in mass fractions by explosive burning in the Si- and O-layers of a 15 $M_\odot$ star, obtained by using a thermal bomb approach. An energy of 1.57×10^{51} erg is injected for 0.02 s into a region containing 0.02 $M_\odot$ above the mass cut at 1.5 $M_\odot$. It can nicely be seen how the ^{28}Si is converted to ^{56}Ni at NSE conditions but with a considerable amount of ^{4}He surviving due to the ineffective triple-α rate at low density (see Section 9.1). In a part of the O-layer, ^{16}O is converted to mainly ^{28}Si and ^{32}S. Further out, the temperature in the shock is not high enough to significantly burn ^{16}O during the time of the shock passage.

The shells experiencing explosive Si- and O-burning show neutron excesses $\eta_n \approx (2\text{–}4) \times 10^{-3}$ for solar metallicity stars and $2 \times 10^{-4} \lesssim \eta_n \lesssim 10^{-3}$ for stars with 0.01 $Z_\odot$. Although QSE abundances are sensitive to η_n, this shows that the resulting abundances are only weakly sensitive to metallicity and the impact of progenitor mass and convection may be more important. In massive stars, incomplete Si-burning occurs with an α-rich freeze-out, see Figure 9.5. The most abundant nucleus under such conditions is ^{56}Ni. Due to the presence of α-particles a fraction of the produced $^{56-58}$Ni is converted to $^{60-62}$Zn by α-captures during the freeze-out. $^{60-62}$Zn is then decaying to $^{60-62}$Ni. Typical amounts of ^{56}Ni ejected by ccSN are

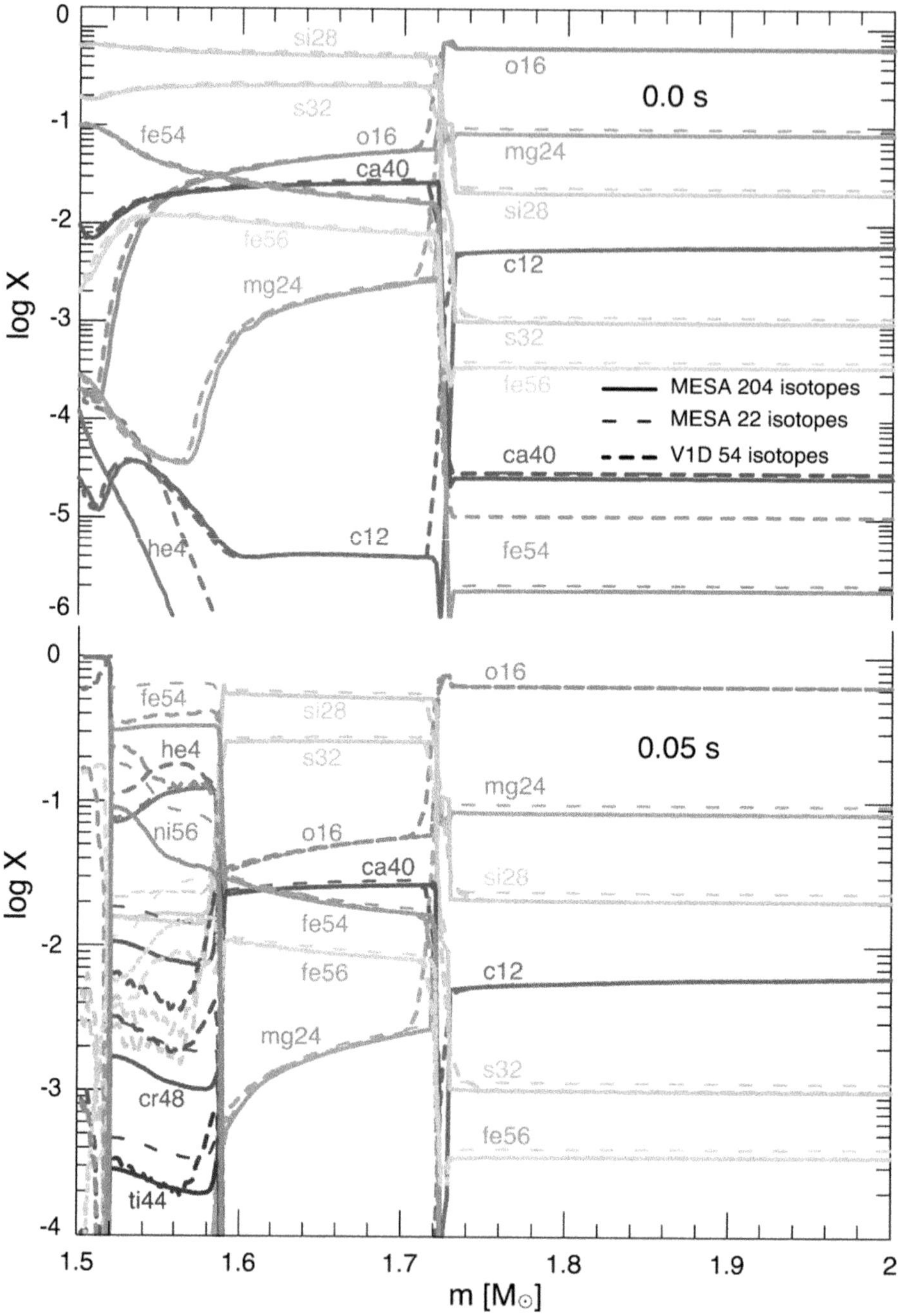

Figure 9.6. Propagation of explosive burning in the layer comprising 0.5 $M_\odot$ above the Fe-core; the composition of the Si- and O-layers at the onset of the shock passage is shown in the top panel and the situation after 0.05 s, when the shock front has moved up about 1.58 $M_\odot$ is shown in the bottom panel. The mass fractions of selected nuclides obtained with two different codes and three different reaction networks are compared. (Reproduced from Paxton et al. 2015. © 2015. The American Astronomical Society. All rights reserved.)

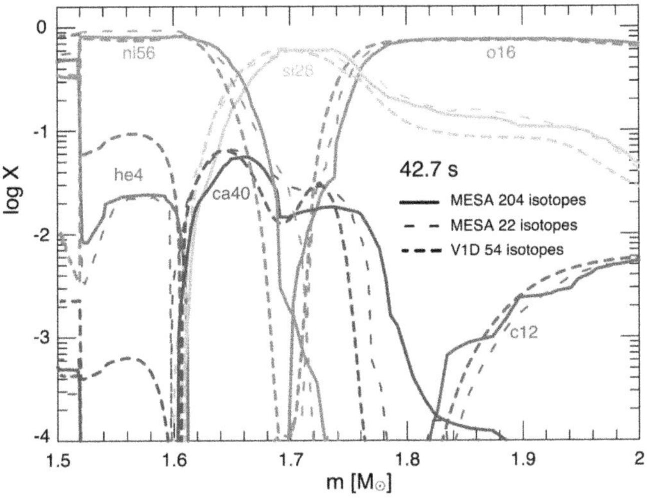

Figure 9.7. Same as Figure 9.6 but 42.7 s, after the shock wave has passed through the outer layers. (Reproduced from Paxton et al. 2015. © 2015. The American Astronomical Society. All rights reserved.)

$(7-8) \times 10^{-2}$ $M_\odot$. Further important products of incomplete Si-burning are unburnt fractions of ^{28}Si, ^{32}S, ^{36}Ar, and ^{40}Ca. Further produced are ^{44}Ca ($\leftarrow$ ^{44}Ti by decay), $^{56,57}Fe$ ($\leftarrow$ $^{56,57}Ni$), ^{59}Co ($\leftarrow$ ^{59}Cu), ^{58}Ni, and traces of ^{43}Ca and $^{64,66}Zn$ ($\leftarrow$ $^{64,66}Ge$). It is to be noted that some of these nuclides are also produced by the weak s-process component in massive stars (see Section 8.2), in particular ^{59}Co, $^{60-62}Ni$, and ^{66}Zn.

Explosive O-burning produces most of the intermediate-mass nuclides in the range from ^{28}Si to ^{42}Ca in a QSE. The main products (ordered by abundance) are ^{28}Si, ^{32}S, ^{36}Ar, ^{40}Ca, ^{38}Ar, ^{34}S. Also produced are ^{33}S, ^{39}K, ^{35}Cl, ^{42}Ca, ^{37}Ar with $X_i < 0.01$. The main products are similar to what is made in hydrostatic O-shell burning and the relative proportion of what is made explosively and hydrostatically depends on the treatment of convection (which may or may not allow for mixing between shells) and the progenitor mass.

Explosive Ne-burning is hard to separate from explosive C-burning and their products are difficult to distinguish from those resulting from the hydrostatic phases before the explosion. The abundances in both shells are only partially affected by the passage of the shock wave, with the Ne-shell more than the C-shell. The C-shell shows almost no explosive burning. The main product of the explosive burning processes is similar to the one of the pre-explosive phases, that is ^{16}O via $^{20}Ne(\gamma,\alpha)^{16}O$. Also importantly made, similar to the hydrostatic phases, are ^{24}Mg and ^{28}Si via $^{20}Ne(\alpha,\gamma)^{24}Mg(\alpha,\gamma)^{28}Si$. At lower level, but still important, ^{23}Na, $^{25,26}Mg$, ^{27}Al, $^{29,30}Si$, and ^{31}P are made. A difference to hydrostatic burning is that a brief, strong burst of protons and neutrons appears due to the peaking temperature in the shock. Neutrons are released by (α,n) reactions on ^{22}Ne and $^{25,26}Mg$, with α-particles (mainly) released from the photodisintegration of ^{20}Ne, leading to the production of neutron-rich isotopes of elements from S to Zr ($36 \leqslant Z \leqslant 88$). This includes production of ^{60}Fe which is important because its decay γ-rays emitted in supernova remnants can be detected with γ-ray satellite telescopes. Another nuclide important for γ-ray astronomy is the proton-rich ^{26}Al, made by ^{25}Mg(p,γ)^{26}Al. The ^{26}Al made by proton capture is partially destroyed by abundant neutrons also released during explosive burning, mainly by ^{26}Al(n,p)^{26}Mg. With respect to ^{26}Al production, it is important to note that the nuclide not only has a ground state with $t_{1/2} = 7.17 \times 10^5$ yr but also an isomeric state with $t_{1/2} = 6.346$ s. The half-life of the isomer is longer than the explosive burning timescale and therefore it acts as a stable nuclide during the burning process and has to be treated as such in the nuclear reaction network (see Section 4.2 for the definition of reaction networks). This is usually done by including the isomer as a separate nuclide. The difficulty in this is that the reactions populating and depopulating the isomer, including the transitions connecting it to the ground state and other excited states, have to be known explicitly. The usual assumption of detailed balance (see Section 4.1.3.2) is not fulfilled in this case. A number of simplifications for treating ^{26}Al in reaction networks have been suggested on the basis that some, although not all, transitions can still be assumed to be equilibrated.

H- and He-shell burning is not affected by the explosion because $\tau_\Delta^{\mathrm{nuc}}$ still is much longer than the shock timescale, even at the elevated shock temperature. These shells are ejected with their pre-explosive composition, except for modifications by the ν-process (see below).

The effect of explosive burning in a 25 $M_\odot$ star with initial solar metallicity is illustrated by Figure 9.8. The ratio of post-explosive to pre-explosive abundances ($Y_{\mathrm{post}}/Y_{\mathrm{pre}}$) are shown as function of mass number A. The change in abundances from Ne up to $A \approx 90$ is due to the explosive burning processes described above. Two additional processes proceeding during the explosion are relevant for further abundance variations at $A < 20$ and $A > 90$, the γ-process and the ν-process. These are described below.

9.3.4.1 The γ-Process in Massive Stars

The abundance increase for proton-rich isotopes with $A > 100$ seen in Figure 9.8 is due to the γ-process, partially photodisintegrating nuclides in the O/Ne-layers of a massive star. The explosive production includes the p-nuclides, which cannot be reached by the s- and r-processes. The definition of p-nuclides is presented in Section 8.4 and the basic operation of the γ-process is explained in Section 8.4.2. In massive stars, the shock wave reaches T_{peak} appropriate for the production of p-nuclides in the O- and Ne-layers (see also Equation (9.9)). The seed nuclei for the γ-process are s- and r-nuclei initially present (inherited by the massive star from its proto-stellar cloud). The weak s-process component produced in the massive star itself does not contribute much as it only produces nuclides with $A \lesssim 90$. Therefore the p-nuclide yield is metallicity dependent and the γ-process in massive stars is a secondary process.

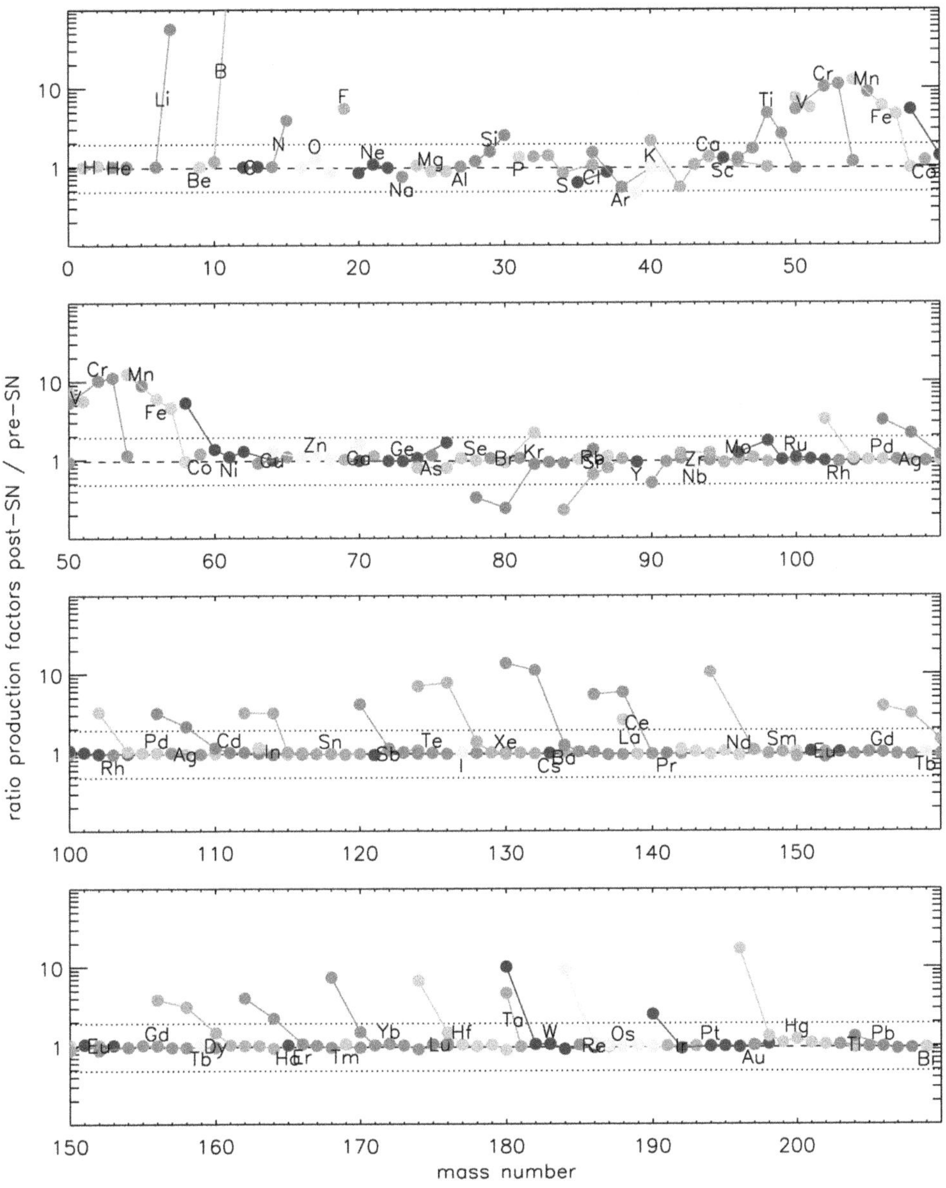

Figure 9.8. Explosive nucleosynthesis in a 25 $M_\odot$ star with initial solar metallicity: only stable nuclides are shown and isotopes of each element are connected by a line and elements are identified by labels and distinctive colors. (Figure taken from Rauscher et al. 2002. © 2002. The American Astronomical Society. All rights reserved.)

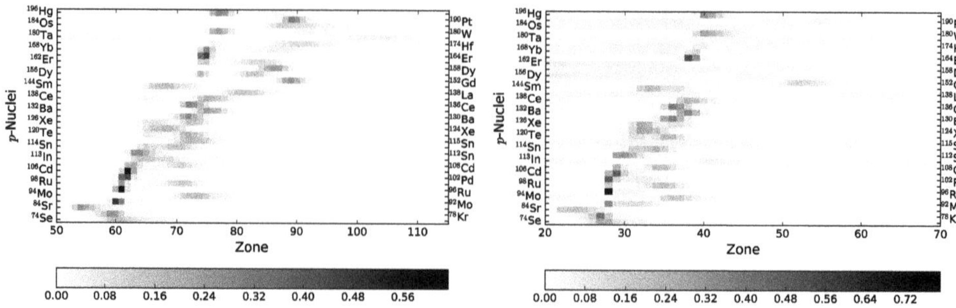

Figure 9.9. Relative abundance change $\Delta^j Y_i$ (given by the color shade) of p-nuclides in each mass shell in the explosion of a 15 $M_\odot$ (left) and a 25 $M_\odot$ (right) star. Zones with a lower number are further inside the star and experience higher T_{peak}. (Reproduced from Rauscher et al. 2016, by permission of Oxford University Press on behalf of the Royal Astronomical Society.)

Figure 9.9 shows the relative abundance change of p-nuclides in the Lagrangian mass zones of a 15 $M_\odot$ and a 25 $M_\odot$ star. The relative abundance change of a nuclide i in the mass shell identified by j is defined as

$$\Delta^j Y_i = \frac{|^j Y_i^{\text{expl}} - {}^j Y_i^{\text{ini}}|}{\sum_j |^j Y_i^{\text{expl}} - {}^j Y_i^{\text{ini}}|}, \tag{9.10}$$

where ${}^j Y_i^{\text{ini}}$ is the initial abundance of nuclide i in shell j and ${}^j Y^{\text{expl}}$ is the value after conclusion of explosive burning. The effect of the decreasing T_{peak} when the shock front moves outwards is seen in the trend that the more tightly bound lighter p-nuclides are affected in layers further inside the star. Most p-nuclides are only produced in a tightly confined region encompassing one or only a few mass layers. This underlines the sensitivity to T_{peak} as explained in Section 8.4.2 and also shown in Figure 8.20. The production of several p-nuclides, on the other hand, is stretched out over a larger number of zones. This indicates that there is not a single most important production process (a single, dominating reaction or reaction chain) but rather a number of different reaction paths to obtain the nuclide, each acting at a different temperature and each rather inefficient by itself.

A feature immediately catching the eye in Figure 9.8 is the missing γ-process production for $A < 100$, which includes the comparatively abundant p-nuclides ^{92,94}Mo and ^{96,98}Ru, as well as ^{84}Sr, ^{78}Kr, and ^{74}Se. This is a persistent problem in p-production in massive stars, independent of stellar mass or stellar model code. In fact, there is not enough seed material present in this range of nuclear mass number to consistently obtain the observed abundances of the lightest p-nuclides in a γ-process. Therefore an additional process would be required, either acting in massive stars or making the p-nuclides (the light ones or all of them) in a different site. Within massive stars, an enhancement of seed abundances with $A < 100$ before the γ-process would help but currently there seems to be no way to accomplish this. Alternatively, perhaps the missing light p-nuclides could be supplied by the νp-process introduced in Section 9.3.3 but the prevailing modeling uncertainties do not allow to draw definitive conclusions on whether this is actually possible in principle and whether the required abundances can be obtained consistently.

Some explosive models have also found a persistent underproduction of p-nuclides in the mass range $150 \leqslant A \leqslant 165$ across a wide range of stellar masses. This underproduction by factors of about 3–4 seems to be solvable by adapting nuclear reaction rates but still no consistent set of reaction rates has been found to remove this problem, with the focus on (γ,α) and (γ,n) reactions in this nuclear mass number range. It may also not appear in other sites, with a γ-process at different conditions (see Section 9.4.3).

Two p-nuclides require special attention, ^{138}La and ^{180m}Ta. They are produced in the γ-process but also receive considerable contributions from the ν-process (see below). For the rarest nuclide among the naturally occurring nuclides, ^{180m}Ta, there is an additional complication. It occurs naturally in its isomeric state with $J^{\pi_q} = 9^-$ at 77.1 keV with $t_{1/2} > 7.15 \times 10^{15}$ yr, not in its $J^{\pi_q} = 1^+$ ground state (g.s.) which has a half-life of only $t_{1/2} = 8.154$ h. All other excited states have much shorter half-lives, not only due to β-decays or particle emission but also because of γ-emission and -absorption. These higher-lying states feed either into the g.s. or the isomeric state via de-excitation chains. Therefore any ^{180}Ta produced in supernovae which does not become ^{180m}Ta will decay and will not contribute to the abundance of this p-nucleus. At the temperatures encountered in explosive O/Ne-burning the excited states within the nucleus are equilibrated by photo-excitation and -deexcitation (see also Section 4.1.3). In equilibrium, the population ratio R_m of the isomer with respect to the g.s. depends on the ratio of the Boltzmann population factors (Equation (4.38)),

$$R_m(T) = \frac{\mathcal{P}_m}{\mathcal{P}_{g.s.}} = \frac{g_m}{g_{g.s.}} e^{-E_m^x/(k_B T)}, \tag{9.11}$$

and therefore sensitively depends on temperature. During the cooling of the ejected plasma, however, the link between ground state and isomeric state via higher-lying intermediate states drops out of equilibrium at a freeze-out temperature T_{freeze}. Except for final de-excitations into the g.s. and the isomer, the population ratio is frozen at $R_m(T_{freeze})$. The value of T_{freeze} depends on the properties (excitation energy, spin, parity, and γ-widths) of the excited states which mediate between the g.s. and the isomer. These mainly comprise levels belonging to a rotational band (see also Section 3.4) built on a state at 592 keV with uncertain spin assignment and including further states at 1.087 MeV, 1.297 MeV, and 1.507 MeV (states at higher excitation energy are not populated sufficiently to appreciably modify the final population ratio). Based on experimental information, the isomeric ratio $\mathcal{P}_m/(\mathcal{P}_g + \mathcal{P}_m)$ was estimated to be 0.38. This translates into $R_m(T_{freeze}) = 0.919$ with $T_{freeze} \simeq 3.831 \times 10^8$ K, well below the γ- and ν-process temperature. The final yield of ^{180m}Ta from stellar model calculations has to be modified accordingly as the usual reaction networks employed in stellar models do not differentiate between ^{180g}Ta and ^{180m}Ta.

9.3.4.2 The ν-Process in Massive Stars

Although not part of the neutrino-driven wind component, the ejected outer layers of the star are expanding and cooling in a stream of neutrinos of all flavors coming from the neutrino spheres. During hydrostatic burning, neutrinos escape freely without interacting with the plasma but in the explosion the neutrino flux becomes so high even in the outer layers that the neutrinos (weakly) interact with the nuclei in the expelled layers and modify the abundances of a few. This is called the *ν-process*. Neutrinos interact in two ways, by charged-current reactions and neutral-current reactions (see Section 7.3.2). Due to the fact that the neutrino energies do not exceed 20 MeV, only ν_e and $\bar{\nu}_e$ interact with nuclides by charged-current processes, transforming them by (anti-)neutrino-induced processes (see also Equation (3.175)). Nevertheless, all neutrino flavors can excite nuclei in inelastic scattering by neutral-current reactions. When the achieved excitation is above the particle emission threshold (i.e., the particle separation energy, see Section 3.2), the nuclei are transformed by emission of protons or neutrons.

The abundance increase seen in Figure 9.8 for ^{7}Li, ^{11}B, ^{15}N, and ^{19}F is due to the ν-process. It also mildly affects the abundances of the radionuclides ^{22}Na and ^{26}Al, not shown in Figure 9.8; and contributes to the abundances of ^{138}La and $^{180\mathrm{m}}$Ta, as mentioned above.

The formation mechanisms for these nuclides are as follows. In the He-shell, neutrinos eject a nucleon from a small fraction of ^{4}He by neutral-current excitation, thus producing ^{3}H and ^{3}He. The abundant ^{4}He produce ^{7}Li by ^{3}H(α,γ)^{7}Li and ^{3}He(α,γ)^{7}Be(β^{+})^{7}Li. This is by far not enough to explain the solar abundance of ^{7}Li and therefore the bulk of ^{7}Li is produced elsewhere (see Chapter 11 and Table 11.1). The bulk of ^{11}B is made in about equal proportions by neutral-current and by charged-current processes in the C-shell. Neutral-current interactions directly produce ^{11}B by ejecting a proton from ^{12}C or make ^{11}C by ejecting a neutron. The unstable ^{11}C then decays to ^{11}B with $t_{1/2}$ = 20.364 min. The relevant charged-current processes are ^{12}C(ν_e,e$^-$p)^{11}C and ^{12}C($\bar{\nu}_e$,e$^+$n)^{11}B. A smaller contribution comes from inelastic neutrino scattering on ^{16}O in the O/Ne-shell, ^{16}O(ν,$\nu'\alpha$p)^{11}B. Neutrino-induced ejection of nucleons from ^{16}O in the C- and O-shells may produce ^{15}N. Modern neutrino spectra with lower average neutrino energies, however, significantly reduced the ^{15}N production in the ν-process compared to earlier works and only marginally increase the previously present ^{15}N abundance. The abundance of ^{19}F is increased by neutral-current and charged-current processes on ^{20}Ne in the O/Ne-layer. It is to be noted that in stars with masses $M \lesssim 17\ M_\odot$ ^{19}F is additionally made during the explosion without any neutrino reactions. In such low-mass stars, the shock wave may still have sufficient energy in the inner He-shell to boost the reaction sequence ^{18}O(p,α)^{15}N(α,γ)^{19}F. Regardless of stellar mass, massive stars are not the main sources of ^{19}F, which primarily comes from AGB stars and Wolf–Rayet stars (see Chapter 6).

Similar to the γ-process, the ν-process produces the nuclides ^{138}La and $^{180\mathrm{m}}$Ta in the O/Ne-layers of a massive star. It dominates over the γ-process in the cooler regions of these layers, where $T < 2$ GK. The two nuclides are made by the charged-current processes ^{138}Ba(ν_e,e$^-$)^{138}La and ^{180}Hf(ν_e,e$^-$)^{180}Ta, respectively. An additional, minor path to ^{180}Ta is ^{179}Ta(n,γ)^{180}Ta, with neutrons released by neutral-current excitation of ^{16}O, ^{20}Ne, and ^{24}Mg. To obtain the surviving amount of ^{180m}Ta the same considerations have to be applied as already explained for the γ-process above.

As pointed out before, there is considerable uncertainty in the neutrino energies and their spectral distribution, as well as their temporal evolution, for the emission during the core collapse and the subsequent explosion. Consequently, also the predictions of abundances from the ν-process carry large uncertainties. Moreover, it has been pointed out that neutrino oscillations may have an impact, both collective oscillations and the MSW effect (see Section 7.2.3) depending on the location of the MSW resonance region (C- or Ne-layer).

9.4 Explosive Burning in Binary Systems

9.4.1 The Roche Lobes

A number of seemingly different phenomena are due to explosive nuclear burning caused by an interaction of two objects in binary systems. About half of all stars are part of a binary with a companion star. The star with higher mass completes its stellar evolution phases (Chapter 6) quicker than the companion and turns into a compact object, a white dwarf or a neutron star. In close binaries, matter from the stellar atmosphere of the remaining companion star can flow toward the compact object and be accreted on its surface.

The gravitational potential of a single, massive object with mass M is, as seen before,

$$V_G(r) = -\frac{GM}{r}. \tag{9.12}$$

In a binary system with two objects with masses $m_1 < m_2$ and position vectors $\vec{r}_1$, $\vec{r}_2$ moving around the common center-of-mass the gravitational potential acting on a third, smaller mass (a "test particle") at position $\vec{r}$ is given by the *Roche potential*

$$V_{\mathrm{Roche}}(\vec{r}) = -G\left[\frac{m_1}{|\vec{r} - \vec{r}_1|} - \frac{m_2}{|\vec{r} - \vec{r}_2|}\right] - \frac{1}{2}(\vec{\Omega}_\omega \times \vec{r})^2. \tag{9.13}$$

The angular momentum $\vec{\Omega}_\omega$ of the binary system is given by

$$\vec{\Omega}_\omega = \sqrt{\frac{G\tilde{M}}{|\vec{r}_1 - \vec{r}_2|}}\,\hat{\vec{n}}, \tag{9.14}$$

with $\hat{\vec{n}}$ being the unit vector perpendicular to the plane of the binary motion and $\tilde{M} = m_2/(m_1 + m_2)$. From a large distance $|\vec{r}\,| = r \rightarrow \infty$, the Roche potential reduces to the spherically symmetrically potential of Equation (9.12) centered on the center-of-mass of m_1 and m_2 and with $M = m_1 + m_2$. At closer distance the potential assumes the complicated shape described by Equation (9.13). An example for the resulting equipotential surface is shown in Figure 9.10. At very close distance to

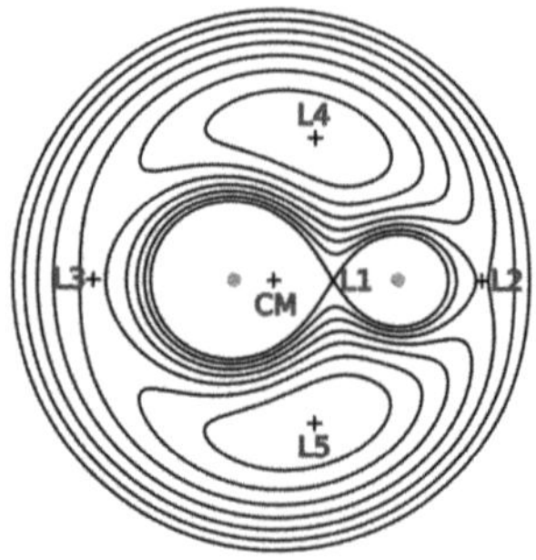

Figure 9.10. Equipotential surface contours of the gravitational potential in the orbital plane of a binary system with $m_2/m_1 = 3$. The two bodies are located at the two red dots, with the more massive body on the left. Also marked are the common center-of-mass (CM) and the Lagrange points, L1–L5. Contours are at equally spaced values of the potential between the value at the inner Lagrange point L1 and its maximum. (This Roche potential image has been modified from an original made available by Philip D. Hall on the Wikimedia website under a CC BY-SA 4.0 licence. It is included within this book on that basis. It is attributed to Philip D. Hall.)

either body the potential again is spherically symmetric because the influence of the second body is negligible. At intermediate distances, however, the equipotential surfaces become deformed. There is a critical surface where two equipotential surfaces surrounding the two bodies are just touching at one point, a saddle point between the gravitational wells of the two bodies. This point is called the first Lagrange point, in short L1. The combined surfaces containing L1 are called the *Roche lobes*. There are further *Lagrange points* where the forces on a test particle cancel. The Lagrange points L2 and L3 are at opposite locations behind the two bodies, the Lagrange points L4 and L5 each form an equilateral triangle with the two bodies.

The radius of the Roche lobe around a body in a binary system can be calculated by

$$R_{\mathrm{Roche}} = |\vec{r}_1 - \vec{r}_2| \frac{0.49 q_{\mathrm{m}}^{2/3}}{0.6 q_{\mathrm{m}}^{2/3} + \ln\left(1 + q_{\mathrm{m}}^{1/3}\right)}, \tag{9.15}$$

where q_{m} is the mass ratio of the two bodies. The radius around the object of mass m_1 is obtained by using $q_{\mathrm{m}} = m_1/m_2$, the one around the object with mass m_2 by $q_{\mathrm{m}} = m_2/m_1$. For $q_{\mathrm{m}} = m_2/m_1$ and $0.1 \leqslant q_{\mathrm{m}} \leqslant 0.8$ the above formula can be reduced to

$$R_{\mathrm{Roche}} = 0.462\,24\,|\vec{r}_1 - \vec{r}_2|\tilde{M}^{1/3}. \tag{9.16}$$

This yields the radius for the object with mass m_2.

For mass transfer in close binaries, matter from one body has to reach the saddle point L1, from where it can spill over into the gravitational well of the second object. This can happen in two ways. The first possibility is that a strong stellar wind (see Chapter 6) is emitted from the surface of one of the stars with sufficient kinetic energy to make it to and over the saddle point. The second possibility is that one of the stars in the binary system expands so much during stellar evolution (for example by reaching the red giant stage) that it fills the volume enclosed by the Roche lobe. Matter from the stellar atmosphere extending beyond the Roche lobe will flow toward the companion object. This is called *Roche lobe overflow*. Such an overflow can also be caused by changing the separation between the two bodies which, in turn, also modifies the Roche lobe and can make it smaller than the extension of one of the stars. A change in distance between the two bodies can, for example, be the result of gravitational wave emission.

The transferred matter does not fall onto the second object in a straight path. Since angular momentum has to be conserved, the infalling matter spirals inwards and forms an *accretion disk*. Material at the inner edge of the accretion disk may finally make it to the surface of the second object when allowed by angular momentum conservation or after it shedded part of its angular momentum by, for example, scattering on other particles or emission of radiation. The transfer of angular momentum between the two bodies may have further interesting consequences. If mass is transferred from the lighter to the more massive body, the lower-mass object has to move away from the more massive body to keep the center-of-mass fixed. This may lead to a cut-off in the mass transfer. On the other hand, if mass is transferred from the more massive body to the lighter one, the orbital separation between the two is reduced. Both possibilities are only important when the transferred mass significantly changes the mass ratio $q_{\rm m}$ between the two bodies.

Thermonuclear explosions caused by mass transfer onto the surface of white dwarfs are thought to be the explanation of novae (Section 9.4.2) and one of the conceivable mechanisms for type Ia supernovae (Section 9.4.3). Thermonuclear explosions in the accreted matter on the surface of a neutron star can explain a certain class of X-ray bursts (Section 9.4.4). Merging neutron stars are another example of an interacting binary system, which was confirmed to be a site of the r-process (Section 9.4.5).

9.4.2 Novae

Classical novae are the second-most frequent type of thermonuclear explosions observed in the Galaxy behind type I X-ray bursts (Section 9.4.4), with an estimated rate of $30 \pm 10\ {\rm yr}^{-1}$ in the Galaxy. The mechanism behind the nova phenomenon is thought to be well understood: a white dwarf accretes matter at a rate of 10^{-10}–$10^{-9}\ {\rm M}_\odot\ {\rm yr}^{-1}$ from a main-sequence companion star by Roche lobe overflow. The accreted matter is mainly composed of hydrogen because it stems from the outer layers of the companion star. The white dwarf matter is degenerate (see Section 2.5 and Chapter 6) and the accreted hydrogen is compressed on the surface of the white dwarf and also becoming degenerate. With more and more accreted matter in a thin layer on the surface of the white dwarf, pressure, density, and temperature (by released gravitational binding energy as well as compressional heating) are rising until H-burning via CNO-cycles (see Section 7.2.2) ignites. Due to the degeneracy of the electron gas as the dominant pressure contribution, pressure cannot adjust to the rising temperature to establish a hydrostatic regime and therefore a thermonuclear runaway ensues. The generated heat triggers strong convection, transporting short-lived, β^+-unstable nuclides (^{13}N, ^{14,15}O, ^{17}F) to the outer and cooler envelope region. The outer region is then heated by the decay of these nuclides and this results in an increase in entropy and a lifting of the degeneracy, which in turn leads to a rapid expansion of the envelope. Matter from the outer 10^{-5}–10^{-4} $M_\odot$ of the white dwarf is ejected with speeds of hundreds to a few thousands km s^{-1}. This leads to a sudden decrease in pressure and temperature which impedes the runaway and the nuclear reactions cease. Depending on the amount of unburnt hydrogen and the accretion rate, another thermonuclear runaway is expected to follow after a few 10–10^5 yr.

The nuclear burning is mainly proceeding through the hot and β-limited CNO-cycles as described in Section 7.2.2 at temperatures above 10^7 K but not exceeding 4×10^8 K. Many observational features of novae have been reproduced in spherical symmetry. Nevertheless, details of the thermonuclear ignition, convection in the envelope, and mixing at the core–envelope interface are less well understood and require sophisticated multi-D approaches. In addition to CO white dwarfs, also ONeMg white dwarfs may accrete matter, with implications for the resulting nucleosynthesis.

In terms of nucleosynthetic contribution to the total budget of nuclides in the Galaxy, novae are of limited importance. The amount of ejected mass per nova is small compared to supernovae. On the other hand, novae are quite frequent events and their ejecta show large yields of certain nuclides, such as ^{13}C, ^{15}N, ^{17}O. They may also provide minor contributions to other nuclides with $A < 40$, such as ^{7}Li, ^{19}F, and ^{26}Al. In addition to ^{26}Al, further radionuclides are ejected: ^{13}N, ^{18}F, and to a lesser extent ^{22}Na. ^{18}F can be directly produced in the CNO-III cycle because ^{16}O is abundant in both CO white dwarfs and ONeMg white dwarfs. The production of ^{7}Li was disputed for many years but recent observations have found ^{7}Li and ^{7}Be in nova outflows. The formation of ^{7}Be proceeds through ^{3}He(α,γ)^{7}Be. The nuclide would be destroyed by ^{7}Be(p,γ)^{8}B but at $T \gtrsim 10^8$ K the reaction equilibrates with its reverse, $^7\text{Be} + \text{p} \leftrightarrows {}^8\text{B} + \gamma$, so that an equilibrium abundance of ^{7}Be survives. ^{7}Li can then appear as a decay product of ^{7}Be provided that an efficient *beryllium transport mechanism* exists, which transports the ^{7}Be from the nucleosynthesis region to the cooler layers of the outer envelope on a timescale shorter than its decay timescale. Otherwise ^{7}Li would be immediately destroyed in the hotter region.

There are two possible reaction sequences to synthesize the radionuclide ^{22}Na, both starting with ^{20}Ne(p,γ)^{21}Na at the interface between core and accreted envelope on ONeMg white dwarfs. This is followed either by ^{21}Na(β^+)^{21}Ne(p,γ)^{22}Na or by ^{21}Na(p,γ)^{22}Mg(β^+)^{22}Na. As of yet, the amount of ^{22}Na produced in nova outbursts has not yet been determined precisely because it crucially depends on mixing between various depths of the core, the outer layers, and the accretion layer on top of the white dwarf. Similarly dependent on this mixing is the ^{26}Al production. Relevant reactions are ^{24}Mg p,γ^{25}Al(β^+)^{25}Mg(p,γ)^{26}Al and ^{25}Mg(p,γ)^{26}Al, as well as the destruction sequence $^{26\text{m}}$Al(β^+)^{26}Mg(p,γ)^{27}Al. Although the process timescale is longer in novae ($\tau_\Delta^{\text{process}} \approx 50$–500 s) than for explosive shell-burning in massive stars, the complications with respect to the comparatively long-lived isomer ^{26m}Al (requiring special treatment in the reaction network with individual reaction rates for ground state and isomer) discussed for explosive Ne-shell burning in massive stars also apply here.

Rapidly accreting white dwarfs have also be suggested to be the site of the i-process, see Section 8.5.

9.4.3 Type Ia Supernovae (Thermonuclear Supernovae)

9.4.3.1 Explosion Mechanism and Nucleosynthesis

The kinetic energy of the type Ia supernova ejecta has been found to be $E_{\mathrm{kin}} \approx 10^{51}$ erg and the energy integrated over the lightcurve $E_{\mathrm{rad}} \approx 10^{49} \ll E_{\mathrm{kin}}$. The lightcurves show extended tails, indicating a large amount of ^{56}Ni to be ejected, more than in other supernova types. Also in contrast to the other supernova types, the type Ia supernovae occur in all types of galaxies, including old elliptical galaxies without active star formation. This points to an involvement of older stars or endpoints of stellar evolution. Although E_{kin} is quite similar to the energy release by core-collapse supernovae, it was realized early on that the explosive burning of about 1 $M_\odot$ of a CO mixture also releases a similar amount of energy. The obvious conclusion was that a type Ia supernova is connected to the complete thermonuclear disruption of a CO white dwarf. Therefore also the term *thermonuclear supernova* is used (contrasting with core-collapse supernova). Despite of many dedicated efforts over the last several decades, how to achieve such a disruption in a manner compatible with the wealth of observational data on lightcurves and spectra basically is a still unsolved problem. Certain details have been clarified and some alternatives have been ruled out but a completely consistent model still has to be completed. As for ccSN, this is also due to the considerable computational effort involved in following the nuclear burning on the surface and in the core of the white dwarf, including detailed energy transport by conduction and convection in multi-D simulations. Providing an account of the historic development of the research in type Ia supernovae is beyond the scope of this book. Likewise, a complete review of the various current propositions and their details cannot be given. Only the most important considerations are presented here, for more details see, for example, Diehl et al. (2018); José (2016).

In terms of nucleosynthesis, observational data suggest ≈0.6–0.7 $M_\odot$ of ^{56}Ni to be ejected per event which implies that type Ia supernova produce about half to two thirds of the ^{56}Fe (as decay product of ^{56}Ni) in the Galaxy (with a type Ia supernova rate of about 0.3 per century). It further implies that a large fraction of the white dwarf material is burned at NSE conditions with Y_e close to 0.5. Therefore mainly nuclides around ^{56}Ni (the Fe-group) are produced. Production details depend on the deviations from $Y_e = 0.5$ and what fraction of the white dwarf is only incompletely burned. This, in turn, depends sensitively on the ignition of nuclear burning and the propagation of the burning front.

Among the proposed solutions to the thermonuclear supernova problem, two basic mechanisms have received the most attention:

- *The single-degenerate mechanism:* A CO white dwarf in a binary system is accreting H- or He-rich material from the outer layers of a low-mass star by Roche lobe overflow. It has been found difficult to obtain full white dwarf disruption by accreting a H-layer from a main-sequence companion star (*cataclysmic variable*). H-burning in the accreted layer either leads to a classical nova or to weak H-shell flashes which also lead to mass loss. *Symbiotic variables* are characterized by a wide binary system of a white dwarf and a red giant star. In such a system, He-rich matter can be accreted on the white dwarf. The question is whether sufficient mass can be accreted to bring a $\approx$1.1 $M_\odot$ white dwarf over the Chandrasekhar mass limit (see Section 2.5). Sub-Chandrasekhar mass explosions have been suggested, where C-ignition is triggered by added mass and heat from He-burning in the accreted layer.
- *The double-degenerate mechanism:* A binary system contains two CO white dwarfs which merge (or collide) due to the loss of energy and angular momentum by gravitational wave emission. Historically, the double-degenerate scenario has been disfavored due to two main reasons: (1) observations suggest that binaries with two white dwarfs close enough to merge in less than the age of the Universe may be scarce; (2) numerical simulations suggest that instead of an explosion such systems would lead to the formation of a neutron star by accretion-induced collapse. It has to be mentioned, though, that also the double-degenerate scenario poses a significant challenge for modelers and that more recent simulations have found explosions for specific assumptions on the progenitor system and by including, for example, magnetic fields.

Despite of the modeling difficulties connected to both mechanisms, the current consensus is that both, the single-degenerate and the double-degenerate channel, may contribute to the total number of type Ia supernovae. This is also driven by the difficulty to obtain the estimated type Ia supernova rate with the single-degenerate channel alone. As of today, the relative contribution of each channel is unknown and estimates vary widely. In the following, a few more details of the single-degenerate mechanism are presented, as past nucleosynthesis studies have focused on this channel.

In single-degenerate Chandrasekhar mass models, the mass of the white dwarf is increased by accretion until it reaches a mass close to M^*_{Ch} (defined in Equation (2.43)). Due to the addition of mass, the white dwarf shrinks according to the mass–radius relation given in Equation (2.31). The central density and temperature of the white dwarf continuously increase due to the contraction during the pile-up of matter on the surface. Additionally, H/He-burning in the accreted layer further heats the white dwarf, with conduction transporting the heat to the center (see Section 2.7.3). At densities exceeding 2×10^9 K (also depending on temperature) explosive C-burning ignites and a burning front runs through the star. One-dimensional simulations assume one ignition point in the center or off-center but also multiple ignition points could be realized in nature and are studied with multi-D codes. There are two basic ways a burning front can propagate: subsonically (*deflagration*) or supersonically (*detonation*). (In layperson's terms, a wood fire or bush fire spreads by deflagration whereas a detonation is what is commonly understood as an explosion.) The flame speed determines whether nuclear burning is complete or remains incomplete, with nuclides only partially consumed in the explosive burning. Whether a deflagration or a detonation occurs depends on density, temperature, and composition. For instance, it is obvious that it is easier to obtain the required overpressure for triggering a detonation at low density ($\approx 3 \times 10^7$ g cm^{-3}) than at high density. The dependence on other parameters is not so straightforward.

For decades, the "standard model" of single-degenerate Chandrasekhar mass supernovae has been the W7 model, one of a series of spherically symmetric models with different accretion rates and internal mixing lengths by Nomoto et al. (1984); Thielemann et al. (1986). In the W7 model a 1 $M_\odot$ white dwarf accretes matter at a rate of 4×10^{-8} $M_\odot$ yr^{-1}. There is steady H/He-burning on its surface and the mass increase of the white dwarf is at the same rate as the accretion. Central C-burning ignites at a central density of $\rho_{\rm ign} = 2.6 \times 10^9$ g cm^{-3} when the white dwarf has reached a mass of 1.378 $M_\odot$. Due to the degeneracy of the white dwarf, a thermonuclear runaway occurs. Initially, the heat can be carried off by convection but above $T \approx 8 \times 10^8$ K a shock front is established. The shock propagates as a deflagration front, with compression waves traveling ahead of the front, expanding the outer layers and thus lowering its density before the arrival of the burning front. The pre-explosive expansion of the white dwarf progressively weakens the nuclear energy release by the burning front as it passes through layers of decreased density. At 1.3 $M_\odot$ and a density of 10^7 g cm^{-3}, still inside the white dwarf, C-burning ceases. Nevertheless, the total energy release so far is 1.8×10^{51} erg which is sufficient to unbind the white dwarf (its gravitational binding energy is $\approx 5 \times 10^{50}$ erg). The kinetic energy of the explosion obtained in the W7 model is about 3×10^{51} erg and a small fraction of the total energy ($\approx 4 \times 10^{49}$ erg) is carried off in a brief neutrino burst of about 1.5 s. In the burning front passage, the white dwarf material undergoes four burning regimes: incomplete C/Ne-burning, incomplete O-burning, incomplete Si-burning, and full NSE. Depending on the respective $T_{\rm peak}$ in a layer, not the full sequence may be completed. About 0.7 $M_\odot$ achieve $T_{\rm peak} = 6 \times 10^9$ K and $\rho = 9 \times 10^7$ g cm^{-3} and therefore a full NSE is established (Section 4.3.2). Due to the high density, also the triple-α rate is efficient in converting lighter species to Fe-group elements (see also Figure 9.5 for a comparison to ccSN densities). The NSE abundances depend on the Y_e of the material, see the discussion in Section 4.3.2 and Figure 4.9. The bulk of the white dwarf matter is ^{12}C and ^{16}O with $Y_e = 0.5$. High densities, however, shift the electron Fermi energies E_F to higher energy (see Equation (1.79)) and this increases the cross sections for electron captures (see also Sections 2.5, 9.3.1), allowing to make the material slightly neutron-rich. In the center of the W7 model the most abundant nuclides are ^{56}Ni, ^{56}Fe, and ^{54}Fe. Further outwards (the region beyond 0.1 $M_\odot$), densities are already lower, electron captures are not efficient, and the abundance is dominated by ^{56}Ni. The region $0.3 \leqslant M \leqslant 0.6$ M$_\odot$ exhibits an admixture of intermediate-mass nuclides, such as ^{40}Ca, ^{36}Ar, and ^{32}S, indicating a slight departure from NSE. The regions further out do not achieve temperatures to allow full NSE and also do not completely burn their contents on the timescale set by the passage of the burning front (see the general discussion in Section 9.1 and the discussion of explosive shell burning in Section 9.3.4). In the region encompassing the mass layers $0.7 \leqslant M \leqslant 0.9$ M$_\odot$ incomplete Si-burning leaves mainly intermediate-mass nuclides, such as ^{28}Si, ^{32}S, ^{36}Ar, ^{40}Ca, and some ^{56}Ni. In the layer with $0.9 \leqslant M \leqslant 1.1$ M$_\odot$ only C- and O-burning occurs and the obtained abundances are dominated by ^{28}Si, ^{32}S, and ^{36}Ar. In the layer with $1.1 \leqslant M \leqslant 1.25$ M$_\odot$ only C/Ne-burning happens, yielding ^{16}O, ^{24}Mg, and ^{28}Si. Only C-burning products appear in the layer $1.25 \leqslant M \leqslant 1.3$ M$_\odot$ whereas in the outermost envelope of the white dwarf the original composition is left unchanged. The model W7 produces 0.58 M$_\odot$ of ^{56}Ni and a total amount of Fe-peak nuclides of 0.86 M$_\odot$.

Parameterized post-processing studies varying the ignition density $\rho_{\rm ign}$ and velocity $v_{\rm def}$ of the deflagration front have indicated that the amount of mass with $Y_{\rm e} < 0.45$ is determined by $\rho_{\rm ign}$ whereas the amount of material with $0.47 \leqslant Y_{\rm e} \leqslant 0.485$ depends on $v_{\rm def}$. Therefore observed abundances of ^{58}Ni, ^{54}Fe and of ^{58}Fe, ^{54}Cr, ^{50}Ti, ^{64}Ni, ^{48}Ca could be used to constrain $v_{\rm def}$ and $\rho_{\rm ign}$, respectively (see Section 4.3.2 for the dependence of NSE abundances on $Y_{\rm e}$).

More recently than W7, another class of models has received particular attention. In these models, the burning front initially starts propagating as a deflagration front but turns into a detonation at some critical density. The prerequisite for a *deflagration–detonation transition* (DDT) is that the deflagration front is running into a preheated region exhibiting fluctuations in temperature, density, and composition sufficient to have pre-burned a fraction of the material before the subsonic shock can reach it. The physical mechanism to induce such fluctuations is not completely clear yet but several possibilities have been suggested. A multi-D treatment is required to catch the details not only of the DDT but also of the detonation wave which can also burn "backwards" into previously unburnt pockets or fingers of the deflagration front. Regardless of the computational difficulties, *delayed-detonation* models may open the road to a better reproduction of observational features, in particular regarding the nucleosynthesis of intermediate-mass nuclides. The comparatively large mass fractions of unburnt C and O in the (one-dimensional) C-deflagration models are processed in the detonation, also providing a slightly higher energy output.

9.4.3.2 p-Nuclides from Thermonuclear Supernovae

Single-degenerate thermonuclear supernovae have been considered as a production site of p-nuclides (see Section 8.4) since the early 1990s when it was realized that a small fraction of the white dwarf material experiences temperatures between 2 and 3.5 GK, required for a γ-process (see Sections 8.4.2 and 9.3.4). Since the γ-process requires seed nuclei for partial photodisintegration, the ad hoc assumption was made that the accreted material on top of the white dwarf is enriched in s-process nuclides. An enrichment by factors 10^3–10^4 was required to obtain p-abundances compatible with solar abundances from galactic chemical evolution. Nevertheless, a persistent problem appeared to be the production of the light p-nuclides, in particular the p-isotopes of Mo and Ru. This switched the interest back to core-collapse supernovae which displayed a similar problem with light p-nuclides but naturally produced the other p-nuclides naturally without the need for further seed enrichment (Section 9.3.4). Recently, however, a study postprocessing tracers (Section 9.1) of a 2D simulation of a white dwarf explosion found co-production of all p-nuclides (with the exception of ^{138}La and ^{180m}Ta) on levels sufficient to explain solar abundances, see Travaglio et al. (2011). The underlying model is a white dwarf having accreted matter from a companion star and exhibiting a DDT in its explosion. The required seed enrichment comes from thermal pulses in the accreted H/He-layer during further accretion. The pulses work similarly to s-processing in the He-shell flashes discussed in Section 7.3.2, with ^{13}C(α,n)^{16}O being the main neutron source for white dwarfs with $M \gtrsim 1.26$ M$_\odot$ and ^{22}Ne(α,n)^{25}Mg for $M < 1.26$ M$_\odot$. Once the white dwarf reaches $M^*_{\rm Ch}$, the thermonuclear burning is ignited in multiple sparks and the subsequent burning with mixing and convection is followed in the 2D hydrodynamic code. For the nucleosynthesis study the white dwarf is subdivided into 51,200 mass elements of equal mass (each having 2.73×10^{-5} M$_\odot$) and the initial composition as well as the temperature and density history of each mass element (tracer) is recorded. The production of p-nuclides and other species is then calculated by post-processing these tracers with an extended nuclear reaction network (Section 4.2). The tracers move around the white dwarf during the explosion. A snapshot of their location 1.45 s after ignition is shown in Figure 9.11. Also seen in this figure are the maximum temperature experienced by each tracer. The resulting nucleosynthesis yields are presented in Figure 9.12.

The peak densities and temperatures of the tracer particles relevant for p-nucleosynthesis are also shown in Figure 8.20. The group of tracers with higher density comes from the interior of the white dwarf as seen in Figure 9.11 whereas the bulk of p-nuclide tracers comes from the surface region with lower density. The lower densities are compatible with the densities obtained in the outer layers of the W7 model, as also shown in Figure 8.20. Generally, p-nuclides are produced in thermonuclear supernovae at densities exceeding those for the γ-process in massive stars by at least two orders of magnitude. This also impacts the reaction flows. Photodisintegration rates are similar to those found in massive stars because the allowed temperature range is set by the nuclear binding of the p-nuclides and is the same for ccSN and SN Ia. Particle-induced rates, on the other hand, scale with the density (see Sections 4.1.2, 4.1.5) and they are therefore considerably faster in SN Ia than in ccSN. This shifts the competition between capture and particle-emission rate in favor of the capture rate (Equation (4.52)). This may have two consequences: (1) In the region of the lightest p-nuclides with the lowest Coulomb barriers proton captures may become an option to create (or destroy) nuclei; (2) the neutrons released by photodisintegration of heavier nuclides may be recaptured on other nuclides

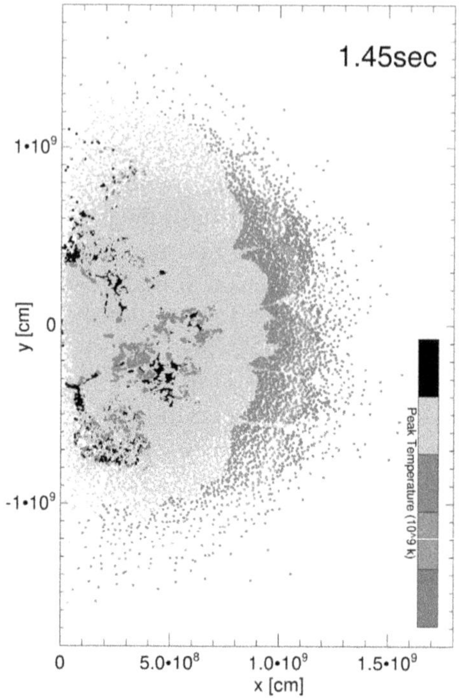

Figure 9.11. Tracer distribution 1.45 s after ignition in a 2D simulation of a white dwarf explosion with deflagration–detonation transition. The color coding refers to T_{peak} a tracer experienced during the entire explosion: $T_{\text{peak}} > 7$ GK (black), $3.7 \leqslant T_{\text{peak}} \leqslant 7$ GK (gray), $3.0 \leqslant T_{\text{peak}} \leqslant 3.7$ GK (red), $2.4 \leqslant T_{\text{peak}} \leqslant 3$ GK (green), $1.5 \leqslant T_{\text{peak}} \leqslant 2.4$ GK. The latter three T_{peak} ranges encompass temperatures for the γ-process. (Adapted from Travaglio et al. 2011.)

(intermediate or heavy ones) and work against the (γ,n) flow within an isotope chain, limiting the flow to very proton-rich isotopes (see Figure 9.13 for an example). This helps to reduce uncertainties caused by reaction rate uncertainties because, for example, neutron capture rates (and by the reciprocity relation in Equation (4.52) also photodisintegrations) are well known at stability.

Another difference between the γ-process in massive stars and in thermonuclear supernovae is the much larger range in density in which the process unfolds, as also seen in Figure 8.20. This results in a much wider range of "paths" to a specific nuclide when integrating over all tracers and justifies the term reaction "boulevard" introduced in Section 8.4.2. As a consequence, although this is not an equilibrium process, the importance of individual rates is diminished and key reactions are hard to find. If one reaction rate in the reaction boulevard is changed by improved theory or a measurement, the overall flow may still only be affected in a minor way as the change can be compensated by an alternative flow path in another tracer. This means that the ensemble of participating rates has to be determined consistently and sufficiently well to obtain a reliable prediction of the total flow summed over all tracers but no single rate can be identified to be crucial for the production of a

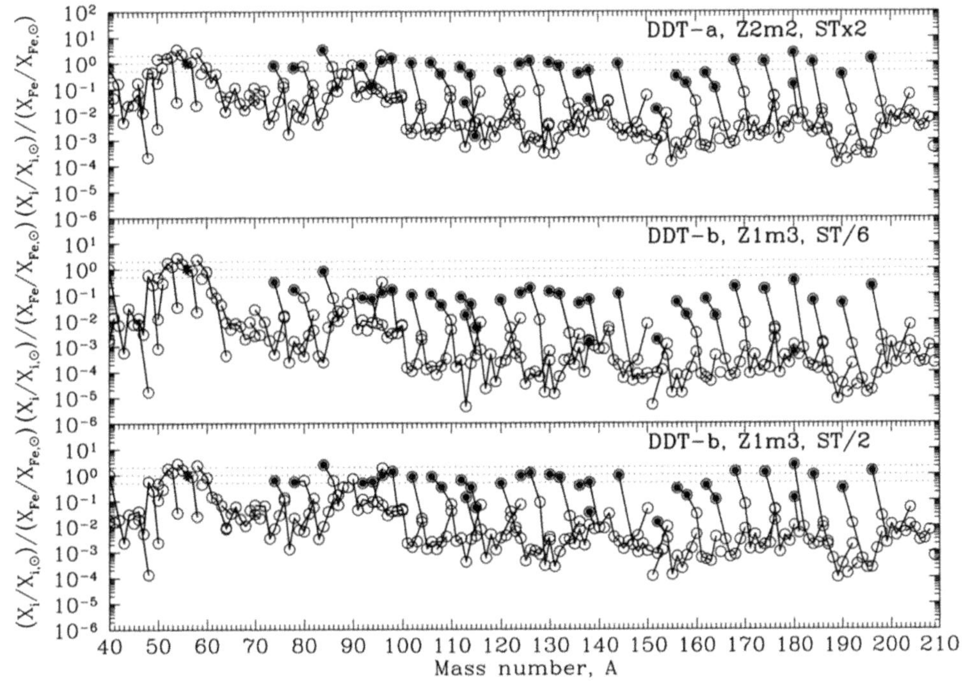

Figure 9.12. Nucleosynthesis yields from the 2D thermonuclear supernova; the filled circles mark p-nuclides. The assumed s-process enrichment in the accreted layer is solar (upper panel) and 1/20 solar metallicity (middle and lower panels) with different strengths of the ^{13}C pocket (see Travaglio et al. 2011 for details). (Reproduced from Travaglio et al. 2011. © 2011. The American Astronomical Society. All rights reserved.)

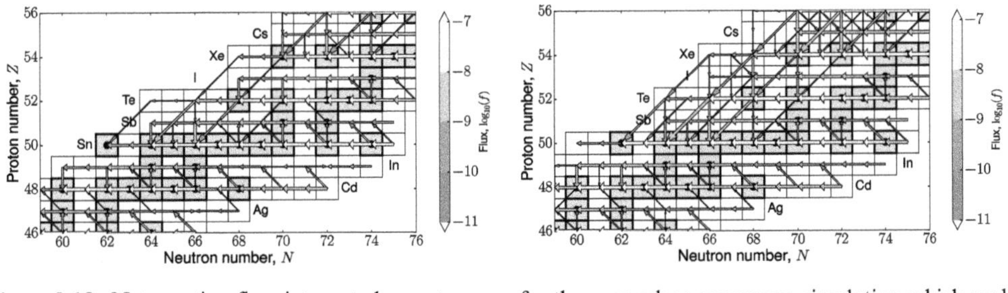

Figure 9.13. Net reaction flow integrated over tracers of a thermonuclear supernova simulation which make p-nuclides. The left panel uses only tracers from the surface region with lower density, whereas the right panel shows the flow summed over tracers at higher density from the interior of the white dwarf. The two tracer groups are also shown in Figures 8.20 and 9.11. (Reproduced from Nishimura et al. 2018, by permission of Oxford University Press on behalf of the Royal Astronomical Society.)

certain p-nuclide. There are only few exceptions. A recent investigation using Monte Carlo variation of reaction rates (Nishimura et al. 2018) and the same tracers as described above found ^{145}Eu(p,γ)^{146}Gd to significantly affect the abundance of ^{146}Sm (and also having a lesser impact on ^{144}Sm). Further reactions of interest, but with minor significance, were ^{70}Ge(α,γ)^{74}Se (affecting ^{74}Se), ^{137}Nd(n,γ)^{138}Nd (^{138}Ce), and

^{170}Hf(α,γ)^{174}W (^{174}Hf). Note that, as common for the γ-process, these reactions proceed at high temperature and therefore strong contributions of reactions on thermally populated excited states dominate the stellar cross section and the ground state cross section only gives a negligible contribution to the stellar rate (see Section 4.1.3.3 for details).

Despite the promise of the above described model regarding the production of p-nuclides, it has to be confirmed by further investigations. Regardless of the considerable uncertainties connected to the accretion, ignition, and burning front propagation, including the DDT, the seed enhancement by thermal pulses in the accreted layer has to be better constrained. Preliminary studies with a self-consistent treatment of the thermal pulses during the accretion phase seem to indicate that, similar to what was found in massive stars, the lighter p-nuclides with $A < 100$ remain underproduced, contrary to what is shown in Figure 9.12. Also, using different progenitors (different mass, composition) would also be of interest. Finally, the idea of having an enhanced s-process seed partially photodisintegrated in a white dwarf explosion only works in a single-degenerate scenario. If the double-degenerate mechanism is found to dominate the white dwarf disruptions in the Galaxy, the amount of p-nuclides would not be sufficient to explain the solar abundance level.

9.4.4 Nuclear Burning on the Surface of Neutron Stars

Brief episodes of X-ray emission in bursts with a duration of seconds to tens of minutes and typical recurrence periods between hours and days are known since the mid-1970s. Two types of such bursts have been identified: type I bursts with a more regular appearance and the more erratic type II bursts which also have very short recurrence times of a few seconds. It is now commonly accepted that type I bursts are the result of a thermonuclear runaway on the surface of a neutron star, accreting material from the Roche lobe overflow of a companion star. The burst lightcurves rise quickly, within a few seconds, to their maxima, followed by an extended tail with a decrease exhibiting a power-law behavior. "Normal" *type I X-ray bursts* (see Figure 9.14) have rise times between 1 and 10 s, last about 10–100 s, output an energy of about 10^{39} erg (peak luminosity $\mathcal{L}^*_{\rm burst} \approx 10^{37}$ erg s^{-1}), and recur within hours to days. About 110 Galactic type I burst sources are known to date. Type II bursts are attributed to accretion instabilities, also connected to a binary with a neutron star. In fact, only two type II sources are known. There is also a subclass of type I bursts which are called *X-ray superbursts*. They have durations of about 1 day, a total energy output of about 10^{42} erg, and recurrence periods of about 1 yr.

In the following the focus is on the nuclear burning causing normal type I bursts. Superbursts are briefly addressed at the end of this section. The basic idea behind the nuclear mechanism behind type I bursts is that the neutron star accretes a layer of H/He-rich material from a companion star. (Accretion of pure H or He as well as the impact of the accretion rate have also been studied.) As more and more material is accreted, density and temperature rise within the layer until hydrogen burning via CNO-cycles (see Section 7.2.2) ignites. While during early accretion the layer is only mildly degenerate and the degeneracy is lifted by a moderate increase in

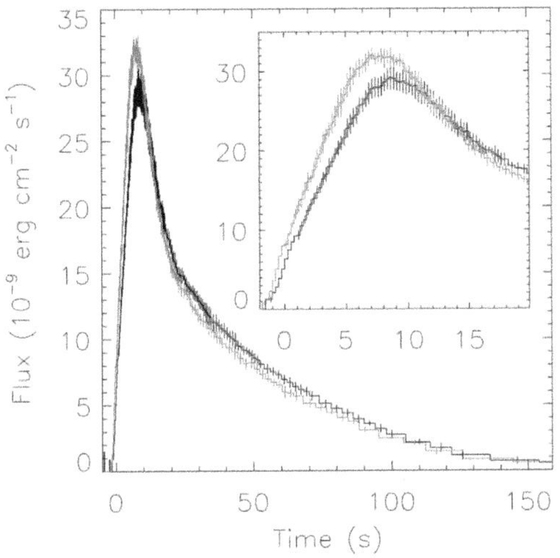

Figure 9.14. Mean profiles of seven X-ray bursts from GS 1826-24, 7 bursts observed 1997–1998 (gray histogram) and 10 bursts observed in 2000. (Reproduced from Galloway et al. 2004. © 2004. The American Astronomical Society. All rights reserved.)

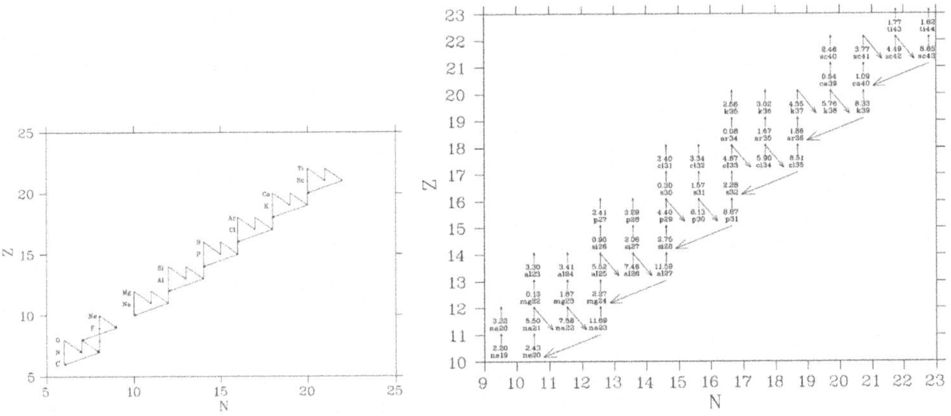

Figure 9.15. Hot CNO-type cycles and their breakup by reaction Q–value. In the right panel, Q–value is given in MeV for each breakout reaction. (Reproduced from Rembges et al. 1997. © 1997. The American Astronomical Society. All rights reserved.)

temperature, once more material is piled up on the neutron star surface combined H/He-burning proceeds under degenerate conditions and results in a thermonuclear runaway which can reach $T_{\rm peak} \approx$ 1–3 GK at densities $\rho \approx 10^5 - 10^6$ g cm^{-3}. In the initial runaway phase, with rising temperature the reaction flow can break out from the normal CNO-I, CNO-II, and CNO-III cycles by accelerated proton-capture reactions and a series of further cycles is then established, showing the same structure as the standard CNO-cycles (see the left panel in Figure 9.15). These generate further energy and finally break up according to their reaction Q–value, as shown in the right panel of Figure 9.15. The reaction flow is then moving away from stability and approaching the proton dripline (see also Section 3.2). In the presence of ^{4}He in the accreted layer, at around 3×10^8 K the triple-α rate dominates the flow from the lightest nuclides to ^{12}C, followed by ^{12}C(p,γ)^{13}N(p,γ)^{14}O. Once a sufficient amount of ^{14}O has been created, there is a strong flow toward heavier nuclides bypassing the CNO-type cycles, starting with ^{14}O(α,p)^{17}F. A sequence of alternating (p,γ) and (α,p) reactions ensues, the so-called *ap-process*, leading up to Ti, as seen in Figure 9.16 (left). In the region of Ca–Ti, sequences of proton-captures and β^+ decays (or electron captures) are taking over. The αp-process switches to the rp-process.

Due to the high densities, proton captures are fast and can compete with their reverse (γ,p) reactions even at a temperature above 1 GK. This would not be possible at lower density, for example in massive stars, where such high temperatures would result in a γ-process. In the accreted layer on the neutron star, instead the *rp-process* (short for rapid proton-capture process) appears. The rp-process bears some resemblance to the r-process (Section 8.3) in terms of reaction types. Instead of

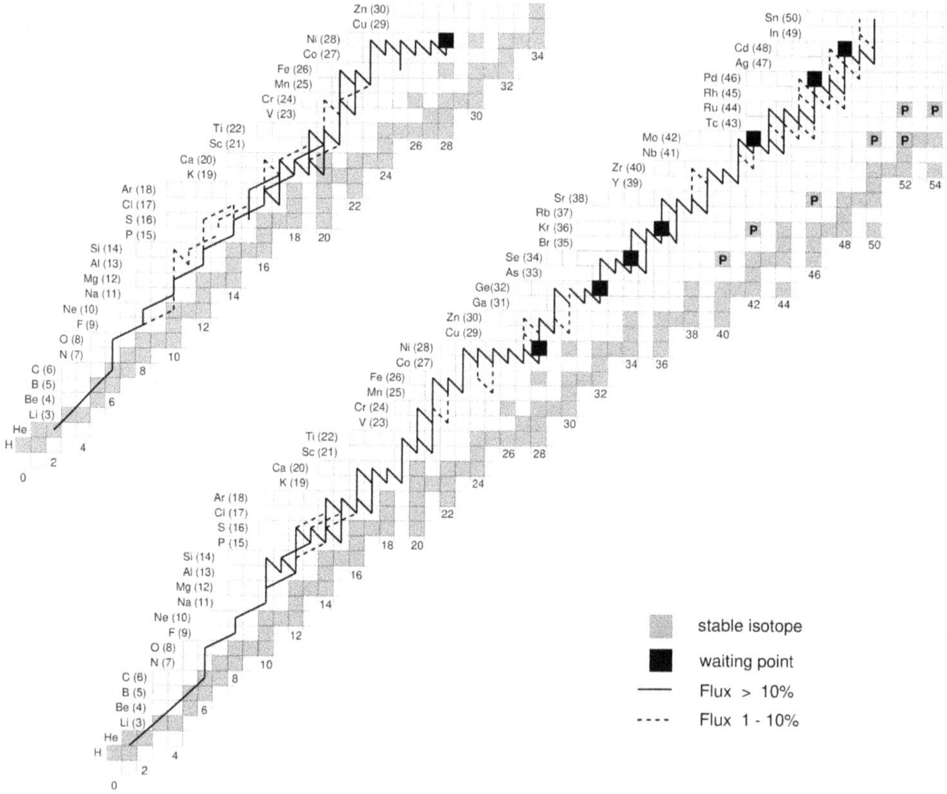

Figure 9.16. Path of the rp-process up to ^{56}Ni (left) and beyond in the freeze-out phase (right). (Republished with permission of The Royal Society, from Wiescher et al. 1998; permission conveyed through Copyright Clearance Center, Inc.)

the (n,γ)–(γ,n) equilibrium within an isotope chain, a (p,γ)–(γ,p) equilibrium in an isotone chain is established. A similar behavior is discussed in Section 9.3.3 for the νp-process. The difference is that the νp-process proceeds at lower density (but making up for it by a higher proton abundance) and starts at ^{56}Ni assembled under NSE conditions. The (p,γ)–(γ,p) equilibrium abundances in each isotone chain can be described with equations similar to Equation (4.150) but with $\bar{n}_n$ replaced by the proton number density $\bar{n}_p$, and S_n^C replaced by the proton separation energy S_p^C. Again, the abundances do not depend on the individual proton-capture rates anymore. Differences to the r-process are that the proton dripline is much closer to stability than the neutron dripline and that proton captures are subject to Coulomb barriers which increase with increasing Z within an isotone chain.

Each isotone chain has a "waiting point" similar to the waiting points discussed for the r-process (Sections 4.3.3, 8.3). These are characterized by low or negative (at the dripline) Q–value, when photodisintegration acts as a barrier to further proton captures, according to the reciprocity relation given in Equation (4.52). Waiting points are always even-Z nuclides because of the weaker binding of the following odd-Z nucleus (see Section 4.2). Just as in the r-process, the processing toward heavier A has to wait until the slow reactions bridge the waiting point. In the νp-process these waiting points are efficiently bridged by (n,p) reactions but there are no free neutrons in the rp-process. In addition to electron captures (or β^+-decays), also two-proton captures help to overcome a waiting point. Two-proton captures are especially interesting where the flow path scratches along the dripline because of the ragged structure of the dripline. The next nucleus with $Z + 1$ may be unbound with respect to protons but the nucleus with $Z + 2$ may again have $S_p > 0$ (Section 3.2).

The two-proton rate can be derived assuming a two-step process in a similar manner as shown for the rate of the triple-α rate in Section 7.3. Due to the fact that even a proton-unbound nucleus has a finite lifetime related to a proton width according to Equation (4.39), a non-zero equilibrium abundance is established for the unbound nucleus even if it has a very short lifetime with respect to proton emission. Assuming an initial nuclide A with (Z, N) and a proton-unbound nucleus B with ($Z + 1$, N), following Section 4.3 the equilibrium abundance is given by (Schatz et al. 1998)

$$Y_B = \frac{\rho N_A \langle \sigma^* v \rangle_{A+p}}{\lambda_p^* + \lambda_{(\gamma,p)}^*} Y_A Y_p, \tag{9.17}$$

where $\langle \sigma^* v \rangle_{A+p}$ is the stellar reactivity to produce the nucleus B. The spontaneous proton-decay rate and the photo-induced proton emission rate of nucleus B are denoted by λ_p^* and $\lambda_{(\gamma,p)}^*$, respectively. In the second step, another proton is captured on the equilibrium abundance of B with the usual rate $\rho^2 N_A^2 \langle \sigma^* v \rangle_{B+p} Y_B Y_p$. Assuming stellar rates (i.e., thermal population of excited states, see Section 4.1.3.3) in each step, the final rate is

$$r_{2p}^* = \frac{1}{2} \rho^3 N_A^3 \langle \sigma^* v \rangle_{2p} Y_A Y_p^2, \tag{9.18}$$

with

$$\langle\sigma^*v\rangle_{2p} = \left(\frac{2\pi\hbar^2}{\mu_{red}^{A+p}k_BT}\right)^{3/2}\frac{G_0^B}{G_0^A}e^{Q_{nuc}^{A+p}/(k_BT)}\langle\sigma^*v\rangle_{B+p}. \tag{9.19}$$

This includes the Q–value of the proton capture on nuclide A and the reactivity of this proton-capture reaction. The G_0 are the (temperature-dependent) normalized nuclear partition functions as defined in Equation (4.39). Since stellar rates obey reciprocity (as shown in Section 4.1.3.2), the reverse reaction B + $\gamma \rightarrow$ A + 2p can be expressed by the reactivity of the two-proton capture rate,

$$\lambda^*_{(\gamma,2p)} = \left(\frac{k_BT}{2\pi\hbar^2}\right)^3(\mu_{red}^{A+p}\mu_{red}^{B+p})^{3/2}\frac{2G_0^A}{G_0^B}e^{-(Q_{nuc}^{A+p}+Q_{nuc}^{B+p})/(k_BT)}\langle\sigma^*v\rangle_{2p}. \tag{9.20}$$

As mentioned above, the lifetime of a waiting point nucleus A in the rp-process depends not only on its lifetime with respect to electron capture or β^+-decay but also on the value of the two-proton capture rate. Therefore $\tau^*_{life} = 1/\lambda^*_{dec}$ (Equations (4.110) and (4.113)) with the total destruction rate

$$\lambda^*_{dec} = \lambda^*_{\beta A} + Y_p^2\rho^2 N_A^2\left(\frac{2\pi\hbar^2}{k_BT}\right)^{3/2}\frac{G_0^B}{2G_0^A}e^{Q_{nuc}^{A+p}/(k_BT)}\langle\sigma^*v\rangle_{B+p}. \tag{9.21}$$

The decay rate of A by electron capture or β^+-decay is included in $\lambda^*_{\beta A}$. It is important to note that this lifetime of the waiting point is not determined by the (p,γ) rate on the waiting point but rather by the proton capture on the next nuclide, B.

At high temperatures, $T > 1.5$–2 GK, another situation has to be considered. The (γ,2p) photodisintegration of nucleus C, made by A(2p,γ)C and thus having (Z + 2, N), will not be negligible and get into equilibrium with the (2p,γ) reaction. Therefore the abundances of A and C will be in equilibrium. Then λ^*_{dec} is given by the decay rates $\lambda^*_{\beta A}$ and $\lambda^*_{\beta C}$ as follows:

$$\lambda^*_{dec} = \lambda^*_{\beta A} + \lambda^*_{\beta C}Y_p^2\rho^2 N_A^2\left(\frac{2\pi\hbar^2}{k_BT}\right)^3\left(\frac{1}{\mu_{red}^{A+p}\mu_{red}^{B+p}}\right)^{3/2}\frac{G_0^C}{4G_0^A}e^{Q_{nuc}^{A+p}/(k_BT)}\langle\sigma^*v\rangle_{B+p}. \tag{9.22}$$

The decay rate $\lambda^*_{\beta B}$ of the intermediate nucleus is negligible because of its low equilibrium abundance.

The upper left side of Figure 9.16 shows the reaction flow reaching ^{56}Ni. The energy release by the reactions leading up to ^{56}Ni together with the CNO-type cycles is so efficient that the temperature is well above 2 GK when this first waiting point is reached. This presents an unfortunate situation because at such high temperature Equation (9.22) has to be used and the strong photodisintegration prevents any flow beyond ^{56}Ni. On the other hand, the decay half-life of ^{56}Ni ($t_{1/2}$ = 6.07 d) is much longer than the typical process timescale of the X-ray burst. This would lead one to think that ^{56}Ni was the endpoint of the combined αp/rp-process. There is an interesting fact, however, with respect to the effective lifetime of ^{56}Ni (and all other waiting points). Considering decays, two-proton captures, and photodisintegration by comparing Equations (9.21) and (9.22) it is found that there is a minimum in the lifetime within a certain temperature window. At low temperature, two-proton captures are not effective and the lifetime is given by the regular decay half-life. At high temperature, photodisintegrations act against one- and two-proton captures and also prevent a conversion to a nucleus with larger Z. At favorable, intermediate temperatures, however, the effective lifetime of a waiting point can become several orders of magnitude shorter than the X-ray burst timescale. Note that the effective lifetime not only depends on temperature but also on density and the respective Q–value, as can be seen in the equations above. For typical conditions, the temperature window is in the range of 1–1.8 GK, with a broader range for ^{56}Ni than for other waiting points. During the initial rise of the burst, ^{56}Ni is not reached before the temperatures exceed the upper limit of the temperature window. Therefore further processing has to wait until after the burst maximum, when the energy generation is reduced due to expansion of the burning layer and/or a significant reduction in protons and α-particles. Once the temperature dropped again to within the favorable temperature window, the rp-process can continue rapidly beyond ^{56}Ni. In fact, it is the energy production of the extended rp-process that determines the tail of the lightcurve after the maximum and prevents a sharper drop in luminosity. The right part of Figure 9.16 shows the extended rp-process path with the additional waiting points ^{64}Ge, ^{68}Se, ^{72}Kr, to name just the first three after ^{56}Ni. It has been found that all of them can be bridged for $\rho \approx 10^6$–10^7 g cm^{-3} and therefore the rp-process is able to efficiently produce nuclides at higher masses once it went beyond ^{56}Ni.

Figure 9.17 shows a possible endpoint region of the rp-process path. It runs along the proton dripline until enters a region of spontaneous α-emitters around Te. It is difficult to bridge this region, as spontaneous α-emission together with (γ,α) reactions act as a barrier against progressing further up in mass. Instead, SnSbTe-cycles made from proton-captures and α-emissions appear, as also illustrated in Figure 9.17. In this figure it is suggested that the cycle is entered at ^{105}Sn. Recent measurements of the proton separation energies in this region, however, suggest that the feeding into the cycle happens closer to stability (Elomaa et al. 2009). This also results in the release of fewer α-particles in the cycle. This endpoint is nevertheless interesting because it limits the production of nuclides to the region $A \lesssim 110$. This has several consequences. Since each X-ray burst is built on the ashes of the previous bursts, the recurrence time will be affected. Part of the ashes is sinking through the accreted layer and eventually settles on the crust of the neutron star, affecting its crust composition (see also Section 9.4.5 for further details on neutron-star properties). Apart from generating the actual burst, this is the reason why it is important to know the nucleosynthesis details connected to the burst. The abundance patterns synthesized and settling on the neutron star crust determine radiative, thermal, mechanical, and electric properties (and their time evolution) of the neutron star crust, affecting the further evolution of the system and probably leading to different observational signatures. Finally, the rp-process was also discussed as a possible

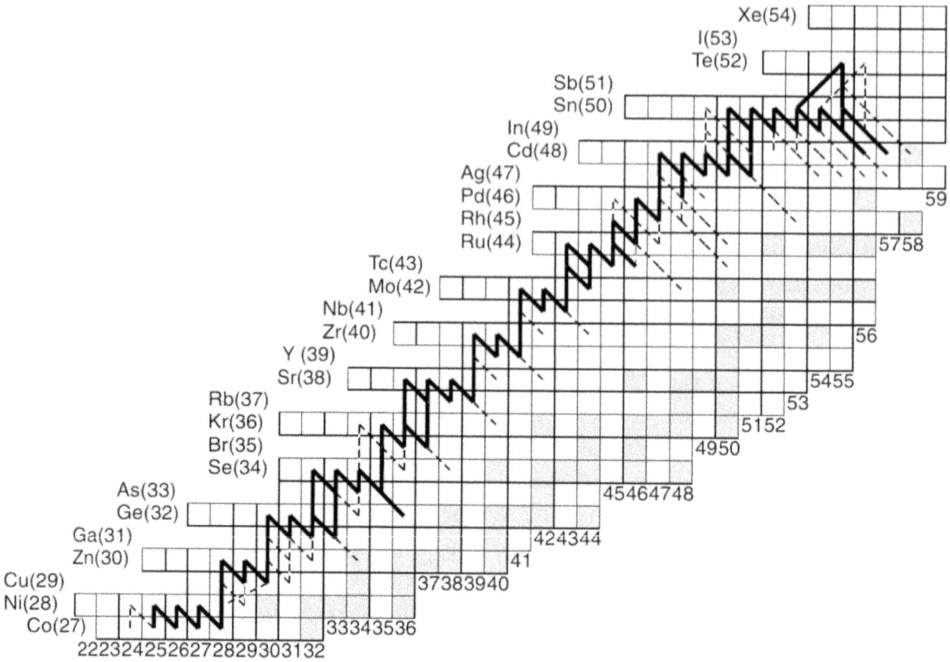

Figure 9.17. Endpoint region of the rp-process. (Reprinted figure with permission from Schatz et al. 2001. Copyright 2001 by the American Physical Society.)

source of p-nuclides (marked by "P" in Figure 9.16). If this were the case (see below), it could only contribute to the lighter p-nuclides.

Contrary to novae, the nuclear energy released in the X-ray burst is not sufficient to unbind the accreted layer due to the larger gravitational potential of the neutron star. The escape velocities (Equation (2.158)) of a typical white dwarf and a typical neutron star are about 6900 km s^{-1} and 1.9×10^5 km s^{-1} (about c/3). This also implies that the envelope expansion is quite limited. Therefore the duration of X-ray bursts is limited by fuel exhaustion rather than by expansion and envelope loss. A debated question has been whether any, even small, loss of material from the top of the accreted (and incinerated) layer is possible. Small amounts of matter could be lost by radiation-driven winds on the order of 10^{-9}–10^{-6} $M_\odot$ (also depending on the accretion rate, as accretion continues during the burst). The lost material can only contain reaction products from the αp- and rp-processes when they are brought to the top of the accretion layer by efficient convection. Currently, this is seen as a major problem for X-ray bursts to contribute to Galactic abundances. For example, recent detailed simulations of convection in the burst layer suggest that the abundance of p-nuclides in the outermost layers of the envelope is too small by orders of magnitude to significantly contribute to the Galactic p-nuclide budget.

About 26 X-ray superbursts from 15 different sources have been discovered so far. The sources also exhibit regular type I bursts but these are quenched for a longer time after the occurrence of a superburst. The energy release and the duration of superbursts suggest that they are ignited deeper inside the envelope, close to the neutron star crust, at densities exceeding 10^9 g cm^{-3}. As most likely there is no ^1H or ^4He at these depths, it was suggested that degenerate C-burning may be the energy source. This faces two problems: (1) the neutron star crust and the layers adjacent to it have to be warm enough to trigger the thermonuclear runaway with C even at high density which requires to understand the detailed heating (by previous type I bursts) and cooling mechanisms; (2) the required C can either be already contained in the accreted matter or synthesized in the rp-process during previous type I bursts. Standard type I burst burning, however, seems to consume all C. Nevertheless, it was shown that only a small amount of C in the deep layers would be sufficient to ignite a superburst.

There are a number of uncertainties in the modeling of regular type I X-ray bursts. The results depend sensitively on the assumed accretion rate and composition of the accreted matter. All nucleosynthesis studies so far have only used one-dimensional, single-zone models. A multi-D treatment would be necessary to understand the detailed impact of convection and mixing between different regions, as well as the detailed energy transport. This would also shed light on the actual ignition phase and the propagation of the burning front. It has been suggested that the thermonuclear runaway is ignited at one place (at a hot spot formed by material falling in from the accretion disk) and the burning front then rapidly fans out and engulfs the whole surface layer. Neutron stars, however, also spin rapidly and the speed of the burning front may be low enough to observe the effect of the rotation during the ignition phase. An unfortunate obstacle to rapid progress in understanding the nuclear processes is the fact that they and their products are hidden from view. The composition and its change during the burst happens below the top layer and the γ-rays emitted by nuclear reactions and decays are thermalized in the neutron star atmosphere above them. Therefore they are seen as X-rays.

In terms of nuclear physics uncertainties, they are difficult to evaluate because there is strong feedback between the energy generation and the hydrodynamic properties of the burning layer. This limits the possibility of post-processing studies. Also because of the hidden nature of the process, estimates of the impact on recurrence times and the lightcurve shape are the only possible observables but these are also affected by, for example, the accretion rate and convection. Similarly to the r-process, individual reactions (one- and two-proton capture) and decay half-lives have only to be known close to waiting points. Proton-separation energies (or alternatively nuclear masses), on the other hand, have to be known along the full rp-path to obtain the equilibrium abundances as shown above. Again, the energy generation is mainly impacted by the separation energies closely behind the waiting points. A number of (α,γ) and (α,p) reactions affect the efficiency of the αp-process, such as $^{12}C(\alpha,\gamma)^{16}O$, $^{18}Ne(\alpha,p)^{21}Na$, $^{25}Si(\alpha,p)^{28}P$, $^{30}P(\alpha,p)^{33}S$, $^{30}S(\alpha,p)^{33}Cl$, $^{32}S(\alpha,\gamma)^{36}Ar$, $^{56}Ni(\alpha,p)^{59}Cu$. For more details see, for example, Diehl et al. (2018); José (2016); Parikh et al. (2008, 2009).

9.4.5 Neutron-star Merger and the Structure of Neutron Stars

Although it is conceivable to have two massive stars in a binary system and both ending their evolution in a core-collapse supernova creating a neutron star, binary neutron star systems are quite rare. To date only eight such systems are known. This is due to several reasons. First, both stars in the system must have masses in the range permitting the final stage to be a neutron star. They must not exchange material from each other because otherwise one of the stars may strip too much mass from the companion. They will most probably not explode simultaneously and therefore also the explosion of the first star should not change the mass of the second star by blowing away outer layers or depositing additional matter. The largest problem seems to be to keep the system bound. If too much mass (more than about 30%) is ejected out of the system by the supernova explosion of one star, the binary may become unbound. Nascent neutron stars get accelerated ("kicked") by strongly asymmetric explosions leading to very asymmetric neutrino emissions. Therefore not only one but both nascent neutron stars must not receive kicks disrupting the binary system. Even if the system is not disrupted, the resulting neutron stars will move on highly eccentric orbits around the common center-of-mass.

Binary neutron stars are of interest for nucleosynthesis because it has been shown, providing a further proof of general relativity, that as orbiting very compact objects they emit gravitational waves and thereby lose energy and angular momentum. This is demonstrated by long-term observation of the first binary neutron star system detected, PSR 1913+16. This is a system with two neutron stars of almost equal mass (about 1.4 $M_\odot$) in elliptical orbits around the center-of-mass. Their maximum distance is 4.8 $R_\odot$, the minimum distance is about 1.1 $R_\odot$, with the semimajor axis of the elliptical orbit being 1,950,100 km and the period of the orbital motion being 7.75 h. According to the measurements the rate of decrease of the orbital period is 76.5 $\mu s\ yr^{-1}$, translating into a shortening of the semimajor axis by 3.5 $m\ yr^{-1}$. This agrees perfectly with the energy loss predicted by general relativity, which calculates the total power in the gravitational wave emission presently to be 7.35×10^{24} W. Therefore the neutron stars are approaching each other. The time until they collide, or less spectacularly expressed, merge has been estimated to 300 million years.

Such an inspiral and merger is the fate of every binary system of two neutron stars. In the merger, the neutron-rich matter is compressed and heated to extreme conditions and part of it is forming an accretion disk around the newly created central object. A fraction of the material may be ejected from the disk but if the combined mass of the remaining material is exceeding the limit for a neutron star, a black hole will be created. Neutron-star mergers have been suggested as one possible mechanism driving high-energy γ-ray bursts. It was also suggested that the ejected matter could be observed as a *blue kilonova*, rich in r-process material. In 2008, indeed a faint and quickly vanishing afterglow was detected at the same position as the preceding short γ-ray burst GRB 080503. The actual presence of r-process material in the ejected matter was spectacularly proven only recently, with the help of the first detected gravitational wave signal from two merging neutron stars (GW170817) in August 2017. Due to the combined data of two gravitational wave detectors (LIGO at two locations on the North American continent and VIRGO in Italy) it was possible to constrain the origin of the signal to a comparatively small region in the sky which could be searched by the combined effort of 70 observatories (on seven continents and in space). This allowed to catch the lightcurve of the event following the merger early on and across the electromagnetic spectrum, providing a wealth of data. The merger event was found to be accompanied by an initial γ-ray burst of about 2 s duration, followed by a detection of the afterglow in the ultraviolet 15.3 h after the event. Over the course of several days the optical color shifted from blue to red as the source of the emission expanded and cooled. The long-term evolution of the lightcurve is determined by the decay heat of intermediate and heavy nuclides coming from an r-process.

Before discussing further details of the observed kilonova, it may be helpful to recall a few properties of neutron stars (neutron star cooling is discussed in Section 9.3.1). What is presented in Section 2.5 regarding the stability of white dwarfs can also partially be applied to neutron stars. Instead of a degenerate electron gas providing the pressure to keep the object stable, in neutron stars the pressure is coming from a degenerate neutron gas. In fact, it is the combined pressure of neutrons, protons, and electrons (and further particles, see below) but the obtained neutron star properties are not very different when considering only neutrons. Why are neutrons appearing at all despite of being unstable as free neutrons? Neutrons decay via $n \rightarrow p + e^- + \nu_e$. To emit an electron, the electron has to find an available energy state because as a fermion the number of electrons occupying the same energy state is limited to two. As outlined before (Equation (1.79)), at high density the Fermi energy E_F of the electrons is shifted to high energy and the electrons become relativistic. This, in turn, allows electron capture on protons to create neutrons. When there are no energy states available for electrons anymore in the range of electron energies appearing in neutron decay, neutrons cannot decay anymore. This is called *Pauli blocking*. They can only decay when an energy "slot" becomes available due to electron capture on a proton. Therefore there is an equilibrium between neutron decay and electron captures, the *β-equilibrium*, which prefers neutrons due to the high E_F of the electrons. The number densities can be obtained according to Equation (1.31) using the total chemical potentials (Equation (1.30)) of neutrons, protons, and electrons and applying the equilibrium condition given in Equation (1.29).

The equation-of-state for a Fermi gas is given in Section 1.5.4. The neutron gas is degenerate but never relativistic due to the neutron mass being much higher than the electron mass. However, there are two important modifications to consider: (1) tightly packed fermions feel the repulsion part of the strong nuclear force (see Section 3.1) and therefore the notion of a *non-interacting* Fermi gas may not be applicable; (2) the stellar structure equations derived in Section 2.2 are based on a Newtonian, non-relativistic treatment which has to be modified in the deep gravitational well of a neutron star. The latter has been addressed by deriving the structure equations within a relativistic framework. This leads to the *Tolman–*

Oppenheimer–Volkoff equations. They can be written in a way corresponding to the structure equations derived in Section 2.2:

$$\frac{\mathrm{d}m}{\mathrm{d}r} = \frac{4\pi r^2 \varepsilon}{\mathrm{c}^2}, \tag{9.23}$$

$$\frac{\mathrm{d}P}{\mathrm{d}r} = -\frac{G\varepsilon m(r)}{\mathrm{c}^2 r^2}\left[1 + \frac{P}{\varepsilon}\right]\left[1 + \frac{4\pi r^3 P}{m(r)\mathrm{c}^2}\right]\left[1 - \frac{2Gm(r)}{\mathrm{c}^2 r}\right]^{-1}. \tag{9.24}$$

Equation (9.23) is exactly the same as the Newtonian relationship between mass and density. The density here is written in terms of the *mass energy density* $\varepsilon = \rho \mathrm{c}^2$ (in complete analogy to $E = m\mathrm{c}^2$). Equation (9.24) extends the Newtonian pressure equation, Equation (2.10), with three correction factors. The first two arise because of special relativistic effects and the third is a correction for general relativity. This third factor can be used to assess whether general relativity is important or not. It is not important when

$$\frac{2GM}{\mathrm{c}^2 R} = \frac{R_\mathrm{S}}{R} \ll 1. \tag{9.25}$$

The length

$$R_\mathrm{S} = \frac{2GM}{\mathrm{c}^2} \tag{9.26}$$

is called the *Schwarzschild radius*. Therefore Equation (9.25) shows that the relation of R_S to the actual physical radius R of an object determines the relevance of general relativistic effects. The Schwarzschild radius would also be the radius of the event horizon of a black hole with mass M. The Schwarzschild radius for the Sun is about 3 km, for a white dwarf with 0.6 $\mathrm{M}_\odot$ and a radius of about 10^4 km one obtains $R_\mathrm{S} = 1.8$ km and $R_\mathrm{S}/R \approx 1.8 \times 10^{-4}$. On the other hand, a neutron star is already close to a black hole with its typically 1.4 $\mathrm{M}_\odot$ and 10 km radius because $R_\mathrm{S} = 4.2$ km and $R_\mathrm{S}/R \approx 0.4$. There is a further interesting point about Equation (9.24). The pressure, providing support against gravitational contraction, appears not only on the left-hand side of the equation but also in the special relativistic terms on the right. Being a form of energy, pressure also is a source of gravity causing additional gravitational force. Regardless of the equation-of-state used (even for a completely incompressible EOS), there is an absolute limit on the mass of a stable object. If the pressure required to stabilize the mass becomes too large, its additional mass contribution leads to the creation of a black hole. It is not straightforward to obtain the absolute mass limit because fundamental principles, such as causality, have to be taken into account when the object contracts under general relativistic conditions. The theoretical maximum mass was shown to be not larger than 3.2 $\mathrm{M}_\odot$ for an object the size of a neutron star. More recent estimates with "realistic" equations-of-state are in the range of 2–3 $\mathrm{M}_\odot$.

The actual mass limit for neutron stars, as well as their mass–radius relation analogous to the one for white dwarfs derived in Section 2.5, is determined by the actual, realistic equation-of-state. Using the equation-of-state of a degenerate neutron-gas together with Equations (9.23) and (9.24) a radius of about 10 km is obtained, not too far from values derived from observation, but a maximum mass of only 0.7 $M_\odot$. This is due to the neglect of interactions between tightly packed nucleons. It is noteworthy that neutron stars with typically 1.4 $M_\odot$ and 10 km radius have an average density of about 10^{15} g cm^{-3}, which is larger than nuclear density ρ_0 by almost an order of magnitude. The canonical neutron star mass of 1.4 $M_\odot$ amounts to about 1.7×10^{57} nucleons. Assuming a sphere with 10 km radius, this results in an average distance between nucleons of only 1.35 fm. Realizing that a nucleon has a radius of about 0.8 fm, it becomes obvious that their interaction cannot be neglected. A realistic equation-of-state has to accurately model the repulsive part of the strong interaction and the detailed mediation through pions. It also has to take into account that further particles contribute at high density (such as muons, hyperons) or pion and kaon condensates may form. Also the appearance of free quarks at very high density in the core of neutron stars has been pondered. Having additional particles softens the equation-of-state because high Fermi momenta of nucleons are replaced by heavier baryons with lower Fermi momenta. A softer equation-of-state yields neutron stars with smaller radii (because it takes higher density to reach the pressure to balance gravity) and a smaller maximum mass. The search for a realistic nuclear equation-of-state is an active field of research at the intersection of nuclear physics, particle physics, and astronomy. The most massive neutron stars detected to date have masses of 1.97 ± 0.04 $M_\odot$ (PSR J1614-2230) and 2.14 ± 0.1 $M_\odot$ (PSR J0740+6620, with a radius of about 30 km), close to the theoretical limit. This already rules out many of the softer equations-of-state that have been suggested but also poses a challenge to the current understanding of nuclear matter.

Realistic neutron star models show that, similar to other stars, a neutron star has an internal structure, which has to be described by variations in the equation-of-state. Three main regions with different properties can be distinguished: the surface, the crust, and the core. Each of them are again subdivided into inner and outer regions. The surface of the neutron star is often subdivided into the atmosphere ($\rho < 10^5$ g cm^{-3}) and the envelope (or ocean) approaching $\rho \approx 10^6$ g cm^{-3}. The atmosphere is only 1 cm thick but already strongly affects the spectrum of the emitted photons (see the nature of X-ray bursts in Section 9.4.4).

Also the crust, with a thickness of 1–2 km can be subdivided further, into an outer crust and an inner crust. The density strongly increases from the outer to the inner crust. The crust is composed of nuclei, mainly ^{56}Fe plus the ashes of the rp-process, if the neutron star is an X-ray burster (see above). With increasing density, electron captures produce more and more neutron-rich nuclides and the NSE abundances shift to more and more massive nuclei with mass numbers of several hundreds. Because of the interplay of nuclear and Coulomb forces, the nuclei become arranged in a crystalline lattice. When reaching a critical density, the *neutron-drip density* $\rho_{drip} = 4 \times 10^{11}$ g cm^{-3}, the chemical potential of the neutrons inside the nuclei becomes zero and they drip out of the nuclei, forming a sea of neutrons in which the remaining nuclei are immersed. If the neutron star temperature is low, the neutron fluid is probably in a superfluid state. With further increasing density, the nuclei dissolve further and instead of a lattice of nuclei, a lattice of voids in a nucleon fluid emerges. In the transition phase, the assemblies of nuclei and voids assume different shapes, according to the surface tension at the boundary between nuclei and voids. The inner crust is often referred to as "nuclear pasta" phase, with increasing density going from spherical nuclear regions ("meatballs") and nuclear slabs ("lasagna") to voids embedded in the nuclear fluid ("swiss cheese"). Finally, the voids and nuclei dissolve completely to a uniform nuclear fluid ("sauce").

The outer and inner core together comprise 99% of the neutron star's mass. The outer core is characterized by the homogeneous neutron–proton fluid, starting at a density of 10^{14} g cm^{-3}. In addition, electrons and muons are present in sufficient amounts to guarantee electric neutrality of the core matter. The composition of the inner core, being the region of the highest density, remains a mystery. Equations-of-state differ mostly in their predictions for the inner core, where also more exotic particle configurations may appear. Since only the radii, masses, and to a certain extent, the temperature of neutron stars can be determined (and oftentimes only inferred indirectly), constraining the inner structure is very challenging.

The equation-of-state also determines how neutron stars behave in a merger event. The softness of the equation-of-state impacts the maximally possible compression and the peak temperatures and densities reached. This affects whether and how much mass can become unbound and ejected from such an event. Indirectly it also determines the nucleosynthesis possible in the ejected matter by shaping temperature, density, and Y_e. Section 8.3 presents the details of r-process nucleosynthesis and the general requirements to obtain a strong r-process. Neutron star mergers are introduced in that section as an interesting alternative to the previously favored core-collapse supernovae because they can provide low Y_e and the entropies required for an r-process with fission cycling.

With the recent detection of the kilonova lightcurve following the gravitational wave event GW170817, the theoretical studies of nucleosynthesis in neutron-star mergers have found a stronger footing. Although individual line spectra of nuclides in the ejecta cannot be identified yet, the combined color and time-evolution of the event's electromagnetic emission indicate that the ejecta contain heavy nuclides produced by rapid neutron captures. Two components in the lightcurve have been identified, a "blue component" attributed to about 0.01 $M_\odot$ lanthanide-rich material and a "red component" attributed to about 0.04 $M_\odot$ of actinide-rich material. Both compositions are the result of an r-process but at different Y_e. The lanthanide-rich ejecta are supposed to come from a close-to relativistic jet ejected perpendicularly from the orbital plane of the merging neutron stars, with Y_e exceeding 0.25. This results in a weak r-process, synthesizing nuclides with $A < 140$. The actinide material is viewed as coming from tidal ejecta from an accretion disk, with a lower $Y_e < 0.25$ and longer processing times, allowing a full r-process running up to its endpoint and probably including fission cycles. The full interpretation of the event and its lightcurve is shown in Figure 9.18. Observations of the remnant are still continuing and promise to yield further details on the composition and origin of the

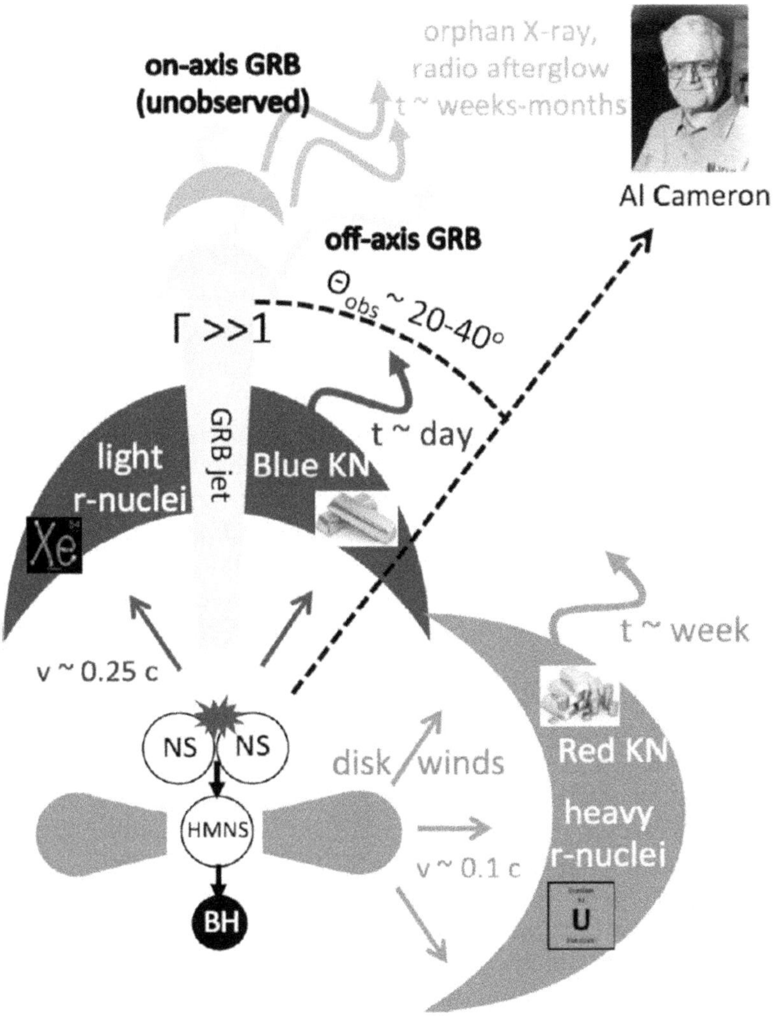

Figure 9.18. Interpretation of the electromagnetic counterpart of GW170817, as viewed by a terrestrial observer (A. G. W. Cameron). The timeline is as follows: (1) Two neutron stars (NS) of comparable mass coalesce. The dynamical stage of the merger ejects a small mass $\lesssim 10^{-3}$–10^{-2} $M_\odot$ at high velocities $v \approx 0.2-0.3$c, part of which are shock-heated polar ejecta with a high electron fraction $Y_e > 0.25$, synthesizing exclusively light r-process nuclei ($A \lesssim 140$; e.g., xenon and silver); (2) The merger product is a temporarily stable high-mass neutron star (HMNS), which generates a large accretion torus ≈ 0.1 $M_\odot$ as it sheds its angular momentum and collapses into a black hole (BH) on a timescale of $\lesssim$0.1–1 s. Winds from this highly-magnetized object also contribute substantially to the fast ejecta with high-Y_e; (3) The torus-BH powers a collimated γ-ray burst (GRB) jet, which burrows through the polar ejecta on a timescale of $\lesssim$2 s; (4) γ-Rays from the core of the GRB jet are relativistically beamed away from our sight line, but a weaker GRB is nevertheless observed from the off-axis jet or the hot cocoon created as the jet breaks through the polar ejecta; (5) On the same timescale, the accretion disk produces a powerful wind ejecting $\approx$0.03–0.06 $M_\odot$ of $Y_e \lesssim 0.25$ matter which expands quasi-spherically at $v \approx 0.1$c and synthesizes also heavier r-process nuclei with $A \lesssim 140$ such as gold and uranium; (6) After several hours of expansion, the fast polar ejecta become diffusive, powering visual wavelength ("blue") kilonova emissions lasting for a few days; (7) Over a longer timescale ($\approx$1 week), the more deeply embedded disk wind ejecta become diffusive, powering red kilonova emission; (8) The GRB/cocoon ejecta decelerate as they run into the interstellar medium, causing a delayed rise of non-thermal X-ray and radio synchrotron afterglow over several weeks to months. (Figure and caption (modified) reprinted from Metzger 2019, Copyright 2019, with permission from Elsevier.)

ejected matter. See, for example, Abbott et al. (2017); Cowperthwaite et al. (2017); Fernández & Metzger (2016); Metzger (2019) for further details.

Neutron star mergers as source of r-process material have been suggested since the mid-1970s, with refined models in the 1980s and 1990s. Hotly debated over the years, however, is their actual contribution to the total r-process content in the Galaxy (and the solar system). The scarcity of the events in itself does not pose much of a problem because a large amount of r-process material is expected to be ejected, much larger than what may be coming from core-collapse supernovae (especially if they can make an r-process only in jet ejecta). Nevertheless, the predicted amount of ejected matter still strongly varies between different numerical simulations and also depends on their input, such as the configuration of the binary and the assumed equation-of-state. The ejection of large amounts of r-process material, on the other hand, posed a problem in tracing the r-process enrichment over time in a galaxy, using galactic chemical evolution models. A large amount of material ejected per event predicts a larger spread in r-process enrichment of metal-poor halo stars than what is observed. An additional problem is that for a binary neutron star merger to occur, two stars have to have completed their stellar evolution and turned into neutron stars. The compact binary then has to merge, taking another 10^8 yr. Therefore such binary mergers can start contributing to the enrichment of interstellar matter with r-process elements only with a time delay. Assuming mergers are the only source of r-process nuclides, they would start to contribute only at higher metallicity than what is observed because r-process elements are already found in very old, metal-poor stars (see Section 8.3.3). In recent, multi-D and multi-zone galactic chemical evolution models, some of them also accounting for accretion of sub-halos into the Milky Way halo, these problems seem to be mitigated and better agreement with the observed distribution of r-process elements also in older stars is found. However, also the results of these models are sensitive to numerical resolution and to their treatment of mixing the interstellar medium. Moreover, there are still large uncertainties in the actual frequency of merger events (in addition to their mass ejection) which enter these calculations.

The emerging picture seems to favor neutron star mergers as the main source of r-process elements, with a smaller contribution from massive stars at early times in the Galaxy. The massive star contribution probably comes from magneto-rotational core-collapse supernovae which form magnetically driven jet outflows providing the conditions for a weak r-process (see, for example, Radice et al. 2016; Thielemann et al. 2017).

Further Reading

Abbott, B. P., et al. 2017, ApJL, 848, L12
Angulo, C., et al. 1999, NuPhA, 656, 3
Argast, D., Samland, M., Thielemann, F.-K., & Qian, Y.-Z. 2004, A&A, 416, 997
Caughlan, G. R., & Fowler, W. A. 1988, ADNDT, 40, 283
Chieffi, A., & Limongi, M. 2013, ApJ, 764, 21
Cowperthwaite, P. S., et al. 2017, ApJL, 848, L17

Cowan, J. J., et al. 2020, RvMP, in press; arXiv:1901.01410
Curtis, S., Ebinger, K., Fröhlich, C., et al. 2019, ApJ, 870, 2
Diehl, R., Hartmann, D. H., & Prantzos, N. (ed) 2018, Astrophysics with Radioactive Isotopes Astrophysics and Space Science Library, Vol. 453 (Berlin: Springer)
Elomaa, V.-V., et al. 2009, PhRvL, 102, 252501
Fernández, R., & Metzger, B. D. 2016, ARNPS, 66, 23
Fröhlich, C., Martínez-Pinedo, G., Liebendörfer, M., et al. 2006, PhRvL, 96, 142502
Fynbo, H. O. U., et al. 2005, Natur, 433, 136
Galloway, D. K., Cumming, A., Kuulkers, E., et al. 2004, ApJ, 601, 466
Janka, H.-T. 2001, A&A, 368, 527
Janka, H.-T. 2017, in Handbook of Supernovae, ed. A. W. Alsabti, & P. Murdin (Cham: Springer)
José, J. 2016, Stellar Explosions (Boca Raton, FL: CRC Press)
Metzger, B. D. 2019, AnPhy, 410, 167923
Langanke, K., Martínez-Pinedo, G., & Sieverding, A. 2019, AAPPSB, 28, 41
Lattimer, J. M., & Prakash, M. 2004, Sci, 304, 536
Limongi, M., & Chieffi, A. 2003, ApJ, 592, 404
Nishimura, N., Rauscher, T., Hirschi, R., et al. 2019, MNRAS, 489, 1379
Nishimura, N., Rauscher, T., Hirschi, R., et al. 2018, MNRAS, 474, 3133
Nomoto, K., Thielemann, F.-K., & Yokoi, K. 1984, ApJ, 286, 644
Parikh, A., José, J., Iliadis, C., Moreno, F., & Rauscher, T. 2009, PhRvC, 79, 045802
Parikh, A., José, J., Moreno, F., & Iliadis, C. 2008, ApJS, 178, 110
Paxton, B., et al. 2015, ApJS, 220, 15
Pruet, J., Hoffman, R. D., Woosley, S. E., Janka, H.-T., & Buras, R. 2006, ApJ, 644, 1028
Rauscher, T., Heger, A., Hoffman, R. D., & Woosley, S. E. 2002, ApJ, 576, 323
Rauscher, T., Nishimura, N., Hirschi, R., et al. 2016, MNRAS, 463, 4153
Radice, D., Galeazzi, F., Lippuner, J., et al. 2016, MNRAS, 460, 3255
Rembges, F., Freiburghaus, C., Rauscher, T., et al. 1997, ApJ, 484, 412
Rosswog, S., & Brüggen, M. 2011, Introduction to High-Energy Astrophysics (Cambridge: Cambridge University Press)
Sallaska, A. L., Iliadis, C., Champagne, A. E., et al. 2013, ApJS, 207, 18
Schatz, H., Aprahamian, A., Barnard, V., et al. 2001, PhRvL, 86, 3471
Schatz, H., et al. 1998, PhR, 294, 167
Shapiro, S. L., & Teukolsky, S. A. 1983, Black Holes, White Dwarfs, and Neutron Stars (New York: Wiley)
Taam, R. E., Woosley, S. E., & Lamb, D. Q. 1996, ApJ, 459, 271
Thielemann, F.-K., Eichler, M., Panov, I. V., & Wehmeyer, B. 2017, ARNPS, 67, 253
Thielemann, F.-K., Hashimoto, M.-A., & Nomoto, K. 1990, ApJ, 349, 222
Thielemann, F.-K., Nomoto, K., & Yokoi, K. 1986, A&A, 158, 17
Travaglio, C., Röpke, F. K., Gallino, R., & Hillebrandt, W. 2011, ApJ, 739, 93
Travaglio, C., Gallino, R., Rauscher, T., et al. 2014, ApJ, 795, 141
Wanajo, S. 2006, ApJ, 647, 1323
Wiescher, M., Schatz, H., & Champagne, A. E. 1998, RSPTA, 356, 2105
Woosley, S. E., Heger, A., & Weaver, T. A. 2002, RvMP, 74, 1015
Woosley, S. E., & Taam, R. E. 1976, Natur, 263, 101

Thomas Rauscher

Chapter 10

Primordial Nucleosynthesis

10.1 Introduction

Nucleosynthesis does not only occur in the hot plasmas of stars and stellar explosions. In its evolution after the Big Bang, the Universe has passed through a phase providing a range of temperatures and densities suited for reactions between the free nucleons present. These reactions built the first nuclei consisting of more than a single proton in what is called Big Bang Nucleosynthesis (BBN). While protons remain the most abundant nuclides also after the completion of BBN, there is a considerable abundance of ^{4}He made, as well as deuterium (^{2}H), ^{3}He, and traces of ^{7}Li. Historically, the success of BBN calculations reproducing observed or inferred primordial abundances of these nuclides over a range of nine orders of magnitude, using a single free parameter, made BBN one of the original three pillars of the hot Big Bang model. The other two are the cosmological expansion, measured by the recession of distant galaxies, and the cosmic microwave background radiation (CMB) as a relic of an early hot phase of the Universe. The role of BBN has changed over time, though. Before the high-resolution measurements of the CMB by the WMAP and PLANCK satellite observatory, BBN was used to constrain the baryon density of the Universe. As will be seen below, the only free parameter is the baryon-to-photon ratio $\eta_{\mathrm{b}} = \bar{n}_{\mathrm{b}}/\bar{n}_{\gamma}$. Over the years, the reaction rates relevant in BBN have been measured with high precision and BBN studies have entered the "precision era." Tension arose with cosmological models favoring a "flat" universe and requiring a higher value of the energy density of the Universe as could be provided by the value obtained with BBN. This indicated the presence of a non-baryonic matter component, termed *dark matter*. The situation was further complicated by the fact that distance measurements using SN Ia as distance indicators had discovered an unexpected acceleration (or deceleration of deceleration) of the

doi:10.1088/2514-3433/ab8737ch10

expansion of the Universe on very large distances. A straightforward way to include this in general relativity is to postulate an additional field giving rise to a repulsive force acting against gravitation and thus causing distant galaxies to be further away than one would expect from original general relativity. This additional component was termed *dark energy* and can be modeled by a non-zero cosmological constant in the Einstein field equations (see below). The sum of the three components—baryonic matter, non-baryonic matter, and dark energy—(plus radiation density, which is negligible since the formation of the CMB) was found to be able to provide a total value of the energy density compatible with a flat universe. BBN had proven to be a reliable source of information regarding the conditions in the early and present universe.

With the availability of the data from WMAP (Komatsu et al. 2009) and PLANCK (Planck Collaboration et al. 2016) cosmology again has entered a new era. The high-resolution data allows to obtain independent measures of the baryon density and the total energy density. These basically confirmed the previous values but provided much higher precision. Having an independent value for the baryon density implies that there is no free parameter left in modeling BBN, once the reaction rates are accurately determined (which they are). This changes the role of BBN to being an excellent tool for probing the physics of the early universe and to motivate (see the description of the ^{7}Li problem below) or constrain extensions to the standard model of particle physics. In the following, standard Big Bang nucleosynthesis (SBBN) is presented without an exhaustive listing of non-standard modifications, except for a few. For further details and more on the history of (S)BBN, the reader is referred to the literature given at the end of this chapter and the references therein.

10.2 Measured Primordial Abundances

While the BBN model itself has reached high precision, there are still considerable uncertainties connected with the determination of primordial abundances for comparison to the model. This is due to the fact that the light element abundances have to be observed in systems with as low metallicity as possible and then extrapolated to zero metallicity. This is complicated by the fact that stars modify abundances, and differently for each type of star and nuclide, and even the observed surface abundances of stars may not be the same as in the molecular cloud from which they formed (see also Chapter 6). The current status on primordial abundances is (ordered by abundance):

- ^{4}He: This abundance is determined from HII (ionized hydrogen) emission lines from low-metallicity extragalactic sources, in particular compact blue dwarf galaxies, supposedly an early type of galaxy. To account for stellar production, a range of (low) metallicities have to be considered and the derived abundance value extrapolated to zero metallicity. The currently recommended value is $Y_{^4\mathrm{He}} = 0.2449 \pm 0.004$ (Pitrou et al. 2018).

- D (^{2}H): Deuterium is weakly bound and can be easily destroyed by nuclear processes in stars already at comparatively low plasma temperature. There are no known significant sources for deuterium production after BBN. Its lines are comparatively pronounced and also detectable as faint absorption lines at high redshift (i.e., from absorbing clouds far away at cosmological distances). The lowest abundance values are obtained from high-redshift clouds which are in the line of sight of distant quasars. The scatter in measurements of different objects at very high redshift is very small, leading to the conclusion that they indeed show primordial values. The recommended primordial abundance (relative to the hydrogen abundance, also written as D/H) is Y_D/Y_p = D/H = $(2.527 \pm 0.03) \times 10^{-5}$ (Pitrou et al. 2018).
- ^{3}He: This nuclide can be produced as well as destroyed in all types of stars and therefore the evolution of its abundance with time (and metallicity) is not easy to track. Observation of He is difficult in general and due to the low $Y_{^3\mathrm{He}}/Y_{^4\mathrm{He}}$ ratio it was only observed in our Galaxy. Only very crude limits, based on the data from a single object and thus not very reliable, can be given: $0.9 \times 10^{-5} \leqslant {}^3\mathrm{He/H} \leqslant 1.3 \times 10^{-5}$ (Pitrou et al. 2018). The next generation of 30 m (and larger) telescopes may allow better constraints from observation of extragalactic HII regions.
- ^{7}Li: This nuclide is special, both in terms of its BBN production and in its later production and destruction. Due to its weak nuclear binding, it can be easily destroyed already at plasma temperatures as low as 2.5 MK. On the other hand, it is produced in at least four sources: BBN, spallation of nuclei in the interstellar medium by cosmic rays, AGB stars (Chapter 6), and novae (Section 9.4.2). Lithium is observed by emission lines from the surface region of low-mass stars in the Milky Way halo. A complication for obtaining an accurate abundance value is that the ionization state of Li has to be known which requires theoretical input from stellar atmosphere models. To minimize effects of convective layers transporting Li from the surface to deeper, hotter regions of the star, where it would be destroyed, usually a selection is made to include only stars with thin surface convection layers. The French astronomers Monique and François Spite discovered that the Li/H abundance levels off in a sample of more than 100 stars with metallicities $-2.7 \lesssim$ [Fe/H] $\lesssim -1.5$. This is called the *Spite plateau*, with ^{7}Li/H = $(1.58 \pm 0.3) \times 10^{-10}$ (Pitrou et al. 2018). Up to now, this has been interpreted as the primordial ^{7}Li value. As shown below, however, this is in discrepancy with current BBN predictions. Apart from more exotic proposed solutions including non-standard physics, it has been suggested that atomic diffusion and pre-main sequence depletion of Li may occur and may also have lowered the ^{7}Li/H of stars in the Spite plateau (see Pitrou et al. 2018 for more details).

In conclusion, probably with the exception of D, primordial abundances are not directly observed but rather inferred. Nevertheless, in the following these determinations are called "observations" for simplicity.

10.3 The Early Universe

10.3.1 Equations for the Expanding Universe

The description of the evolution of the Universe from the Big Bang to today is based on the nonlinear field equations of general relativity. The Einstein field equations can be written very compactly in tensor form:

$$R_{\mu\nu} - \frac{1}{2}g_{\mu\nu}R_{\mathrm{GR}} = \frac{8\pi \mathrm{G}}{\mathrm{c}^4}\tilde{F}_{\mu\nu} - g_{\mu\nu}\Lambda, \tag{10.1}$$

where $R_{\mu\nu}$ is the Ricci curvature tensor, $g_{\mu\nu}$ the metric tensor, R_{GR} the scalar curvature, $\tilde{F}_{\mu\nu}$ the energy–momentum tensor, and Λ the cosmological constant. The tensors are 4×4 matrices according to the three spatial dimensions and one time dimension and therefore this is actually a set of four coupled, nonlinear equations. Each tensor has 10 independent components. There are four Bianchi identities related to curvature in multi-dimensional space, which reduce the number of independent equations from 10 to 6, leaving the metric with four degrees of freedom. This corresponds to the freedom to choose a coordinate system. Equation (10.1) relates space curvature (left-hand side) to energy content (right-hand side). It reduces to Newton's gravitation law for weak gravitational fields and velocities much slower than the speed of light.

The metric determines the large-scale geometry of spacetime. In order to solve the tensor equation a metric (an abstract definition of distance in multi-dimensional space) appropriate for the problem of interest has to be defined. Exact solutions can only be obtained with certain simplifying assumptions. The most general metric for a universe fulfilling the *cosmological principle* of being a homogeneous and isotropic universe is the *Friedmann–Lemaître–Robertson–Walker metric* (FLRW),

$$\mathrm{d}s^2 = \mathrm{c}^2\mathrm{d}t^2 - \tilde{r}^2(t)\mathrm{d}l^2, \tag{10.2}$$

with $\mathrm{d}l^2$ being the metric (distance measure) of an isotropic and homogeneous three-dimensional space, and $\tilde{r}(t)$ is a scale factor describing the change in the co-moving length scale of the Universe, that means the relative expansion or contraction. To fulfill the cosmological principle, the metric $\mathrm{d}l^2$ in Equation (10.2) has to describe a space with constant curvature and therefore being either flat, a sphere of constant positive curvature, or a hyperbolic (saddle-shaped) space with constant negative curvature. Only then the scale factor is a meaningful definition. In spherical coordinates (r, θ, φ), the FLRW becomes

$$\mathrm{d}s^2 = \mathrm{c}^2\mathrm{d}t^2 - \tilde{r}^2(t)\left[\frac{\mathrm{d}r^2}{1 - \tilde{k}r^2} + r^2\mathrm{d}\theta^2 + r^2\sin^2\theta\mathrm{d}\varphi^2\right]. \tag{10.3}$$

The curvature in the three spatial dimensions is specified by the curvature constant $\tilde{k}$, assuming one of three possible values: 0 (flat, Euclidic space), +1 (spherical, closed space), −1 (hyperbolic, open space). The time dependence of the scale factor, describing the expansion or contraction of the Universe, can be derived by using the

FLRW with the field equations and solving the resulting set of coupled differential equations. This leads to the *Friedmann–Lemaître equations*,

$$\left[\frac{1}{\tilde{r}(t)}\frac{\mathrm{d}\tilde{r}(t)}{\mathrm{d}t}\right]^2 = \frac{8\pi G\rho}{3} + \frac{\Lambda c^2}{3} - \frac{\tilde{k}c^2}{\tilde{r}(t)}, \tag{10.4}$$

$$\frac{1}{\tilde{r}(t)}\frac{\mathrm{d}^2\tilde{r}(t)}{\mathrm{d}t^2} = -\frac{4\pi G}{3}\left[\rho + \frac{3P}{c^2}\right] + \frac{\Lambda c^2}{3}. \tag{10.5}$$

These two equations can be combined to obtain

$$\frac{\mathrm{d}}{\mathrm{d}t}[\rho\tilde{r}^3(t)] + \frac{P}{c^2}\frac{\mathrm{d}}{\mathrm{d}t}\tilde{r}^3(t) = 0. \tag{10.6}$$

This expresses mass–energy conservation and can also be interpreted as the first law of thermodynamics ($\mathrm{d}U + P\mathrm{d}V = 0$) in an adiabatic expansion (which is implicitly assumed in the FLRW). Equivalently, this can also be recast as

$$\frac{\mathrm{d}\rho}{\mathrm{d}t} = -\frac{3}{\tilde{r}(t)}\frac{\mathrm{d}\tilde{r}(t)}{\mathrm{d}t}\left[\rho + \frac{P}{c^2}\right]. \tag{10.7}$$

So far, only the time dependence of the scale factor was made explicit. In fact, also the (co-moving) mass density $\rho = \rho(t)$ and the pressure $P = P(t)$ vary with time. Thus, there are three variables but only two linearly independent equations, Equations (10.4) and (10.5). Just as is the case with the stellar structure equations presented in Section 2.2, an equation-of-state has to be used as a third, independent equation. In cosmology, the simple equation-of-state of a perfect fluid is used,

$$P(t) = \tilde{\omega}\rho c^2. \tag{10.8}$$

Inserting Equation (10.8) into Equation (10.7) and integrating over time yields

$$\frac{\mathrm{d}\rho}{\mathrm{d}t} = -\frac{3}{\tilde{r}(t)}\frac{\mathrm{d}\tilde{r}(t)}{\mathrm{d}t}\rho[1 + \tilde{\omega}], \tag{10.9}$$

implying

$$\rho \propto \tilde{r}^{-3(1+\tilde{\omega})}. \tag{10.10}$$

The proportionality factor $\tilde{\omega}$ is chosen according to the type of fluid:

- $\tilde{\omega} = 0$ for a matter-dominated universe (gas of non-interacting, non-relativistic particles);
- $\tilde{\omega} = 1/3$ for a radiation-dominated universe; this means that the pressure is dominated by radiation pressure and the density to be used is the radiation density, $\rho = \rho_{\mathrm{rad}}$;
- $\tilde{\omega} = -1$ when the dynamics of the Universe are dominated by the cosmological constant. This implies a constant density $\rho = \rho_\Lambda = \Lambda c^2/(8\pi G) = \mathrm{const}$;

- Further values of $\tilde{\omega}$ can accommodate exotic states of matter and energy, such as the hypothetical "quintessence" with varying $\tilde{\omega}(t)$ and/or $\tilde{\omega} < -1$ ("phantom energy").

For a mixture of different fluids, Equation (10.9) holds separately for each and therefore a simple linear combination can be solved straightforwardly, see also below. For most applications, however, it is sufficient to consider only one dominating pressure contribution. The early universe until the formation of the CMB is radiation dominated and it is matter dominated since then. The contribution of the cosmological constant may take over at very late epochs of the Universe. With the help of Equation (10.6) and an equation-of-state, the relation between density and scale factor is found to be $\rho \propto 1/\tilde{r}^3$ for a matter-dominated universe and $\rho \propto 1/\tilde{r}^4$ for a radiation-dominated universe. This also shows that, when both contributions from matter and radiation are present, an initially radiation-dominated universe turns into a matter-dominated one with increasing scale factor.

Assuming a flat universe ($\tilde{k} = 0$) the solution of the Friedmann–Lemaître equations (it is again advantageous to use Equation (10.6)) with an equation-of-state as described by Equation (10.8) yields the temporal evolution of the scale factor as

$$\tilde{r}(t) = \tilde{r}_0 t^{2/(3(1+\tilde{\omega}))}, \tag{10.11}$$

with $\tilde{r}_0$ being an integration constant fixed by the choice of initial conditions. Therefore $\tilde{r}(t) \propto \sqrt{t}$ for a radiation-dominated universe. This sets the expansion rate to be used during SBBN. For a matter-dominated universe, $\tilde{r}(t) \propto t^{2/3}$.

Several important cosmological parameters are defined via the Friedmann–Lemaître equations. The *Hubble–Lemaître parameter* is defined as

$$H(t) = \frac{1}{\tilde{r}(t)} \frac{\mathrm{d}\tilde{r}(t)}{\mathrm{d}t}. \tag{10.12}$$

The *Hubble–Lemaître constant* H_0 is today's value of the Hubble–Lemaître parameter, $H_0 = H(t_{\mathrm{today}})$. It is related to the *Hubble–Lemaître law* specifying the recession velocity of galaxies from an observer, $v_{\mathrm{rec}} = H_0 d$, where d is the distance of the Galaxy. Note that over cosmological distances the distance has to be corrected for the curvature of spacetime (as it is done for the luminosity distance when using SN Ia to determine cosmological distances). Another useful parameter is the *deceleration parameter*,

$$q_{\mathrm{dec}}(t) = -\tilde{r}(t) \frac{\mathrm{d}^2\tilde{r}(t)}{\mathrm{d}t^2} \left[\frac{\mathrm{d}\tilde{r}(t)}{\mathrm{d}t} \right]^{-2} = -1 - \frac{1}{H(t)^2} \frac{\mathrm{d}H(t)}{\mathrm{d}t}, \tag{10.13}$$

which is a measure of the change in the expansion or contraction of the Universe. It is called *deceleration* parameter because the action of gravity decelerates the expansion after the Big Bang. With the discovery of dark energy, however, it is also possible that the expansion accelerates over cosmological distances and timescales.

Rewriting Equation (10.4) with the Hubble–Lemaître parameter and considering the combined densities of matter (comprised of baryonic and non-baryonic matter, $\rho_M = \rho_b + \rho_{non-b}$), radiation, and the cosmological constant yields

$$H(t)^2 = \frac{8\pi G}{3}\left[\rho_M(t) + \rho_{rad}(t) + \rho_\Lambda\right] - \frac{\tilde{k}c^2}{\tilde{r}(t)^2}. \tag{10.14}$$

Assuming a flat, matter-dominated universe ($\tilde{k} = 0$, $\rho_{rad} \approx 0$) and also $\Lambda = 0$ this transforms to

$$\rho_M = \frac{3H(t)^2}{8\pi G} = \rho_{M,crit}(t). \tag{10.15}$$

This defines the *critical density* $\rho_{M,crit}$ which leads to an asymptotically stopped expansion for $t \to \infty$ in such a universe. With the definition of the dimensionless density parameter

$$\Omega_M(t) = \frac{\rho_M(t)}{\rho_{M,crit}(t)} \tag{10.16}$$

it can be seen that $\Omega_M = 1$ has to be fulfilled in such a flat, matter-dominated universe.

Equation (10.16) can easily be generalized by keeping all the terms of Equation (10.14) which again have to sum up to

$$\rho_{crit}(t) = \frac{3H(t)^2}{8\pi G} = \rho_M(t) + \rho_{rad}(t) + \rho_\Lambda - \frac{3\tilde{k}c^2}{8\pi G\tilde{r}(t)^2}. \tag{10.17}$$

Density parameters can be defined for each term on the right-hand side of Equation (10.14) by division by the generalized critical density $\rho_{crit} = 3H/(8\pi G)$ as defined in Equation (10.17). This is straightforward for Ω_M, Ω_{rad}, Ω_Λ and defines a "curvature density parameter" as

$$\Omega_k(t) = -\frac{\tilde{k}c^2}{H(t)^2\tilde{r}(t)^2}. \tag{10.18}$$

Finally, this leads to

$$\Omega_{tot} = \Omega_M(t) + \Omega_{rad}(t) + \Omega_\Lambda(t) + \Omega_k(t) = 1 \tag{10.19}$$

for the general case. Note that the individual density parameters depend on time but Ω_{tot} does not. Equation (10.14) is sometimes expressed by using today's values of the density parameters:

$$\frac{H(t)^2}{H_0^2} = \frac{\Omega_{0,rad}}{\tilde{r}(t)^4} + \frac{\Omega_{0,M}}{\tilde{r}(t)^3} + \frac{\Omega_{0,k}}{\tilde{r}(t)^2} + \Omega_\Lambda, \tag{10.20}$$

with today's value of the critical density $\rho_{0,crit} = 3H_0/(8\pi G)$. Today radiation density is negligible, $\Omega_{0,rad} = 0$. For a flat universe, obviously $\Omega_{0,k} = 0$. The remaining dimensionless density parameters and H_0 are derived from various types of observations, partially complementary. They define the overall geometry of space-time. To obtain the temporal evolution of the scale parameter, additionally the cosmological equation-of-state has to be known. Assuming the form given in Equation (10.8) and combining data for the various density parameters, the fluid parameter $\tilde{\omega}$ for dark energy may also be determined.

The unfortunate choice of the term "dark matter" often gives rise to confusion. Occasionally, in astronomy the term dark matter has been used to describe any non-luminous matter which cannot be observed by their electromagnetic wave emission. This would include cold white dwarfs, brown dwarfs, and cold interstellar dust made of "normal" baryonic matter. Most widely used nowadays is the term dark matter in the context of galaxy dynamics and galaxy formation where an explanation of observational data seems to require an additional component (additional to the luminous matter) contributing to the gravitational attraction. At least a part of this additional matter component has to interact only weakly or not at all with baryonic matter. In the context of Friedmann–Lemaître cosmology a contribution of non-baryonic matter is required because the value of Ω_b obtained from SBBN or the CMB, together with the value for Ω_Λ, is not sufficient to obtain $\Omega_b + \Omega_\Lambda = 1$, as required by Equation (10.19). This has become even more convincing since WMAP and PLANCK provided also tight constraints on Ω_{tot}. Therefore a non-baryonic matter contribution, not impacting SBBN, is required and $\Omega_M = \Omega_b + \Omega_{non-b} = \Omega_{lum} + \Omega_{non-lum}$, where Ω_{non-b} and $\Omega_{non-lum}$ do not necessarily fully coincide. In fact, the observationally inferred value (from stars, dust, galaxies, and galaxy clusters) for the baryonic matter content of the Universe is lower by more than an order of magnitude than Ω_b inferred from SBBN and the CMB. This is known as the *missing baryon problem*.

The cosmological parameter values obtained from PLANCK data (within 1σ errors) are (Planck Collaboration et al. 2016):

$$\begin{aligned}
\Omega_{tot} &= 1.0023^{+0.0056}_{-0.0054}, \\
\Omega_{0,b} &= 0.0486 \pm 0.0010, \\
\Omega_{0,non-b} &= 0.2589 \pm 0.0057, \\
\Omega_{0,M} &= 0.3089 \pm 0.0062, \\
\Omega_\Lambda &= 0.6911 \pm 0.0062, \\
\rho_{0,crit} &= (8.62 \pm 0.12) \times 10^{-27}\ \mathrm{kg\ m^{-3}}, \\
\tilde{\omega}_\Lambda &= 0.980 \pm 0.053.
\end{aligned} \tag{10.21}$$

It is to be noted that today's values of the density parameters and the critical density implicitly depend on the adopted value of the Hubble–Lemaître constant H_0, as they are derived from CMB data and the Universe has expanded since the freeze-out of the CMB. The value determined from PLANCK data is $H_0 = 67.74 \pm 0.46$ km s^{-1} Mpc^{-1}. With these parameters, the age of the Universe is determined as $t_{today} = (13.799 \pm 0.021) \times 10^9$ yr. Within uncertainties, the value for $\Omega_{0,b}$ agrees excellently with the value obtained by SBBN before WMAP and

PLANCK. It implies that less than 5% of the content of the Universe is composed of the baryonic matter familiar to us from everyday experience. This is a modern version of the Copernican revolution.

10.3.2 Evolution until the Nucleosynthesis Epoch

Using the cosmological parameters defined above and extrapolating backwards in time it is easy to see that the very early universe had to have an extremely high relativistic energy density. Going back to $t = 0$ reveals a singularity in the Friedmann–Lemaître equations because they would give infinite density and temperature. It is well known that general relativity breaks down on the Planck scale and this is why there is currently no physical model available to describe the Planck era of the Universe, the first 10^{-43} s. After 10^{-43} s the Friedmann–Lemaître equations can be used to predict the expansion and cooling of the Universe. Nevertheless, up until about 10^{-11} s the picture remains still speculative because there is no direct observational evidence and also the energies are still way beyond what is achievable in particle accelerators. Based on certain suggestions, it is assumed that at approximately 10^{-37} s a phase transition caused cosmic inflation, i.e., a sudden, exponential growth of the scale factor $\tilde{r}$ during an extremely short period of time. After inflation, the Universe reheats, attaining conditions allowing a plasma of free quarks and gluons (the *quark–gluon plasma*) and further postulated heavy elementary particles and their antiparticles. All plasma constituents move at relativistic speeds. At some instance, an unknown process must have caused *baryogenesis*, i.e., the violation of baryon number conservation resulting in a slight excess of quarks and leptons over antiquarks and antileptons, of the order of one part in 30 million. Still particle–antiparticle pairs can be constantly created due to the relativistic energies involved but the basic asymmetry continues to exist. This pair production ceases in order of decreasing particle masses (only particles with $m_0c^2 < k_BT/2$ can be created) as the Universe further expands and cools. After about 10^{-11} s particle energies have dropped to values testable with particle accelerators and the following processes are better grounded in experimental knowledge.

At about 10^{-6} s and temperatures dropping below $k_BT \approx 100$–200 MeV (10^{12} K), freeze-out from the quark–gluon plasma creates baryons, such as neutrons and protons. This marks the beginning of the hadron epoch. As long as $k_BT \gtrsim 1$ MeV, the hot plasma is composed of nucleons, photons, e^-, e^+, ν_e, $\bar{\nu}_e$, ν_μ, $\bar{\nu}_\mu$, ν_τ, and $\bar{\nu}_\tau$ in thermal equilibrium. The pressure is dominated by the contribution of the ultra-relativistic particles (all particles except nucleons) and therefore according to Equations (1.94) and (1.95) the equation-of-state can be written as

$$P = \frac{1}{3}\frac{g}{2}\tilde{a}T^4, \tag{10.22}$$

with the statistical degrees of freedom given by

$$g = \sum_{i \in \text{bosons}} g_i + \sum_{j \in \text{fermions}} g_j. \tag{10.23}$$

The pressure contribution of non-relativistic nucleons is the one of an ideal gas, $P = \bar{n}k_BT$, and is negligible due to the linear dependence on T. Neutrinos and antineutrinos have $g = 1$, all other leptons and photons have $g = 2$. Therefore the statistical factor in Equation (10.23) evaluates to $g = 43/4$.

Because of complete thermal equilibrium, all plasma constituents can be transformed into each other. For example, protons and neutrons are converted into each other by $e^- + p \leftrightarrows n + \nu_e$ and $e^+ + n \leftrightarrows p + \bar{\nu}_e$. The equilibrium abundances can be calculated as outlined in Section 1.4, making use of chemical potentials as defined in Equation (1.30). Since $\mu_c^{tot} \approx 0$ for ultra-relativistic particles, only the non-relativistic neutrons and protons have to be considered and therefore in equilibrium $\mu_{c,n}^{tot} = \mu_{c,p}^{tot}$. Using Equations (1.56) and (1.57), the ratio of neutron and proton number-densities can be obtained as

$$\frac{\bar{n}_n}{\bar{n}_p} = \frac{X_n}{X_p} = e^{(\mu_{c,n}^{tot} - m_nc^2 - \mu_{c,p}^{tot} + m_pc^2)/(k_BT)}\left(\frac{m_n}{m_p}\right)^{3/2} \simeq e^{-\Delta_{np}c^2/(k_BT)}, \tag{10.24}$$

where Δ_{np} is the neutron–proton mass difference and the last step used $(m_n/m_p)^{3/2} \approx 1$. As long as the equilibrium is upheld, the ratio in Equation (10.24) adjusts according to the temperature. There are always fewer neutrons than protons because of the larger mass of neutrons.

At about 1 s and for $k_BT < 1$ MeV ($T < 10$ GK) electrons are not energetic enough anymore to efficiently bridge Δ_{np} and photons do not have enough energy to create positrons via pair production. Therefore also the production of ν_e and $\bar{\nu}_e$ in the nucleon transformations stops. At this point the neutrinos cannot keep up thermal equilibrium with the particles and they decouple thermally. This is called *weak freeze-out* and *weak decoupling*. The neutron-to-proton ratio of Equation (10.24) therefore also freezes out at the temperature of the weak freeze-out T_{weak}. Below T_{weak}, the ratio only changes by the decay of free neutrons. The temperature T_{weak} is not well defined because freeze-out is not instantaneous due to the high energy tails of the thermal energy distributions. A reasonable value is $k_BT_{weak} \approx 0.8$ MeV but weak interactions may affect the ratio down to $k_BT_{weak} \approx 0.28$ MeV. In analogy to the CMB for photons, there is also a (as of yet undetectable) *cosmic neutrino background radiation* from the decoupled neutrinos. The redshifted freeze-out temperature corresponds to $\approx$1.95 K, lower than the temperature of the CMB.

After weak decoupling, the Universe still is radiation dominated and photons are in thermal equilibrium with nucleons, electrons, and positrons. Around $k_BT \approx 0.5$ MeV the photon temperature is raised by the complete annihilation of positrons by electrons. The neutrino temperature remains largely unaffected because neutrinos have already decoupled. This also affects the expansion of the Universe because of the change in g in Equation (10.23) at weak freeze-out. To be able to compute the evolution of the photon temperature (which is also the nucleon temperature because it is still a radiation-dominated phase) during BBN it is necessary to determine the new g after weak freeze-out. Since the expansion is adiabatic, the entropy has to be conserved before and after the weak freeze-out. Again using Equations (1.94) and (1.95), the entropy before freeze-out is

$$S_{\text{before}} = \frac{g_{\text{before}}}{2}\frac{4}{3}\tilde{a}T_{\nu}^{3} \tag{10.25}$$

and after freeze-out it is

$$S_{\text{after}} = S_{\gamma} + 6S_{\nu} = \frac{4}{3}\tilde{a}T^{3} + 6\frac{7}{8}\frac{g_{\nu}}{2}\frac{4}{3}T_{\nu}^{3}. \tag{10.26}$$

The factor 6 is for the six neutrino species (neutrinos and antineutrinos) and $g_{\nu} = 1$. The temperature before freeze-out is taken to be the neutrino temperature as all plasma components are in thermal equilibrium and $T = T_{\nu}$. After freeze-out, a distinction has to be made between the neutrino temperature and the temperature of the remaining plasma components. Setting $S_{\text{before}} = S_{\text{after}}$ yields a relation between T and T_{ν},

$$\frac{T}{T_{\nu}} = \left(\frac{11}{4}\right)^{1/3}. \tag{10.27}$$

The energy density (internal energy) after weak freeze-out has to be calculated with the two different temperatures but Equation (10.27) can be used to express one by the other. With Equation (1.95) the two contributions are written as

$$u = \tilde{a}T^{4} + \frac{7}{8}\frac{6}{2}\tilde{a}T_{\nu}^{4} = \left[1 + \frac{21}{8}\left(\frac{4}{11}\right)^{4/3}\right]\tilde{a}T^{4} \simeq \frac{3.3626}{2}\tilde{a}T^{4}. \tag{10.28}$$

Therefore g changes from $g_{\text{before}} = 43/4 = 10.75$ to 3.3626 after weak freeze-out. With Equation (10.11) it was shown above that for a radiation-dominated universe $\tilde{r} \propto \sqrt{t}$. Combining this with inserting $\rho = u = (3.3626/2)\tilde{a}T^{4}$ into Equation (10.4) allows to relate time and temperature,

$$t = \sqrt{\frac{3c^{2}}{16\pi G\tilde{a}}}\frac{1}{\sqrt{3.3626}}\frac{1}{T^{2}}. \tag{10.29}$$

The temperature in GK is then found with

$$T_{9}(t) = \frac{13.336}{\sqrt{t}}. \tag{10.30}$$

This expression is used to track the temperature during SBBN.

The change in the number of degrees-of-freedom at weak decoupling determines the subsequent expansion and temperature evolution. This is why BBN is also sensitive to the number and type of particles present in the plasma before and at the weak freeze-out. That is why historically BBN was used to constrain the number of neutrino families N_ν. The statistical factor before freeze-out depends on N_ν as given by

$$g_{\text{before}} = \frac{43}{4}\left[1 + \frac{7}{43}(N_\nu - 3)\right]. \tag{10.31}$$

Increasing the number of neutrinos (or other ultra-relativistic particles) increases the pressure and leads to a faster expansion. In turn, this results in an earlier weak decoupling at higher T_{weak}. According to Equation (10.24), a higher T_{weak} causes a higher neutron-to-proton ratio. As will be shown below, the primordial ^{4}He abundance is determined by this ratio and a change of less than 20% in the weak rates or the expansion rate would already destroy the good agreement of calculated to measured primordial ^{4}He abundances. The BBN constraint of $N_\nu < 3.3$ was later confirmed by measurements of particle decays at CERN. Also WMAP and PLANCK provide tight constraints on N_ν and confirmed $N_\nu = 3$.

The solution of the Friedmann–Lemaître equations is an initial value problem but without choosing the boundary condition, the solution is not unique. The equations describe an adiabatic expansion but the value of the entropy can be chosen freely when lacking further information. This proved to be the original strength of the SBBN model because the nuclear reaction rates during SBBN depend on ρ and this is set by the entropy condition. The global baryon-to-photon ratio

$$\eta_{\text{b}} = \frac{\bar{n}_{\text{b}}}{\bar{n}_\gamma} \tag{10.32}$$

in a radiation-dominated universe is proportional to an inverse entropy per baryon, as can be seen when comparing Equations (1.64), (1.72), and (1.77). In SBBN calculations before WMAP and PLANCK, η_{b} was used as the only free parameter and it was constrained by comparison to primordial abundance data. Knowing the temperature from Equation (10.30), the photon number density can be obtained as

$$\bar{n}_\gamma = \frac{2.404}{\pi^2}\left(\frac{k_{\text{B}}T}{\hbar c}\right)^3 \tag{10.33}$$

according to Equation (1.72). Choosing a value for η_{b}, the baryon number density is then given by $\bar{n}_{\text{b}} = \eta_{\text{b}}\bar{n}_\gamma$ and the baryon density by $\rho_{\text{b}} = \bar{n}_{\text{b}}m_{\text{u}}$ (the latter relation neglects the neutron–proton mass difference which is only a small correction here). The neutron and proton number-densities before the onset of BBN can be obtained from this value by application of Equation (10.24). In practical units, the baryon number density in cm^{-3} can be calculated by $\bar{n}_{\text{b}} = 2.029 \times 10^{28}\eta_{\text{b}}T_9^3$ and the baryon density in g cm^{-3} by $\rho_{\text{b}} = 3.376 \times 10^4\eta_{\text{b}}T_9^3$, where T_9 is the temperature given in GK. The baryon density, and therefore also η_{b}, is directly related to Ω_{b}. In practical units the relation is $\eta_{\text{b}} = 2.7377 \times 10^{-8}\Omega_{0,\text{b}}h_{\text{H}}^2$, where h_{H} is the Hubble–Lemaître constant H_0 in units of 100 km s^{-1} Mpc^{-1}. This number is derived by considering Equation (10.20) and using today's temperature of the CMB (T_0) to obtain $\bar{n}_\gamma = 410.73(T/T_0)^3$ cm^{-3} (Coc et al. 2014).

The above relations are sufficient to specify the conditions to run a reaction network calculation to follow the synthesis of the first nuclides. This nucleosynthesis process is described in detail in Section 10.4. During BBN, the Universe keeps expanding and cooling, and the nuclear reactions cease below $T \lesssim 10^8$ K ($k_B T \lesssim 0.01$ MeV) when the age of the Universe is about 1000 s.

Much later, after about 47,000 yr, the shift from a radiation-dominated to a matter-dominated universe occurs when the pressure contribution of the relativistic radiation becomes equal and less than the contribution from (non-relativistic) nuclei. At an age of the Universe of 377,700 ± 3200 yr (Planck Collaboration et al. 2016), *recombination* occurred. At this time the temperature has sufficiently dropped so that electrons can combine with the nuclei synthesized by BBN and form neutral atoms. The term “recombination” has its origin in the history of the field and is misleading because electrons and nuclei actually combine for the first time. The formation of neutral atoms removes charged scatterers for the photons and the photons decouple. This is similar to the weak decoupling of neutrinos much earlier in the evolution of the Universe. From that moment on, photons and matter are not in thermal equilibrium anymore and the photons retain the temperature they had at decoupling. This was about 3000 K but due to the further expansion of the Universe the photons have been redshifted until today they are observed at a temperature of 2.7255 ± 0.0006 K corresponding to a thermal microwave spectrum. This is the cosmic microwave background radiation (CMB). The photon pattern in the CMB still carries the signatures of the surface of last scattering (similar to the photosphere of a star). It is the details of these patterns (tiny temperature fluctuations and their geometric distribution, polarization) which allow to determine cosmological parameters with high precision.

10.4 Standard Big Bang Nucleosynthesis (SBBN)

SBBN can be understood as the result of H-burning in an NSE (see Section 4.3.2) at high temperature and low density, i.e., at high entropy, with a rapid, extremely α-rich freeze-out from NSE. The electron abundance Y_e is set by the neutron-to-proton ratio and is slightly larger than 0.5. After weak decoupling the temperature drops from an initial value of 10 GK, when neutrons and protons exist as free nucleons. First nuclides can only form after the temperature dropped below about $k_B T \lesssim 0.1$ MeV, corresponding to the binding energy of the deuteron. The synthesis of deuterons is the bottleneck of SBBN, the synthesis of further nuclides has to wait until deuterons are not significantly photodisintegrated anymore by energetic photons. This delays the onset for another about 9–10 s and leaves enough time for a fraction of neutrons to decay. At weak decoupling the neutron-to-proton ratio given by Equation (10.24) is 1/6 and changes to 1/7 (which is $X_n = 0.125$, $X_p = 0.875$) by neutron decay until the start of BBN. The change of neutron mass fraction with temperature (and therefore with time, see Equation (10.30)) is shown in Figure 10.1. Coming from high temperature, the neutron mass fraction is set by Equation (10.24) until weak decoupling (labeled A in the figure), after which neutrons decay until the onset of nucleosynthesis (labeled B in the figure). This illustrates that SBBN is

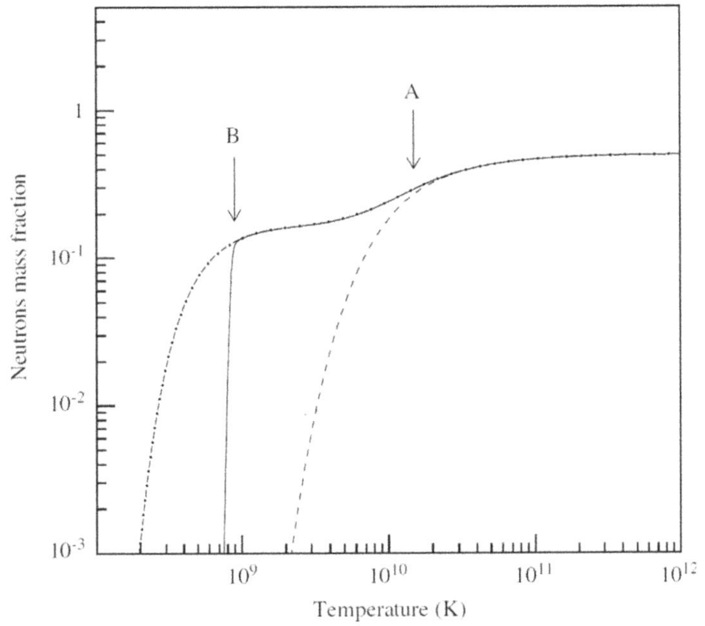

Figure 10.1. Neutron mass fraction as function of temperature; the dashed line is the n–p equilibrium value, free neutron decay is shown by the dash-dotted line, and the full line is the SBBN network calculation. The freeze-out of the weak reactions occurs at the temperature labeled A, and the deuterium bottleneck is overcome at label B. (Reprinted from Coc 2009, Copyright 2009, with permission from Elsevier.)

sensitive to the half-life of the neutron in a similar manner as it is to changing, for example, the number of neutrino families (Equation (10.31)). A longer half-life has the same effect as increasing N_ν, giving higher X_α. The currently adopted value of the neutron half-life for SBBN is $t_{1/2} = 880.2 \pm 1.0$ s, although there seems to be a systematic discrepancy between in-beam and trap measurements, for example, as addressed in Czarnecki et al. (2018).

Once deuterons have been created the equilibrium quickly adapts to favor the most strongly bound nuclide reachable at high T. Almost instantaneously, all remaining neutrons combine with protons to form ^{4}He, as also clearly seen in Figure 10.1. As two neutrons are required for each ^{4}He, $Y_\alpha = 0.5Y_n = 0.5X_n$ (see Section 1.3 for the definitions of mass fractions and abundances). Therefore the mass fraction of ^{4}He becomes $X_\alpha = 4Y_\alpha = 2X_n$, yielding $X_\alpha = 0.25$. More generally, the ^{4}He mass fraction can be expressed in terms of the ratio X_n/X_p, directly obtained from Equation (10.24), as

$$X_\alpha \simeq \frac{2X_n/X_p}{1 + X_n/X_p}, \tag{10.34}$$

assuming $X_n + X_p = 1$, i.e., mass fractions of any other nuclides are zero or negligible. Since $Y_p > Y_n$, there are protons left over. Therefore the main nucleosynthesis products of SBBN are ^{1}H (protons) with a mass fraction of about 0.75 and ^{4}He with a mass fraction of about 0.25.

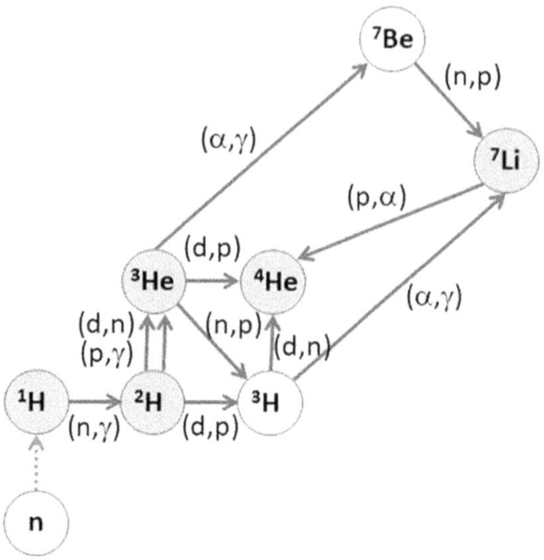

Figure 10.2. The SBBN reaction network with the 12 main reactions (including neutron decay). The nuclides ^{3}H and ^{7}Be are unstable but have so long half-lives that they can be considered stable during SBBN. They decay before the formation of the CMB, though.

In reactions leading from protons to ^{4}He further nuclides are only produced with several orders of magnitude lower mass fractions. For example, the next most abundant nuclide is the bottleneck ^{2}H, with a mass fraction about a factor of 0.001 lower than X_α. Figure 10.2 shows the minimal reaction network for SBBN. All reaction cross sections have been measured to high precision. In Figure 10.3 the mass fractions of the most abundant nuclides are plotted as a function of time. Striking the eye are the steep increase of ^{4}He once ^{2}H is present, the rapid neutron consumption, and the relatively modest decrease in the mass fraction of free protons. The half-lives of ^{3}H and ^{7}Be are long enough that they do not decay significantly during SBBN and the change in their mass fractions is due to nuclear reactions. Later on, but long before the formation of the CMB, they decay to ^{3}He and ^{7}Li, respectively.

The final primordial abundances as a function of η_b are presented in Figure 10.4. Such calculations were used to derive η_b by the requirement to simultaneously reproduce the observed primordial abundances of D (^{2}H), ^{3}He, ^{4}He, and ^{7}Li. Nowadays, η_b is set by WMAP and PLANCK data and BBN is used to test potential modifications of the standard model.

The dark blue lines in Figure 10.4 give the calculated abundances or mass fractions with an extended network (see Figure 10.5 and the discussion further below). The line widths corresponds to the uncertainty in the nuclear reaction rates. The results with the simpler network shown in Figure 10.2 are indicated by dashed light-blue lines close to or coinciding with the dark blue lines. The general dependence on η_b is easy to understand for D, ^{3}He, and ^{4}He. An increase in η_b also increases the particle number densities, causing faster reaction rates (Equation (4.9)) toward ^{4}He. This increases the amount of produced ^{4}He and decreases the abundances of D and ^{3}He. The sensitivity of ^{4}He to a variation in the number of neutrino families N_ν

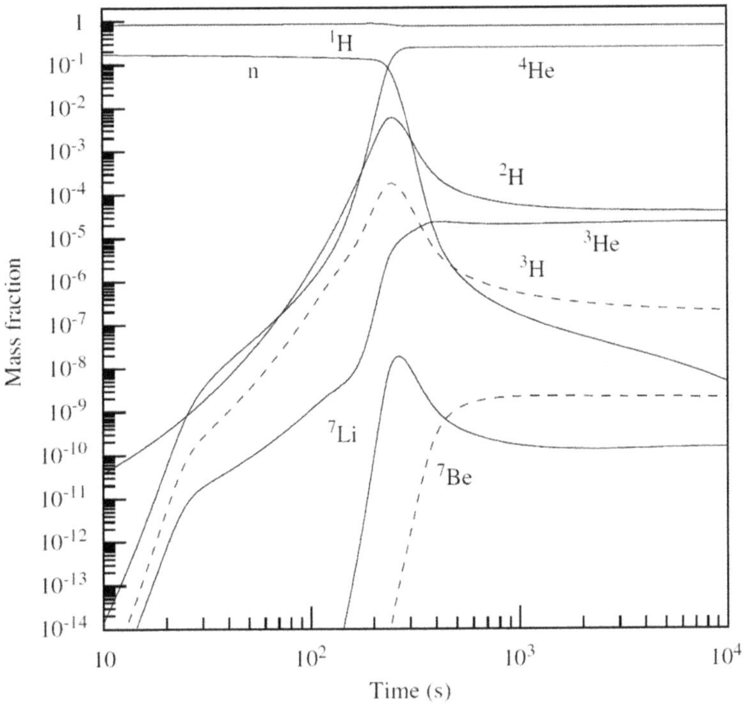

Figure 10.3. Time-evolution of the mass fractions of light nuclides at WMAP Ω_b; dashed lines indicate mass fractions of unstable nuclides. (Reprinted from Coc 2009, Copyright 2009, with permission from Elsevier.)

(discussed above) is shown by the red dot-dashed lines in the ^{4}He plot. These lines give the production range when varying N_ν in the range 3.30 ± 0.27 as suggested by CMB data. For the default SBBN calculation $N_\nu = 3$ was used.

The behavior of the ^{7}Li abundance is more complicated, exhibiting a minimum. At low η_b (low density) ^{7}Li is created by ^{3}H(α,γ)^{7}Li. With increasing density, however, the destruction by ^{7}Li(p,α)^{4}He increases more than this creation rate and the abundance declines. At even higher density the destruction is then offset by an increased production of ^{7}Be through ^{3}He(α,γ)^{7}Be. The produced ^{7}Be later decays to ^{7}Li. The quickly vanishing neutron abundance allows only a small fraction of ^{7}Be to be directly converted to ^{7}Li by ^{7}Be(n,p)^{7}Li.

The current observational abundance limits are given by horizontal dashed green lines and hatched areas in Figure 10.4. (Previously used limits are shown by horizontal, black dotted lines.) Despite of the large production of ^{4}He it is not a good indicator of η_b because its dependence on η_b is much weaker than for other nuclides. The abundance of ^{3}He is also not very useful to constrain η_b due to the very uncertain observational limits. The minimum in the ^{7}Li abundance as function of η_b does not permit a single, unique solution for η_b. Therefore the best determination of η_b is provided by D.

Within limits, the η_b limits obtained by WMAP and PLANCK agree and they also agree, within limits, to the value derived from SBBN before the availability of

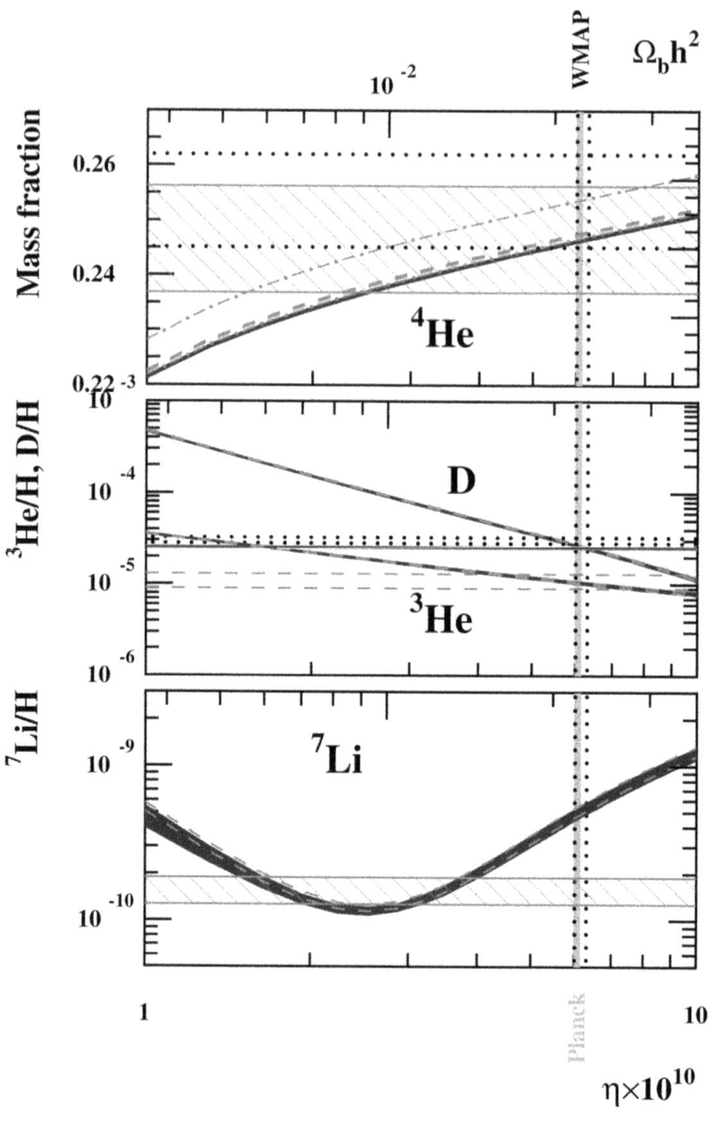

Figure 10.4. Calculated SBBN abundances as function of η_b (Ω_b) compared to measured primordial abundances (horizontal, green dashed lines and hatched areas). Note that a linear scale is used for the ^{4}He mass fraction whereas the other scales are logarithmic. The width of the dark blue lines indicates the impact of the reaction rate uncertainties. Also indicated is the value of η_b (Ω_b) determined from the cosmic microwave background by the WMAP and Planck satellites. For further details, see text. (Reproduced from Coc et al. 2014. © 2014. Published by IOP Publishing Ltd on behalf of SISSA Medialab srl. All rights reserved.)

these precise CMB measurements. As seen in Figure 10.4, the observed primordial values agree well with the SBBN values for D, ^{4}He, and maybe also ^{3}He. The calculated ^{7}Li abundance, however, does not agree within uncertainties. It is higher by a factor of about 3.5. This is the well-known *Li-problem*, which started to emerge already before the CMB measurements and became more pronounced since. A large

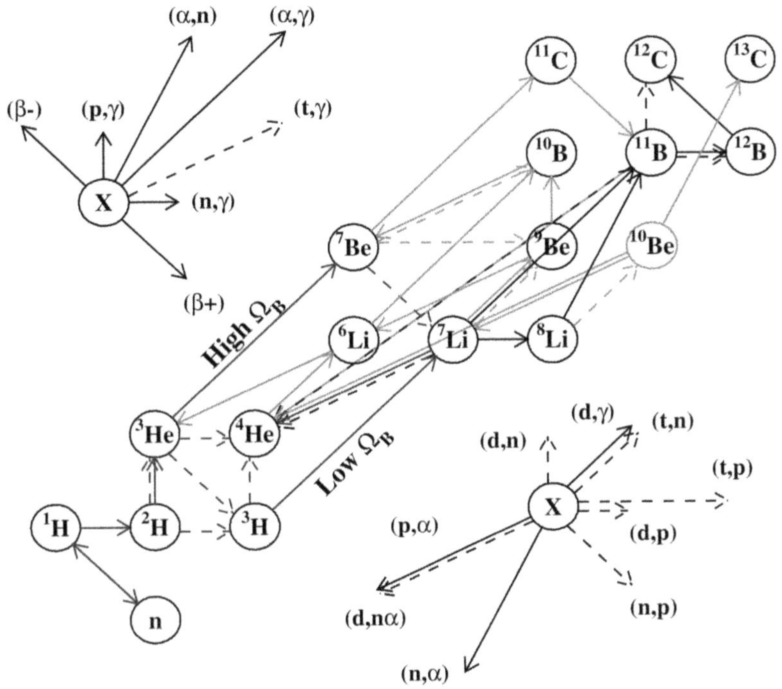

Figure 10.5. Extended SBBN reaction network to include CNO production. Shown are important reactions for production of ^{4}He, ^{2}H (D), ^{3}He, and ^{7}Li (blue), ^{6}Li (green), ^{9}Be (pink), ^{10,11}B (cyan), and CNO (black). Note that CNO production is via ^{11}B but follows a different path than primordial ^{11}B formation through the late-time ^{11}C decay. (Reproduced from Coc et al. 2014. © 2014. Published by IOP Publishing Ltd on behalf of SISSA Medialab srl. All rights reserved.)

number of possible solutions, some more exotic than others, have been proposed over the past years. Conservative solutions suggested to modify nuclear reaction cross sections. The current consensus, also supported by additional measurements and many variation studies, is that there is no nuclear solution of the Li-problem. Non-standard solutions include the addition of new particles but this is difficult because it also modifies the abundances of other nuclides. A recently suggested solution modifies the thermal energy distributions of the interacting nuclei (Hou et al. 2017). Given the unclear situation in the observations (see above in Section 10.2), however, it may be premature to resort to exotic solutions. The adopted uncertainty in observed values of the primordial Li abundance may be too optimistic and it is still not ruled out that ^{7}Li has been reduced from its primordial value even in the oldest stars presently observed and in the Spite plateau.

The triple-α rate (see Sections 7.3 and 9.1) is ineffective at the given density on the timescale set by the adiabatic freeze-out and does not produce appreciable amounts of heavier nuclides. It would be interesting to have production of, for example, CNO nuclei because this would drastically change the life of the first generation of stars. Without CNO nuclei, H-burning in stars proceeds via the slow pp-chains (explained in detail in Section 7.2.1). For stars with more than about 2 $M_\odot$, and especially for

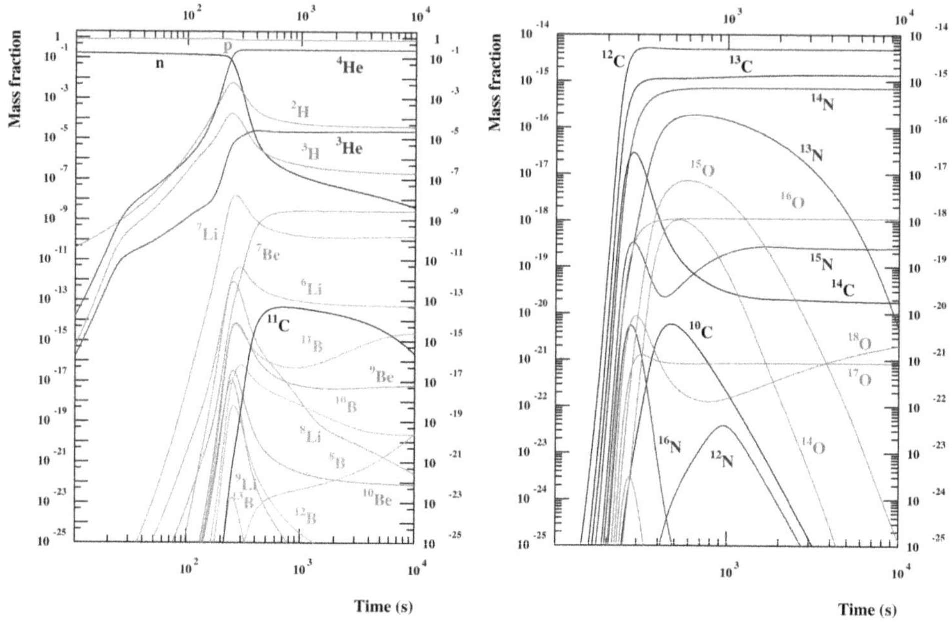

Figure 10.6. Time-evolution of the mass fractions of CNO nuclides. (Reproduced from Coc et al. 2012.

massive stars with more than 8 $M_\odot$, the pp-chains do not generate energy at a rate sufficient to keep the star stable. Since also not enough ^{4}He is produced for the next burning phase (Chapter 7) massive stars would not enter hydrostatic equilibrium but directly collapse. With a presence of CNO nuclides, these more massive stars use the more efficient CNO-cycles (Section 7.2.2) for energy generation and therefore can pass through regular, hydrostatic burning phases. Carbon production has been recently studied (for example, Coc et al. 2012, 2014 and references therein) making use of the extended reaction network shown in Figure 10.5 but otherwise standard conditions. The time-evolution of the mass fractions of the lighter and the CNO nuclides is shown in Figure 10.6. Carbon is produced only at a very low level, with the highest production in the unstable ^{11}C, too low to be of relevance for energy generation in the first generation of stars. Because the reactions leading to C are not as well constrained experimentally as the reactions involving the lighter nuclides, however, there may still be room for improvement.

Production of very heavy elements has been studied in non-standard BBN nucleosynthesis (Matsuura et al. 2005, Rauscher et al. 1994). The investigations were motivated by the idea that large density fluctuations before BBN may lead to regions with different neutron-to-proton ratios, even with $\bar{n}_{\rm n} \gg \bar{n}_{\rm p}$. This could lead to production of nuclides beyond C and even a primordial r-process with fission cycling (see Section 8.3 for a discussion of the r-process). It was found that this was not feasible, however, because of the constraints put by the light elements and CMB limits on fluctuations. Moreover, the initially proposed mechanism for creating the density fluctuations was found to be overestimating the effect.

Further Reading

Coc, A. 2009, NIMPA, 611, 224

Coc, A., Goriely, S., Xu, Y., Saimpert, M., & Vangioni, E. 2012, ApJ, 744, 158

Coc, A., Uzan, J.-P., & Vangioni, E. 2014, JCAP, 10, 050

Cyburt, R. H., Fields, B. D., Olive, K. A., & Yeh, T.-H. 2016, RvMP, 88, 015004

Czarnecki, A., Marciano, W. J., & Sirlin, A. 2018, PhRvL, 120, 202002

Fields, B. D., Olive, K. A., Yeh, T.-H., & Young, C. 2020, JCAP, 3, 10

Hou, S. Q., He, J. J., Parikh, A., et al. 2017, ApJ, 834, 165

Kolb, E. W., & Turner, M. 2018, The Early Universe (Boca Raton, FL: CRC Press)

Komatsu, E., et al. 2009, ApJs, 180, 330

Liddle, A. 2003, Introduction to Modern Cosmology (Chichester: Wiley)

Matsuura, S., Fujimoto, S.-I., Nishimura, S., Hashimoto, M.-A., & Sato, K. 2005, PhRvD, 72, 123505

Pitrou, C., Coc, A., Uzan, J.-P., & Vangioni, E. 2018, PhR, 754, 1

Planck Collaboration, et al. 2016, A&A, 594, A13

Rauscher, T., Applegate, J. H., Cowan, J. J., Thielemann, F.-K., & Wiescher, M. 1994, ApJ, 429, 499

Thomas Rauscher

Chapter 11

Galactic Origin of the Elements

11.1 Production Sites

Chapters 7–10 present the details of nuclear processes of single and binary systems. Their nucleosynthesis output is summarized in Table 11.1. Not included in the table are the nuclei above Fe made in the main s-process component (Section 8.2), the r-process (Section 8.3), the γ-process (Section 8.4), and the i-process (Section 8.5). The site producing the main s-process component is well known, it comes from AGB stars (Chapter 6). The site or sites of the r-process are not well determined, although there are strong indications that a main component is made in neutron star mergers and a weak component comes from magneto-hydrodynamically (MHD) driven jets from core-collapse supernovae (see Sections 8.3, 9.4.5). The symbols used in Table 11.1 for the origin of the nuclides are: BB (SBBN, Chapter 10), GCR (Galactic cosmic rays, see below), L (low-mass stars, below 8 $M_\odot$, including AGB and solar-type stars Chapters 6, 7), M (hydrostatic and explosive burning in massive stars with $M \gtrsim 8\ M_\odot$, ccSN, see Chapter 7, Section 9.3), Ms (weak s-process component from massive stars), Mν (ν-process in massive stars, Section 9.3.4), nov (novae, Section 9.4.2), SNIa (type Ia supernovae, Section 9.4.3), α-fo (α-rich freeze-out from massive stars or other sources, Section 9.3.4), ν-wind (ν-driven wind in ccSN or other sources).

Galactic cosmic rays (GCR) have not been discussed in any of the other chapters. They are made of extremely high-energy particle radiation up to $\approx 10^{19}$ eV permeating interstellar space. The particle flux is strongly declining with energy, however, following a power law. The integrated energy density is low despite of the high energies. In fact, it is comparable to other sources of interstellar radiation. For example, the energy density of GCR is about 1 eV cm^{-3} and the energy density of visible starlight is 0.3 eV cm^{-3}. It is not yet completely clear where GCR originate. A definite connection to supernova remnants is found, where charged particles perhaps

doi:10.1088/2514-3433/ab8737ch11

Table 11.1. Origin of Nuclides

Nuclide	Origin	Nuclide	Origin	Nuclide	Origin	Nuclide	Origin
^{1}H	BB	^{22}Ne	M	^{39}K	M,Mν	^{53}Cr	M
^{2}H	BB	^{23}Na	M	^{40}K	Ms,M	^{54}Cr	SNIa
^{3}He	BB,L	^{24}Mg	M	^{41}K	M	^{55}Mn	SNIa,M,Mν
^{4}He	BB,L,M	^{25}Mg	M	^{40}Ca	M	^{54}Fe	SNIa,M
^{6}Li	GCR	^{26}Mg	M	^{42}Ca	M	^{56}Fe	M,SNIa
^{7}Li	BB,Mν,L,GCR, nov	^{27}Al	M	^{43}Ca	M,α-fo	^{57}Fe	M,SNIa
^{9}Be	GCR	^{28}Si	M	^{44}Ca	α-fo,SNIa	^{58}Fe	Ms,SNIa
^{10}B	GCR	^{29}Si	M	^{46}Ca	M	^{59}Co	Ms,α-fo,SNIa, Mν
^{11}B	Mν	^{30}Si	M	^{48}Ca	SNIa	^{58}Ni	α-fo
^{12}C	L,M	^{31}P	M	^{45}Sc	α-fo,M,Mν	^{60}Ni	α-fo,Ms
^{13}C	L,M	^{32}S	M	^{46}Ti	M,SNIa	^{61}Ni	Ms,α-fo,SNIa
^{14}N	L,M	^{33}S	M	^{47}Ti	SNIa,M	^{62}Ni	Ms,α-fo
^{15}N	nov,Mν	^{34}S	M	^{48}Ti	M,SNIa	^{64}Ni	Ms
^{16}O	M	^{36}S	Ms,M	^{49}Ti	M	^{63}Cu	Ms,M
^{17}O	nov,L	^{35}Cl	M,Mν	^{50}Ti	SNIa,Ms	^{65}Cu	Ms
^{18}O	M	^{37}Cl	Ms,M	^{50}V	M	^{64}Zn	ν-wind,α-fo,Ms
^{19}F	Mν,M,L	^{36}Ar	M	^{51}V	α-fo,SNIa,M, Mν	^{66}Zn	Ms,α-fo,SNIa
^{20}Ne	M	^{38}Ar	M	^{50}Cr	M,α-fo,SNIa	^{67}Zn	Ms
^{21}Ne	M	^{40}Ar	Ms,M	^{52}Cr	M,α-fo,SNIa	^{68}Zn	Ms

could be accelerated in the shocks interacting with the interstellar medium, but the acceleration mechanism is not completely understood. Also the Galactic center has been found to contribute to GCR fluxes.

The composition of GCR is very close to the solar abundance, with mainly protons and ^{4}He and a small content of heavier nuclides. The exact composition depends on the GCR energy. A striking deviation from the solar abundance pattern is the extreme overabundance of Li, Be, and B. They are about 500,000 times more abundant than in the Sun. They are thought to be produced in *spallation reactions* in which C, N, and O nuclei are broken into fragments by interaction with the high-energy protons.

Figure 11.1 illustrates the (estimated) contributions of various nucleosynthesis sites to the solar system abundances of each element. The color shades represent the fraction of the contribution to the total element abundance. The elemental abundances comprise the sum of the abundances of each isotope. As explained in the previous chapters and also shown in Table 11.1 above, each nucleosynthesis process and site contributes differently to each isotope. Not shown in Figure 11.1 are additional contributions by the νp-process (Section 9.3.3) and the rp-process (Section 9.4.4), which are speculated to maybe also contribute to proton-rich

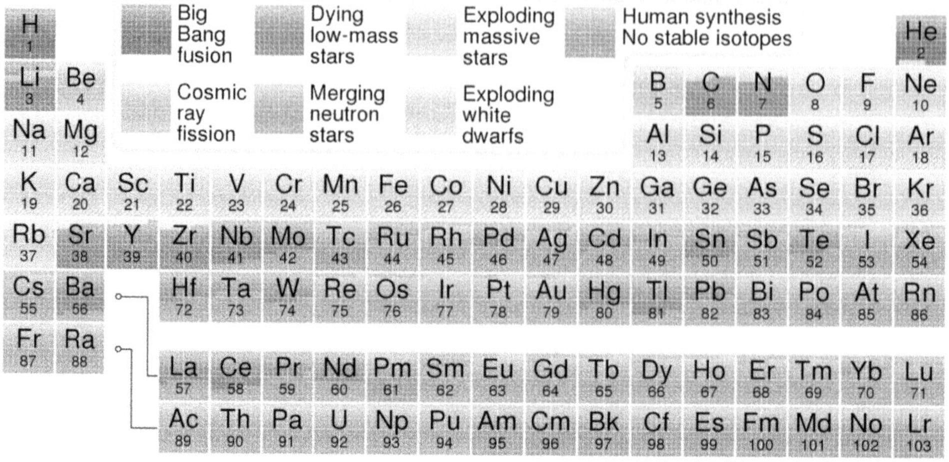

Figure 11.1. Contributions of various sources to the solar system abundances of elements. (This nucleosynthesis periodic table image has been obtained by the author from the Wikimedia website where it was made available by Cmglee under a CC BY-SA 3.0 licence. It is included within this book on that basis. It is attributed to Cmglee.)

isotopes of elements above Fe, as well as the r-process component from MHD jet outflows in ccSN explosions (Section 8.3). They are mainly important, however, to understand abundance patterns in old, metal-poor stars and their contribution to solar system abundances probably is very small.

11.2 Galactic Chemical Evolution Models

Stellar models allow to follow the evolution of a star with given mass and initial composition and also allow to calculate (perhaps in combination with further modeling, for example, of an explosion) the nucleosynthesis yield of the star. Nucleosynthesis yield means the amount of different nuclides ejected into the interstellar medium (ISM). The integrated contribution of stars over the history of a galaxy enriches the ISM with heavier and heavier elements over time, see Figure 11.2. This is called *galactical chemical evolution* (GCE). Since stars are formed from the ISM containing the nucleosynthesis products of previous stellar generations and stars on the main sequence of the HRD (see Chapter 6) keep their original composition in the surface layers, the spectra of old main-sequence stars provide a view of the composition of the early Galaxy. To understand the compositions of stars, from very old ones to the solar composition, it is necessary to construct GCE models.

The GCE models currently in use range from simple, homogeneous models, assuming instantaneous mixing of the ejecta of dying stars with the ISM within the full Galaxy and neglecting different zones and gas flows, over inhomogeneous models relaxing the instantaneous mixing assumption in a semi-analytical way and accounting for the dispersion of nuclide abundances in stars and in the ISM, to chemo-dynamical models also treating dynamic gas flows, for example, in smoothed-particle hydrodynamics (SPH, see also Section 2.9.3). Apart from the numerical treatment, there are two fundamental problems with GCE models. The

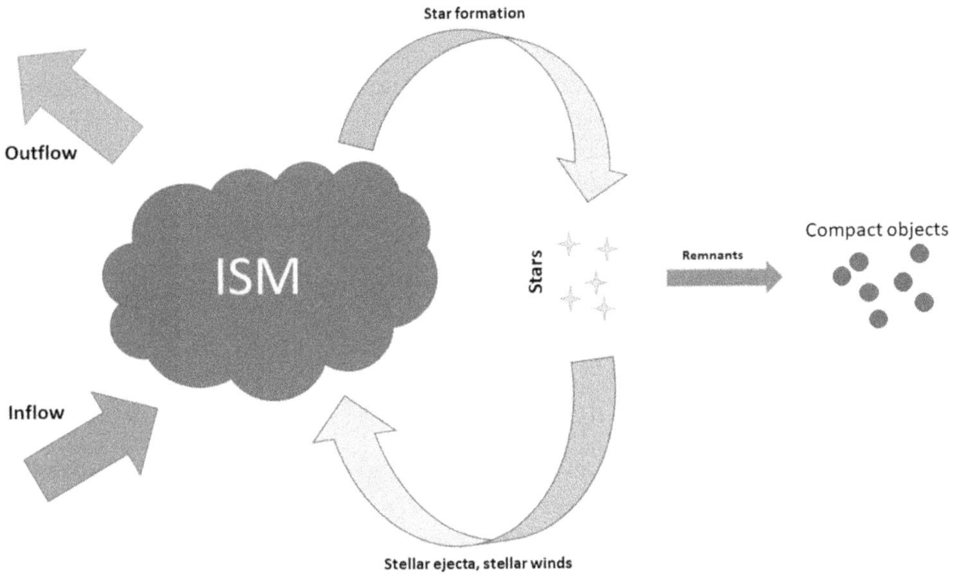

Figure 11.2. Sketch of the matter cycle within a galaxy.

first is the insufficient number and type of observables to constrain the parameters of the models. Only already somehow processed abundance distributions (and mostly elemental abundances and not isotopic abundances) for a range of stellar ages are available. Information on abundances in the early Galaxy is only available from low-mass stars and not from massive stars. Constraints for modeling large-scale gas flows and galaxy formation processes are scarce and indirect. This problems worsens for the more complex models because they also have more adjustable parameters. The second problem is the uncertainties in the input to the models. The stellar abundance yields bear some uncertainties, especially for elements made in explosive nucleosynthesis (Chapter 9) because in certain cases not even the actual production site is known (for example, for the r-process, Section 8.3, or the p-elements, Section 8.4). Moreover, the star formation rate (SFR) and the initial mass function (IMF) (see Chapter 6) are not well understood or constrained, especially for early times, and lowest and highest stellar masses, respectively. This has a major impact in all GCE models. Finally, the initial composition of a galaxy has to be known at the start of the calculation. This could be the primordial composition (inferred from observations or calculated from SBBN, see Chapter 10) but could also be already affected by the galaxy formation process and by inflows from outside a galaxy during formation or even later times. At the current state, GCE models therefore can provide a statistical understanding of how abundances evolve with time in a galaxy but cannot yet make sound predictions for a given system or fully model a specific galaxy. Nevertheless, GCE models are useful for helping to understand stellar nucleosynthesis and the yields of stars, and for establishing a time sequence of events, for example, tracking the overall metallicity evolution over time or inferring at what time certain stellar sites or events have significantly contributed to the abundances of specific elements. Studies of the latter, however, are again sensitive to model details and are often debatable (for example, the current discussion on whether there is need for an additional r-process source in the early Galaxy, see also Section 9.4.5).

As an example for how a GCE model works, the equations for a simple model are presented below. For further details on and reviews of more sophisticated models, see the literature (and references therein) in the "Further Reading" section at the end of this chapter.

Consider a system (galaxy or galactic region) containing the mass $M_{\rm gal} = M_{\rm gal}(t)$. Its time dependence is given by

$$\frac{\mathrm{d}M_{\rm gal}}{\mathrm{d}t} = f_{\rm gal}^{\rm in} - f_{\rm gal}^{\rm out}, \tag{11.1}$$

where $f_{\rm gal}^{\rm in}$ and $f_{\rm gal}^{\rm out}$ are the rates of mass inflows and outflows, respectively. Without any mass addition or mass loss over time, the right-hand side of Equation (11.1) is zero, of course, and this is then called the *closed box model*. For the closed box model and for special functional dependences of inflows or outflows it is possible to derive analytical solutions when making additional assumptions, see below. In the general case, Equation (11.1) and the following equations form a set of differential equations that has to be solved numerically.

The system is composed of a gas mass (the ISM) and the objects distributed within, that is the stars of different mass and composition as well as their final states, such as white dwarfs, neutron stars, and black holes. The time dependence of the gas mass is

$$\frac{\mathrm{d}M_{\rm gas}}{\mathrm{d}t} = -\Psi_{\rm SFR}(t) + M_{\rm ej}(t) + f_{\rm gal}^{\rm in}(t) - f_{\rm gal}^{\rm out}(t), \tag{11.2}$$

where $\Psi_{\rm SFR}$ is the star formation rate (SFR), specifying the rate at which gas is incorporated into stars. The SFR is one of the most uncertain parameters in a GCE model. The quantity $M_{\rm ej}$ is the mass ejection rate by dying stars. It is calculated as

$$M_{\rm ej}(t) = \int_{M_{\rm low}(t)}^{M_{\rm upper}} [M - M_{\rm c}]\Psi_{\rm SFR}(t - \tau_{\rm M})\mathcal{F}_{\rm IMF}(M)\,\mathrm{d}M, \tag{11.3}$$

with the initial mass function (IMF) as introduced in Equation (6.5). The integral runs from the minimal mass of stars dying at time t (i.e., a star with mass M created at time $t - \tau_{\rm M}$ and with lifetime $\tau_{\rm M} < t$) to the upper limit of the IMF (see Chapter 6). Stars with masses lower than about 1 $M_\odot$, however, live too long to affect mass ejection (and enrichment of the ISM with heavier nuclides, see Equation (11.7) below), they only block mass from recycling in the ISM. A star of mass M leaves behind a compact object of mass $M_{\rm c}$. Therefore the net ejection back into the ISM is $M - M_{\rm c}$. Obviously, the masses in stars and in final states $M_{\rm non-gas}$, and in the gas are related to the total mass by $M_{\rm gal}(t) = M_{\rm gas}(t) + M_{\rm non-gas}(t)$. With progressing time, more and more stars have ended their lives and in $M_{\rm non-gas}(t) = M_{\rm stars}(t) + M_{\rm fin}(t)$ the contribution of mass caught up in a final state M_{fin} will dominate more and more over the mass in stars, M_{stars}. The mass in final, compact objects is

$$M_{\text{fin}}(t) = \int_0^t c_{\text{fin}}(t)\,\mathrm{d}t, \tag{11.4}$$

with the mass creation rate of the final states given by

$$c_{\text{fin}}(t) = \int_{M_{\text{low}}(t)}^{M_{\text{upper}}} M_{\text{c}}\Psi_{\text{SFR}}(t - \tau_{\text{M}})\mathcal{F}_{\text{IMF}}(M)\,\mathrm{d}M. \tag{11.5}$$

Equation (11.2) is closely related to the evolution of the composition of the ISM. According to Equation (1.11), the mass composed of a specific nuclide i is simply $M_{\text{gas}}X_i$ and therefore

$$\frac{\mathrm{d}(M_{\text{gas}}X_i)}{\mathrm{d}t} = -\Psi_{\text{SFR}}(t)X_i + {}^{i}M_{\text{ej}}(t) + f_{\text{gal}}^{\text{in}}(t)X_{\text{in},i} - f_{\text{gal}}^{\text{out}}(t)X_{\text{out},i}, \tag{11.6}$$

where all quantities now relate to ejection into and absorption out of the ISM of nuclide i. The mass fractions $X_{\text{in},i}$ and $X_{\text{out},i}$ are for the infalling and outflowing gas, respectively. Usually, $X_{\text{in},i}$ is taken from a primordial composition and $X_{\text{out},i} = X_i$ with an average ISM composition. The *stellar yield* M_i^{y} for a nuclide i, that means the mass ejected as nuclide i, is entering the calculation of the mass ejection rate,

$${}^{i}M_{\text{ej}}(t) = \int_{M_{\text{low}}(t)}^{M_{\text{upper}}} M_i^{\text{y}}(M)\Psi_{\text{SFR}}(t - \tau_{\text{M}})\mathcal{F}_{\text{IMF}}(M)\,\mathrm{d}M. \tag{11.7}$$

The stellar yield may also depend on time because the abundance of a nuclide produced in a star may depend on the star's metallicity. It is to be noted that also primary nuclides (produced in H- or He-burning, see the definition at the end of Section 8.2.1) may also be affected by metallicity because metallicity also impacts the mass loss rate of a star during its lifetime. While the stellar yield M_i^{y} is to be used in Equation (11.7), other quantities are more useful to describe the actual production of a nuclide in a star. Since M_i^{y} contains also a part of the nuclide mass fraction already initially present when the star is formed, which is then re-ejected, it is obvious to define a *net yield*

$$M_i^{\text{net}}(M) = M_i^{\text{y}}(M) - M_i^{\text{ini}}(M), \tag{11.8}$$

where the initially present mass composed of nuclei i is subtracted from the yield,

$$M_i^{\text{ini}}(M) = X_{0,i}(M - M_{\text{c}}), \tag{11.9}$$

with $X_{0,i}$ being the mass fraction of nuclide i in the cloud from which the star formed. While stellar yields can only be positive or zero, net yields can also be negative when a nuclide is effectively destroyed in a star. Another widely used quantity is the *production factor*,

$$f_{\text{prod}} = \frac{M_i^{\text{y}}(M)}{M_i^{\text{ini}}(M)}. \tag{11.10}$$

Obviously, $0 \leqslant f_{\text{prod}} < 1$ implies destruction and $f_{\text{prod}} > 1$ implies production. For instance, massive stars are the only producers of ^{16}O with $f_{\text{prod}} \approx 10$. The stellar yields are derived from a variety of stellar models for stars with different masses and metallicities and other nucleosynthesis sites (see Chapters 6 and 9).

The simplest model makes the following assumptions: closed box, instant mixing assumption (IMA, all the ISM has the same composition), the original composition is primordial, the IMF is constant in time. A further important assumption is the *instantaneous recycling approximation* (IRA). It poses that all stars above a certain mass die instantaneously ($\tau_M = 0$), whereas all other stars live eternally. The appropriate mass limit depends on the age of the studied system. Usually, the division is made at 1 $M_\odot$, corresponding to a galactic age of $T_{\text{gal}} \approx 12$ Gyr. With the IRA, $\Psi_{\text{SFR}}(t - \tau_M)$ becomes $\Psi_{\text{SFR}}(t)$ and therefore can be taken out of all the integrations over mass. Then the above equations can be solved analytically. A convenient definition is the return function

$$\tilde{\mathcal{R}} = \int_{M_{\text{low}}(T_{\text{gal}})}^{M_{\text{upper}}} [M - M_{\text{c}}]\mathcal{F}_{\text{IMF}}(M)\, \mathrm{d}M, \tag{11.11}$$

which is the fraction of the mass of a stellar generation that returns to the ISM. Further useful is the yield per stellar generation and per mass caught in “eternal” objects (low-mass stars and final states of stellar evolution),

$$\mathcal{F}_i^{\text{y}} = \frac{1}{1 - \tilde{\mathcal{R}}} \int_{M_{\text{low}}(T_{\text{gal}})}^{M_{\text{upper}}} M_i^{\text{net}}(M)\mathcal{F}_{\text{IMF}}(M)\, \mathrm{d}M. \tag{11.12}$$

The mass fraction of nuclide i in the gas then evaluates to

$$X_i - X_{0,i} = \mathcal{F}_i^{\text{y}} \ln\left(\frac{M_{\text{gal}}}{M_{\text{gas}}}\right). \tag{11.13}$$

The ratio $M_{\text{gas}}/M_{\text{gal}}$ is called the *gas fraction.* These mass fractions are independent of time and of the shape of the SFR. Thus, the IRA relates the chemical enrichment to the amount of gas left in the system. It is useful to study gas flows and can be used to derive metallicity distributions in stars.

For the SFR, a power law depending on the density of the gas mass (and therefore on M_{gas}) has been suggested. An often used, simple dependence is

$$\Psi_{\text{SFR}} = f_{\text{SFR}}\, M_{\text{gas}}^k, \tag{11.14}$$

with f_{SFR} being the efficiency of star formation and k being an empirical parameter. Assuming $k = 1$ results in a time dependence of the gas mass given by

$$M_{\text{gas}} = M_{\text{gal}} \mathrm{e}^{-f_{\text{SFR}}(1-\tilde{\mathcal{R}})t} \tag{11.15}$$

and Equation (11.13) becomes

$$X_i - X_{0,i} = \mathcal{F}_i^{\text{y}} f_{\text{SFR}}[1 - \tilde{\mathcal{R}}]t. \tag{11.16}$$

This establishes a unambiguous relation between metallicity and time and thus demonstrates that stellar ages are connected to the (surface) metallicity of stars.

For the relation of the mass fraction of a secondary nuclide i_2 to a primary one i_1, the closed box model with IRA finds

$$X_{i_2} = \alpha_{21} X_{i_1} \ln\left(\frac{M_{\text{gal}}}{M_{\text{gas}}}\right) = \frac{\alpha_{21}}{\mathcal{F}^{\text{y}}_{i_1}} X_{i_1}^2 \tag{11.17}$$

where α_{21} is a proportionality factor relating the secondary yield to a primary mass fraction, $\mathcal{F}^{\text{y}}_{i_2} = \alpha_{21} X_{i_1}$. Therefore the mass fraction of the secondary nuclide increases much faster than the one of the primary.

The IRA turns out to be a surprisingly good approximation for nuclides produced exclusively in massive stars, such as O. On the other hand, it is a poor approximation for elements produced on long timescales, such as C, N, and Fe. When the restriction of the closed box is lifted and gas flows into and/or out of the system are allowed, the IRA still allows analytical solutions for very specific assumptions on the functional form of the modeled gas flows. More realistic flows, however, require numerical solutions of the equations laid out above.

For modeling the solar neighborhood, usually models using two or three separate regions, connected by flows between them, are introduced. The *solar cylinder* is defined as cylindrical region perpendicular to the plane of the Milky Way, centered on the Sun (which is 8 kpc away from the Galactic center) and usually assumed to have a radius of 0.5 kpc. Within the cylinder, roughly three regions with different compositions and kinematic properties can be distinguished. The *Galactic halo* with an age of about 12–13 Gyr contains the most metal-poor stars with a large velocity dispersion perpendicularly to the disk. The *thick disk* region has an age of about 10 Gyr, exhibits higher metallicity than the halo and smaller velocity dispersion. Finally, the *thin disk* is younger (5–6 Gyr) than both halo and thick disk, is more concentrated with maximal stellar motions perpendicularly to the plane of about 100 pc for young stars and about 300 pc for old stars, and it contains three quarters of the total mass of the Galaxy. It is to be noted that the three regions have not necessarily evolved out of each other, despite of the age sequence. In fact, modern hierarchical models of galaxy formation suggest that they may have formed somewhat independent from each other. Therefore they can be treated independently in simple GCE models.

One observable is the number of stars per metallicity $\mathcal{Z}$ (see Equation (1.13) for the definition of metallicity). Applying the closed box model with IRA to the local thin disk, too many stars at low metallicity are predicted. This is known as the *G-dwarf problem.* (G dwarfs are bright enough to be easily sampled and live long enough to reflect the early composition of the disk. The same problem would appear in stars with the spectral classes F and K.) A reasonable hypothesis to cure this problem is that the disk cannot be treated as a closed box but was built from infall of almost zero metallicity material. The infall rate is not well constrained by observations but an exponentially decreasing rate $f^{\text{in}}_{\text{gal}}(t) \propto \exp(-t/\tau_{\text{in}})$ with a long timescale of $\tau_{\text{in}} \approx 7$ Gyr can provide a reasonably good fit to the data.

This concludes the description of the most simple GCE models. Inhomogeneous models lift the assumption of instantaneous mixing of the ISM, subdivide the gas mass into cells, in which the random birth and death of stars is tracked. Some mixing between cells also has to be invoked at certain time intervals. This results in a model with discrete time steps and amounts to a book keeping exercise. Chemo-dynamical models follow the gas flows with hydrodynamic models and couple these to the differential equations of GCE.

11.3 Nucleocosmochronology

The fact that certain nuclides decay with very long half-lives comparable to astronomical timescales, such as the age of the solar system or the Galaxy, can be used to determine the time since their synthesis in stellar events. Accordingly, research in this area is termed *nucleocosmochronology*. Depending on their half-lives, different nuclides are suited to explore different timescales. The half-life has to be sufficiently long for some amount to have been present at least at the closure of the solar system (in the early stage of its formation when the pre-solar cloud solidifies and the abundances in solid bodies receive no further contribution from outside sources) and short enough that measurable decay happens. The latter rules out naturally occurring nuclides with very long half-lives, such as ^{144}Nd ($t_{1/2} = 2.3 \times 10^6$ Gyr), ^{148}Sm ($t_{1/2} = 7 \times 10^6$ Gyr), ^{152}Gd ($t_{1/2} = 1.1 \times 10^5$ Gyr), or ^{186}Os ($t_{1/2} = 2 \times 10^6$ Gyr), much longer than the age of the Universe with about 13.8 Gyr (see Section 10.3). Nuclides with half-lives 10^{11}–10^9 yr probe timescales on the order of the galactic age, shorter half-lives of 10^8–10^7 yr may relate to events close to the formation of the solar system. Table 11.2 provides a list of unstable nuclides potentially useful as cosmochronometers (their actual utility, however, may be affected by further limitations, see below). Apart from unstable nuclides still found on Earth naturally, most of the nuclides listed in

Table 11.2. Potential Cosmochronometers Sorted by Decreasing Half-life (in Myr)

Nuclide	Decay Product	$t_{1/2}$	Nuclide	Decay Product	$t_{1/2}$
^{147}Sm^{n}	^{143}Nd	1.06×10^5	^{205}Pb	^{205}Tl	15.3
^{87}Rb^{n}	^{87}Sr	4.89×10^4	^{182}Hf	^{182}W	8.9
^{187}Re^{n}	^{187}Os	4.5×10^4	^{207}Pd	^{107}Ag	6.5
^{232}Th^{n}	^{208}Pb	1.4×10^4	^{97}Tc	^{97}Mo	4.2
^{238}U^{n}	^{206}Pb	4.47×10^3	^{98}Tc	^{98}Ru	4.2
^{235}U^{n}	^{207}Pb	704	^{154}Dy	^{142}Nd	3
^{244}Pu	^{208}Pb	80.8	^{60}Fe^{n}	^{60}Ni	2.6
^{146}Sm	^{142}Nd	69	^{135}Cs	^{135}Ba	2.3
^{92}Nb	^{92}Zr	34	^{150}Gd	^{142}Nd	1.8
^{129}I	^{129}Xe	16	^{93}Zr	^{93}Nb	1.5
^{247}Cm	^{235}U	15.6	^{26}Al	^{26}Mg	0.72

Notes. Nuclides occurring naturally on Earth are marked by superscript "n" (see text for the ^{60}Fe case).

Table 11.2 are *extinct radionuclides*. They are not found in natural terrestrial abundances but there are indications that they have been present in the early solar system. Their decay products are found in meteorites which have protected them from chemically fractionating, that means from being moved around and mixed with other material according to their physical and chemical properties. These captured decay products then produce isotopic anomalies with respect to the average terrestrial isotope distribution. The first indication for the existence of such extinct radioactivities was the discovery of Xe-gas enclosed in meteoritic material, composed of ^{129}Xe, the decay product of ^{129}I. A special case is ^{60}Fe which has a comparatively short half-life but was nevertheless found in deep-sea sediments. The geological dating of the sediments indicates that this ^{60}Fe was deposited between 1.7 and 3.2 Myr ago, after the closure of the solar system bodies, and hints at the arrival of ejecta from a nearby supernova (Wallner et al. 2016). Any ^{26}Al from the same supernova was below the detection level, probably due to the short half-life (Feige et al. 2018). This finding is consistent with the predicted ^{60}Fe/^{26}Al production ratio in ccSN. Up to that discovery, ^{60}Fe was identified in meteorites by an increased ^{60}Ni abundance and classified as extinct.

The most straightforward use of half-lives for time measurements would be the direct application of the decay law (see Equation (4.107)) but this requires knowledge of the initial amount. One of the few direct applications is shown in Figure 8.11 of Section 8.3.3. Although the stellar surface abundances of CS22892-052 may have received contributions of a few r-process events, it is the current understanding that they always produce the same relative elemental abundances for the heavy r-nuclides (see Section 8.3) and thus their superposition does not destroy the typical r-process pattern. Therefore it is possible to determine the r-process conditions required to produce the pattern found for stable nuclides and then to compare the Th abundance obtained with these conditions to the observed Th abundance. From the ratio of observed and calculated abundance the age of the star can be determined. While the decay half-life of ^{232}Th is well known, its predicted abundance, of course, is vulnerable to all uncertainties connected to the r-process production of actinides.

In meteoritic material, in which it can be assumed that the decay products have not been separated from the progenitor nuclides, also the ratio between progenitor and product abundances can be used as a chronometer. This also requires further assumptions, though. The initial abundance of the decay products has to be estimated or the progenitor should not be extinct. Moreover, the abundances should not have been affected by irradiation since the formation of the solar system. Finally, the abundance can only be extrapolated back to the closure of the solar system, not back to the original production site because solar system abundances are already a superposition of contributions from many sites. It helps to combine several of such age estimates to remove some of the uncertainty regarding the initial abundances. At least, consistent age values should be obtained.

Any further analyses of the decay of radionuclides have to be connected to GCE models which track the production and ISM distribution of these nuclides until the formation of the solar system. The nuclides i in Equation (11.6) are assumed to be stable. The equation can be extended by adding a term $-\lambda_{\mathrm{dec}} X_i M_{\mathrm{gas}}$ to the right-hand-side, accounting for the decay of the nuclides in the ISM (decay within the star is included in the net yield). Again, all the complications of following GCE would enter here, in principle. Therefore widespread use was and is made with a toy model, using the IRA, neglecting inflows and outflows, and assuming an exponentially decreasing SFR (and thus also exponentially decreasing nuclide production). Often, in addition to the exponentially declining production the possibility of a late-time production spike is also taken into account. The superposition of the exponential and the sudden component leads to an abundance after a galactic nucleosynthesis time interval Δ_{syn} of

$$Y_i(\Delta_{\text{syn}}) = \Psi^0_{\text{SFR}} e^{-f\left(\Delta_{\text{syn}},\, \tau^{\text{R}}_{\text{life}}\right)} P_i \left[\frac{1}{\tau^{\text{R}}_{\text{life}} - \tau^{i}_{\text{life}}} \left(e^{-\tau^{i}_{\text{life}}\Delta_{\text{syn}}} - e^{-\tau^{\text{R}}_{\text{life}}\Delta_{\text{syn}}}\right) + \frac{S_0}{\tau^{\text{R}}_{\text{life}}(1 - S_0)} \left(1 - e^{-\tau^{\text{R}}_{\text{life}}\Delta_{\text{syn}}}\right) e^{-\tau^{i}_{\text{life}}\Delta_{\text{syn}}} \right], \tag{11.18}$$

where τ^{i}_{life} is the decay lifetime of nuclide i, P_i is its stellar production by mass, $\tau^{\text{R}}_{\text{life}}$ is a parameter describing the exponential decline of the star formation as given by the term $\Psi^0_{\text{SFR}} e^{-f(\Delta_{\text{syn}}, \tau^{\text{R}}_{\text{life}})}$ with an initial star formation rate Ψ^0_{SFR}. The abundance contributed by the sudden component is S_0. The number of free parameters can be reduced by considering abundance ratios of two unstable nuclide i, j with their abundances evolving on the galactic timescale,

$$\frac{Y_i(\Delta_{\text{syn}})}{Y_j(\Delta_{\text{syn}})} = \frac{P_i}{P_j} f\left(\tau^{i}_{\text{life}}, \tau^{j}_{\text{life}}, S_0, \tau^{\text{R}}_{\text{life}}, \Delta_{\text{syn}}\right). \tag{11.19}$$

The terms with $\Psi^0_{\text{SFR}} e^{-f(\Delta_{\text{syn}}, \tau^{\text{R}}_{\text{life}})}$ cancel out. A further advantage is that the absolute amounts produced do not have to be known, production ratios can be used. Abundance ratios can be predicted more reliably. In order to determine the three parameters S_0, $\tau^{\text{R}}_{\text{life}}$, Δ_{syn} three chronometric pairs have to be used (the half-lives of the nuclides are known). If the sudden component is neglected, two pairs are sufficient.

Various combinations of the nuclides listed in Table 11.2 have been used for investigating ages and timescales. In the selection of appropriate chronometric pairs it is advantageous to use nuclides originating from the same nucleosynthesis process in the same sites. Otherwise, additional complications arise because the contributions of different processes with different timing have to be considered and a simple model may not suffice anymore. For example, with respect to r-process nucleosynthesis the actinide pairs ^{232}Th/^{238}U, ^{235}U/^{238}U, and ^{244}Pu/^{238}U have been used and compared to meteoritic abundance ratios. The first two pairs can provide information on galactic timescales whereas the third pair suffers from the comparatively short half-life of ^{244}Pu and can only be used for times shortly before the closure of the solar system.

Chronometric pairs involving nuclides receiving contributions from different nucleosynthesis process (or sites) are called *mixed chronometers*. These are better addressed by more sophisticated GCE models, accounting for the different production frequencies for different processes. The mixed chronometers include, for example, the ^{187}Re/^{187}Os pair. While ^{187}Re is produced in the r-process, ^{187}Os is produced in the s-process but additionally receives a contribution from the decay of ^{187}Re. This contribution (the cosmoradiogenic osmium) has to be used in the chronometer pair. This necessitates a sound understanding of the s-process production and its evolution with Galactic time. Similarly, ^{87}Rb/^{87}Sr receive contributions from both the s- and the r-process, as well as ^{147}Sm. For any pairs involving Pb isotopes, the s-process has to be considered. The pairs ^{206}Pb/^{238}U and ^{207}Pb/^{235}U in addition have a third contribution from α-decay chains of nuclides above Pb, also produced in the r-process.

Nuclides with shorter half-lives can also be used in connection with other production processes, such as the γ-process (see Section 8.4.2). For example, the ^{142}Nd/^{144}Sm abundance ratio in meteorites correlates with the ^{146}Sm/^{144}Sm production ratio in the γ-process. Unfortunately, there are still considerable uncertainties in the nuclear reactions involved and therefore neither this production ratio is well known nor is the Galactic production time interval (which also depends on whether the γ-process occurs in ccSN or SNIa, see Sections 8.4.2, 9.3.4, and 9.4.3). Were this interval known, then the production ratio could be inferred. Such an attempt was made to infer the ^{92}Nb/^{92}Mo production ratio using a GCE model (with infall) tuned to simultaneously reproduce the early solar system values of extinct radioactive p-nuclides. As mentioned in Section 9.3.4, p-nuclides with $A < 100$ are notoriously underproduced in massive stars. As an alternative, proton-rich nucleosynthesis processes, such as the rp-process (Section 9.4.4) or the νp-process (Section 9.3.3) have been suggested. These processes, however, do not produce ^{92}Nb which is shielded from decay from the proton-rich side by the stable ^{92}Mo. According to Dauphas et al. (2003), therefore, the contribution of such proton-rich processes is strongly constrained by the ^{92}Nb/^{92}Mo ratio inferred from early solar system abundances.

Finally, an important source of abundance information, including isotopic abundances, has to be mentioned: *presolar grains*. These are miniscule grains enclosed in meteoritic material. They are thought to have condensed in the stellar winds of AGB stars and in supernova ejecta. This means that they are indeed pristine "star dust," not contaminated by any subsequent GCE effects. Therefore the original composition ejected from a star is "frozen" into such a grain and the decay products of any radionuclides are also enclosed therein. Not having to deal with complications from GCE obviously is a huge advantage but new complications are introduced because the exact mechanism for grain formation (and possible chemical fractionation and biasing in this formation process) is not well understood. Furthermore, the mixing of matter ejected from different layers of a star in, for example, a ccSN before grain formation also still leaves room for interpretation. Nevertheless, in the last decades, also with the advent of improved equipment allowing highly precise analyses of the grain material, the study of presolar grains has become one of the most important source for obtaining detailed information on isotopic abundances and the processes creating them within stars. More about this exciting field and the different types of grains can be found, for example, in Lugaro (2005).

Further Reading

Argast, D., Samland, M., Thielemann, F.-K., & Qian, Y.-Z. 2004, A&A, 416, 997
Boezio, M., & Mocchiutti, E. 2012, APh, 39, 95
Chiappini, C., Matteucci, F., & Gratton, R. 1997, ApJ, 477, 765
Clayton, D. D. 1983, ApJ, 268, 381
Clayton, D. D. 1988, MNRAS, 234, 1
Clayton, D. D., & Nittler, L. R. 2004, ARA&A, 42, 39
Côté, B., O'Shea, B. W., Ritter, C., Herwig, F., & Venn, K. A. 2017, ApJ, 835, 128
Cowan, J. J., Thielemann, F.-K., & Truran, J. W. 1991, PhR, 208, 267
Dauphas, N., & Chaussidon, M. 2011, AREPS, 39, 351
Dauphas, N., Rauscher, T., Marty, B., & Reisberg, L. 2003, NuPhA, 719, C287
Dauphas, N., et al. 2008, ApJ, 686, 560
Diehl, R., Hartmann, D. H., & Prantzos, N. (ed) 2018, Astrophysics with Radioactive Isotopes, Astrophysics and Space Science Library, Vol. 453 (Berlin: Springer)
Feige, J., et al. 2018, PhRvL, 121, 221103
Gibson, B. K., Fenner, Y., Renda, A., Kawata, D., & Lee, H. 2003, PASA, 20, 401
Gibson, B. K., Pilkington, K., Brook, C. B., Stinson, G. S., & Bailin, J. 2013, A&A, 554, A47
Kubryk, M., Prantzos, N., & Athanassoula, E. 2015, A&A, 580, A127
Lugaro, M. 2005, Stardust from Meteorites: An Introduction to Presolar Grains (Singapore: World Scientific)
Matteucci, F. 2012, Chemical Evolution of Galaxies (Berlin, Heidelberg: Springer)
Prantzos, N. 2008, EASPS, 32, 311
Prantzos, N., Abia, C., Cristallo, S., Limongi, M., & Chieffi, A. 2020, MNRAS, 491, 1832
Prialnik, D. 2009, An Introduction to the Theory of Stellar Structure and Evolution (2nd ed.; Cambridge: Cambridge University Press)
Qian, Y.-Z., & Wasserburg, G. J. 2000, PhR, 333, 77
Rauscher, T., Dauphas, N., Dillmann, I., et al. 2013, RPPh, 76, 066201
Rolfs, C. E., & Rodney, W. S. 1988, Cauldrons in the Cosmos (Chicago, IL: University of Chicago Press)
Romano, D., Matteucci, F., Zhang, Z.-Y., Papadopoulos, P. P., & Ivison, R. J. 2017, MNRAS, 470, 401
Rizzuti, F., Cescutti, G., Matteucci, F., et al. 2019, MNRAS, 489, 5244
Thielemann, F.-K., Eichler, M., Panov, I. V., & Wehmeyer, B. 2017, ARNPS, 67, 253
Tinsley, B. M. 1980, FCPh, 5, 287
Travaglio, C., Galli, D., & Burkert, A. 2001, ApJ, 547, 217
Travaglio, C., Galli, D., Gallino, R., et al. 1999, ApJ, 521, 691
Travaglio, C., Gallino, R., Rauscher, T., et al. 2014, ApJ, 795, 141
Travaglio, C., Gallino, R., Rauscher, T., Röpke, F. K., & Hillebrandt, W. 2015, ApJ, 799, 54
Wallner, A., et al. 2016, Natur, 532, 69
Wehmeyer, B., Pignatari, M., & Thielemann, F.-K. 2015, MNRAS, 452, 1970
Woosley, S. E., Heger, A., & Weaver, T. A. 2002, RvMP, 74, 1015

www.ingramcontent.com/pod-product-compliance
Ingram Content Group UK Ltd.
Pitfield, Milton Keynes, MK11 3LW, UK
UKHW051849210426
5322IPUK00024B/618

9 780750 318280